Statistics of Extremes

Statistics of Extremes

Theory and Applications

Jan Beirlant, Yuri Goegebeur, and Jozef Teugels
*University Center of Statistics, Katholieke Universiteit Leuven,
Belgium*

Johan Segers
Department of Econometrics, Tilburg University, The Netherlands

with contributions from:

Daniel De Waal
*Department of Mathematical Statistics, University of the Free State,
South Africa*

Chris Ferro
Department of Meteorology, The University of Reading, UK

John Wiley & Sons, Ltd

Copyright © 2004   John Wiley & Sons Ltd, The Atrium, Southern Gate, Chichester,
West Sussex PO19 8SQ, England

Telephone (+44) 1243 779777

Email (for orders and customer service enquiries): cs-books@wiley.co.uk
Visit our Home Page on www.wileyeurope.com or www.wiley.com

Reprinted August 2005

This publication is designed to provide accurate and authoritative information in regard to the subject
matter covered. It is sold on the understanding that the Publisher is not engaged in rendering professional
services. If professional advice or other expert assistance is required, the services of a competent
professional should be sought.

Other Wiley Editorial Offices

John Wiley & Sons Inc., 111 River Street, Hoboken, NJ 07030, USA

Jossey-Bass, 989 Market Street, San Francisco, CA 94103-1741, USA

Wiley-VCH Verlag GmbH, Boschstr. 12, D-69469 Weinheim, Germany

John Wiley & Sons Australia Ltd, 33 Park Road, Milton, Queensland 4064, Australia

John Wiley & Sons (Asia) Pte Ltd, 2 Clementi Loop #02-01, Jin Xing Distripark, Singapore 129809

John Wiley & Sons Canada Ltd, 22 Worcester Road, Etobicoke, Ontario, Canada M9W 1L1

Wiley also publishes its books in a variety of electronic formats. Some content that
appears in print may not be available in electronic books.

Library of Congress Cataloging-in-Publication Data

Statistics of extremes : theory and applications / Jan Beirlant ... [et al.], with contributions
    from Daniel De Waal, Chris Ferro.
        p.cm. — (Wiley series inprobability and statistics)
    Includes bibliographical references and index.
    ISBN 0-471-97647-4 (acid-free paper)
        1. Mathematical statistics. 2. Maxima and minima. I. Beirlant, Jan. II. Series.
    QA276.S783447 2004
    519.5–dc22   2004051046

British Library Cataloguing in Publication Data

A catalogue record for this book is available from the British Library

ISBN 10: 0-471-97647-4 (H/B)
ISBN 13: 978-0-471-97647-9 (H /B)

This book is printed on acid-free paper responsibly manufactured from sustainable forestry
in which at least two trees are planted for each one used for paper production.

Contents

Preface

The key result obtained by Fisher and Tippett in 1928 on the possible limit laws of the sample maximum has seemingly created the idea that extreme value theory was something rather special, very different from classical central limit theory. In fact, the number of publications dealing with statistical aspects of extremes dated before 1970 is at most a dozen. The book by E. J. Gumbel, published by Columbia University Press in 1958, has for a long time been considered as the main referential work for applications of extreme value theory in engineering subjects. A close look at this seminal publication shows that in the early stages one tried to approach extreme value theory via central limit machinery. During the decade following its appearance, no change occurred in the lack of interest among probabilists and statisticians who contributed only a very limited number of relevant papers.

From the theoretical point of view, the 1970 doctoral dissertation by L. de Haan *On Regular Variation and its Applications to the Weak Convergence of Sample Extremes* seems to be the starting point for theoretical developments in extreme value theory. For the first time, the probabilistic and stochastic properties of sample extremes were developed into a coherent and attractive theory, comparable to the theory of sums of random variables. The statistical aspects had to wait even longer before they received the necessary attention.

In Chapter 1, we illustrate why and how one should look at extreme values in a data set. Many of these examples will reappear as illustrations and even as case studies in the sequel. The next five chapters deal with the univariate theory for the case of independent and identically distributed random variables. Chapter 2 covers the probabilistic limiting problem for determining the possible limits of sample extremes together with the connected domain of attraction problem. The extremal domain of attraction condition is, however, too weak to use to fully develop useful statistical theories of estimation, construction of confidence intervals, bias reduction, and so on. The need for second order information is illustrated in Chapter 3. Armed with this information, we attack the tail estimation problem for the Pareto case in Chapter 4 and for the general case in Chapter 5. All the methods developed so far are then illustrated by a number of case studies in Chapter 6.

The last five chapters deal with topics that are still in full and vigorous development. We can only try to give a picture that is as complete as possible at the time of writing. To broaden the statistical machinery in the univariate case, Chapter 7 treats a variety of alternative methods under a common umbrella of regression-type methods. Chapters 8 and 9 deal with multivariate extremes and repeat some of the methodology of previous chapters, in more than one dimension. In the first of these two chapters, we deal with the probabilistic aspects of multivariate extreme value theory by including the possible limits and their domains of attraction; the next chapter is then devoted to the statistical features of this important subject. Chapter 9 gives an almost self-contained survey of extreme value methods in time series analysis, an area where the importance of extremes has already long been recognized. We finish with a separate and tentative chapter on Bayesian methods, a topic in need of further and deep study.

We are aware that it is a daring act to write a book with the title *Statistics of Extremes*, the same as that of the first main treatise on extremes. What is even more daring is our attempt to cope with the incredible speed at which statistical extreme value theory has been exploding. More than half of the references in this book appeared over the last ten years. However, it is our sincere conviction that over the last two decades extreme value theory has matured and that it should become part of any in-depth education in statistics or its applications. We hope that this slight attempt of ours gets extreme value theory duly recognized.

Here are some of the main features of the book.

1. The probabilistic aspects in the first few chapters are streamlined to quickly arrive at the key conditions needed to understand the behaviour of sample extremes. It would have been possible to write a more complete and rigorous text that would automatically be much more mathematical. We felt that, for practical purposes, we could safely restrict ourselves to the case where the underlying random variables are sufficiently continuous. While more general conditions would be possible, there is little to gain with a more formal approach.

2. Under this extra condition, the mathematical intricacies of the subject are usually quite tractable. Wherever possible, we provide insight into why and how the mathematical operations lead to otherwise peculiar conditions. To keep a smooth flow in the development, technical details within a chapter are deferred to the last section of that chapter. However, statements of theorems are always given in their fullest generality.

3. Because of the lively speed at which extreme value theory has been developing, thoroughly different approaches are possible when solving a statistical problem. To avoid single-handedness, we therefore included alternative procedures that boast sufficient theoretical and practical underpinning.

4. Being strong believers in graphical procedures, we illustrate concepts, derivations and results by graphical tools. It is hard to overestimate the role of the latter in getting a quick but reliable impression of the kind and quality of data.

5. Examples and case studies are amply scattered over the manuscript, some of them reappearing to illustrate how a more advanced technique results

in better insight into the data. The wide variety in areas of application as covered in the first chapter beautifully illustrates how extreme value theory has anchored itself in various branches of applied statistics.

6. An extensive bibliography is included. This material should help the reader find his or her way through the bursting literature on the subject. Again, as the book is statistical in nature, many important contributions to the probabilistic and stochastic aspects of extreme value theory have not been included.

The book has been conceived as a graduate or advanced undergraduate course text where the instructor has the choice of including as much of the theoretical development as he or she desires. We expect the reader to be familiar with basic probability theory and statistics. A bit of knowledge about Poisson processes would also be helpful, especially in the chapters on multivariate extreme value theory and time series. We have attempted to make the book as self-contained as possible. Only well-known results from analysis, probability or statistics are used. Sometimes they are further explained in the technical details.

Software that was used in the calculations is available at http://www.wis. kuleuven.ac.be/stat/extreme.html#programs

We take pleasure in thanking first Daan de Waal, Chris Ferro and Björn Vandewalle who have contributed substantially to the topics covered in the book. We are also very grateful to Tertius de Wet, Elisabeth Joossens, Alec Stephenson and Björn Vandewalle for furnishing additional material. A large number of colleagues provided the data on which most of the practical applications and the case studies have been based. We thank in particular Jef Caers, Johan Fereira, Philippe Delongueville (SECURA), Robert Verlaak (AON), Robert Oger and Viviane Planchon and Jozef Van Dyck (Probabilitas).

We are very obliged to our universities as well. Especially, the members of and visitors from the University Center for Statistics, Catholic University, Leuven, deserve our great appreciation. We are particularly grateful for the constructive input from the graduate students who have followed courses partially based on material from the book.

Our contact with J. Wiley through Sian Jones and Rob Calver has always been pleasant and constructive.

Our main gratitude goes to our loving families, who have endured our preoccupation with this ever-growing project for so long.

Jan BEIRLANT
Yuri GOEGEBEUR
Johan SEGERS
Jozef L. TEUGELS

1

WHY EXTREME VALUE THEORY?

1.1 A Simple Extreme Value Problem

Many statistical tools are available in order to draw information concerning specific measures in a statistical distribution. In this textbook, we focus on the behaviour of the extreme values of a data set. Assume that the data are realizations of a sample $X_1, X_2, \ldots, X_n$ of n independent and identically distributed random variables. The ordered data will then be denoted by $X_{1,n} \leq \cdots \leq X_{n,n}$. The sample data are typically used to study properties about the distribution function

$$F(x) = P(X \leq x),$$

or about its inverse function, the *quantile function* defined as

$$Q(p) := \inf\{x : F(x) \geq p\}.$$

Suppose we would like to examine the daily maximal wind speed data in the city of Albuquerque shown in Figure 1.1 (taken from Beirlant *et al.* (1996a)). In the classical theory, one is often interested in the behaviour of the mean or average. This average will then be described through the expected value $E(X)$ of the distribution. On the basis of the law of large numbers, the sample mean $\bar{X}$ is used as a consistent estimator of $E(X)$. Furthermore, the central limit theorem yields the asymptotic behaviour of the sample mean. This result can be used to provide a confidence interval for $E(X)$ in case the sample size is sufficiently large, a condition necessary when invoking the central limit theorem. For the Albuquerque wind speed data, these techniques lead to an average maximum daily wind speed of 21.65 miles per hour, whereas (21.4–21.9) is a 95% confidence interval for $E(X)$ based on the classical theory.

Statistics of Extremes: Theory and Applications J. Beirlant, Y. Goegebeur, J. Segers, and J. Teugels
© 2004 John Wiley & Sons, Ltd ISBN: 0-471-97647-4

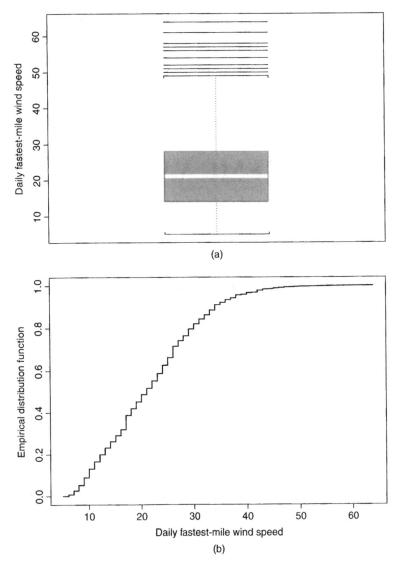

Figure 1.1   (a) Boxplot and (b) empirical distribution function of the daily maximal wind speed data in the city of Albuquerque.

In case of wind speeds, it can be just as important to estimate tail probabilities.
Suppose a shed breaks down if the wind speed is larger than 30 miles per hour,
then it is interesting to estimate the tail probability $p = P(X > 30)$. To this end,
one can use the empirical distribution function defined by

$$\hat{F}_n(x) = \frac{i}{n} \text{ if } x \in [x_{i,n}, x_{i+1,n})$$

where $x_{i,n}$ is the i-th ordered sample value. For the Albuquerque data, this leads to $\hat{p} = 1 - \hat{F}_n(30) = 0.18$.

However, we should add some critical remarks to these considerations. What if the second moment $E(X^2)$ or even the mean $E(X)$ is not finite? Then the central limit theorem does not apply and the classical theory, dominated by the normal distribution, is no longer relevant. Or, what if one wants to estimate $p = P(X > x)$, where $x > x_{n,n}$ and the estimate $\hat{p}$ defined above yields the value 0? Such questions concerning the shed are important since the damage caused by extreme wind speeds can be substantial, perhaps even catastrophical. Clearly, we cannot simply assume that such x-values are impossible. However, the traditional technique based on the empirical distribution function, does not yield any useful information concerning this type of question. In terms of the *empirical quantile function*

$$\hat{Q}_n(p) := \inf\{x : \hat{F}_n(x) \geq p\},$$

problems arise when we consider high quantiles $\hat{Q}_n(1 - p)$ with $p < \frac{1}{n}$.

These observations show that it is necessary to develop special techniques that focus on the extreme values of a sample, on extremely high quantiles or on small tail probabilities. In practical situations, these extreme values are often of key interest. The wind speed example provides just one illustration but there are numerous other situations where extreme value reasoning is of prime importance.

1.2 Graphical Tools for Data Analysis

Given data, a practitioner wants to use graphics that will show in a clear and efficient way the features of the data that are relevant for a given research question. In this section, we concentrate on visually oriented statistical techniques that provide as much information as possible about the tail of a distribution. In later chapters, these graphical tools will help us to decide on a reasonable model to describe the underlying statistical population. Our emphasis will not be on *global* models that aim at describing the data in their entirety or the distribution on its full support. Rather, we perform statistical fits above certain (high) thresholds. Motivation for this was provided in Section 1.1.

We will not recapitulate common statistical graphics such as histograms, smooth density estimates and boxplots. Instead we will focus on *quantile-quantile* (QQ) and *mean excess* (or mean residual life) plots, which are often more informative for our purposes. Moreover, many popular estimation methods from extreme value theory turn out to be directly based on these graphical tools.

1.2.1 Quantile-quantile plots

The idea of quantile plots, or more specifically *Quantile-Quantile plots* (shortly *QQ-plots*), has emerged from the observation that for important classes of distributions, the quantiles $Q(p)$ are *linearly related* to the corresponding quantiles of a standard example from this class. Linearity in a graph can be easily checked by eye and can further be quantified by means of a correlation coefficient. This tool could therefore be ideally used when trying to answer the classical *goodness-of-fit* question: does a particular model provide a plausible fit to the distribution of the random variable at hand?

Historically, the normal distribution has provided the prime class of models where QQ-plots constitute a powerful tool in answering this question. As will be shown in the sequel, the exponential distribution plays a far more important role for our purposes. The rationale for QQ-plots remains the same but the calculations are even easier. We start by explaining and illustrating the QQ-plot idea for the exponential model $Exp(\lambda)$ (see Table 1.1). This very same methodology can then be explored and extended in order to provide comparisons of empirical evidence available in the data when fitting models such as the log-normal, Weibull or others.

Restricting our attention first to the $Exp(\lambda)$ model, we can propose the standard exponential distribution

$$1 - F_1(x) := \exp(-x), \quad x > 0$$

as the standard example from the class of distributions with general survival function

$$1 - F_\lambda(x) = \exp(-\lambda x).$$

We want to know whether the real population distribution F belongs to this class, parametrized by $\lambda > 0$. The answer has to rely on the data $x_1, \ldots, x_n$ that we have at our disposal. It is important to note that this parameter value can be considered as a nuisance parameter here since its value is not our main point of interest at this

Table 1.1 QQ-plot coordinates for some distributions.

Distribution	$F(x)$	Coordinates
Normal	$\int_{-\infty}^{x} \frac{1}{\sqrt{2\pi}\sigma} \exp\left(-\frac{(u-\mu)^2}{2\sigma^2}\right) du$ $x \in \mathbb{R}; \mu \in \mathbb{R}, \sigma > 0$	$(\Phi^{-1}(p_{i,n}), x_{i,n})$
Log-normal	$\int_{0}^{x} \frac{1}{\sqrt{2\pi}\sigma u} \exp\left(-\frac{(\log u-\mu)^2}{2\sigma^2}\right) du$ $x > 0; \mu \in \mathbb{R}, \sigma > 0$	$(\Phi^{-1}(p_{i,n}), \log x_{i,n})$
Exponential	$1 - \exp(-\lambda x)$ $x > 0; \lambda > 0$	$(-\log(1 - p_{i,n}), x_{i,n})$
Pareto	$1 - x^{-\alpha}$ $x > 1; \alpha > 0$	$(-\log(1 - p_{i,n}), \log x_{i,n})$
Weibull	$1 - \exp(-\lambda x^\tau)$ $x > 0; \lambda, \tau > 0$	$(\log(-\log(1 - p_{i,n})),$ $\log x_{i,n})$

moment. We might even wonder whether this parameter has any relevance at all for modelling reality since the whole parametric model itself is still at question. The quantile function for the exponential distribution has the simple form

$$Q_\lambda(p) = -\frac{1}{\lambda} \log(1 - p), \quad \text{for } p \in (0, 1).$$

Hence, there exists a simple linear relation between the quantiles of any exponential distribution and the corresponding standard exponential quantiles

$$Q_\lambda(p) = \frac{1}{\lambda} Q_1(p) \text{ for } p \in (0, 1).$$

Starting with a given set $x_1, x_2, \ldots, x_n$, the practitioner replaces the unknown population quantile function Q by the empirical approximation $\hat{Q}_n$ defined below. In an orthogonal coordinate system, the points with values

$$(-\log(1 - p), \hat{Q}_n(p))$$

are plotted for several values of $p \in (0, 1)$. *We then expect that a straight line pattern will appear in the scatter plot if the exponential model provides a plausible statistical fit for the given statistical population.* When a straight line pattern is obtained, the slope of a fitted line can be used as an estimate of the parameter λ^{-1}. Indeed, if the model is correct, then the equation

$$Q_\lambda(p) = \frac{1}{\lambda}(-\log(1 - p))$$

holds. Remark that the intercept for the given model should be 0 as $Q_\lambda(0) = 0$. In general,

$$\hat{Q}_n(p) = x_{i,n}, \quad \text{for } \frac{i - 1}{n} < p \le \frac{i}{n}.$$

A very practical choice of values of p is given by

$$p \in \left\{ \frac{1}{n}, \frac{2}{n}, \ldots, \frac{n - 1}{n}, 1 \right\}.$$

The alternative choice

$$p \in \left\{ \frac{1 - .5}{n}, \frac{2 - .5}{n}, \ldots, \frac{n - 1 - .5}{n}, \frac{n - 0.5}{n} \right\}$$

applies a *continuity correction* in that we compare a discontinuous function $\hat{Q}_n$ with the continuous function $Q_1(x) := -\log(1 - x)$. Moreover, this choice avoids overflow problems at $p = 1$. The same holds for the choice

$$p \in \left\{ \frac{1}{n + 1}, \frac{2}{n + 1}, \ldots, \frac{n - 1}{n + 1}, \frac{n}{n + 1} \right\}.$$

While other choices are found in the literature, the latter option $p_{i,n} := i/(n + 1)$, $i = 1, 2, \ldots, n$, will be used in the sequel. The empirical quantiles are plotted on the vertical axis and the standard exponential quantiles on the horizontal axis.

A straight line can be fitted through the scatter plot using a classical least-squares algorithm. This straight line with slope a and intercept 0 is obtained from minimizing the sum of squares

$$\sum_{i=1}^{n} \left(x_{i,n} + a \log(1 - p_{i,n})\right)^2.$$

This procedure yields the well-known formula for the least-squares fit

$$\hat{a} = \frac{\sum_{i=1}^{n} x_{i,n} q_{i,n}}{\sum_{i=1}^{n} q_{i,n}^2}$$

where we have put

$$q_{i,n} := -\log(1 - p_{i,n}), \quad i = 1, 2, \ldots, n.$$

The fitted straight line can then be used as a tool to visually check the linear structure of the scatter plot. Moreover, we get an estimate of the parameter value λ if the linearity has been satisfactorily fulfilled.

The exponential QQ-plot has a further important interpretation. The function that is approximated when plotting

$$\left(x_{i,n}, -\log\left(1 - p_{i,n}\right)\right), \quad i = 1, \ldots, n$$

is given by

$$x \mapsto -\log(1 - F(x)).$$

This is precisely the function that maps a random variable X with a continuous distribution function F into the standard exponential distribution. Indeed, the distribution function of $-\log(1 - F(X))$ is given by

$$P\left(-\log(1 - F(X)) \le x\right) = P\left(X \le Q\left(1 - \exp(-x)\right)\right) =$$

$$= F(Q(1 - \exp(-x))) = 1 - \exp(-x)$$

and so $-\log(1 - F(X)) \sim Exp(1)$.

Often data are only available above a certain threshold t. For instance, a reinsurance company might only receive information about claims larger than a priority t. In case of an exponential model, this operation of conditioning on the event $(X > t)$ leads to a shifted exponential model. Evaluation of the distribution of data that are larger than t coincides with the conditional distribution of X, given $(X > t)$. Now, in case of an exponential model,

$$P(X > x \mid X > t) = \frac{P(X > x)}{P(X > t)} = \exp(-\lambda(x - t)), \quad x > t.$$

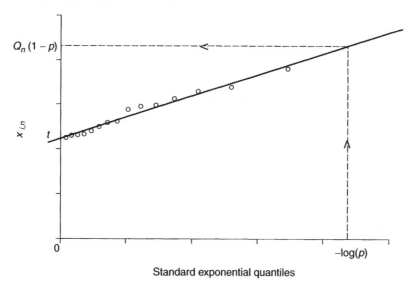

Figure 1.2   Exponential QQ-plot: estimation of extreme quantiles.

The corresponding quantile function is then equal to

$$Q(p) = t - \frac{1}{\lambda} \log(1 - p), \quad 0 < p < 1.$$

As a result, the exponential QQ-plot introduced above will show an intercept t at the value $p = 0$.

Suppose that on the basis of an exponential QQ-plot, a *global* exponential fit appears appropriate. Then we can answer a crucial question in extreme value analysis that has been mentioned before: the estimation of an extreme quantile $Q(1 - p)$ with p small is given by

$$\hat{q}_p = t - \frac{1}{\hat{\lambda}} \log(p).$$

Conversely, a small exceedance probability $p = P(X > x | X > t)$ will be estimated by

$$\hat{p}_x = \exp\left(-\hat{\lambda}(x - t)\right).$$

Here $\hat{\lambda}$ can be obtained from the least-squares regression on the exponential QQ-plot, or just by taking the maximum likelihood estimator $\hat{\lambda} = 1/(\bar{x} - t)$. This is illustrated in a graphical way in Figure 1.2.

Example 1.1 For a practical example of the above method, let us look at the daily maximal wind speed measurements obtained in Zaventem, Belgium (Figure 1.3). We restrict ourselves to data above $t = 82$ km/hr. We see that the histogram shows

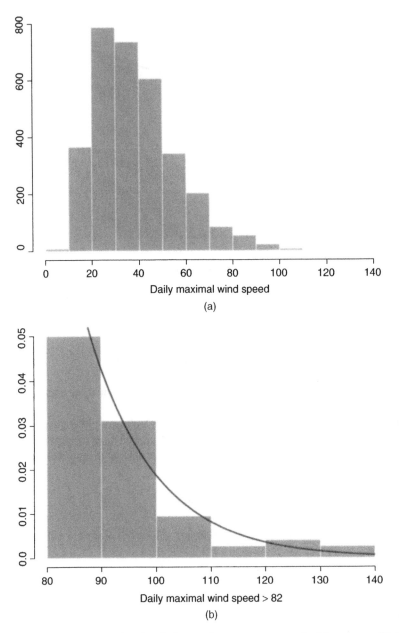

Figure 1.3   Daily maximal wind speed measurements in Zaventem (Belgium) from 1985 till 1992. (a) histogram for all data, (b) conditional histogram for wind speeds larger than 82 km/hr with fitted exponential density superimposed ($\hat{\lambda} = 1/(\bar{x} - 82)$) and (c) exponential QQ-plot for wind speeds larger than 82 km/hr.

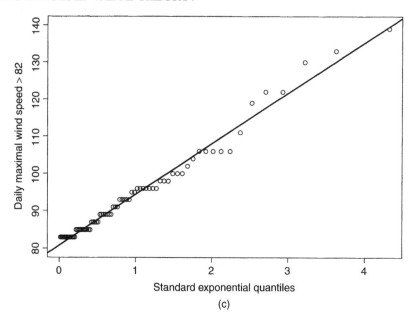

Standard exponential quantiles

(c)

an exponentially decreasing form. The global exponential quantile plot shows a straight line pattern with intercept 82. A similar behaviour but with different values of λ has been shown to hold for most cities in Belgium.

The amount of evidence for the *global* goodness-of-fit can be measured by means of the correlation coefficient

$$r_Q = \frac{\sum_{i=1}^{n}(x_{i,n} - \bar{x})(q_{i,n} - \bar{q})}{\sqrt{\sum_{i=1}^{n}(q_{i,n} - \bar{q})^2 \sum_{i=1}^{n}(x_i - \bar{x})^2}},$$

where

$$\bar{x} = n^{-1}\sum_{i=1}^{n} x_{i,n}$$

and where

$$\bar{q} = n^{-1}\sum_{i=1}^{n} q_{i,n}.$$

The quantity r_Q always satisfies the inequality $0 \leq r_Q \leq 1$. Indeed, since the $x_{i,n}$ and the $q_{i,n}$ are increasing, the correlation coefficient will be non-negative. Moreover, $r_Q = 1$ if and only if all the points lie perfectly on a straight line. Therefore, r_Q can be used as a measure of *global* fit of the exponential model to the data. A formal significance test can be based on the statistic r_Q and rejects the hypothesis of exponentiality when the number obtained differs too much from the value 1.

Equivalently, the value obtained can be compared with that of some tabulated critical value.

We summarize our findings for the case of an exponential QQ-plot as this will help us in pinning down our objectives for the case of general QQ-plots. We denote by Q_s the quantile function of the standard distribution from a given parametric model.

In order to accept a proposed model as a plausible population model:

(i) start from a characterizing linear relationship between (an increasing function of) the theoretical quantiles $Q(p)$ from the proposed distribution and the computable quantiles $Q_s(p)$;

(ii) replace the theoretical quantiles $Q(p)$ by the corresponding empirical quantiles $\hat{Q}_n(p)$;

(iii) plot the (increasing function of the) empirical quantiles $\hat{Q}_n\left(\frac{i}{n+1}\right) = x_{i,n}$ against the corresponding specific quantiles $Q_s\left(\frac{i}{n+1}\right)$;

(iv) inspect the linearity in the plot, for instance, by performing a linear regression on the QQ-plot and by investigating the regression residuals and the correlation coefficient.

Strong linearity implies a good fit. Quantiles and return periods can then be estimated from the linear regression fit $y = \hat{b} + \hat{a}x$ on the QQ-plot:

$$\hat{a} = \frac{\sum_{i=1}^{n}(x_{i,n} - \bar{x})Q_s(p_{i,n})}{\sum_{i=1}^{n}(Q_s(p_{i,n}) - \bar{q})^2},$$

$$\hat{b} = \bar{x} - \hat{a}\bar{q},$$

where $\bar{q} = 1/n \sum_{i=1}^{n} Q_s(p_{i,n})$. Indeed, $\hat{q}_p = \hat{b} + \hat{a}Q_s(1 - p)$ can be used for the estimation of the extreme quantiles. Further, $\hat{p}_x = \bar{F}_s((x - \hat{b})/\hat{a})$ with F_s, the inverse function of Q_s, serves as an estimate for the exceedance probability.

QQ-plots can be used in cases more general than the exponential distribution discussed above. In fact, they can be used to assess the fit of any statistical model. Some other important cases are given below (see Table 1.1 for the QQ-plot coordinates):

- The normal distribution. The coordinates of the points on a normal QQ-plot follow immediately from the representation for normal quantiles

$$Q(p) = \mu + \sigma \Phi^{-1}(p)$$

where Φ^{-1} denotes the standard normal quantile function.

- The log-normal distribution. Since log-transformed log-normal random variables are normally distributed, log-normality can be assessed by a normal QQ-plot of the log-transformed data

$$(\Phi^{-1}(p_{i,n}), \log x_{i,n}), \quad i = 1, \dots, n.$$

- The Pareto distribution. The coordinates of the points on a Pareto QQ-plot follow immediately from the exponential case since a log-transformed Pareto random variable is exponentially distributed.

- The Weibull distribution. The quantile function of the Weibull distribution (cf. Table 1.1) is given by

$$Q(p) = \left(-\frac{1}{\lambda}\log(1-p)\right)^{\frac{1}{\tau}},$$

or equivalently, after a log transformation by

$$\log Q(p) = \frac{1}{\tau}\log\frac{1}{\lambda} + \frac{1}{\tau}\log(-\log(1-p)).$$

This then yields as coordinates for the Weibull QQ-plot

$$\left(\log(-\log(1-p_{i,n})), \log x_{i,n}\right), \quad i = 1, \ldots, n.$$

Example 1.2 Our next example deals with the Norwegian fire insurance data already treated in Beirlant *et al.* (1996a). Together with the year of occurrence, we know the values ($\times$ 1000 Krone) of the claims for the period 1972–1992. A priority of 500 units was in force. The time plot of all the claim values is given in Figure 1.4(a). We will concentrate on the data from the year 1976 for which Figure 1.4(b) shows a histogram. To assess the distributional properties for 1976, we constructed exponential and Pareto QQ-plots, see Figures 1.4(c) and (d) respectively. The points in the exponential QQ-plot bend upwards and exhibit a convex pattern indicating that the claim size distribution has a heavier tail than expected from an exponential distribution. Apart from the last few points, the Pareto QQ-plot is more or less linear indicating a reasonable fit of the Pareto distribution to the tail of the claim sizes. At the three largest observations the Pareto model does not fit so well.

The distribution functions considered so far share the property that QQ-plots can be constructed without knowledge of the correct model parameters. In fact, parameter estimates can be obtained as a pleasant side result. This is particularly true for location-scale models where the intercept of the line fitted to the QQ-plot represents location while the slope represents scale. Unfortunately, this property does not extend to all distributions. In such cases, the construction of QQ-plots involves parameter estimation.

To deal with this more general case, consider a random variable X with distribution function F_θ, where θ denotes the vector of model parameters. To evaluate the fit of F_θ to a given sample $X_1, \ldots, X_n$ using QQ-plots, several possibilities exist. A straightforward approach is to compare the ordered data with the corresponding quantiles of the fitted distribution, i.e., plotting

$$\left(F_{\hat{\theta}}^{\leftarrow}(p_{i,n}), X_{i,n}\right), \quad i = 1, \ldots, n,$$

where $\hat{\theta}$ denotes an estimator for θ based on $X_1, \ldots, X_n$.

Alternatively, one can construct *probability-probability* or *PP*-plots which refer to graphs of the type

$$\left(F_{\hat{\theta}}(X_{i,n}), p_{i,n}\right), \quad i = 1, \ldots, n,$$

or

$$\left(1 - F_{\hat{\theta}}(X_{i,n}), 1 - p_{i,n}\right), \quad i = 1, \ldots, n.$$

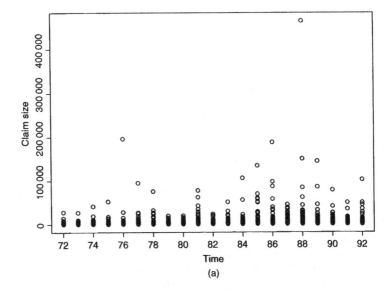

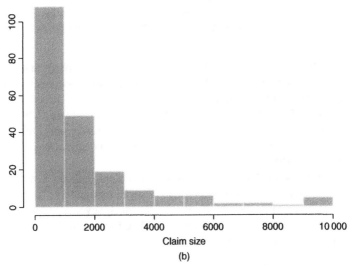

Figure 1.4   (a) Time plot for the Norwegian fire insurance data, (b) histogram, (c) exponential *QQ*-plot and (d) Pareto *QQ*-plot for the 1976 data.

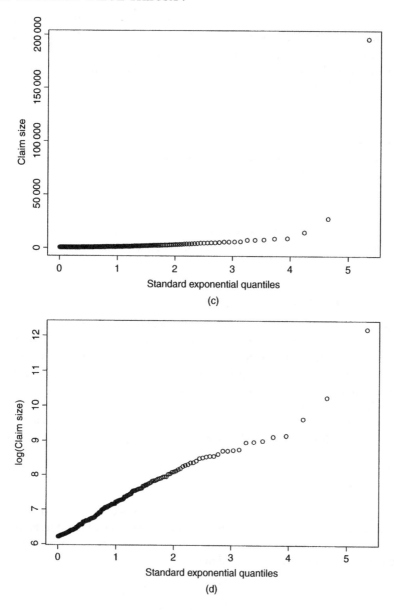

(c)

(d)

For instance, a normal *PP*-plot then consists of

$$\left(\Phi \left(\frac{X_{i,n} - \hat{\mu}}{\hat{\sigma}} \right), p_{i,n} \right), \quad i = 1, \ldots, n.$$

The underlying principle here is that

$$F_\theta(X_{i,n}) \overset{\mathcal{D}}{=} U_{i,n}, \quad i = 1, \ldots, n,$$

where $U_{i,n}$, $i = 1, \ldots, n$ denotes the set of order statistics of a random sample of size n from the $U(0, 1)$ distribution and where $\overset{\mathcal{D}}{=}$ denotes equality in distribution.

One can then return to the exponential framework by transforming the data first to the exponential case followed by a subsequent assessment of the exponential quantile fit. The quantities

$$E_{i,n} = -\log(1 - F_\theta(X_{i,n})), \quad i = 1, \ldots, n,$$

are then the order statistics associated with a random sample of size n from the standard exponential distribution. Hence, another natural procedure to assess the fit of F_θ to $X_1, \ldots, X_n$ is to construct

$$\left(-\log(1 - p_{i,n}), -\log(1 - F_{\hat{\theta}}(X_{i,n}))\right), \quad i = 1, \ldots, n,$$

and to inspect the closeness of the points to the first diagonal. Such a plot is sometimes referred to as a W-plot. Of course, in all these plots the coordinates can be reversed.

1.2.2 Excess plots

The probabilistic operation of conditioning a random variable X on the event $(X > t)$ is of major importance in actuarial practice, especially in reinsurance. Take an excess-of-loss treaty with a retention t on any particular claim in the portfolio. The reinsurer has to pay a random amount $X - t$ but only if $X > t$. When an actuary wants to decide on a priority level t through simulation, he needs to calculate the *expected amount to be paid out per client* when a given level t is chosen. This then is an important first step in deciding on the premium. For instance, the net premium principle depends on the mean claim size $E(X)$. For the overshoot, the actuary will calculate *the mean excess function* or *mean residual life function* e

$$e(t) = E(X - t \mid X > t)$$

assuming that for the proposed model, $E(X) < \infty$. In the whole of extreme value methodology, it is natural to consider data above a specified high threshold.

In practice, the mean excess function e is estimated by $\hat{e}_n$ on the basis of a representative sample $x_1, \ldots, x_n$. Explicitly,

$$\hat{e}_n(t) = \frac{\sum_{i=1}^n x_i 1_{(t,\infty)}(x_i)}{\sum_{i=1}^n 1_{(t,\infty)}(x_i)} - t,$$

where $1_{(t,\infty)}(x_i)$ equals 1 if $x_i > t$, and 0 otherwise. This expression is obtained by replacing the theoretical average by its empirical counterpart, i.e., by averaging the data that are larger than t and subtracting t.

Often the empirical function $\hat{e}_n$ is plotted at the values $t = x_{n-k,n}$, $k = 1, \ldots,$ $n - 1$, the $(k + 1)$-largest observation. Then the numerator equals $\sum_{i=1}^n x_i 1_{(t,\infty)}$ $(x_i) = \sum_{j=1}^k x_{n-j+1,n}$, while the number of x_i larger than t equals k. The estimates of the mean excesses are then given by

$$e_{k,n} := \hat{e}_n(x_{n-k,n}) = \frac{1}{k} \sum_{j=1}^{k} x_{n-j+1,n} - x_{n-k,n}. \tag{1.1}$$

In this section, variations on the mean excess function and its empirical counterparts will be examined from the viewpoint of their statistical applications. But first we need to understand better the behaviour of the theoretical mean excess function. This will help us link empirical shapes with specific theoretical models. In Chapter 4, we will see how the statistic $\hat{e}_{n,\log X}(\log X_{n-k,n})$ with $\hat{e}_{n,\log X}$ denoting the empirical mean excess function of log-transformed data, appears as the *Hill estimator*.

The calculation of e for a random variable with survival function $1 - F$ starts from the formula

$$e(t) = \frac{\int_t^{x_*}(1 - F(u))\, du}{1 - F(t)} \tag{1.2}$$

where $x_* = \sup\{x : F(x) < 1\}$ is the right endpoint of the support of F. The derivation of this alternative formula goes as follows. Apply Fubini's theorem to write

$$\int_t^{x_*}(x - t)\, dF(x) = \int_t^{x_*} \int_t^{x} dy\, dF(x) \tag{1.3}$$

$$= \int_t^{x_*} dy \int_y^{x_*} dF(x) = \int_t^{x_*}(1 - F(y))\, dy.$$

One can also derive an inverse relationship, indicating how one calculates F from e. This then shows that e uniquely determines F. Indeed, from relation (1.2)

$$\int_0^t \frac{1}{e(u)}\, du = \int_0^t \frac{1 - F(u)}{\int_u^{x_*}(1 - F(v))\, dv}\, du$$

$$= -\int_0^t d_u \log \int_u^{x_*}(1 - F(v))\, dv$$

$$= \log \int_0^{x_*}(1 - F(v))\, dv - \log \int_t^{x_*}(1 - F(v))\, dv$$

$$= \log(e(0)(1 - F(0))) - \log((1 - F(t))e(t))$$

so that

$$\frac{1 - F(t)}{1 - F(0)} = \frac{e(0)}{e(t)} \exp\left(-\int_0^t \frac{1}{e(u)}\, du\right).$$

When considering the shapes of mean excess functions, again the exponential distribution plays a central role. A characteristic feature of the exponential distribution is its *memoryless property*, meaning that whether the information $X > t$ is given or not, the outcome for the average value of $X - t$ is the same as if one started at $t = 0$ and calculated $E(X)$. The mean excess function e for the exponential distribution is constant and given by

$$e(t) = \frac{1}{\lambda} \quad \text{for all } t > 0.$$

When the distribution of X has a heavier tail than the exponential distribution (*HTE*), then we find that the mean excess function ultimately increases, while for lighter tails (*LTE*) e ultimately decreases. For example, the Weibull distribution with $1 - F(x) = \exp(-\lambda x^\tau)$ satisfies the asymptotic expression

$$e(t) = \frac{t^{1-\tau}}{\lambda \tau}(1 + o(1))$$

yielding an ultimately decreasing (resp. increasing) e in case $\tau > 1$ (resp. $\tau < 1$). Hence, the shape of e yields important information on the *LTE* or *HTE* nature of the tail of the distribution at hand. The graphs of e for some well-known distributions are sketched in Figure 1.5.

Plots of empirical mean excess values $e_{k,n}$ as introduced in (1.1) can be constructed in two alternative ways, i.e., $e_{k,n}$ versus k, or $e_{k,n}$ versus $x_{n-k,n}$. Remembering the discussion in the preceding subsection, one looks for the behaviour of the plotted $e_{k,n}$ values for decreasing k values or for increasing $x_{n-k,n}$ values. In case of the wind speed data from Zaventem, the constant behaviour becomes apparent from the plot in Figure 1.6. On the other hand, data from the 1976 Norwegian fire insurance example show a *HTE* pattern (see Figure 1.7).

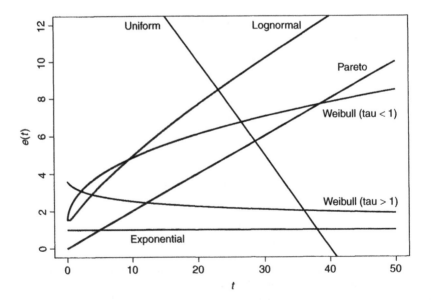

Figure 1.5   Shapes of some mean excess functions.

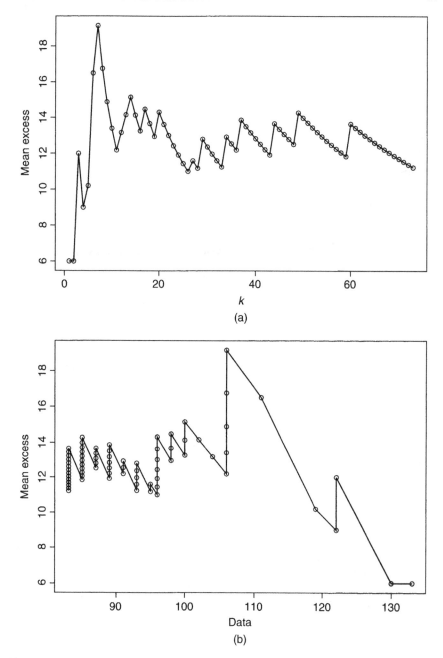

Figure 1.6 Mean excess plots for the daily maximal wind speed measurements larger than 82 km/hr in Zaventem: (a) $e_{k,n}$ versus k and (b) $e_{k,n}$ versus $x_{n-k,n}$.

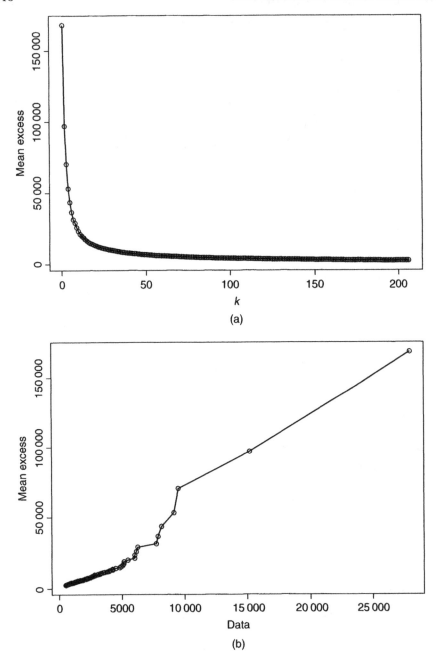

Figure 1.7   Mean excess plots for the 1976 data from the Norwegian fire insurance example: (a) $e_{k,n}$ versus k and (b) $e_{k,n}$ versus $x_{n-k,n}$.

Apart from being an estimate of the function e at a specific value $x_{n-k,n}$, the quantity $e_{k,n}$ can also be interpreted as an estimate of the slope of the exponential QQ-plot to the right of a reference point with coordinates $\left(-\log\left(\frac{k+1}{n+1}\right), x_{n-k,n}\right)$. Here, we make use of the continuity correction in the QQ-plot. Indeed, this slope can be estimated by the ratio of the differences in the vertical and horizontal coordinates between the remaining points and the reference point itself:

$$\tilde{E}_{k,n} = \frac{\frac{1}{k}\sum_{j=1}^{k} X_{n-j+1,n} - X_{n-k,n}}{-\frac{1}{k}\sum_{j=1}^{k}\log\left(\frac{j}{n+1}\right) + \log\left(\frac{k+1}{n+1}\right)}.$$

In this expression, the denominator can be interpreted as an estimator of the mean excess function of the type $e_{k,n}$ taken at $-\log\left(\frac{k+1}{n+1}\right)$ and based on the standard exponential (theoretical) quantiles $-\log(1-p)$ with $p = 1 - p_{j,n}$, $j = 1, \ldots, k$. The denominator hence is an approximation of the mean excess function of the standard exponential distribution $Exp(1)$ as in (1.1) and hence is approximately equal to 1. Using Stirling's formula, one can even verify that this is a very precise approximation even for small values of k. Hence we find that $E_{k,n}$ constitutes an approximation of $\tilde{E}_{k,n}$.

The above discussion explains the ultimately increasing (respectively, decreasing) behaviour of the mean excess function for *HTE* distributions (respectively, *LTE* distributions). In case of a *HTE* distribution, the exponential QQ-plot has a convex shape for the larger observations and the slopes continue to increase near the higher observations. This then leads to an increasing mean excess function. A converse reasoning holds for *LTE* distributions. Illustrations for this principle are sketched in Figure 1.8.

1.3 Domains of Applications

1.3.1 Hydrology

The ultimate interest in flood frequency analysis is the estimation of the *T-year flood discharge* (water level), which is the level exceeded every T years on average. Here, a high quantile of the distribution of discharges is sought. Usually, a time span of 100 years is taken, but the estimation is mostly carried out on the basis of flood discharges for a shorter period. Consequences of floods exceeding such a level can be disastrous. For example, the 100-year flood levels were exceeded by the American flood of 1993 and caused widespread devastation in the states in the Mid-West. The floods in the Netherlands in 1953 were really catastrophic and triggered the Delta Plan of dike constructions, still of interest in that country. By law, dikes in the low-land countries, Belgium and the Netherlands, should be built as high as the 10^4-year flood discharge.

Another hydrological parameter for which the tail of the corresponding distribution is of special interest is rainfall intensity. This parameter is important in modelling water course systems, urban drainage and water runoff. Clearly, the effective capacity of such systems is determined by the most extreme intensities.

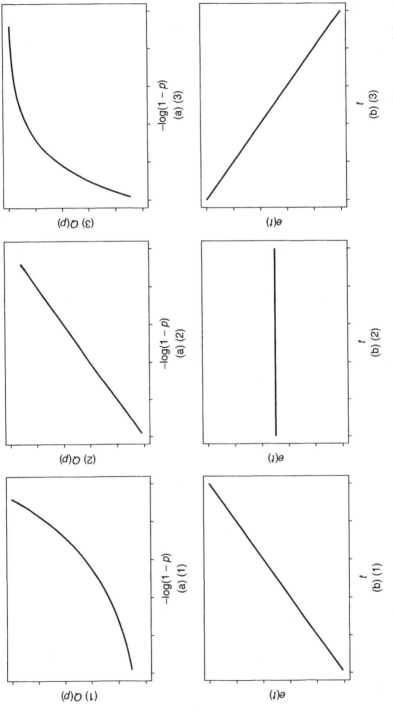

Figure 1.8 (a) Exponential QQ-plot and (b) mean excess plot for simple examples from (1) *HTE*-type, (2) exponential-type ($\bar{F}(x) \approx \exp(-x)$ for large) and (3) *LTE*-type distributions.

Quite often only periodic, even annual, maxima are available. Then, alternative to T-year water levels, the conclusions of an extreme value analysis are requested in terms of *a return period*. The latter is expressed in terms of the reciprocal of the survival function of the periodic maxima, say Y,

$$T(x) = \frac{1}{P(Y > x)}.$$

Later, we will describe how the concept of return period can easily be adapted to cases where the distribution of values of Y is studied.

Case study:

Annual maximal river discharges of the Meuse river from 1911 till 1996 at Borgharen in Holland. The time plot of the annual maximal discharges is given in Figure 1.9(a). In order to get an idea about the tail behaviour of the annual maxima distribution, an exponential QQ-plot was constructed, see Figure 1.9(b). As is clear from this QQ-plot, the annual maxima distribution does not exhibit a HTE tail behaviour. This is also confirmed by the mean excess plots given in Figures 1.9(c) and (d).

1.3.2   Environmental research and meteorology

Meteorological data generally have no alarming aspects as long as they are situated in a narrow band around the average. The situation changes for instance when concentrations occur that overshoot a specific ecological threshold like with ozone concentration. Rainfall and wind data provide other illustrations with tremendous impact on society as they are among the most common themes for discussion. Just recall the questions concerning global warming and climate change. Typically, one is interested in the analysis of maximal and minimal observations and records over time (often attributed to global warming) since these entail the negative consequences.

Case studies:

(i) *Wind speed database provided by NIST, Gaithersburg, consisting of daily fastest-mile speeds measured by anemometers situated 10 m above the ground. The data have been filed for a period of 15 to 26 years and concern 49 airports in the US over a period between 1965 and 1992. Wind speeds from hurricanes and tornadoes have not been incorporated. We select three cities from the study. Table 1.2 gives the length of the data and their observation periods. Figure 1.10 represents the corresponding boxplots.*

(ii) *Daily maximum temperatures at Uccle, Belgium. The data plotted in Figure 1.11 are daily maximum surface air temperatures recorded in degrees Celsius at Uccle, Belgium. These data were gathered as part of the European*

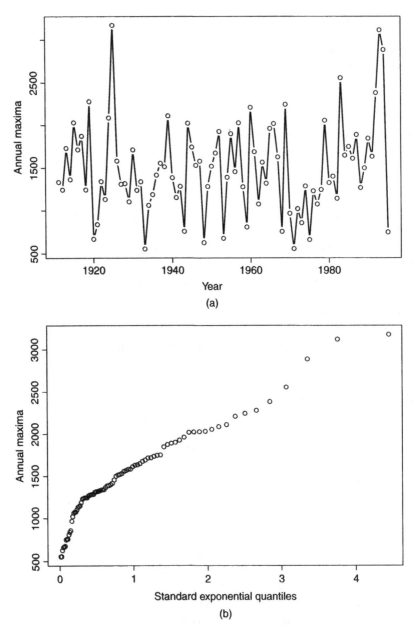

Figure 1.9   (a) Time plot, (b) exponential QQ-plot, (c) $e_{k,n}$ versus k and (d) $e_{k,n}$ versus $x_{n-k,n}$ for the annual maximal discharges of the Meuse.

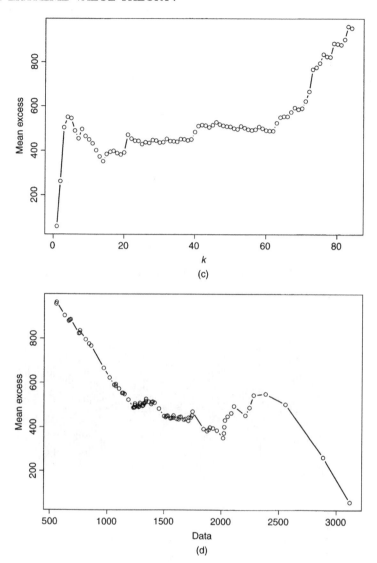

(c)

(d)

Table 1.2 Wind speed database.

City	State	Length	Period
Albuquerque	New Mexico	6939	1965–'83
Des Moines	Iowa	5478	1965–'79
Grand Rapids	Michigan	5478	1965–'79

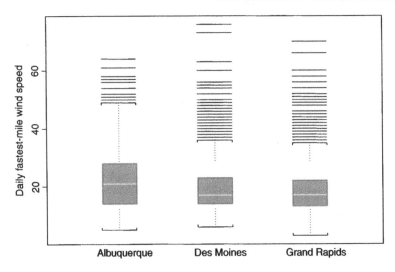

Figure 1.10   Boxplots of the daily fastest-mile wind speeds.

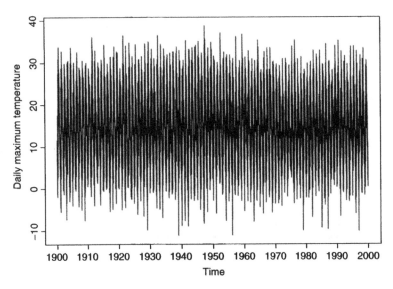

Figure 1.11   Time plot of the daily maximum temperature at Uccle, Belgium.

*Climate Assessment and Dataset project (Klein Tank and co-authors (2002))
and are freely available at* www.knmi.nl/samenw/eca.

1.3.3   Insurance applications

One of the most prominent applications of extreme value thinking can be found
in non-life insurance. Some portfolios seem to have a tendency to occasionally

include a large claim that jeopardizes the solvency of a portfolio or even of a substantial part of the company. Apart from major accidents such as earthquakes, hurricanes, airplane accidents and so on, there is a vast number of occasions where large claims occur. Once in a while, automobile insurance leads to excessive claims. More often, fire portfolios encounter large claims. Industrial fires, especially, cause a lot of side effects in loss of property, temporary unemployment and lost contracts.

An insurance company will always safeguard itself against portfolio contamination caused by claims that should be considered as extreme rather than average. In an excess-of-loss reinsurance contract, the reinsurer pays for the claim amount in excess of a given retention. The claim distribution is therefore truncated to the right, at least from the viewpoint of the ceding company. The estimation of the upper tail of the claim size distribution is of major interest in order to determine the net premium of a reinsurance contract. Several new directions in extreme value theory were influenced by methods developed in the actuarial literature.

Case studies:

(i) *The Secura Belgian Re data set depicted in Figure 1.12(a) contains 371 automobile claims from 1988 till 2001 gathered from several European insurance companies, which are at least as large as 1,200,000 Euro. These data were corrected among others for inflation. The ultimate goal is to provide the participating reinsurance companies with an objective statistical analysis in order to assist in pricing the unlimited excess-loss layer above an operational priority R. These data will be studied in detail in Chapter 6; here we use them only to illustrate some concepts introduced above. The exponential QQ-plot of the claim sizes is given in Figure 1.12(b). From this plot, a point of inflection with different slopes to the left and the right can be detected. This becomes even more apparent in the mean excess plots given in Figures 1.12(c) and (d): behind 2,500,000, the rather horizontal behaviour changes into a positive slope. As we will see later, the mean excess function is an important ingredient for establishing the net premium of a reinsurance contract.*

(ii) *The SOA Group Medical Insurance Large Claims Database. This database records, among others, all the claim amounts exceeding 25,000 USD over the period 1991–92 and is available at http://www.soa.org. There is no truncation due to maximum benefits. The study conducted by Grazier and G'Sell Associates (1997), where a thorough description of these data can be found, collects information from 26 insurers. The 171,000 claims recorded are part of a database including about 3 million claims over the years 1991–92. Here we deal with the 1991 data. The histogram of the log-claim amounts shown in Figure 1.13(a) gives evidence of a considerable right-skewness. Further, the convex shape of the exponential quantile plot (Figure 1.13(b)) and the increasing behaviour of the mean excess plots (Figures 1.13(c) and (d)) in the largest observations indicate a HTE nature of the claim size distribution.*

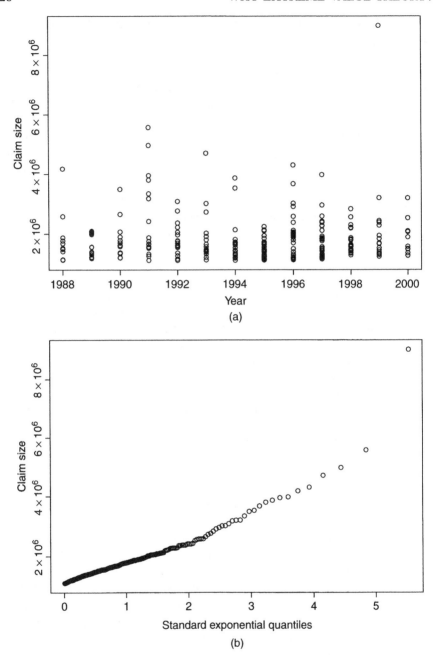

Figure 1.12   Secura Belgian Re data: (a) Time plot, (b) exponential QQ-plot,
(c) $e_{k,n}$ versus k and (d) $e_{k,n}$ versus $x_{n-k,n}$.

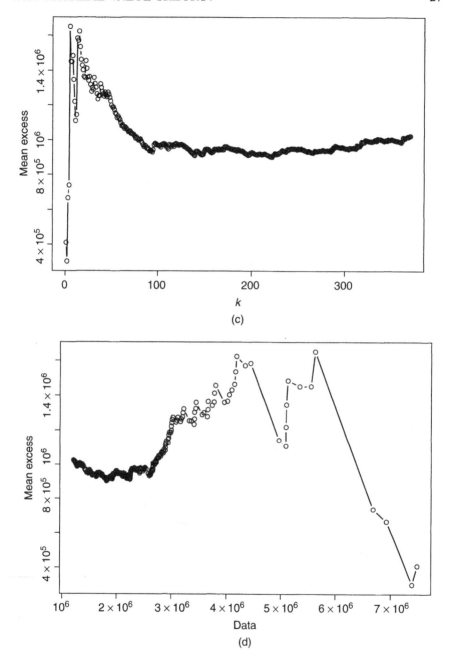

(c)

(d)

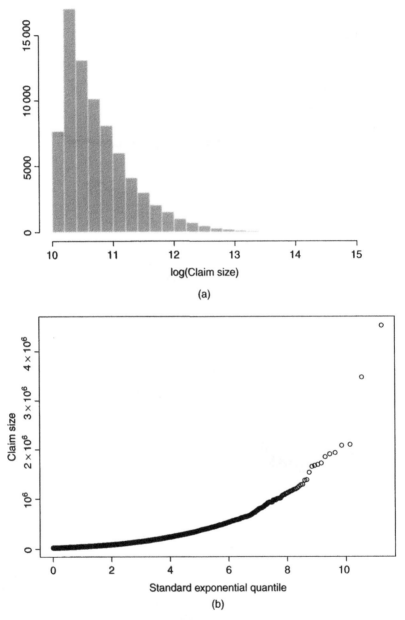

Figure 1.13   SOA Group Medical Insurance data: (a) histogram of log-claim amount, (b) exponential QQ-plot, (c) $e_{k,n}$ versus k and (d) $e_{k,n}$ versus $x_{n-k,n}$.

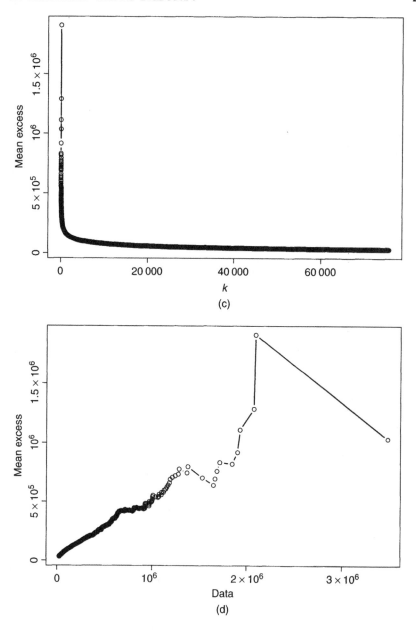

(iii) *Claim data from a fire insurance portfolio provided by the reinsurance broker Aon Re Belgium. The data contain 1668 observations on the claim size, the sum insured and the type of building; see Beirlant et al. (1998). Claim sizes are expressed as a fraction of the sum insured. Figure 1.14 shows the scatter plots of the log(claim size) versus sum insured for three types of building.*

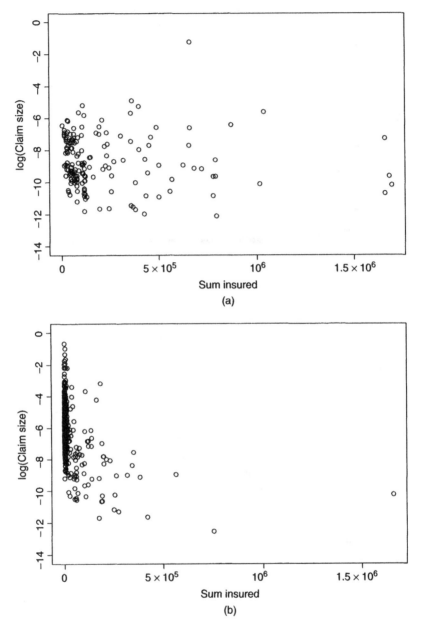

Figure 1.14   Aon Re Belgium data: log(claim size) versus sum insured for three types of buildings.

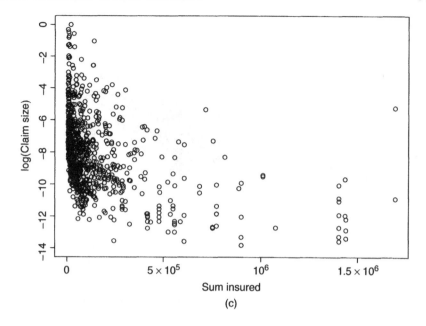

(c)

For reinsurers, the possible influence of covariate information like the sum insured and the type of building is of prime importance for premium differentiation according to the risk involved.

(iv) *Loss-ALAE data studied by Frees and Valdez (1998) and Klugman and Parsa (1999). The data shown in Figure 1.15 comprise 1,500 general liability claims (expressed in USD) randomly chosen from late settlement lags and were provided by Insurance Services Office, Inc. Each claim consists of an indemnity payment (loss) and an allocated loss adjustment expense (ALAE). Here, ALAE are types of insurance company expenses that are specifically attributable to the settlement of individual claims such as lawyers' fees and claims investigation expenses. In order to price an excess-of-loss reinsurance treaty when the reinsurer shares the claim settlement costs, the dependence between losses and ALAE's has to be accounted for. Our objective is to describe the extremal dependence.*

1.3.4 Finance applications

Financial time-series consist of speculative prices of assets such as stocks, foreign currencies or commodities. Risk management at a commercial bank is intended to guard against risks of loss due to a fall in prices of financial assets held or issued by the bank. It turns out that returns, that is, the relative differences of consecutive prices or differences of log-prices, are the appropriate quantities to be investigated; see for instance, Figures 1.16(a) and (b), where we show the time plot of the

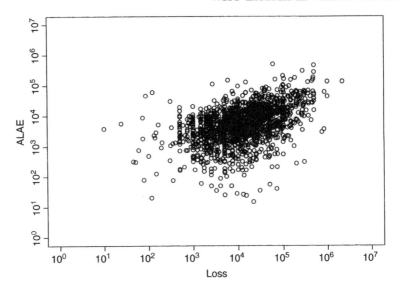

Figure 1.15   Loss-ALAE data: scatterplot of loss vs ALAE.

Standard & Poors 500 closing values and daily percentage returns respectively, from January 1960 up to 16 October 1987, the last market day before the big crash of Black Monday, 19 October 1987. For more about financial applications, we refer to the book by Embrechts *et al.* (1997).

The Value-at-Risk (VaR) of a portfolio is essentially the level below which the future portfolio will drop with only a small probability. VaR is one of the important risk measures that have been used by investors or fund managers in an attempt to assess or predict the impact of unfavourable events that may be worse than what has been observed during the period for which relevant data are available.

1.3.5   Geology and seismic analysis

Applications of extreme value statistics in geology can be found in the magnitudes of and losses from earthquakes, in diamond sizes and values, in impact crater size distributions on terrestrial planets, and so on; see for instance Caers *et al.* (1999a, 1999b). The importance of tail characteristics of such data can be linked to the interpretation of the underlying geological process.

Case studies:

(i) Pisarenko and Sornette (2003) analysed shallow earthquakes (depth <70 km) in the Harvard catalog over the period 1977–2000. In this study, the tails of the seismic moment distributions for subduction and mid-ocean ridge zones are compared. The database contains seismic moment measurements (in dyne-cm) of 6458 earthquakes in subduction zones and 1665 earthquakes in mid-ocean ridge zones. For both zones, the seismic moment distributions are of

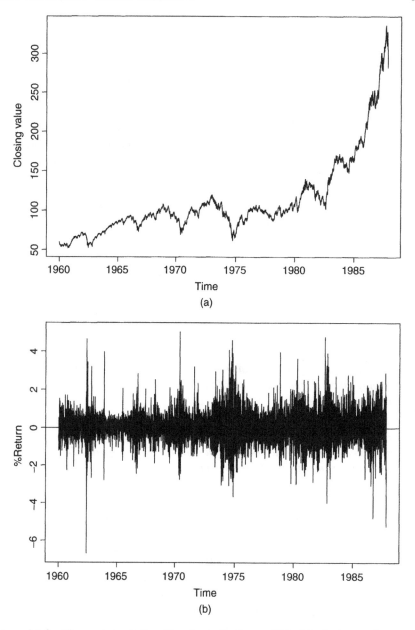

Figure 1.16 Time plot of the Standard & Poors 500 (a) closing values and (b) daily % returns.

HTE *type as indicated by the exponential quantile and mean excess plots given in Figure 1.17.*

(ii) *In agriculture, soil analysis is the basis of fertilizer and amendment recommendations in the context of managing soil fertility and crop performance. Fertilizers are used to meet crop demand for nutrients while amendments are necessary to stabilize and improve both soil structure and water infiltration, and to optimize pH levels. Recently, a new concept of crop management, called precision farming has emerged. It permits within-field variation of crop techniques, for instance, to adjust fertilizer inputs on the basis of soil sampling and soil analysis. As the development of these techniques increased the demand for soil data, laboratories are now burdened with large datasets. In this context, the Belgian non-profit organization REQUASUD (Réseau Qualité Sud i.e. South Quality Network) was created in 1989 to put efficient advices and analysis services at the practitioner's disposal. REQUASUD developed a centralized soil database that contains more than 150,000 soil chemical composition (phKCl, K, Mg, Ca, etc.) records. It also has information about sample origin (zip code), soil texture, soil occupation, previous and recent cultures. The Unit of Geopedology (Gembloux Agricultural University, Belgium) is the reference laboratory for soil analyses and the database is centralized at the Unit of Biometry, Data Management and Agrometeorology (Agricultural Research Centre of Gembloux). Detailed studies of the data allow extension services to study physical and chemical properties of agricultural soils and to manage them according to their fertility potential and their ability to support cultures.*

The Condroz database contains calcium content and pH level measurements of 19,516 soil samples originating from different cities in the Condroz, a geographical region in the southern part of Belgium. Figure 1.18 shows a map of Belgium in which the area covered by the data is grey coloured. For a detailed description of these data, we refer to Goegebeur et al. (2004). The data have been analysed with emphasis on the development of an automatic procedure for highlighting suspicious calcium measurements in order to guarantee database quality. Our focus will be on the related issue of modelling extreme calcium measurements in terms of the covariates pH level and city. As is clear from the calcium versus pH scatter plot given in Figure 1.19(a), both variables are positively associated. Moreover, note that extreme calcium measurements tend to occur more often at the higher pH levels, indicating the need for describing the tail of the calcium distribution in terms of the covariate pH. Here, we comment on the tail behaviour of the calcium content distribution conditional on pH = 6.5. The convex shape of the exponential QQ-plot and (hence) the increasing mean excess function near the largest observations give evidence of a HTE-type tail behaviour, see Figures 1.19(b), (c) and (d).

(iii) *Diamond data. The profitability of a diamond exploration heavily depends on the quality of the stones found in a particular area. In turn, the overall value*

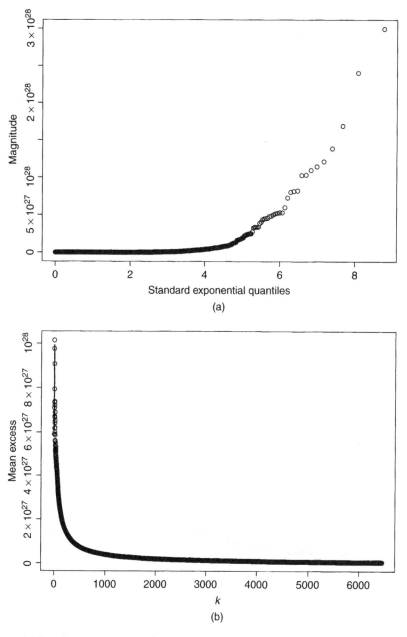

Figure 1.17   Pisarenko and Sornette data: exponential QQ-plots ((a) and (d) respectively) and mean excess plots ((b), (c) and (e), (f) respectively) of seismic moments for subduction respectively mid-ocean ridge zones.

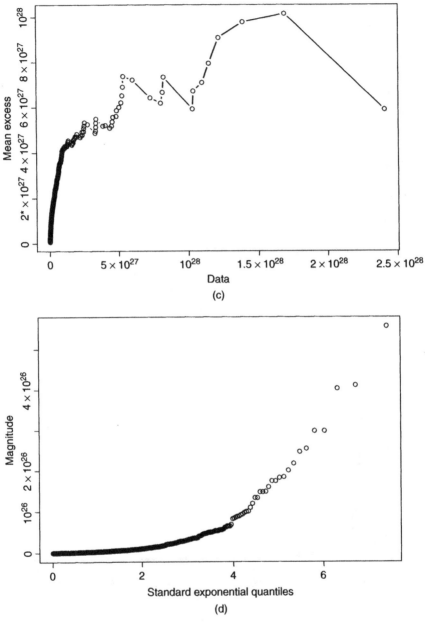

Figure 1.17   *(continued)*

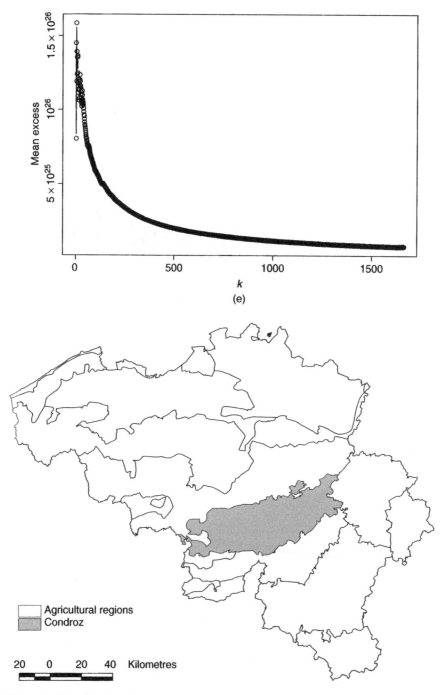

Figure 1.18 Condroz data: geographical area covered by the Condroz database.

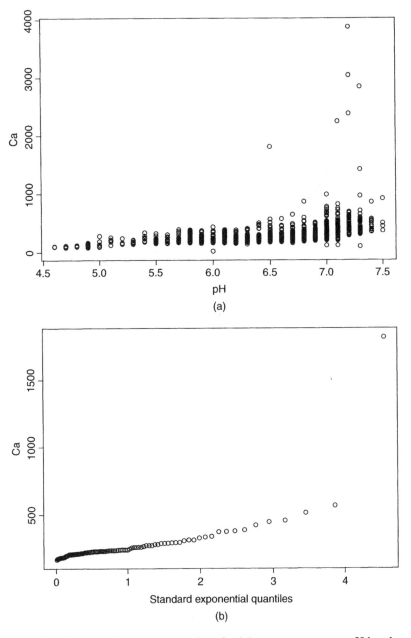

Figure 1.19   Condroz data: (a) scatterplot of calcium content versus pH level measured in soil samples, (b) exponential QQ-plot of the calcium levels at pH $= 6.5$,
(c) $e_{k,n}$ of the calcium levels at pH $= 6.5$ as a function of k and (d) $e_{k,n}$ of the
calcium levels at pH $= 6.5$ as a function of $x_{n-k,n}$.

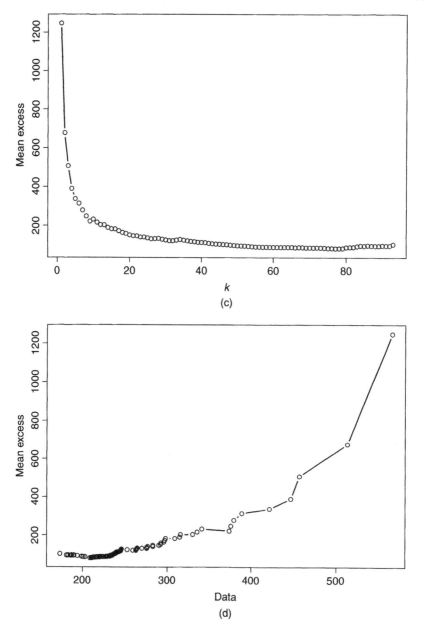

(c)

(d)

of a diamond is influenced by factors such as carat, colour, clarity and cut. This is illustrated in Figure 1.20 in which the value (in USD) versus size (in carat) scatterplot is given for a sample of 1914 diamonds obtained from a kimberlite deposit.

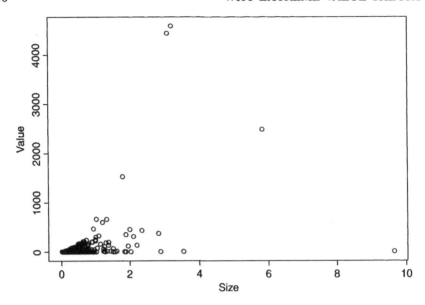

Figure 1.20   Diamond data.

1.3.6 Metallurgy

An important problem from the area of metallurgy that received wide attention is the estimation of the size of the largest inclusions in a metal as metal fatigue typically originates at very large inclusions. See, for instance, the special issue of Extremes, 1999, dedicated to this subject (Bomas *et al.* 1999, Murakami and Beretta 1999, Svensson and de Maré 1999). Here an interesting connection exists with Wicksell's corpuscle problem when only the sizes of vertical sections of such 'grains' are measured. Wicksell (1925) gave the integral relation between the distribution of sizes of spheric objects and the distribution of vertical sections.

Applying a cyclic loading on a metallic component may cause its failure even if the maximum stress is below the static strength limit of the material. This phenomenon is termed *fatigue*. Any material has a minimum stress range—called the fatigue strength—below which it can endure an indefinite number of cycles. However, fatigue properties of steel are strongly influenced by the presence of microscopic particles of oxides or foreign material known as inclusions. Fatigue strength increases with decreasing defect size and therefore, the size of the maximum inclusion is an important indicator of the quality of a particular metallic component. It is infeasible to completely destruct a component in order to find its largest inclusion. Inference about this inclusion has to be based on a representative sample. In general, observations are taken on polished plane surfaces of samples of steel resulting in sizes of two-dimensional cross-sections of those inclusions that intersect the surface. This raises the additional problem to infer about three-dimensional sizes of large inclusions from data in two-dimensional surface

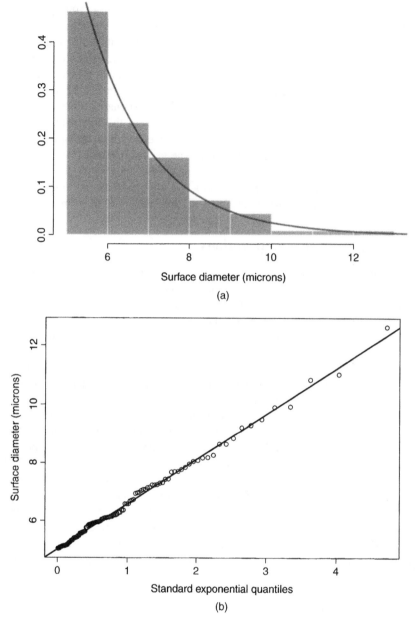

Figure 1.21   (a) Histogram of surface diameters and (b) exponential QQ-plot of surface diameters.

sections. For recent contributions, we refer to Drees and Reiss (1992), Takahashi and Sibuya (1996), Takahashi and Sibuya (1998), Anderson and Coles (2000) and Beretta and Anderson (2002). As an example we use the data from Anderson and Coles (2000). The data contain 112 surface diameters of inclusions on a polished surface above the threshold of 5μm. The units of measurement are taken to insure that the area of the measured surface would be 1. In Figure 1.21(a), we show the histogram of the surface diameters with a fitted exponential density function $(\hat{\lambda} = 1/(\bar{x} - 5) = 1/1.548152)$ superimposed. The fit of the exponential distribution can be further evaluated on the basis of the exponential QQ-plot given in Figure 1.21(b) where the straight line shows the least-squares fit.

Another important problem from metallurgy is the study of pit corrosion. Corrosion can lead to the failure of metal structure such as tanks or tubes. Extreme value analysis becomes relevant since pits of large depth are of primary interest.

1.3.7 Miscellaneous applications

Network traffic data exhibit properties that are inconsistent with traditional queueing models. In fact, next to several other unusual properties, the distribution of quantities such as transmission lengths, transmission rates, file sizes and CPU job completion times can be well modelled by Pareto-type laws which will be discussed in Chapter 4. Some references include Guerin et al. (2000), Resnick (1997), Resnick and Rootzén (2000).

Recently, models for old age mortality data received renewed attention. In fact, there is a debate on whether or not there is a fixed upper limit to the length of human life (see Thatcher (1999)).

We can also refer to Zipf's (1941, 1949) classic study of the dynamics of community sizes where Pareto-type distributions are found again, see, for instance, Feuerverger and Hall (1999). In fact, Pareto laws were also observed for the distribution of biological genera, ranked by the number of species they contain (Willis (1922)), and for the distribution of word usage frequencies in numerous linguistic and literary contexts (Zipf (1935)). The book of Zipf (1941, 1949) presents an incredible variety of phenomena, including examples from economics, business, commerce, economical geography, industry, travel, communication, traffic, sociology, psychology, music, politics and warfare. In many of these examples, an approximate fit of Pareto distribution is rather convincing, particularly, in the tail.

1.4 Conclusion

As a conclusion, we find that the area of extreme value statistics, as statistics in general, offers a wide variety of problems. Aside from the classical problem of analysing the distribution of a single random variable on the basis of a random sample, we find data structures for which time-series models, regression and multivariate settings are appropriate. After parametric and non-parametric approaches in a frequentist approach, Bayesian parameter estimation techniques are also now in use. The goal of this text is to provide an introduction to each of those models and methods. To reach this goal, we provide in Part I the basic theoretical probabilistic and statistical background. In addition, we elaborate on some case studies from the list above.

2

THE PROBABILISTIC SIDE
OF EXTREME VALUE THEORY

Consider a random sample $\{X_i, 1 \leq i \leq n\}$ from a distribution F. In the preceding chapter, it was mentioned that in many situations, extreme value analysis is often (to be) built on a sequence of data that are block maxima, for instance, yearly *maxima*. A traditional statistical discussion on the *mean* is based on the central limit theorem and hence often returns to the normal distribution as a basis for statistical inference. The classical central limit theorem states that the distribution of

$$\sqrt{n}\left(\frac{(X_1 + \cdots + X_n)/n - E(X)}{\sqrt{\text{var}(X)}}\right) = \frac{X_1 + \cdots + X_n - nE(X)}{\sqrt{n\,\text{var}(X)}}$$

converges for $n \to \infty$ to a standard normal distribution. In general, the *central limit problem* deals with the sum $S_n := X_1 + X_2 + \cdots + X_n$ and tries to find constants $a_n > 0$ and b_n such that $Y_n := a_n^{-1}(S_n - b_n)$ tends in distribution to a non-degenerate distribution. Once the limit is known, it can be used to approximate the otherwise cumbersome distribution of the quantity Y_n.

A first question is to determine what distributions can appear in the limit. Then comes the question for which F any such limit is attained. The answer reveals that typically the normal distribution is attained as a limit for this sum (or average) S_n of independent and identically distributed random variables, except when the underlying distribution F possesses a too heavy tail; in the latter case, a *stable distribution* appears as a limit. Specifically, Pareto-type distributions F with infinite variance will yield non-normal limits for the average: the extremes produced by such a sample will corrupt the average so that an asymptotic behaviour different from the normal behaviour is obtained.

In this chapter, we will be mainly concerned with the corresponding problem for the sample maximum rather than the average: we will consider both the possible

---

Statistics of Extremes: Theory and Applications   J. Beirlant, Y. Goegebeur, J. Segers, and J. Teugels
© 2004 John Wiley & Sons, Ltd   ISBN: 0-471-97647-4

limits and the different ways to describe the sets of distributions from which sample maxima are converging to these limits.

2.1   The Possible Limits

In what follows, we will replace the sum S_n by the maximum

$$X_{n,n} = \max\{X_1, X_2 \dots, X_n\}.$$

Of course, we could just as well study the minimum rather than the maximum. Clearly, results for one of the two can be immediately transferred to the other through the relation

$$X_{1,n} = -\max\{-X_1, -X_2, \dots, -X_n\}.$$

It is natural to consider the probabilistic problem of finding the possible limit distributions of the maximum $X_{n,n}$. Hence, the main mathematical problem posed in extreme value theory concerns the search for distributions of X for which there exist a sequence of numbers $\{b_n; n \geq 1\}$ and a sequence of positive numbers $\{a_n; n \geq 1\}$ such that for all real values x (at which the limit is continuous)

$$P\left(\frac{X_{n,n} - b_n}{a_n} \leq x\right) \to G(x) \tag{2.1}$$

as $n \to \infty$. The standardization with b_n and a_n appears natural since otherwise $X_{n,n} \to x_*$ a.s. It is required that the limit G should be a non-degenerate distribution; in fact any number can appear as the degenerate limit of $(X_{n,n} - a_n)/b_n$ whatever the underlying distribution. Again, the problem is twofold: (i) find all possible (non-degenerate) distributions G that can appear as a limit in (2.1); (ii) characterize the distributions F for which there exist sequences $\{a_n; n \geq 1\}$ and $\{b_n; n \geq 1\}$ such that (2.1) holds for any such specific limit distribution.

   The first problem is the *(extremal) limit problem*. It has been solved in Fisher and Tippett (1928), Gnedenko (1943) and was later revived and streamlined by de Haan (1970). Once we have derived the general form of all possible limit laws, we need to solve the second part of the problem, which is called the *domain of attraction problem*. This can be described more clearly in the following manner. Assume that G is a possible limit distribution for the sequence $a_n^{-1}(X_{n,n} - b_n)$. What are the necessary and sufficient conditions on the distribution of X to get precisely that limiting distribution function G. General and specific examples can easily illustrate the variety of distributions attracted to the different limits. The set of such distributions will be called the *domain of attraction* of G and is often denoted by $\mathcal{D}(G)$.

   Trying to avoid an overly mathematical treatment, we will not solve the above problem in its full generality. We rather provide a direct and partial approach to this problem, which works under the assumption that the underlying distribution possesses a continuous, strictly increasing distribution function F.

In contrast with the central limit problem, the normal distribution does not appear as a limiting distribution owing to the inherent skewness that is observed in a distribution of maxima. In this section, we will show that all *extreme value distributions*

$$G_\gamma(x) = \exp\left(-(1 + \gamma x)^{-1/\gamma}\right), \quad \text{for } 1 + \gamma x > 0,$$

with $\gamma \in \mathbb{R}$ can occur as limits in (2.1). The real quantity γ is called the *extreme value index (EVI)*. It is a key quantity in the whole of extreme value analysis.

In order to solve this general limit problem for extremes, we rely on a classical concept from probability theory. Suppose that $\{Y_n\}$ is a sequence of random variables. Then we say that Y_n *converges in distribution* or *converges weakly* to Y, if the distribution function of Y_n converges pointwise to the distribution function of Y, at least in all points where the latter is continuous. We write $Y_n \overset{\mathcal{D}}{\to} Y$. In probability theory, one often proves weak convergence by looking at the corresponding convergence of the characteristic functions. However, for our purposes, we will rely on another well-known result from probability theory, that is, *the Helly-Bray theorem*, see Billingsley (1995). This result transfers the convergence in distribution to the convergence of expectations.

Theorem 2.1 *Let* Y_n *have distribution function* F_n *and let* Y *have distribution function* F. *Then* $Y_n \overset{\mathcal{D}}{\to} Y$ *iff for all real, bounded and continuous functions* z, $E(z(Y_n)) \to E(z(Y))$.

For example, the sequence $\{Y_n\}$ satisfies a *weak law of large numbers* if the random variable Y is degenerate. Alternatively, by the Helly-Bray theorem $Y_n \overset{\mathcal{D}}{\to} Y$ degenerate in the constant c, iff $E(z(Y_n)) \to z(c)$. We will then write $Y_n \overset{P}{\Rightarrow} c$.

For the case of normalized maxima, the limit laws will depend on the crucial parameter γ. For this reason, we include that parameter into the notation. So, put $Y_n := a_n^{-1}(X_{n,n} - b_n)$ and $Y = Y_\gamma$. We then have that $Y_n \overset{\mathcal{D}}{\to} Y_\gamma$ iff for all real, bounded and continuous functions z,

$$E\left\{z\left(a_n^{-1}(X_{n,n} - b_n)\right)\right\} \to \int_{-\infty}^{\infty} z(v) dG_\gamma(v)$$

as $n \to \infty$ where $G_\gamma(v) := P(Y_\gamma \leq v)$.

The idea of the above equivalence is that convergence in distribution can be translated into the convergence of expectations for a sufficiently broad class of functions z. We go through the derivation of the extreme value laws as some of the intermediate steps are crucial to the whole theory of extremes. Therefore, let z be as above, a real, bounded and continuous function over the domain of F. First, note that

$$P(X_{n,n} \leq x) = P(\cap_{i=1}^n (X_i \leq x)) = \prod_{i=1}^n P(X_i \leq x) = F^n(x).$$

Therefore, we find that

$$E\left\{z\left(a_n^{-1}(X_{n,n} - b_n)\right)\right\} = n \int_{-\infty}^{\infty} z\left(\frac{x - b_n}{a_n}\right) F^{n-1}(x)dF(x).$$

We could restrict the domain of the distribution F to the genuine interval of support. This is determined by the *left boundary* $_*x := \sup\{x : F(x) = 0\}$ and the *right boundary* $x_* := \inf\{x : F(x) = 1\}$. Unless this is important, we will not specify these endpoints explicitly.

Recall that F was supposed to be continuous. Therefore, we can set $F(x) = 1 - \frac{v}{n}$. We solve this equation for x. The solution can be put in terms of the *inverse function* $F^{\leftarrow}(y) = \inf\{x : F(x) \geq y\}$, which of course equals the *quantile function* $Q(y)$, or the *tail quantile function* $U(y) = F^{\leftarrow}\left(1 - \frac{1}{y}\right)$. Most often, we will use the prescription

$$U(y) = Q\left(1 - \frac{1}{y}\right) = x \qquad \text{and} \qquad F(x) = 1 - \frac{1}{y}. \tag{2.2}$$

Note in particular that $_*x = U(1)$ and that $x_* = U(\infty)$ while U is non-decreasing over the interval $[1, \infty)$. The integral above therefore equals

$$\int_0^n z\left(\frac{U\left(\frac{n}{v}\right) - b_n}{a_n}\right)\left(1 - \frac{v}{n}\right)^{n-1} dv.$$

Now observe that $\left(1 - \frac{v}{n}\right)^{n-1} \to e^{-v}$ as $n \to \infty$ while the interval of integration extends to the positive half-line. The only place where we still find elements of the underlying distribution is in the argument of z. Therefore, we conclude that a limit for $E\left\{z\left(a_n^{-1}(X_{n,n} - b_n)\right)\right\}$ can be obtained when for some sequence a_n we can make $a_n^{-1}(U(n/v) - b_n)$ convergent for all positive v. It seems natural to think of $v = 1$, which suggests that $b_n = U(n)$ is an appropriate choice. The natural condition to be imposed and crucial to all that follows is that for some positive function a and any $u > 0$,

$$\lim_{x \to \infty} \{U(xu) - U(x)\}/a(x) =: h(u) \text{ exists} , \tag{$\mathcal{C}$}$$

with the limit function h not identically equal to zero.

Let us pause to prove the following basic limiting result.

Proposition 2.2 *The possible limits in* $(\mathcal{C})$ *are given by*

$$ch_\gamma(u) = c\int_1^u v^{\gamma-1}dv = c\frac{u^\gamma - 1}{\gamma}$$

where $c \geq 0$, γ is real and where we interpret $h_0(u) = \log u$.

The case where $c = 0$ is to be excluded since it leads to a degenerate limit for $(X_{n,n} - b_n)/a_n$. Next, the case $c > 0$ can be reduced to the case $c = 1$ by incorporating c in the function a. Hence, we replace the condition $(\mathcal{C})$ by the more informative extremal domain of attraction condition

$$\lim_{x \to \infty} \{U(xu) - U(x)\} / a(x) =: h_\gamma(u) \text{ for } u > 0. \qquad (\mathcal{C}_\gamma)$$

indicating that the possible limits are essentially described by the one-parameter family h_γ. If necessary, we will even explicitly refer to the auxiliary function a by referring to $\mathcal{C}_\gamma(a)$. We will say that the underlying distribution F satisfies the *extreme value condition* $\mathcal{C}_\gamma(a)$ if condition $(\mathcal{C}_\gamma)$ holds with the *auxiliary function* a. At instances, we will also assume functions other than the tail quantile function U to satisfy $\mathcal{C}_\gamma(a)$.

Let us prove the above proposition. Let $u, v > 0$, then

$$\frac{U(xuv) - U(x)}{a(x)} = \frac{U(xuv) - U(xu)}{a(xu)} \frac{a(ux)}{a(x)} + \frac{U(xu) - U(x)}{a(x)}. \qquad (2.3)$$

If we accept that the above convergence condition $(\mathcal{C})$ is satisfied, then automatically the ratio $a(ux)/a(x)$ has to converge too. Let us call the limit $g(u)$. Then the mere existence of a limit can be translated into a condition on the function g. Indeed, for $u, v > 0$, we have

$$\frac{a(xuv)}{a(x)} = \frac{a(xuv)}{a(xv)} \frac{a(xv)}{a(x)}$$

and therefore, the function g satisfies the classical *Cauchy functional equation*

$$g(uv) = g(u)g(v).$$

The solution of this equation follows from

Lemma 2.3 *Any positive measurable solution of the equation* $g(uv) = g(u)g(v)$, $u, v > 0$ *is automatically of the form* $g(u) = u^\gamma$ *for a real* γ.

If one writes $a(x) = x^\gamma \ell(x)$, then the limiting relation $a(xu)/a(x) \to u^\gamma$ leads to the condition $\ell(xu)/\ell(x) \to 1$. This kind of condition is basic within the *theory of regular variation* and will be discussed in more detail in a later section. In particular, any measurable function $\ell(x)$, positive for large enough x, that satisfies $\ell(xu)/\ell(x) \to 1$ will be called a *function of slow variation* or a *slowly varying function* (s.v.). The function $a(x) = x^\gamma \ell(x)$ is then of *regular variation* or *regularly varying*, with index of regular variation γ.

We continue our derivation of Proposition 2.2. Take $g(u) = u^\gamma$. Then it seems natural to clarify the notation in (2.3) by assuming that the right-hand side in the limit relation is given by h_γ. The latter function satisfies the functional equation

$$h_\gamma(uv) = h_\gamma(v) u^\gamma + h_\gamma(u). \qquad (2.4)$$

When $\gamma = 0$, then we immediately find that there exists a constant c such that $h(u) = c \log u$. If $\gamma \neq 0$, then by symmetry we find that for $u, v > 1$

$$h_\gamma(uv) = h_\gamma(v) u^\gamma + h_\gamma(u) = h_\gamma(u) v^\gamma + h_\gamma(v).$$

From this it follows that for some constant d, $h_\gamma(u) = d(u^\gamma - 1)$. We can incorporate the case $\gamma = 0$ if we replace the constant d by the constant $c := \gamma d$. We therefore have derived that the right-hand side of an expression of the form

$$\lim_{x \to \infty} \frac{U(ux) - U(x)}{a(x)} = h(u)$$

is necessarily of the form $h(u) = ch_\gamma(u)$ for some constant c, where the auxiliary function a is regularly varying with index γ.

For the case where U is a tail quantile function, we can do even better. For, U is monotonically non-decreasing. So if $\gamma > 0$, then the constant d is also non-negative since $h_\gamma(u)$ is non-decreasing, while if $\gamma < 0$, then also $d < 0$. Therefore, in both cases $c = \gamma d > 0$ and so the quantity c can be incorporated in the non-negative auxiliary function a. This solves the equation (2.4) and proves the expression in the statement of Proposition 2.2.

Let us return to the derivation of the explicit form of the limit laws in Theorem 2.1. Under $C_\gamma(a)$, we find that with $b_n = U(n)$ and $a_n = a(n)$

$$E\left\{z\left(a_n^{-1}(X_{n,n} - b_n)\right)\right\} \to \int_0^\infty z\left(h_\gamma(1/v)\right) e^{-v} dv =: \int_{-\infty}^\infty z(u) \, dG_\gamma(u)$$

as $n \to \infty$. This in particular shows that (up to scale and location) the class of limiting distributions is a *one-parameter* family, indexed by γ. To get a more precise form, we rewrite the right-hand side of the above equation.

It is easy to derive the three standard extremal types. Put $h_\gamma(1/v) = u$. Then

- if $\gamma > 0$

$$E\left\{z\left(a_n^{-1}(X_{n,n} - b_n)\right)\right\} \to \int_{-\gamma^{-1}}^\infty z(u)d\left(\exp\left(-(1 + \gamma u)^{-1/\gamma}\right)\right) ,$$

- if $\gamma = 0$

$$E\left\{z\left(a_n^{-1}(X_{n,n} - b_n)\right)\right\} \to \int_{-\infty}^\infty z(u)d\left(\exp(-e^{-u})\right) ,$$

- if $\gamma < 0$

$$E\left\{z\left(a_n^{-1}(X_{n,n} - b_n)\right)\right\} \to \int_{-\infty}^{-\gamma^{-1}} z(u)d\left(\exp\left(-(1 + \gamma u)^{-1/\gamma}\right)\right) .$$

Note that the range of G_γ depends on the sign of γ. For $\gamma > 0$, the carrier contains the positive half-line but has a negative left endpoint $-1/\gamma$. For $\gamma < 0$, the distribution contains the whole negative half-line and has a positive right endpoint $-1/\gamma$. Finally, for the *Gumbel* distribution G_0, for $\gamma = 0$, the range is the whole real line. For convenience, we will sometimes write S_γ for the range of the extremal law G_γ. Also, we will write $\eta_\gamma(u)$ for the solution of the equation $h_\gamma(1/v) = u$ in terms of u. Explicitly

$$\eta_\gamma(u) = (1 + \gamma u)^{-1/\gamma} \tag{2.5}$$

with the understanding that $\eta_0(u) = e^{-u}$.

The above analysis entails that under $(\mathcal{C}_\gamma)$, the extreme value distributions can be obtained as limit distributions for the maximum of a simple random sample. It suffices to take $b_n = U(n)$ and $a_n = a(n)$ to have that $a_n^{-1}(X_{n,n} - b_n) \overset{\mathcal{D}}{\to} Y_\gamma$. Another way of stating this result is to write that $F \in \mathcal{D}(G_\gamma)$ if F satisfies $\mathcal{C}_\gamma(a)$. It can also be derived that they are also *the only possible* limits that can be obtained. For more details on this, we refer the reader to the work of Gnedenko (1943) and de Haan (1970); see also Beirlant and Teugels (1995). Densities of some extreme value distributions are sketched in Figure 2.1.

The above result implies that, after a location and scale transformation $x \to (x - b)/a$, the sample distribution of maxima $(Y - b)/a$ can be approximated by an extreme value distribution G_γ if the size n of the pure random samples from which these maxima are computed is sufficiently large.

Now we should turn to the domain of attraction problem: for what kind of distributions are the maxima attracted to a specific extreme value distribution, or which satisfy the central condition $(\mathcal{C}_\gamma)$? It will turn out that the sign of the EVI is the dominating factor in the description of the tail of the underlying distribution F. For that reason, we distinguish between the three cases where $\gamma > 0$, $\gamma < 0$ and the intermediate case where $\gamma = 0$. Because of their intrinsic importance, we treat these cases in separate sections. But before doing that, let us include a concrete example.

2.2   An Example

We incorporate an example to illustrate the above procedure.

The annual maximal discharges of the Meuse river in Belgium consist of maxima $Y_1, \ldots, Y_m$ where m denotes the number of years available. Using a QQ-plot, we can attempt to fit the distribution of $Y = \max\{X_1, \ldots, X_{365}\}$ with an extreme value distribution. In this practical example, a right-skewness is apparent from an explorative data analysis (Figure 2.2).

The quantile function of an extreme value distribution is given by

$$Q_\gamma(p) = \frac{\left(\frac{1}{\log(1/p)}\right)^\gamma - 1}{\gamma}, \quad p \in (0, 1).$$

The case $\gamma = 0$ corresponds to the Gumbel distribution with quantile function

$$Q_0(p) = \log\left(\frac{1}{\log(1/p)}\right), \quad p \in (0, 1).$$

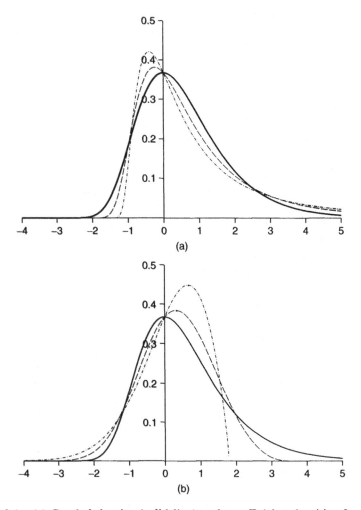

Figure 2.1   (a) Gumbel density (solid line) and two Fréchet densities for parameters $\gamma = 0.28$ (broken line) and $\gamma = 0.56$ (broken-dotted line) and (b) Gumbel
density (solid line) and two extreme value Weibull densities for $\gamma = -0.28$ (broken
line) and $\gamma = -0.56$ (broken-dotted line).

Note that the extreme value quantiles are obtained from standard Fréchet(1) quan-
tiles $\frac{1}{\log(1/p)}$ (see Table 2.1) by using Box-Cox transformations $x \to \frac{x^{\gamma}-1}{\gamma}$. Except
for the special case of a Gumbel quantile plot, a quantile plot for an extreme
value distribution can only be obtained after specifying a value for γ. We start by
considering the simple case of a Gumbel QQ-plot:

$$\left(-\log\left(-\log\frac{i}{n+1} \right), x_{i,n} \right), \quad i = 1, \ldots, n.$$

The Gumbel quantile plot for the annual maximum discharges of the Meuse river is given in Figure 2.3(a). The plot indicates that a Gumbel distribution fits the data quite well and supports the common use in hydrology of this simplified model for annual river discharge maxima. Performing a least squares regression plot on the graph given in Figure 2.3(a), we obtain an intercept value $\hat{b} = 1247$ and a slope $\hat{a} = 446$, which yield estimates of location and scale parameters of the fitted Gumbel model.

When constructing an extreme value QQ-plot, we look for the value of γ in the neighbourhood of 0, which maximizes the correlation coefficient on the QQ-plot.

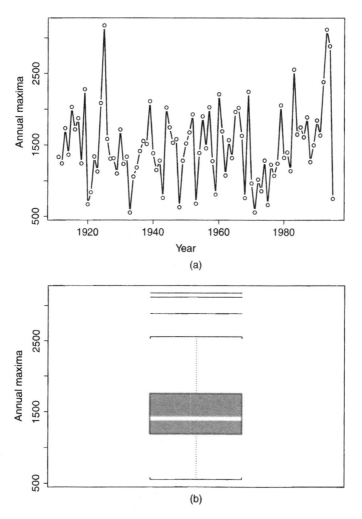

Figure 2.2   (a) Time plot, (b) boxplot, (c) histogram and (d) normal quantile plot for the annual maximal discharges of the Meuse.

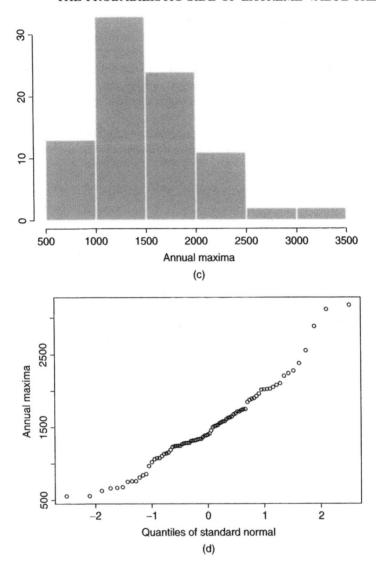

Figure 2.2   (*continued*)

This is obtained for a value $\hat{\gamma} = -0.034$. The corresponding QQ-plot is given in Figure 2.3(b). Here, the least squares line is fitted with $\hat{b} = 1252$ and $\hat{a} = 462$. Of course, γ, a and b can also be estimated by other methods. Here, we can mention the maximum likelihood method and the method of probability-weighted moments. In fact, these estimation methods are quite popular in this context. They will be discussed in Chapter 5.

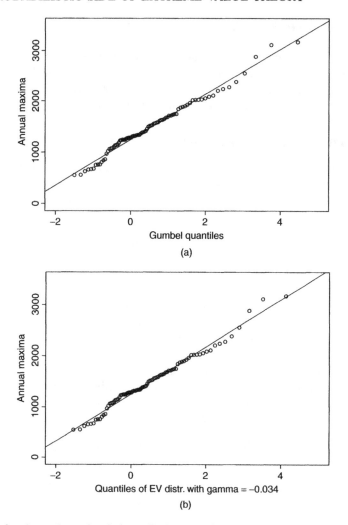

Figure 2.3 Annual maximal river discharges of the Meuse: (a) Gumbel quantile plot and (b) extreme value quantile plot.

These first estimates of γ, b and a can now be used to estimate the 100-year return level for this particular problem. Indeed,

$$\hat{U}(100) = \hat{Q}\left(1 - \frac{1}{100}\right)$$

$$= 1251.86 + 461.55\frac{\left(-\log\left(1 - \frac{1}{100}\right)\right)^{0.034} - 1}{-0.034} = 3217.348 .$$

2.3   The Fréchet-Pareto Case: $\gamma > 0$

We begin with a simple example that can easily be generalized. Then we look for the connection between the tail quantile function U and the underlying distribution F. After rewriting the limiting result in its historical form, we give some sufficient conditions. We finish with a number of examples.

2.3.1   The domain of attraction condition

As *the* prime example, we mention the *strict Pareto distribution* $Pa(\alpha)$ with survival function $\bar{F}(x) = x^{-\alpha}$, $x > 1$. Here, α is a positive number, called *the Pareto index*. For this distribution, $Q(p) = (1 - p)^{-1/\alpha}$ and hence $U(x) = x^{\gamma}$ with $\gamma = 1/\alpha$. Then

$$\{U(xu) - U(x)\}/a(x) = \left((xu)^{\gamma} - x^{\gamma}\right)/a(x)$$

$$= \frac{x^{\gamma}}{a(x)}\left(u^{\gamma} - 1\right)$$

so that the auxiliary function $a(x) = \gamma x^{\gamma}$ leads to

$$\{U(xu) - U(x)\}/a(x) = \frac{u^{\gamma} - 1}{\gamma} = h_{\gamma}(u),$$

and hence $\mathcal{C}_{\gamma}(a)$ is clearly satisfied, actually with equality.

However, there is a much broader class of distributions that satisfies $\mathcal{C}_{\gamma}(a)$ with $\gamma > 0$. Indeed, we can take $U(x) = x^{\gamma}\ell_{U}(x)$ where ℓ_{U} is a slowly varying function. Then $(\mathcal{C}_{\gamma})$ is also satisfied. Indeed, for $x \uparrow \infty$,

$$\{U(xu) - U(x)\}/a(x) = \left((xu)^{\gamma}\ell_{U}(xu) - x^{\gamma}\ell_{U}(x)\right)/a(x)$$

$$= \frac{\ell_{U}(x)x^{\gamma}}{a(x)}\left(\frac{\ell_{U}(xu)}{\ell_{U}(x)}u^{\gamma} - 1\right)$$

$$\sim \left(u^{\gamma} - 1\right)/\gamma$$

when choosing $a(x) = \gamma x^{\gamma}\ell_{U}(x) = \gamma U(x)$, or even more flexibly,

$$\lim_{x \uparrow \infty} a(x)/U(x) = \gamma. \tag{2.6}$$

Distributions for which $U(x) = x^{\gamma}\ell_{U}(x)$ are called *Pareto-type distributions*. Remark that for these distributions, U is regularly varying with index γ since

$$\lim_{x \to \infty} \frac{U(xt)}{U(x)} = \lim_{x \to \infty} \frac{(xt)^{\gamma}\ell_{U}(xt)}{x^{\gamma}\ell_{U}(x)} = t^{\gamma} \text{ for all } t > 0.$$

Also note that once the domain of attraction condition $\mathcal{C}_{\gamma}(a)$ is satisfied, we can choose the normalizing constants by the expressions

$$b_{n} = U(n) = n^{\gamma}\ell_{U}(n) \qquad \text{and} \qquad a_{n} = a(n).$$

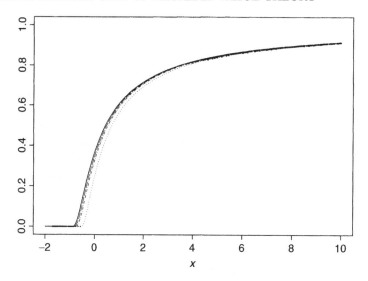

Figure 2.4   Plot of $P\{(X_{n,n} - U(n))/(\gamma U(n)) \le x\} = \left(1 - \frac{(1+x)^{-1}}{n}\right)^{n}$ for $n = 2$ (dotted line), $n = 5$ (broken-dotted line), $n = 10$ (broken line) and its limit, for $n \to \infty$, $\exp(-(1+x)^{-1})$ (solid line).

The convergence of $(X_{n,n} - U(n))/(\gamma U(n))$ to its extreme value limit is illustrated in Figure 2.4 for the case of the strict Pareto distribution with $\gamma = 1$. Here, the convergence appears to be quite fast. In Chapter 3, this will be shown not to be the case overall.

2.3.2   Condition on the underlying distribution

We will show later that the condition (C_{γ}) for $\gamma > 0$ is equivalent to the condition that for $w > 0$,

$$\frac{1 - F(xw)}{1 - F(x)} \to w^{-1/\gamma} \text{ as } x \to \infty.$$

This is precisely saying that $x^{1/\gamma}(1 - F(x))$ is s.v. Hence, there exists a s.v. function $\ell_F(x)$ such that $1 - F(x) = x^{-\alpha}\ell_F(x)$ where $\alpha := \frac{1}{\gamma}$. It will then follow that the definition of a *Pareto-type distribution* can be formulated in terms of the distribution F as well as in terms of the tail quantile function. The term Pareto-type refers to the tail of the distribution, and roughly spoken, it means that as $x \to \infty$, the survival function $1 - F(x)$ tends to zero at a polynomial speed, that is, as $x^{-\alpha}$ for some unknown index α.

The link between the two s.v. functions depends on the concept of the *de Bruyn conjugate* introduced in Proposition 2.5 in section 2.9.3. In its neatest form, we can state that there is full equivalence between the statements

$$1 - F(x) = x^{-1/\gamma} \ell_F(x) \qquad \text{and} \qquad U(x) = x^\gamma \ell_U(x) \qquad (2.7)$$

where the two slowly varying functions ℓ_F and ℓ_U are linked together via the de Bruyn conjugation. Remark that with increasing value of γ the tail becomes heavier, that is, the dispersion is larger; otherwise stated, large outliers become even more likely. When $\gamma > 1$, the expected value $E(X)$ of X does not exist as can be readily checked for the strict Pareto distribution. For $\gamma > 0.5$, even the variance is infinite. For this reason, Pareto-type distributions are often invoked to model data with extremely heavy tails. More specifically, denoting the positive part of X by X_+, one finds that

$$E(X_+^c) = \begin{cases} \infty, & c\gamma > 1, \\ < \infty, & c\gamma < 1. \end{cases}$$

2.3.3   The historical approach

We remark that our approach follows a different path than in the classical theory. From the general approach, we know that $\{X_{n,n} - U(n)\}/a(n) \overset{\mathcal{D}}{\to} Y_\gamma$. However, as derived in Theorem 2.3, we also know that $a(n)/U(n) \to \gamma$. If we write

$$\frac{X_{n,n}}{U(n)} = \frac{a(n)}{U(n)} \left\{ \frac{X_{n,n} - U(n)}{a(n)} + \frac{U(n)}{a(n)} \right\}$$

then $\frac{X_{n,n}}{U(n)} \overset{\mathcal{D}}{\to} Z_\gamma$ where

$$P\{Z_\gamma \le z\} = P\left\{ \gamma \left(Y_\gamma + \frac{1}{\gamma} \right) \le z \right\} = G_\gamma \left(\frac{z-1}{\gamma} \right) = \exp(-z^{-1/\gamma}). \qquad (2.8)$$

In the early literature on the subject, the latter distribution is often abbreviated by $\Phi_{1/\gamma}(z)$ and called the *extreme value distribution of type II*. The limit law in terms of Z_γ looks simpler than the one for Y_γ. We write $F \in \mathcal{D}(G_\gamma) = \mathcal{D}(\Phi_{1/\gamma})$. From the statistical point of view, it is advisable to work with the condition in terms of the quantile function U as the investigator usually does not know in advance that $\gamma > 0$.

As the first example of (2.8) appeared in Fréchet (1927), the class $\mathcal{D}(\Phi_{1/\gamma})$ should be attributed to him. On the other hand, an equivalent description is given by the Pareto-type distributions. We therefore opted for the *Fréchet-Pareto-class* terminology.

2.3.4   Examples

Examples of distributions of Fréchet-Pareto type are given in Table 2.1. We note that the condition in (2.7) can easily be used to derive explicit examples for the Fréchet case.

Table 2.1 A list of distributions in the Fréchet domain.

Distribution	$1 - F(x)$	Extreme value index	$\ell_F(x)$
Pa(α)	$x^{-\alpha}$, $x > 1; \alpha > 0$	$\frac{1}{\alpha}$	1
GP(σ, γ)	$\left(1 + \frac{\gamma x}{\sigma}\right)^{-\frac{1}{\gamma}}$, $x > 0; \sigma, \gamma > 0$	γ	$\left(\frac{\sigma}{\gamma}\right)^{\frac{1}{\gamma}}\left(1 + \frac{\sigma}{\gamma x}\right)^{-\frac{1}{\gamma}}$
Burr(η, τ, λ) (type XII)	$\left(\frac{\eta}{\eta + x^{\tau}}\right)^{\lambda}$, $x > 0; \eta, \tau, \lambda > 0$	$\frac{1}{\lambda\tau}$	$\left(\frac{\eta}{1 + \frac{\eta}{x^{\tau}}}\right)^{\lambda}$
Burr(η, τ, λ) (type III)	$1 - \left(\frac{\eta}{\eta + x^{-\tau}}\right)^{\lambda}$, $x > 0; \eta, \tau, \lambda > 0$	$\frac{1}{\tau}$	$\frac{\lambda}{\eta}\left(1 - \frac{1}{2}\frac{\lambda+1}{\eta}x^{-\tau} + o(x^{-\tau})\right)$
F(m, n)	$\int_x^\infty \frac{\Gamma\left(\frac{m+n}{2}\right)}{\Gamma\left(\frac{m}{2}\right)\Gamma\left(\frac{n}{2}\right)}\left(\frac{m}{n}\right)^{m/2} w^{m/2-1}\left(1 + \frac{m}{n}w\right)^{-(m+n)/2} dw$, $x > 0; m, n > 0$	$\frac{2}{n}$	$\frac{\Gamma\left(\frac{m+n}{2}\right)}{\Gamma\left(\frac{m}{2}\right)\Gamma\left(\frac{n}{2}+1\right)}\left(\frac{m}{n}\right)^{m/2}\left(\frac{m}{n} + \frac{1}{x}\right)^{-(m+n)/2}(1 + o(1))$
InvΓ(λ, α)	$\int_x^\infty \frac{\lambda^{\alpha}}{\Gamma(\alpha)}\exp(-\lambda/w)w^{-\alpha-1}dw$, $x > 0; \lambda, \alpha > 0$	$\frac{1}{\alpha}$	$\frac{\lambda^{\alpha}}{\Gamma(\alpha+1)}\exp(-\lambda/x)(1 + o(1))$
logΓ(λ, α)	$\int_x^\infty \frac{\lambda^{\alpha}}{\Gamma(\alpha)}w^{-\lambda-1}(\log w)^{\alpha-1}dw$, $x > 1; \lambda, \alpha > 0$	$\frac{1}{\lambda}$	$\frac{\lambda^{\alpha-1}}{\Gamma(\alpha)}(\log x)^{\alpha-1}\left(1 + \frac{\alpha-1}{\lambda}\frac{1}{\log x} + o\left(\frac{1}{\log x}\right)\right)$
Fréchet(α)	$1 - \exp(-x^{-\alpha})$, $x > 0; \alpha > 0$	$\frac{1}{\alpha}$	$1 - \frac{x^{-\alpha}}{2} + o(x^{-\alpha})$
$\lvert T_n \rvert$	$\int_x^\infty \frac{2\Gamma\left(\frac{n+1}{2}\right)}{\sqrt{n\pi}\Gamma\left(\frac{n}{2}\right)}\left(1 + \frac{w^2}{n}\right)^{-\frac{n+1}{2}} dw$, $x > 0; n > 0$	$\frac{1}{n}$	$\frac{2\Gamma\left(\frac{n+1}{2}\right)}{\sqrt{n\pi}\Gamma\left(\frac{n}{2}\right)}n^{\frac{n-1}{2}}\left(1 - \frac{n^2(n+1)}{2(n+2)}x^{-2} + o(x^{-2})\right)$

A well-known sufficient condition can be given in terms of the *hazard function*

$$r(x) = \frac{f(x)}{1 - F(x)}$$

where it is assumed that F has a derivative f.

Proposition 2.1 Von Mises' theorem *If $x_* = \infty$ and $\lim_{x \uparrow \infty} xr(x) = \alpha > 0$, then $F \in \mathcal{D}(\Phi_\alpha)$.*

If we use Theorem 2.3 (*ii*) below and define $\epsilon(x) := xr(x) - \alpha$, then it easily follows by integration that

$$1 - F(x) = \{1 - F(1)\} \exp\left\{-\int_1^x \frac{\alpha + \epsilon(u)}{u} du\right\} =: Cx^{-\alpha}\ell(x) \text{ for } x \geq 1.$$

The function $\ell(x)$ defined in this way can easily be shown to be slowly varying.

The von Mises result is readily applied when for some $\beta > 1$, the tail of the density f is of the form $f(x) \sim x^{-\beta}\ell(x)$ for some s.v. $\ell(x)$. It then follows that also the tail of F is regularly varying. Actually,

$$1 - F(x) = \int_x^\infty f(y)\, dy = xf(x) \int_1^\infty \frac{f(xv)}{f(x)}\, dv \sim (\beta - 1)^{-1}xf(x)$$

and thus $xr(x) \to \beta - 1$ so that the von Mises sufficient condition is satisfied with $\alpha = \beta - 1$.

Examples of Pareto-type distributions are the Burr distribution, the Generalized Pareto distribution, the log-gamma distribution and the Fréchet distribution (see Table 2.1).

Note that all *t-distributions* are in the Fréchet-Pareto domain. A *t*-density f with k degrees of freedom has the form

$$f(x) = \frac{\Gamma\left(\frac{k+1}{2}\right)}{\sqrt{k\pi}\,\Gamma\left(\frac{k}{2}\right)} \left(1 + \frac{x^2}{k}\right)^{-\frac{k+1}{2}}, \quad x \in \mathbb{R}.$$

Then

$$1 - F(x) \sim \frac{1}{k\sqrt{\pi}} \frac{\Gamma\left(\frac{k+1}{2}\right)}{\Gamma\left(\frac{k}{2}\right)} \left(1 + \frac{x^2}{k}\right)^{-\frac{k}{2}} \in \mathcal{R}_{-k}.$$

Hence, $F \in \mathcal{D}(\Phi_k)$. In particular, the Cauchy distribution is an element of $\mathcal{D}(\Phi_1)$.

But also all *F-distributions* are in the Fréchet-Pareto domain. An *F*-density with m and n degrees of freedom is given by the expression

$$f(x) = \frac{\Gamma\left(\frac{m+n}{2}\right)}{\Gamma\left(\frac{m}{2}\right)\Gamma\left(\frac{n}{2}\right)} \left(\frac{m}{n}\right)^{m/2} x^{\frac{m}{2}-1}\left(1 + \frac{m}{n}x\right)^{-\frac{m+n}{2}}, \quad x > 0.$$

As before $1 - F(x) \sim C_{m,n} x^{-n/2}$ as $x \to \infty$ with $C_{m,n}$ a constant and hence $F \in \mathcal{D}(\Phi_{n/2})$.

Most of the above examples have a slowly varying function ℓ_U of the type

$$\ell_U(x) = C\left(1 + Dx^{-\beta}(1 + o(1))\right) \quad x \to \infty$$

for some constants $C > 0$, $D \in \mathbb{R}$, $\beta > 0$. The corresponding subclass of the Pareto-type distributions is often named *the Hall class of distributions* referring to Hall (1982). This class plays an important role in the discussion of estimators of a positive EVI γ.

2.3.5 Fitting data from a Pareto-type distribution

Here, we recall and extend some of the basic notions on QQ-plots discussed earlier in section 1.2.1.

Case 1: The strict Pareto case

The strict Pareto distribution $Pa(\alpha)$ with survival function $\bar{F}(x) = x^{-\alpha}$ $(x > 1)$ has quantile function $Q(p) = (1 - p)^{-1/\alpha}$. Hence, $\log Q(p) = -\frac{1}{\alpha} \log(1 - p)$. A Pareto QQ-plot is obtained from an exponential QQ-plot after taking the logarithm of the data:

$$\left(-\log(1 - p_{i,n}), \log x_{i,n}\right), \quad i = 1, \ldots, n.$$

Indeed, taking a log-transformation from a strict Pareto random variable results in an exponential random variable. We expect to see a linear shape with intercept zero and slope approximately equal to $1/\alpha$ in case a strict Pareto distribution fits well.

Case 2: The bounded Pareto case

We put a left bound on the Pareto distribution. To do that, we consider the conditional distribution of X given $X > t$ where the data are censored by a lower retention level t. The conditional distribution is obtained from

$$\bar{F}_t(x) = P\left(X > x | X > t\right) = \left(\frac{x}{t}\right)^{-\alpha}, \quad x > t$$

and its quantile function is given by

$$Q_t(p) = t\,(1 - p)^{-1/\alpha}, \quad p \in (0, 1).$$

Hence, performing the same procedure described as for the strict Pareto case, we plot

$$\left(-\log(1 - p_{i,n}), \log x_{i,n}\right), \quad i = 1, \ldots, n,$$

and we obtain a straight line pattern with intercept $\log t$ and a slope again approximately equal to $1/\alpha$.

Case 3: Pareto-type distributions

Pareto-type distributions were introduced in 2.3.1. The effect obtained by using a Pareto QQ-plot to a *Pareto-type* distribution is illustrated in Figure 2.5.

In all the examples in 2.3.4, the leading factor in the expression of the survival function is of the form $x^{-\frac{1}{\gamma}}$ for some positive number γ. The value of γ is also

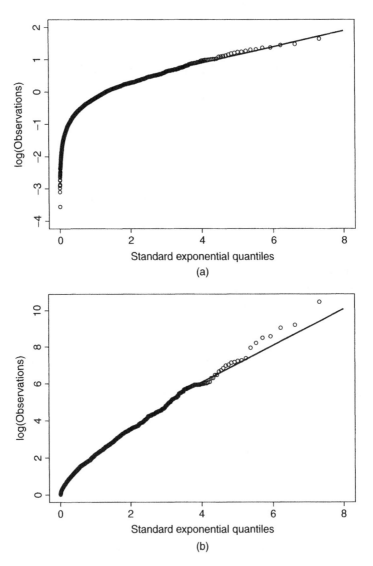

Figure 2.5   Pareto QQ-plot for simulated data of size $n = 1500$ from (a) the Burr(1,2,2) distribution ($\gamma = 0.25$) and (b) the $\log\Gamma(1, 2)$ distribution ($\gamma = 1$).

given in Table 2.1 in terms of the model parameters. In Chapter 4, we will present an estimation procedure for α based on the Pareto QQ-plot.

As we have shown, Pareto-type distributions were found to be the precise set of distributions for which the sample maxima are attracted to an extreme value distribution with EVI $\gamma = 1/\alpha > 0$. However, it now appears that this set of distributions can be used in a broader statistical setting than the one considered in section 3.1 where the statistical practitioner had only access to data from independent block maxima. Indeed, Pareto-type distributions are characterized by the specification

$$U(x) = x^\gamma \ell_U(x)$$

for the tail quantile function U. Here, ℓ_U denotes an s.v. function as indicated in (2.7). So,

$$Q(1 - p) = p^{-\gamma} \ell_U \left(\frac{1}{p} \right), \quad p \in (0, 1).$$

Then

$$\log Q(1 - p) = -\gamma \log p + \log \ell_U \left(\frac{1}{p} \right)$$

and, since for every s.v. function ℓ_U

$$\frac{\log \ell_U \left(\frac{1}{p} \right)}{\log p} \to 0, \quad \text{as } p \to 0,$$

(see (2.19) below) we have that

$$\frac{\log Q(1 - p)}{- \log p} \to \gamma, \quad \text{as } p \to 0$$

which explains the ultimate linear appearance in a Pareto QQ-plot in case of an underlying Pareto-type distribution. This is illustrated in Figure 2.5.

In Figure 2.5, the Pareto QQ-plot has been constructed for simulated data sets of size $n = 1500$ (a) from the Burr distribution and (b) from the log-gamma distribution. It is clear that the straight line pattern only appears at the right end of the plot. This has been suggested by superimposing a line segment on the right tail of the QQ-plot. The fitting of this line segment has been performed manually. In both cases, the slope of the fitted line is set equal to the reference value $1/\alpha$.

Pareto-type behaviour can also be deduced from *probability-probability plots* or *PP-plots*. For the strict Pareto distribution, the coordinates of the points on the *PP*-plot are given by

$$\left(x_{i,n}^{-1/\gamma}, 1 - p_{i,n} \right), \quad i = 1, \ldots, n.$$

In case the Pareto distribution fits well, we expect the points to be close to the first diagonal. This 'classical' PP-plot requires knowledge of γ, or at least of an estimate of γ. Alternatively, log-transforming the above coordinates and changing signs leads to the plot

$$\left(\log x_{i,n}, -\log(1 - p_{i,n})\right), \quad i = 1, \dots, n, \tag{2.9}$$

which is obtained from the Pareto QQ-plot by interchanging the coordinates. For the strict Pareto distribution $-\log(1 - F(x)) = \frac{1}{\gamma} \log x$, so the Pareto probability

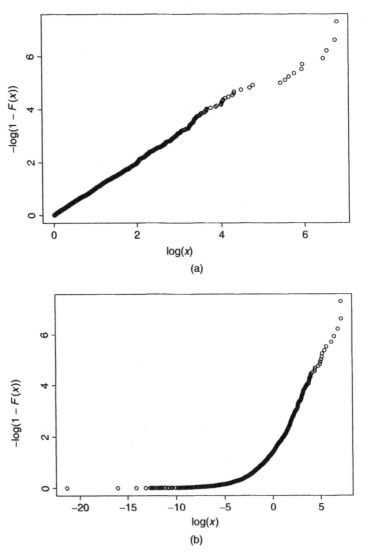

Figure 2.6   Pareto *PP*-plot for simulated data of size $n = 1500$ from (a) the Pa(1) distribution, (b) the Burr(1,0.5,2) distribution and (c) the log $\Gamma(1, 1.5)$ distribution.

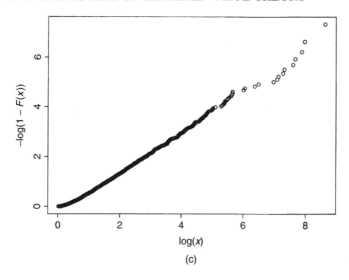

(c)

plot defined by (2.9) will be approximately linear with slope $\frac{1}{\gamma}$. Using arguments similar to the ones in the discussion of the Pareto quantile plot, it is easy to show that for Pareto-type distributions, the plot (2.9) will be ultimately linear with slope $\frac{1}{\gamma}$. This is illustrated in Figure 2.6.

2.4 The (Extremal) Weibull Case: $\gamma < 0$

We follow the same pattern as in the previous case. It will turn out that there is a full equivalence between the cases $\gamma > 0$ and $\gamma < 0$.

2.4.1 The domain of attraction condition

Let us again start with a simple example, slightly more complicated than the uniform distribution. Let $0 < x_* < \infty$ and look at the survival function on $(0, x_*)$:

$$1 - F(x) = (1 - x/x_*)^\beta,$$

where $\beta > 0$. It follows that $U(x) = x_*(1 - x^{-1/\beta})$ on $[1, \infty)$. Then

$$\{U(xu) - U(x)\}/a(x) = \frac{x_*}{a(x)}\left((1 - (xu)^{-\frac{1}{\beta}}) - (1 - x^{-\frac{1}{\beta}})\right)$$

$$= \frac{x_* x^{-\frac{1}{\beta}}}{a(x)}\left(1 - u^{-\frac{1}{\beta}}\right)$$

$$= \frac{x_* x^{-\frac{1}{\beta}}}{\beta a(x)}\, h_{-\frac{1}{\beta}}(u)\,.$$

Therefore, we recover the condition $C_\gamma(a)$ if we make the choice $\gamma = -\frac{1}{\beta} < 0$ and $a(x) = (1/\beta)x_*x^{-1/\beta}$ for the auxiliary function. Note that $a \in \mathcal{R}_\gamma$ as it should be and that $a(x) = (-\gamma)(x_* - U(x))$.

Again, there is a much broader class of distributions where the above reasoning holds. Let $x_* < \infty$, put $1 - F(x) = (x_* - x)^{-1/\gamma}\ell_F(1/(x_* - x))$ as $x \uparrow x_*$, and put $\ell(v) = \ell_F^\gamma(v)$. It then follows by Proposition 2.5 that $U(y) = x_* - y^\gamma \ell_U(y)$ as $y \uparrow \infty$ where $\ell_U(y) = (\ell^*(y^{-\gamma}))^{-1}$ with ℓ^* denoting the de Bruyn conjugate of ℓ (defined in section 2.9.3 below). Then

$$\{U(xu) - U(x)\}/a(x) = \frac{x^\gamma \ell_U(x)}{a(x)}\left(1 - u^\gamma \frac{\ell_U(xu)}{\ell_U(x)}\right)$$

$$\sim -\gamma \frac{x^\gamma \ell_U(x)}{a(x)}h_\gamma(u)$$

which is of the required $C_\gamma(a)$ form if we choose

$$\frac{a(x)}{x_* - U(x)} \to -\gamma . \tag{2.10}$$

Figure 2.7 shows the convergence of $(X_{n,n} - U(n))/(-\gamma(x_+ - U(n)))$ to its extreme value limit in case of the $U(0, 1)$ distribution.

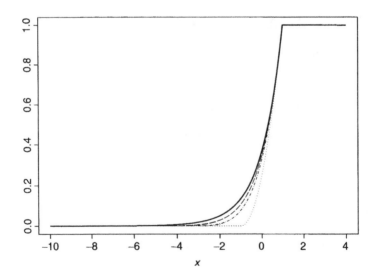

Figure 2.7 Plot of $P\{(X_{n,n} - U(n))/a(n) \leq x\} = \left(1 + \frac{x-1}{n}\right)^n$ for $n = 2$ (dotted line), $n = 5$ (broken-dotted line), $n = 10$ (broken line) and its limit, for $n \to \infty$, $\exp(-(1 - x))$ (solid line).

2.4.2 Condition on the underlying distribution

Note that again all the previous steps can be reversed so that there is full equivalence between the statements

$$1 - F\left(x_* - \frac{1}{x}\right) = x^{\frac{1}{\gamma}}\ell_F(x), \quad x \uparrow \infty$$

and   (2.11)

$$U(x) = x_* - x^{\gamma}\ell_U(x), \quad x \uparrow \infty.$$

The proof is similar to the one given for $\gamma > 0$ in section 2.9.3.

2.4.3 The historical approach

We give an indication of the earlier derivation for the case where $\gamma < 0$. We continue to take $x_* < \infty$ and $(x_* - U(x))/a(x) \to -\gamma^{-1}$. So, write

$$\frac{X_{n,n} - x_*}{x_* - U(n)} = \frac{a(n)}{x_* - U(n)}\left\{\frac{X_{n,n} - U(n)}{a(n)} + \frac{U(n) - x_*}{a(n)}\right\}.$$

If we are in the domain of attraction $\mathcal{D}(G_\gamma)$ for $\gamma < 0$, then also the left-hand side of the above equation converges in distribution, say to Z_γ and we have

$$Z_\gamma \stackrel{\mathcal{D}}{=} (-\gamma)\left(Y_\gamma + \frac{1}{\gamma}\right)$$

with distribution

$$P(Z_\gamma \leq z) = \exp\left(-|z|^{-\frac{1}{\gamma}}\right)$$

for $z < 0$.

Again, the latter limit relation has been treated differently in the older literature. One used to write for $z < 0$, $\Psi_\alpha = \exp(-|z|^\alpha)$ for the *extreme value distribution of type III*. We then see that $F \in \mathcal{D}(G_\gamma) = \mathcal{D}(\Psi_{-1/\gamma})$. It is true that the above limit distribution is again a bit simpler than the one using Y_γ. However, usually the statistician has no prior information on the sign of the EVI γ. Moreover, splitting the one-parameter set of limit distributions into three apparently different subcases spoils the unity of extreme value theory.

2.4.4 Examples

Put $Y := (x_* - X)^{-1}$. As mentioned before, the extreme value Weibull case and the Fréchet case are easily linked through the identification:

$$F_X \in \mathcal{D}(\Psi_\alpha) \Leftrightarrow F_Y \in \mathcal{D}(\Phi_\alpha).$$

Table 2.2 A list of distributions in the extreme value Weibull domain.

Distribution	$1 - F\left(x_+ - \frac{1}{x}\right)$	Extreme value index	$\ell_F(x)$
Uniform	$\frac{1}{x}$ $x > 1$	-1	1
Beta(p,q)	$\int_{1-\frac{1}{x}}^{1} \frac{\Gamma(p+q)}{\Gamma(p)\Gamma(q)} u^{p-1}(1-u)^{q-1}du$ $x > 1; p, q > 0$	$-\frac{1}{q}$	$\frac{\Gamma(p+q)}{\Gamma(p)\Gamma(q+1)}\left(1-\frac{1}{x}\right)^{p-1}\left\{1 + \frac{p-1}{q+1}\left(1-\frac{1}{x}\right)^{-1}\frac{1}{x} + o(x^{-1})\right\}$
Reversed Burr	$\left(\frac{\beta}{\beta+x^\tau}\right)^\lambda$ $x > 0; \lambda, \beta, \tau > 0$	$-\frac{1}{\lambda\tau}$	$\beta^\lambda(1 - \lambda\beta x^{-\tau} + o(x^{-\tau}))$
Extreme value Weibull	$1 - \exp(-x^{-\alpha})$ $x > 0; \alpha > 0$	$-\frac{1}{\alpha}$	$1 - \frac{x^{-\alpha}}{2} + o(x^{-\alpha})$

This equivalence follows from simple algebra in that

$$1 - F\left(x_* - \frac{1}{x}\right) = P\left(X > x_* - \frac{1}{x}\right) = P\left(\frac{1}{x_* - X} > x\right) = 1 - F_Y(x).$$

Some examples of distributions in the extreme value Weibull domain are given in Table 2.2.

Apart from the determination of the right endpoint x_*, the cases $\gamma < 0$ and $\gamma > 0$ are fully equivalent. Note in particular that in case a density exists, $f_Y(x) = x^{-2} f_X(x_* - x^{-1})$. Therefore, there exists a sufficient von Mises condition in terms of the hazard function is $r = f/(1 - F)$.

Proposition 2.1 Von Mises' theorem *If $x_* < \infty$ and $\lim_{x \uparrow x_*} (x_* - x)\, r(x) = \alpha > 0$, then $F \in \mathcal{D}(\Psi_\alpha)$.*

The proof is similar to the one for the Fréchet-Pareto case. From the condition, it follows that

$$\frac{f\left(x_* - \frac{1}{t}\right)}{1 - F\left(x_* - \frac{1}{t}\right)} \sim \alpha t$$

which leads to $f_X(x_* - t) \sim t^{\alpha-1}\ell(1/t)$ when $t \to 0$ and where ℓ is slowly varying.

Explicit examples are known as well. *Beta-distributions* are among the most popular elements from the extreme value Weibull domain. Recall the *beta-density* with parameters p and q

$$f(x) = \frac{1}{B(p, q)} x^{p-1} (1 - x)^{q-1} .$$

Here, $x_* = 1$ and $1 - F(1 - x) \sim \{q B(p, q)\}^{-1} x^q$ making the distribution an element from $\mathcal{D}(\Psi_q)$. In particular, the uniform distribution is an element of $\mathcal{D}(\Psi_1)$.

A graphical description of tails of models with right-bounded support as a function of the value of γ is given in Figure 2.8. Remark especially the different shapes near the endpoint x_* around the values $\gamma = -1/2$ and -1.

2.5 The Gumbel Case: $\gamma = 0$

This case, often called extremal type I, is more diverse than the two previous ones: this set of tails turns out to be quite complex as can be seen from Table 2.3, which contains a list of distributions in this domain.

2.5.1 The domain of attraction condition

The class (C_0) is called the *Gumbel class* as here the maxima are attracted to the Gumbel distribution function $\Lambda(x) := G_0(x) = \exp(-e^{-x})$. The domain of attraction is denoted by $\mathcal{D}(\Lambda)$.

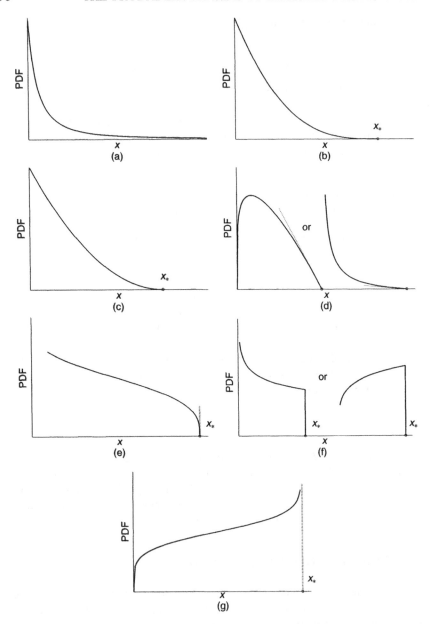

Figure 2.8 Tails of distributions: different cases along the value of the *EVI*. (a) $\gamma \geq 0$, no upper bound, (b) $-1 < \gamma < 0$, finite endpoint x_*, zero density, (c) $\gamma = -1/k$ ($k \leq 3$, integer), zero density at x_*, first $(k-2)$ derivatives of the density function are zero, (d) $\gamma = -1/2$, zero density, finite first derivative at x_*, (e) $-1 < \gamma < -1/2$, zero density at x_*, infinite first derivative, (f) $\gamma = -1$, non-zero finite density at x_* and (g) $\gamma < -1$, infinite density at x_*.

The most central example is the exponential distribution with survival function $1 - F(x) = e^{-\lambda x}$, $x > 0$, with $\lambda > 0$. Then $Q(p) = -\frac{1}{\lambda}\log(1 - p)$, and so $U(x) = \frac{1}{\lambda}\log x$. Then

$$\{U(xu) - U(x)\}/a(x) = \frac{1}{\lambda}(\log(xu) - \log(x))/a(x)$$

$$= \frac{1}{\lambda a(x)}\log(u)$$

so that the constant function $a(x) = 1/\lambda$ leads to

$$\{U(xu) - U(x)\}/a(x) = \log(u),$$

and $(\mathcal{C}_0)$ is satisfied.

The convergence of $(X_{n,n} - U(n))/a(n)$ to the Gumbel limit is illustrated in Figure 2.9 for the $Exp(\lambda)$ distribution.

As opposed to the other two domains of attraction, the elements of $\mathcal{D}(\Lambda)$ cannot be considered as being from the same *type* as this prime example. Validating that all the other examples from Table 2.3 belong to $(\mathcal{C}_0)$ can be a tedious job. We provide some alternative conditions for $\mathcal{D}(\Lambda)$ next.

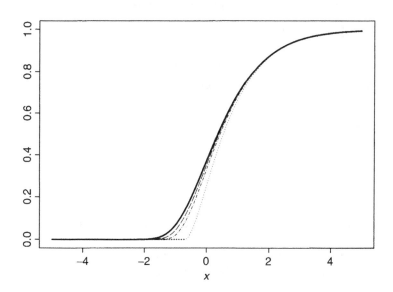

Figure 2.9   Plot of $P\{(X_{n,n} - U(n))/a(n) \le x\} = \left(1 - \frac{\exp(-x)}{n}\right)^n$ for $n = 2$ (dotted line), $n = 5$ (broken-dotted line), $n = 10$ (broken line) and its limit, for $n \to \infty$, $\exp(-\exp(-x))$ (solid line).

Table 2.3   A list of distributions in the Gumbel domain.

Distribution	$1 - F(x)$
Benktander II	$x^{-(1-\beta)} \exp\left(-\frac{\alpha}{\beta} x^{\beta}\right),$ $x > 0; \alpha, \beta > 0$
Weibull	$\exp(-\lambda x^{\tau}),$ $x > 0; \lambda, \tau > 0$
Exponential $\exp(\lambda)$	$\exp(-\lambda x),$ $x > 0; \lambda > 0$
Gamma	$\frac{\lambda^m}{\Gamma(m)} \int_x^{\infty} u^{m-1} \exp(-\lambda u) du,$ $x > 0; \lambda, m > 0$
Logistic	$1/(1 + \exp(x)),$ $x \in \mathbb{R}$
Log-normal	$\int_x^{\infty} \frac{1}{\sqrt{2\pi}\sigma u} \exp\left(-\frac{1}{2\sigma^2}(\log(u) - \mu)^2\right) du,$ $x > 0; \mu \in \mathbb{R}, \sigma > 0$

2.5.2   Condition on the underlying distribution

The characterization of $\mathcal{D}(\Lambda)$ in terms of the distribution function is also more complex than in the other two cases. The pioneering thesis of de Haan (1970) gave a solution to this problem, revitalizing the interest in extreme value analysis.

Proposition 2.1 *The distribution F belongs to $\mathcal{D}(\Lambda)$ if and only if for some auxiliary function b for every $v > 0$*

$$\frac{1 - F(y + b(y)v)}{1 - F(y)} \to e^{-v} \tag{2.12}$$

as $y \to x_*$. Then

$$\frac{b(y + vb(y))}{b(y)} \to 1 .$$

This result being of quite different nature than for the Fréchet-Pareto and the extreme value Weibull case, the question can now be raised if the formulation in (2.12) can be generalized to the general case of $(\mathcal{C}_\gamma)$. This will be done in section 2.6. A sketch of proof for this general case will be given in section 2.9.4.

2.5.3   The historical approach and examples

The case $\gamma = 0$ has been responsible for the seemingly awkward treatment of the three extreme value domains. If one tries to find the appropriate centering and norming constants for the most classical distribution from statistics, the normal distribution, the calculations are by no means trivial, on the contrary.

The von Mises sufficiency condition is a bit more elaborate than before.

Proposition 2.2 Von Mises' theorem *If $r(x)$ is ultimately positive in the neighbourhood of x_*, is differentiable there and satisfies $\lim_{x \uparrow x_*} \frac{dr(x)}{dx} = 0$, then F belongs to $\mathcal{D}(\Lambda)$.*

The calculations involved in checking the attraction condition to Λ are often tedious. In this respect, the von Mises criterion can be very handy, particularly as the Gumbel domain is very wide. This can be illustrated with the *normal distribution* as well as with the (classical) *Weibull distribution* $F(x) = 1 - e^{-x^\alpha}$ with $\alpha > 0$ and $x > 0$. Also, the *logistic distribution* has an explicit expression $F(x) = \{1 + \exp(-(x-a)/b)\}^{-1}$, which is easily shown to satisfy the von Mises condition. The calculations for the *log-normal distribution* are somewhat tedious.

2.6 Alternative Conditions for $(\mathcal{C}_\gamma)$

We return to the general case. In view of statistical issues for arbitrary values of the EVI γ, we need alternative conditions for the general domain of attraction condition $(\mathcal{C}_\gamma)$. The proofs of the results in this section are probably among the most technical points in the discussion of the attraction problem for the maximum. Then again, we do not treat the most general case since we started out from the added restriction that F is continuous. Proofs are deferred to the end of the chapter.

(*i*) A first and equivalent condition is given in terms of the distribution function. The result comes from de Haan (1970) and extends Proposition 2.1 for the case $\gamma = 0$ to the general case. The derivation is postponed to section 2.9.4.

Proposition 2.1 *The distribution F belongs to $\mathcal{D}(G_\gamma)$ if and only if for some auxiliary function b and $1 + \gamma v > 0$*

$$\frac{1 - F(y + b(y)v)}{1 - F(y)} \to (1 + \gamma v)^{-1/\gamma} \quad (\mathcal{C}_\gamma^*) \tag{2.13}$$

as $y \to x_*$. Then

$$\frac{b(y + vb(y))}{b(y)} \to u^\gamma = 1 + \gamma v \ .$$

As will be shown in section 2.9.4, the auxiliary function b can be taken as $b(y) = a(U^{\leftarrow}(y))$.

Another equivalent condition is closely linked to the above. Instead of letting $y \to \infty$ in an arbitrary fashion, we can restrict y by putting $1 - F(y) = n^{-1}$ or equivalently $y = U(n)$.

Proposition 2.2 *The distribution F satisfies* (C_γ) *if and only if*

$$n\{1 - F(U(n) + b_n v)\} \to H(v) \qquad (2.14)$$

for a positive sequence b_n *and a positive, non-constant function H.*

As before, the mere existence of the limit is enough for the explicit form $(1 + \gamma v)^{-1/\gamma}$ for a real γ.

(*ii*) Here is a first necessary condition for (C_γ) that will be crucial in the statistical chapters. The condition (C_γ) entails that for $x \to \infty$

$$\frac{U(x)}{a(x)}\{\log U(xu) - \log U(x)\} \to \begin{cases} \log u, & \text{if } \gamma \geq 0, \\ \frac{u^\gamma - 1}{\gamma}, & \text{if } \gamma < 0 \\ & \text{provided } x_* > 0 . \end{cases} \qquad (\tilde{C}_\gamma)$$

(*iii*) The relationship between U and a as appearing in conditions (C_γ) and $(\tilde{C}_\gamma)$ is different in the three cases. This becomes clear from the following result.

Theorem 2.3 *Let* (C_γ) *hold.*

(*i*) *Fréchet-Pareto case:* $\gamma > 0$. *The ratio* $a(x)/U(x) \to \gamma$ *as* $x \to \infty$ *and U is of the same regular variation as the auxiliary function a; moreover,* (C_γ) *is equivalent with the existence of a s.v. function* ℓ_U *for which* $U(x) = x^\gamma \ell_U(x)$.

(*ii*) *Gumbel case:* $\gamma = 0$. *The ratios* $a(x)/U(x) \to 0$ *and* $a(x)/\{x_* - U(x)\} \to 0$ *when* x_* *is finite.*

(*iii*) *(Extremal) Weibull case:* $\gamma < 0$. *Here* x_* *is finite, the ratio* $a(x)/\{x_* - U(x)\} \to -\gamma$ *and* $\{x_* - U(x)\}$ *is of the same regular variation as the auxiliary function a; moreover,* (C_γ) *is equivalent with the existence of an s.v. function* ℓ_U *for which* $x_* - U(x) = x^\gamma \ell_U(x)$.

However, the function a can be linked to the mean excess function of the log-transformed data over all cases. We first consider the case where $\gamma > 0$. In a later section, we will see how Hill's estimator can be motivated from the fact that

$$e_{\log X}(\log x) := E\left\{\log \frac{X}{x} \Big| X > x\right\} = \int_x^\infty \frac{\log(u/x)}{1 - F(x)}\, dF(u)$$

$$= \int_x^\infty \frac{1 - F(u)}{1 - F(x)}\frac{du}{u} = \int_1^\infty \frac{1 - F(xy)}{1 - F(x)}\frac{dy}{y}$$

for $x \to \infty$ tends to $\int_1^\infty y^{-1/\gamma}\frac{dy}{y} = \gamma$ (in Theorem 2.3 we show why the limit and integral can be interchanged).

In the sequel, we will show that under $(\tilde{C}_\gamma)$

$$\frac{U(x)e_{\log X}(\log U(x))}{a(x)} \to \begin{cases} 1, & \text{if } \gamma \geq 0, \\ \int_1^\infty \frac{u^\gamma - 1}{\gamma} \frac{du}{u^2} = \frac{1}{1-\gamma}, & \text{if } \gamma < 0 \\ & \text{provided } x_* > 0, \end{cases} \quad (2.15)$$

when $x \to \infty$. It then follows that the function b appearing in (C_γ^*) can be taken as

$$b(t) = (1 - \gamma^-)te_{\log X}(\log t) \quad (2.16)$$

where

$$\gamma^- = \begin{cases} 0, & \text{if } \gamma \geq 0, \\ \gamma, & \text{if } \gamma < 0. \end{cases}$$

2.7   Further on the Historical Approach

As illustrated in the previous sections, the literature offers other forms that are more customary as they refer to the distribution F rather than to the tail quantile function U. We feel the need to formulate the result in its historic form. We include the domain of attraction conditions as well. The next theorem contains the historical results derived by Fisher and Tippett (1928) and Gnedenko (1943).

Distributions that differ only in location and scale are called *of the same type*.

Theorem 2.1 Fisher-Tippett-Gnedenko Theorem. *The extremal laws are exactly those that agree to within type with one of the following* ($\alpha > 0$)

(i) *Fréchet-Pareto-type:*

$$\Phi_\alpha(x) = \begin{cases} 0, & \text{if } x \leq 0, \\ e^{-x^{-\alpha}}, & \text{if } x \geq 0. \end{cases}$$

Moreover, $F \in \mathcal{D}(\Phi_\alpha)$ *if and only if for* $x \to \infty$

$$\frac{1 - F(\lambda x)}{1 - F(x)} \to \lambda^{-\alpha}, \text{ for all } \lambda > 0 .$$

(ii) *Gumbel-type:*

$$\Lambda(x) = e^{-e^{-x}} \quad x \in \mathbb{R}.$$

Moreover $F \in \mathcal{D}(\Lambda)$ *if and only if for some auxiliary function b for* $x \to \infty$

$$\frac{1 - F(x + tb(x))}{1 - F(x)} \to e^{-t}, \text{ for all } t > 0.$$

(iii) (Extremal) Weibull-type:

$$\Psi_\alpha(x) = \begin{cases} e^{-|x|^\alpha}, & if \ x \le 0, \\ 1, & if \ x \ge 0. \end{cases}$$

Moreover $F \in \mathcal{D}(\Psi_\alpha)$ *if and only if* $x_* < \infty$ *and for* $x \to \infty$

$$\frac{1 - F(x_* - \frac{1}{\lambda x})}{1 - F(x_* - \frac{1}{x})} \to \lambda^{-\alpha}, \ for \ all \ \lambda > 0 \ .$$

2.8   Summary

We summarize the most important results of this chapter concerning conditions on $1 - F$ or the tail quantile function $U(y) = Q(1 - 1/y)$ for a distribution to belong to the maximum domain of attraction of an extreme value distribution.

	Conditions on U	Conditions on $1 - F$
General case $\gamma \in \mathbb{R}$	$(\mathcal{C}_\gamma)$	$(\mathcal{C}_\gamma^*)$
	$\lim_{x\to\infty} \frac{U(ux)-U(x)}{a(x)} = h_\gamma(u)$	$\lim_{y\to x_*} \frac{1-F(y+b(y)v)}{1-F(y)} = \eta_\gamma(v)$
		where $b(y)$ can be taken as
	$\Downarrow$	$(1 - \gamma^-)y e_{\log X}(\log y)$
	$(\tilde{\mathcal{C}}_\gamma)$	
	$\lim_{x\to\infty} \frac{U(x)}{a(x)} \log \frac{U(ux)}{U(x)}$	
	$= \begin{cases} \log u, & if \ \gamma \ge 0, \\ \frac{u^\gamma - 1}{\gamma}, & if \ \gamma < 0, \end{cases}$	
	where $a(x)$ can be taken as	
	$(1 - \gamma^-)U(x)e_{\log X}(\log U(x))$	
Pareto-type distributions $\gamma > 0$	$\lim_{x\to\infty} \frac{U(ux)}{U(x)} = u^\gamma, \ u > 0$	$\lim_{y\to\infty} \frac{1-F(yv)}{1-F(y)} = v^{-\frac{1}{\gamma}}, \ v > 0$

2.9   Background Information

In this section, we collect a number of results that provide most of the background necessary to fully understand the mathematical derivations. Some of these results are proved while for the proofs of other statements, we refer the reader to the literature. We begin with information on inverse functions that is useful to see what kind of conditions we are actually imposing on the underlying distributions. We then collect information on functions of regular variation: Here Bingham *et al.* (1987) is an appropriate reference. In the last part, we return to alternative forms of the condition (C_γ) and the relation to the underlying distribution F.

2.9.1 Inverse of a distribution

Start with a general distribution F. Denote by ${}_*x := \inf\{x : F(x) > 0\}$ the left-end boundary of F while similarly $x_* := \sup\{x : F(x) < 1\}$. We define the *tail quantile function* U by $U(t) := \inf\{x : F(x) \geq 1 - 1/t\}$. Note that the tail quantile function U and the quantile function Q are linked via the relation $U(t) = Q(1 - 1/t)$.

From this definition, we get the following inequalities:

(i) If $z < U(t)$, then $1 - F(z) > 1/t$;

(ii) for all $t > 0$, $1 - F(U(t)) \leq 1/t$;

(iii) for all $x < x_*$, $U\left(1/(1 - F(x))\right) \leq x$.

Note that $U(1) = {}_* x$ while $U(\infty) = x_*$. It is easy to prove that *if F is continuous*, then the equality $1 - F(U(t)) = 1/t$ is valid while if U is continuous, $U\left(1/(1 - F(x))\right) = x$. In any case, U will be left-continuous while F is right-continuous. The easy transition from F to U and back is the main reason for always assuming that F and U are continuous.

Even more special is the important case where F has a strictly positive derivative f on its domain of definition. For then

$$1 - F(U(t)) = \int_{U(t)}^{x_*} f(y)\, dy = \frac{1}{t}\,.$$

But then also U has a derivative u and both derivatives are linked by the equation

$$f(U(t))u(t) = t^{-2} \tag{2.17}$$

which could help in calculating u or U if F is known through its density function f.

2.9.2 Functions of regular variation

In this section, we treat a class of functions that shows up in a vast number of applications in the whole of mathematics and that is intimately related to the class of power functions. We first give some generalities. Then we state a number of fundamental properties. We continue with properties that are particularly important for us.

Definition 2.1 *Let f be an ultimately positive and measurable function on $\mathbb{R}_+$. We will say that f is regularly varying if and only if there exists a real constant ρ for which*

$$\lim_{x \uparrow \infty} \frac{f(xt)}{f(x)} = t^\rho \quad \text{for all } t > 0. \tag{2.18}$$

We write $f \in \mathcal{R}_\rho$ and we call ρ the index of regular variation. In the case $\rho = 0$, the function will be called slowly varying (s.v.) or of slow variation. We will reserve the symbol ℓ for such functions.

The class of all regularly varying functions is denoted by $\mathcal{R}$.

It is easy to give examples of s.v. functions. Typical examples are

- $\ell(x) = (\log x)_+^\alpha$ for arbitrary α. Further on, we will always drop the $+$ sign; regular variation indeed is an asymptotic concept and hence does not depend on what happens at fixed values.

- $\ell(x) = \prod_1^k (\log_k x)^{\alpha_k}$ where $\log_1 x = \log x$ while for $n \geq 1$, $\log_{n+1} x :=$ $\log(\log_n x)$.

- ℓ satisfying $\ell(x) \to c \in (0, \infty)$.

- $\ell(x) = \exp\{(\log x)^\beta\}$ where $\beta < 1$.

The class $\mathcal{R}_0$ has a lot of properties that will be constantly used. Some of the proofs are easy. For others, we refer to the literature.

Proposition 2.2 *Slowly varying functions have the following properties:*

(i) *$\mathcal{R}_0$ is closed under addition, multiplication and division.*

(ii) *If ℓ is s.v. then ℓ^α is s.v. for all $\alpha \in \mathbb{R}$.*

(iii) *If $\rho \in \mathbb{R}$, then $f \in \mathcal{R}_\rho$ iff $f^{-1} \in \mathcal{R}_{-\rho}$.*

Mathematically, the two most important results about functions in $\mathcal{R}_0$ are given in the following theorem due to Karamata.

Theorem 2.3

(i) **Uniform Convergence Theorem.** *If $\ell \in \mathcal{R}_0$, then the convergence*

$$\lim_{x \uparrow \infty} \frac{\ell(xt)}{\ell(x)} = 1$$

is uniform for $t \in [a, b]$ where $0 < a < b < \infty$.

(ii) **Representation Theorem.** *$\ell \in \mathcal{R}_0$ if and only if it can be represented in the form*

$$\ell(x) = c(x) \exp\left\{ \int_1^x \frac{\epsilon(u)}{u} du \right\}$$

where $c(x) \to c \in (0, \infty)$ and $\epsilon(x) \to 0$ as $x \to \infty$.

The second part of the theorem above allows direct construction of elements of $\mathcal{R}_0$. For later reference, we collect some more properties in the next proposition.

Proposition 2.4 *Let ℓ be slowly varying. Then*

(i)

$$\lim_{x \uparrow \infty} \frac{\log \ell(x)}{\log x} = 0. \tag{2.19}$$

(ii) *For each $\delta > 0$ there exists a x_δ so that for all constants $A > 0$ and $x > x_\delta$*

$$Ax^{-\delta} < \ell(x) < Ax^{\delta}.$$

(iii) *If $f \in \mathcal{R}_\rho$ with $\rho > 0$, then $f(x) \to \infty$, while for $\rho < 0$, $f(x) \to 0$ as $x \uparrow \infty$.*

(iv) **Potter Bounds.** *Given $A > 1$ and $\delta > 0$ there exists a constant $x_o(A, \delta)$ such that*

$$\frac{\ell(y)}{\ell(x)} \leq A \, max \left\{ \left(\frac{y}{x}\right)^{\delta}, \left(\frac{y}{x}\right)^{-\delta} \right\}, \quad x, y \geq x_o.$$

Statements (i) and (ii) follow easily from the representation theorem. Unfortunately, (ii) does not characterize slow variation. For (iii) and $\rho > 0$, use (ii) to see that for x large enough $f(x) \sim x^\rho \ell(x) > x^{\rho/2}$ which tends to ∞ with x. Relation (iii) shows that regularly varying functions with $\rho \neq 0$ are akin to monotone functions. Further, Potter bounds are often employed when estimating integrals with slowly varying functions in the integrand.

2.9.3 Relation between F and U

The link between the tail of the distribution F and its tail quantile function U depends on an inversion result from the theory of s.v. functions. We introduce the concept of the *de Bruyn conjugate* whose existence and asymptotic uniqueness is guaranteed by the following result.

Proposition 2.5 *If $\ell(x)$ is s.v., then there exists an s.v. function $\ell^*(x)$, the de Bruyn conjugate of ℓ, such that*

$$\ell(x)\ell^*(x\ell(x)) \to 1, \quad x \uparrow \infty.$$

The de Bruyn conjugate is asymptotically unique in the sense that if also $\tilde{\ell}$ is s.v. and $\ell(x)\tilde{\ell}(x\ell(x)) \to 1$, then $\ell^* \sim \tilde{\ell}$. Furthermore $(\ell^*)^* \sim \ell$.

As a simple example, check that if $\ell = \log$, then $\ell^* \sim 1/\log$. Let us illustrate how (2.7) can be obtained from this proposition. Put $y := 1/(1 - F(x))$. Then $1 - F(x) = x^{-\alpha}\ell_F(x)$ translates into

$$y = x^{\alpha}\ell_F^{-1}(x) = (x\ell(x))^{\alpha}$$

where

$$\ell(x) := \ell_F^{-1/\alpha}(x).$$

By Proposition 2.5, one can solve the equation $y^{1/\alpha} = x\ell(x)$ for x yielding $x = y^{1/\alpha}\ell^*(y^{1/\alpha})$ where ℓ^* is the de Bruyn conjugate of ℓ. The direct connection between the two functions F and U is given by $U(1/(1 - F(x))) \sim x$ (check this!), or by

$$x \sim U(y) = y^{\gamma}\ell_U(y) = y^{1/\alpha}\ell^*(y^{1/\alpha}).$$

This yields that indeed $\gamma = \alpha^{-1}$ and that $\ell_U(x) \sim \ell^*(x^{\gamma})$. We see how the link between ℓ_F and ℓ_U runs through ℓ and its de Bruyn conjugate ℓ^*. Note that all the previous steps can be reversed so that there is full equivalence between the statements.

   Example. Assume that $\ell_F(x) = (\log x)^{\beta}$; then $\ell(x) = (\log x)^{-\beta/\alpha} = (\log x)^{-\beta\gamma}$ and we need to solve the equation $y = (x\ell(x))^{\alpha}$ for x. This means that

$$y^{\frac{1}{\alpha}} = x(\log x)^{-\beta\gamma} \sim y^{\frac{1}{\alpha}}\ell^*(y^{\frac{1}{\alpha}})\{\log y^{\frac{1}{\alpha}} + \log \ell^*(y^{\frac{1}{\alpha}})\}^{-\beta\gamma}.$$

This in turn leads to

$$1 \sim \ell^*(y)(\log y)^{-\beta\gamma}\left(1 + \frac{\log \ell^*(y)}{\log y}\right)^{-\beta\gamma}.$$

Now use Proposition 2.5 to deduce that $\ell^*(y) = (\log y)^{\beta\gamma}$ and hence that $\ell_U(x) \sim (\gamma \log x)^{\beta\gamma}$.

2.9.4   Proofs for section 2.6

We start with the proof of Proposition 2.1. To this end, define $k(x, u)$ by the relation $U(ux) - U(x) = k(x, u)a(x)$ where $u > 0$ has been fixed. By definition, the inverse function $Q(1 - \frac{1}{ux})$ equals $U(x) + k(x, u)a(x)$ where $k(x, u) \to h_{\gamma}(u)$ as $x \uparrow \infty$. But then

$$\frac{1}{ux} = 1 - F(U(x) + k(x, u)a(x))$$

while at the same time $1/x = 1 - F(U(x))$. Put $y = U(x)$. We find that

$$\frac{1}{u} = \frac{1 - F(y + \tilde{k}(y, u)a(U^{\leftarrow}(y)))}{1 - F(y)}$$

where we have put $\tilde{k}(y, u) = k(U^{\leftarrow}(y), u) = k(x, u)$. Replace $a(U^{\leftarrow}(y))$ by $b(y)$. Further change u into v by the relation $v = h_\gamma(u)$. Equivalently, by (2.5) $u = (1 + \gamma v)^{1/\gamma}$; for $\gamma = 0$, $v = \log u$ immediately leads to $u = \exp v$. Replace this u by v in $\check{k}(y, u)$ to obtain $\check{k}(y, v)$. We have therefore transformed the original equation into

$$\frac{1 - F(y + b(y)\check{k}(y, v))}{1 - F(y)} = (1 + \gamma v)^{-1/\gamma} = \eta_\gamma(v).$$

However, when $x \to \infty$, $k(x, u) \to h_\gamma(u)$ translates into $y \uparrow x_*$ and $\check{k}(y, v) \to v$. By the monotonicity of F we see that then also

$$\frac{1 - F(y + b(y)v)}{1 - F(y)} \to (1 + \gamma v)^{-1/\gamma}. \qquad (\mathcal{C}_\gamma^*) \qquad (2.20)$$

If also U is continuous, then the converse derivation holds as well.

Note that the substitutions $y = U(x)$ and $v = h_\gamma(u)$ yield

$$\frac{b(y + vb(y))}{b(y)} = \frac{a(U^{\leftarrow}(y + vb(y)))}{a(U^{\leftarrow}(y))}$$

$$= \frac{a(U^{\leftarrow}(U(x) + vb(U(x))))}{a(x)} \to u^\gamma = 1 + \gamma v \; .$$

Indeed, the last step follows again from the basic relation $(\mathcal{C}_\gamma)$ since for $x \to \infty$

$$U(x) + vb(U(x)) = U(xu) + (v - k(x, u))a(x) = U(xu) + o(1)a(x)$$
$$= U(xu)(1 + o(1))$$

together with the uniform convergence theorem 2.3(i).

The statement of Proposition 2.2 needs no proof in one direction as the choice of y provides a necessary condition immediately. That the condition is also sufficient is less obvious and needs interpolation type arguments as can be found in de Haan (1970).

To prove the statement of (ii) recall that when $x \to 1$, then $\log x \sim x - 1$. If $\gamma \leq 0$, then $U(xu)/U(x) \to 1$ and hence $\log(U(xu)/U(x)) \sim U(xu)/U(x) - 1$. Multiplying by $U(x)/a(x)$, we can use $(\mathcal{C}_\gamma)$ to see that

$$\frac{U(x)}{a(x)} \{\log U(xu) - \log U(x)\} \sim \frac{U(xu) - U(x)}{a(x)} \to \frac{u^\gamma - 1}{\gamma} \qquad (2.21)$$

as $x \to \infty$. In case $\gamma > 0$, the quantity on the left tends to $\log u$ since $U(x)/a(x) \to \gamma^{-1}$ while $U(tx)/U(x) \to u^\gamma$. Combination of the two statements leads to the condition for $x \to \infty$

$$\frac{U(x)}{a(x)} \{\log U(xu) - \log U(x)\} \to \begin{cases} \log u, & \text{if } \gamma \geq 0, \\[2mm] \frac{u^\gamma - 1}{\gamma}, & \text{if } \gamma < 0 \\ & \text{provided } x_* > 0. \end{cases} \qquad (\tilde{\mathcal{C}}_\gamma)$$

(*iii*) In case $\gamma > 0$, $U(x)/a(x) \to 1/\gamma$. Also, in section 2.6, it was outlined that when $\gamma > 0$ $e_{\log X}(\log x) \to \gamma$ and hence $e_{\log X}(\log U(x)) \to \gamma$ as $x \to \infty$. Indeed, limit and integral can be interchanged there using the dominated convergence theorem based on the Potter bounds 2.4(iv). So we find indeed that in this case

$$\frac{U(x)}{a(x)} e_{\log X}(\log U(x)) \to 1.$$

On the other hand, when $\gamma \leq 0$, we have

$$
\begin{aligned}
e_{\log X}(\log U(x)) &:= \int_{U(x)}^{\infty} \frac{\log(y/U(x))}{(1 - F)(U(x))} dF(y) \\
&= -x \int_{U(x)}^{\infty} \log \frac{y}{U(x)} d(1 - F(y)) \\
&= x \int_{x}^{\infty} \log \frac{U(v)}{U(x)} \frac{dv}{v^2} \\
&= \int_{1}^{\infty} (\log U(wx) - \log U(x)) \frac{dw}{w^2},
\end{aligned}
$$

where a first substitution $1 - F(y) = \frac{1}{v}$ was followed by a second $v = wx$.

Now, when $\gamma \leq 0$, apply (2.21) to see that under $(\tilde{C}_\gamma)$ the requested result follows. This final limit can be taken using the dominated convergence theorem with the help of Potter type bounds on the (2.21), which can be found, for instance, in Dekkers *et al.* (1989).

3

AWAY FROM THE MAXIMUM

3.1 Introduction

We start with a simple observation. It would be unrealistic to assume that only the maximum of a sample contains valuable information about the tail of a distribution. Other large order statistics could do this as well. In this chapter, we investigate how far we can move away from the maximum. If we stay close to the maximum, then only few order statistics are used and our estimators will show large variances. Running away from the maximum will decrease the variation as the number of useful order statistics increases; however, as a side result, the bias will become larger. If we plan to use the subset $\{X_{n-k+1,n}, X_{n-k+2,n}, \ldots, X_{n,n}\}$ of the order sample values

$$X_{1,n} \leq X_{2,n} \leq \cdots \leq X_{n-k+1,n} \leq \cdots \leq X_{n-1,n} \leq X_{n,n}$$

then we need to find out how to choose the rank k in the order statistic $X_{n-k+1,n}$ optimally. What is clear is that k should be allowed to tend to ∞ together with the sample size. But whether k/n needs to be kept small is not so obvious.

In this chapter, we offer an intuitive guideline that is developed from a closer look at the asymptotics of the larger order statistics. Again, for a practitioner, it would look awkward to only use the largest value in a sample to estimate tail quantities, especially as this value may look so large that it seems hardly related to the sample. From this analysis, it will follow that the choice of the number of useful larger order statistics will also depend on the second-order behaviour in the relations $(\mathcal{C}_\gamma)$ and $(\mathcal{C}_\gamma^*)$. These will be developed in this chapter too. Mathematical details are deferred to a final section.

Statistics of Extremes: Theory and Applications   J. Beirlant, Y. Goegebeur, J. Segers, and J. Teugels
© 2004 John Wiley & Sons, Ltd   ISBN: 0-471-97647-4

3.2   Order Statistics Close to the Maximum

There is a great variety of possible limit laws for the individual order statistics. If the index k is small, we can expect that the limit behaviour of $X_{n-k+1,n}$ is similar to that of the maximum $X_{n,n}$. We derive the corresponding theorems first.

To do anything explicit with the k-th largest order statistic, we need its distribution. This can be rather easily obtained from a combinatorial argument. We look for the probability that $X_{n-k+1,n}$ does not overshoot the value x. To do that, take any of the n elements from the sample and force it to have a value u at most x. The remaining $n-1$ values have to be distributed binomially so that $k-1$ of the other sample values lie to the right of u, while the remaining $n-k$ values sit to the left of u. Therefore,

$$P\{X_{n-k+1,n} \le x\} = \frac{n(n-1)!}{(k-1)!(n-k)!} \int_{-\infty}^{x} F^{n-k}(u)\{1-F(u)\}^{k-1}\,\mathrm{d}F(u). \quad (3.1)$$

Case 1: k fixed

We follow the procedure that gave us the solution of the extremal limit problem. We again use Theorem 2.1 and take z, a real, bounded and continuous function. Then

$$E\{z(a_n^{-1}(X_{n-k+1,n} - b_n))\}$$
$$= \frac{n!}{(k-1)!(n-k)!} \int_{-\infty}^{\infty} z\left(\frac{x-b_n}{a_n}\right) F^{n-k}(x)\{1-F(x)\}^{k-1}\,\mathrm{d}F(x).$$

The substitution $F(x) = 1 - \frac{u}{n}$ turns this into the form

$$E\{z(a_n^{-1}(X_{n-k+1,n} - b_n))\}$$
$$= \frac{1}{(k-1)!}\frac{n!n^{-k}}{(n-k)!} \int_{0}^{n} z(a_n^{-1}(U(n/u) - b_n))u^{k-1}\left(1 - \frac{u}{n}\right)^{n-k}\,\mathrm{d}u.$$

From what we learnt about the maximum, where $k = 1$, the argument of z does not depend upon k. Therefore, we still can take $b_n = U(n)$ and $a_n = a(n)$ as in the case of the maximum. Without much trouble, we see that if F satisfies $C_\gamma(a)$, then, as $n \to \infty$

$$E\left\{z\left(\frac{X_{n-k+1,n} - U(n)}{a(n)}\right)\right\} \to \frac{1}{\Gamma(k)} \int_{0}^{\infty} z(h_\gamma(1/u))e^{-u}u^{k-1}\,\mathrm{d}u \quad (3.2)$$

since for large n and fixed k, $n!/(n-k)! \sim n^k$. We can interpret the right-hand side as an expectation with respect to a gamma density. If we denote by $\{E_j\}_{j=1}^{\infty}$ a sequence of i.i.d. exponential random variables with mean 1, then (3.2) can be written in the form

$$\frac{X_{n-k+1,n} - U(n)}{a(n)} \xrightarrow{\mathcal{D}} h_\gamma\left(\left(\sum_{j=1}^{k} E_j\right)^{-1}\right). \quad (3.3)$$

Case 2: $k \to \infty, n - k \to \infty$

In what follows, we assume that k, $n - k$ and therefore n all tend to ∞. Of course, at this point, we do not know in advance what kind of centering and normalization should be used. So, take a centering sequence $\{b_n\}$ of real numbers and a normalizing sequence $\{a_n\}$ of positive reals for which $a_n^{-1}(X_{n-k+1,n} - b_n)$ will converge in distribution. Recall the formula (3.1) and let z again be any real-valued bounded and continuous function on $\mathbb{R}$. Then we want to investigate the limiting behaviour of

$$Z_n := E \left\{ z \left(\frac{X_{n-k+1,n} - b_n}{a_n} \right) \right\}$$

$$= \frac{n!}{(k-1)!(n-k)!} \int_{-\infty}^{\infty} z \left(\frac{x - b_n}{a_n} \right) F^{n-k}(x)\{1 - F(x)\}^{k-1} \, dF(x)$$

where $\{a_n\}$ and $\{b_n\}$ are norming, respectively centering constants.

As happened in the previous analysis, it is not surprising that, apart from the argument of z, all other ingredients can be controlled, regardless of the properties of the underlying distribution. This is done by changing x into v by the transformation

$$1 - F(x) = \frac{k}{n} + \sqrt{\frac{k(n-k)}{n^3}} \, v. \tag{3.4}$$

Why this specific transformation works is fully explained in section 3.4.1. If we solve the above equation for x and replace $x = x(k, n, v)$ in the expression Z_n, then all we have to do is to find appropriate choices of a_n and b_n such that

$$\tau_n(v) := \frac{x(k, n, v) - b_n}{a_n} \to \tau(v) \tag{3.5}$$

for v bounded. Indeed, as shown in section 3.4.1, we have the following covering result.

Proposition 3.1 *If (3.5) holds for v bounded, then*

$$E \left\{ z \left(\frac{X_{n-k+1,n} - b_n}{a_n} \right) \right\} \to \frac{1}{\sqrt{2\pi}} \int_{-\infty}^{\infty} z(\tau(v)) \, e^{-\frac{1}{2}v^2} \, dv. \tag{3.6}$$

Since the sum $\sum_{j=1}^{k} E_j$ in (3.3) will be ultimately approximated by a normal variable when $k \uparrow \infty$, the appearance of the normal distribution should not be surprising in view of the central limit theorem.

The explicit value of $x(k, n, v)$ is easily obtained from (3.4) and is

$$x(k, n, v) = U \left(\left(\frac{k}{n} + \sqrt{\frac{k(n-k)}{n^3}} \, v \right)^{-1} \right)$$

$$= U \left(\frac{n}{k} \left(1 - \sqrt{\frac{1 - \frac{k}{n}}{k}} \, v + o \left(\frac{1}{\sqrt{k}} \right) \right) \right).$$

The choice of b_n in (3.5) is almost automatic, since 0 is a bounded value for v. Therefore, we put $b_n = x(k, n, 0) = U(n/k)$. The choice of a_n can then be made by requiring the convergence of

$$\tau_n(v) = \frac{U\left(\frac{n}{k}\left(1 - \sqrt{\frac{1-\frac{k}{n}}{k}}\, v + o\left(\frac{1}{\sqrt{k}}\right)\right)\right) - U\left(\frac{n}{k}\right)}{a_n} \to \tau(v)$$

for v bounded. That the choice of a_n has to be determined by the limiting behaviour of the ratio k/n will then become clear. Here are a few illustrations on how conditions on the distribution F or on the tail quantile function influence the possible choice of k and as such of a_n.

(i) We immediately get a *weak law*. Assume $k/n \to 0$. If F satisfies $C_\gamma(a)$, then the choice $a_n = a(n/k)$ yields

$$\tau_n(v) = \frac{U\left(\frac{n}{k}\left(1 - \frac{v}{\sqrt{k}}(1 + o(1))\right)\right) - U\left(\frac{n}{k}\right)}{a\left(\frac{n}{k}\right)} \to 0 = \tau(v).$$

Hence $Z_n \to z(0)$ and

$$\frac{X_{n-k+1,n} - U\left(\frac{n}{k}\right)}{a\left(\frac{n}{k}\right)} \overset{P}{\Rightarrow} 0.$$

In the case $\gamma > 0$, we can go one step further. Indeed, since $\frac{U(x)}{a(x)} \to \gamma^{-1}$, we also have

$$\frac{X_{n-k+1,n}}{a\left(\frac{n}{k}\right)} \overset{P}{\Rightarrow} \frac{1}{\gamma}.$$

We illustrate the result that the k-th largest order statistic leads to an asymptotically consistent estimator for the extreme value index. In Figure 3.1, we show $X_{n-k+1,n}/a(n/k)$ as a function of n, $n = 1, \ldots, 5,000$, with $k = \lfloor n^{0.25} \rfloor$ (solid line), $k = \lfloor n^{0.5} \rfloor$ (broken line) and $k = \lfloor n^{2/3} \rfloor$ (broken-dotted line) for data generated from the Burr$(1, 1, 1)$ distribution.

(ii) Getting a closer look at what happens around n/k, we note that the fundamental relation (C_γ) suggests that as $k \to \infty$

$$\frac{U\left(\frac{n}{k}\left(1 - \frac{v}{\sqrt{k}}(1 + o(1))\right)\right) - U\left(\frac{n}{k}\right)}{a\left(\frac{n}{k}\right)} \sim \frac{1}{\gamma}\left(\left(1 - \frac{v}{\sqrt{k}}\right)^\gamma - 1\right) \sim -\frac{v}{\sqrt{k}}.$$

Under appropriate continuity conditions, we expect that $a_n = k^{-1/2}a(n/k)$ is a good choice since then $\tau_n(v) \to -v$. Since the normal distribution is symmetric, this choice leads for the limiting operation $\frac{k}{n} \to 0$ to

$$\sqrt{k}\frac{X_{n-k+1,n} - U\left(\frac{n}{k}\right)}{a\left(\frac{n}{k}\right)} \overset{D}{\to} Y \sim N(0, 1).$$

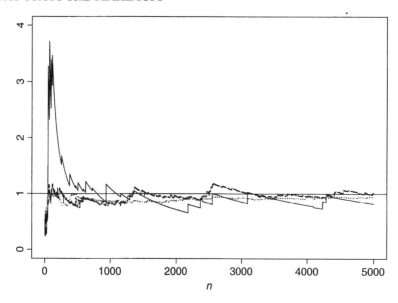

Figure 3.1   Plot of $X_{n-k+1,n}/a(n/k)$ versus n with $k = \lfloor n^{0.25} \rfloor$ (solid line), $k = \lfloor n^{0.5} \rfloor$ (broken line) and $k = \lfloor n^{2/3} \rfloor$ (broken-dotted line) for data generated from the Burr(1, 1, 1) distribution.

(iii) Take $\gamma > 0$. This last result is interesting since it suggests that we might be able to construct asymptotic confidence intervals for the unknown index γ based on this normal approximation. To try this, write the above result in the form

$$\sqrt{k}\left\{\frac{X_{n-k+1,n}}{a(\frac{n}{k})} - \frac{1}{\gamma}\right\} - \sqrt{k}\left\{\frac{U(\frac{n}{k})}{a(\frac{n}{k})} - \frac{1}{\gamma}\right\} \xrightarrow{\mathcal{D}} Y \sim N(0,1). \qquad (3.7)$$

If we hope to get normal convergence of the random quantity on the left, then the second quantity needs to have a limit as well. In other words, if we want normal convergence of the estimator on the left, we need *second-order information* on how fast $U(x)/a(x)$ tends to $1/\gamma$ as $x \to \infty$. This speed of convergence will then tell us at what speed k is allowed to tend to ∞ with n. Let us give a simple example from the *Hall-class* where $U(x) = Cx^{\gamma}(1 + Dx^{-\beta}(1 + o(1)))$ for some $\beta > 0$ and real constants $C > 0$ and D. Then it easily follows that $a(x) = \gamma C x^{\gamma}$ and the condition asks the quantity $\sqrt{k}\frac{D}{\gamma}\left(\frac{n}{k}\right)^{-\beta}$ to converge, or that $k \sim const \times n^{2\beta/(1+2\beta)}$. We illustrate the convergence of $\sqrt{k}(X_{n-k+1,n} - U(n/k))/a(n/k)$ to its normal limit using the Burr(η, τ, λ) distribution. This distribution belongs to the Hall-class with $\gamma = (\lambda\tau)^{-1}$, $C = \eta^{1/\tau}$, $D = -1/\tau$ and $\beta = 1/\lambda$. In Figure 3.2, we show the histogram of $\sqrt{k}(X_{n-k+1,n} - U(n/k))/a(n/k)$ with $k = \lfloor n^{2/3} \rfloor$ for simulated data sets of size (a) $n = 100$, (b) $n = 500$, (c) $n = 1000$ and (d) $n = 2000$ from the Burr(1, 1, 1) distribution.

(iv) Our last choice has nothing to do with extremal behaviour. Nevertheless, we include it to show how the basic condition (C_γ) for the extremal behaviour is replaced by a continuity condition on the tail quantile function. In the central part of the sample, the behaviour of $X_{n-k+1,n}$ is well approximated by a normal distribution as can be shown as follows. Assume that k tends to ∞ in such a way

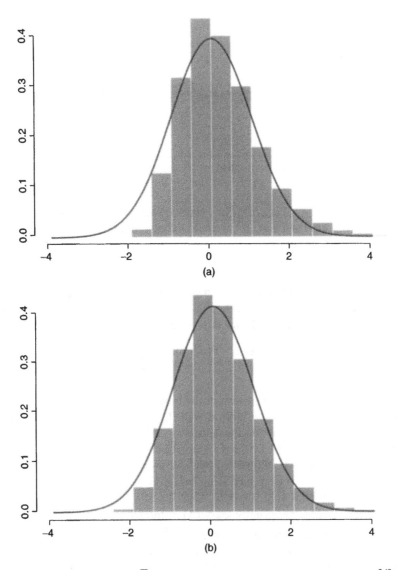

(a)

(b)

Figure 3.2   Histogram of $\sqrt{k}(X_{n-k+1,n} - U(n/k))/a(n/k)$ with $k = \lfloor n^{2/3} \rfloor$ for simulated data from the Burr$(1, 1, 1)$ distribution: (a) $n = 100$, (b) $n = 500$, (c) $n = 1000$ and (d) $n = 2000$.

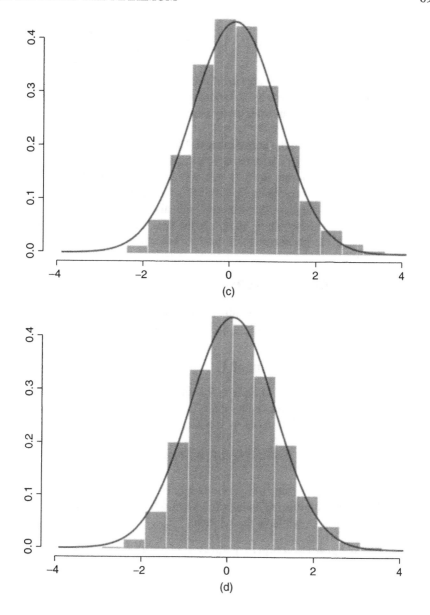

(c)

(d)

that the ratio $k/n \to \lambda \in (0, 1)$. Since now the argument in $U(n/k) \sim b_n$ does no longer tend to ∞, the condition on F or U refers to finite values of the argument. This time, we use the continuity of U. For, one can then show that if F has a density f that is continuous and strictly positive at $U(\lambda^{-1})$, then

$$\sqrt{\frac{n}{\lambda(1-\lambda)}} f\left(U\left(\frac{1}{\lambda}\right)\right) \left\{X_{n-k+1,n} - U\left(\frac{1}{\lambda}\right)\right\} \overset{\mathcal{D}}{\to} N(0, 1)$$

or

$$\sqrt{n}\left\{X_{n-k+1,n} - U\left(\frac{1}{\lambda}\right)\right\} \xrightarrow{\mathcal{D}} N\left(0, \frac{\lambda(1-\lambda)}{f^2\left(U\left(\frac{1}{\lambda}\right)\right)}\right). \tag{3.8}$$

This last result can be fruitfully applied if one needs to estimate central quantiles. However, note that there is hardly any relationship between the behaviour of the largest order statistics and quantiles away from the tails.

What did we learn from the above analysis? That we are entitled to use more of the larger order statistics, that asymptotic consistency is not too difficult to obtain, but that the bias as it appears in (3.7) will only be controllable if we have some additional second-order information on the tail of U. Alternatively, the choice of the number of *useful* larger order statistics will also be decided by the second-order behaviour of the functions F and U. In the next step, we will use more order statistics, but the same phenomena will play a fundamental role.

3.3   Second-order Theory

This section covers second-order results on the condition $(\mathcal{C}_\gamma)$. The discussion of their equivalent versions for distributions is given in section 3.3.3.

3.3.1   Remainder in terms of U

Assume that F satisfies $\mathcal{C}_\gamma(a)$ where $a \in \mathcal{R}_\gamma$. In this section, we derive a remainder of the limiting operation expressed by the $(\mathcal{C}_\gamma)$-condition. Let there exist a second (ultimately) positive function $a_2(x) \to 0$ when $x \to \infty$ such that

$$\frac{U(ux) - U(x)}{a(x)} - h_\gamma(u) \sim a_2(x)k(u), \quad x \uparrow \infty \tag{3.9}$$

for all $u > 0$. We know from the previous section that for some s.v. ℓ, $a(x) = x^\gamma \ell(x)$. Moreover, with this real value of γ we also have that $h_\gamma(u) = \int_1^u v^{\gamma-1}\, dv$. Therefore, for $u, v > 0$, we have the relations

$$h_\gamma(uv) = u^\gamma h_\gamma(v) + h_\gamma(u) = v^\gamma h_\gamma(u) + h_\gamma(v), \tag{3.10}$$

$$h_{-\gamma}\left(\frac{1}{u}\right) = -h_\gamma(u) \tag{3.11}$$

and

$$u^\gamma h_{-\gamma}(u) = h_\gamma(u). \tag{3.12}$$

We first derive an equation for k in (3.9). To get the latter, we follow the usual approach of replacing u in the equation by uv and subsequent rewriting. We get

$$U(uvx) - U(x) = \{U(uvx) - U(ux)\} + \{U(ux) - U(x)\}.$$

A somewhat tedious calculation yields the fundamental equation

$$a_2^{-1}(x) \left\{ \frac{U(uvx) - U(x)}{a(x)} - h_\gamma(uv) \right\}$$

$$= a_2^{-1}(ux) \left\{ \frac{U(uvx) - U(ux)}{a(ux)} - h_\gamma(v) \right\} \left\{ \frac{a(ux)}{a(x)} \frac{a_2(ux)}{a_2(x)} \right\}$$

$$+ a_2^{-1}(x) \left\{ \frac{U(ux) - U(x)}{a(x)} - h_\gamma(u) \right\} + a_2^{-1}(x) \left\{ \frac{a(ux)}{a(x)} h_\gamma(v) + h_\gamma(u) - h_\gamma(uv) \right\}.$$

The third term on the right can be simplified by using (3.10) and the fact that a is regularly varying with index γ. It yields

$$a_2^{-1}(x) h_\gamma(v) \left\{ \frac{a(ux)}{a(x)} - u^\gamma \right\} = u^\gamma a_2^{-1}(x) h_\gamma(v) \left\{ \frac{\ell(ux)}{\ell(x)} - 1 \right\}.$$

The above equation leads to an equation for k when $x \to \infty$ since from (3.9) (in case k is not a multiple of h_γ) we have either the convergence of $a_2(ux)/a_2(x)$ or, equivalently, the convergence of $a_2^{-1}(x) \left\{ \frac{\ell(ux)}{\ell(x)} - 1 \right\}$ for all $u > 0$.

The first implies regular variation of a_2. So, we assume from now on that

$$a_2 \in \mathcal{R}_\rho, \quad \rho \leq 0. \tag{3.13}$$

The alternative condition on ℓ can be written in the form

$$\frac{\ell(ux) - \ell(x)}{a_2(x)\ell(x)} \to m(u). \tag{3.14}$$

However, this is precisely a limiting relation as discussed in section 2.1. Clearly, the auxiliary function $a_2(x)\ell(x)$ is of ρ-regular variation and therefore $m(u) = ch_\rho(u)$ for some constant c. This result is in accordance with what we know from the theory of slow variation with remainder as it has been developed by Goldie and Smith (1987). Alternatively, we can say that $\log \ell$ satisfies $C_\rho(a_2)$.

Using this additional information in the fundamental equation, we arrive at the following functional equation for the function k

$$k(uv) = u^{\gamma+\rho} k(v) + k(u) + cu^\gamma h_\gamma(v) h_\rho(u) \tag{3.15}$$

which is valid for all $u, v > 0$. The derivation of the solution k in (3.15) is given in the last section of this chapter. Of course, one could consider a number of subcases resulting from the possible values of both γ and ρ. The derivation in section 3.4.3 shows this to be unnecessary. The next result is basically due to de Haan and Stadtmüller (1996).

Theorem 3.1 *Assume U satisfies $C_\gamma(a)$; assume further that for all $u > 0$*

$$\frac{U(ux) - U(x)}{a(x)} - h_\gamma(u) \sim a_2(x) k(u), \quad x \uparrow \infty$$

where $a_2 \to 0$ belongs to $\mathcal{R}_\rho$ with $\rho \le 0$. Then for some real constant c

$$a_2^{-1}(x) \left\{ \frac{a(ux)}{a(x)} - u^\gamma \right\} \to c\, u^\gamma h_\rho(u).$$

Furthermore, for this constant c and an arbitrary constant $A \in \mathbb{R}$,

$$k(u) = A\, h_{\gamma+\rho}(u) + c \int_1^u t^{\gamma-1}\, h_\rho(t)\, dt.$$

If $\rho < 0$, then an appropriate choice of the auxiliary function a results in a simplification of the limit function.

Proposition 3.2 *Under the conditions of Theorem 3.1, if $\rho < 0$, then*

$$\lim_{x \to \infty} \frac{1}{a_2(x)} \left(\frac{U(ux) - U(x)}{\tilde{a}(x)} - h_\gamma(u) \right) = \tilde{c} h_{\gamma+\rho}(u),$$

where $\tilde{c} = \rho^{-1}c + A$ and $\tilde{a}(x) = a(x)\left(1 - \rho^{-1}ca_2(x)\right)$ for all x such that $a_2(x) < |\rho/c|$.

3.3.2   Examples

We give a couple of distributions that often appear in applications and for which the second-order quantities can be easily derived.

Weibull distributions

A distribution that is often employed in reliability theory is the *Weibull distribution* defined on $\mathbb{R}_+$ by the tail expression

$$1 - F(x) = \exp\{-Cx^\beta\}$$

where both C and β are positive. This distribution is to be distinguished from the extreme value Weibull type discussed in Chapter 2. For $\beta = 1$, one finds the exponential distribution while for $\beta = 2$, one gets the *Rayleigh distribution*. The tail quantile function is easily found and equals $U(y) = (C^{-1}\log y)^{1/\beta}$. It then easily follows that $a(x) = \beta^{-1}C^{-1/\beta}(\log x)^{(1-\beta)/\beta}$. Hence, F satisfies $C_0(a)$. Moreover

$$\frac{U(xu) - U(x)}{a(x)} - \log u = \beta \log x \left(\left(1 + \frac{\log u}{\log x}\right)^{1/\beta} - 1 \right) - \log u$$

$$\sim \frac{1-\beta}{2\beta} \frac{(\log u)^2}{\log x}$$

so that $\rho = 0$, $a_2(x) = (\log x)^{-1}$, $c = \frac{1-\beta}{\beta}$ and $k(u) = \frac{c}{2}(\log u)^2$. The case where $\beta = 1$ is particularly simple as it should be.

Hall-class

We have already hinted at the *Hall-class of distributions* that is defined in terms of the tail quantile function by an expansion of the form

$$U(x) = Cx^{\gamma}\left(1 + \frac{D_1}{x^{\beta}} + \frac{D_2}{x^{\eta}} + \cdots\right)$$

where $C > 0$, $\gamma > 0$ while $0 \le \beta \le \eta \le \cdots$. The calculations for the second-order quantities are particularly simple and lead to $a(x) = \gamma C x^{\gamma}$ and therefore

$$\frac{U(xu) - U(x)}{a(x)} - h_{\gamma}(u) = \frac{D_1(\gamma - \beta)}{\gamma} x^{-\beta} h_{\gamma-\beta}(u) + \cdots$$

so that $\rho = -\beta$, $c = 0$, $a_2(x) = x^{-\beta}$ while $k(u) = \frac{D_1(\gamma-\beta)}{\gamma} h_{\gamma-\beta}(u)$.
As a special example, we look at the Burr(η, τ, λ) distribution where

$$1 - F(x) = \left(\frac{\eta}{\eta + x^{\tau}}\right)^{\lambda}.$$

With $C = \eta^{\frac{1}{\tau}}$, $\gamma = (\lambda\tau)^{-1}$, $D_1 = -\tau^{-1}$ and $\beta = \lambda^{-1}$ this distribution turns out to be a special case of the general Hall-class.

3.3.3 Remainder in terms of F

We transform the remainder condition that has been stated above in terms of U towards a statement based on the distribution function F. This can sometimes be useful in verifying such remainder condition on specific examples. Also, in the subsequent statistical chapters, these will be of use.

Condition (3.9) can be written in the form

$$U(ux) = U(x) + a(x)h_{\gamma}(u) + a_1(x)k_x(u)$$

where we put $a_1(x) = a(x)a_2(x)$ and where $k_x(u) \to k(u)$ as $x \to \infty$.

We operate on the above equation with the function $1 - F$. By continuity, the left-hand side turns into $(ux)^{-1}$. We can also replace x^{-1} by $(1 - F)(U(x))$. We then obtain

$$\frac{1}{u} = \frac{(1 - F)\left(U(x) + a(x)h_{\gamma}(u) + a_1(x)k_x(u)\right)}{(1 - F)(U(x))}$$

We now define $y = U(x)$ and replace $a(x)$ by $h(U(x))$. The above relation turns into the form

$$\frac{1}{u} = \frac{(1 - F)\left(y + h(y)h_{\gamma}(u) + h(y)\chi(y)\kappa_y(u)\right)}{(1 - F)(y)}$$

where we have defined

$$\chi(y) := a_2\left(U^{\leftarrow}(y)\right) \tag{3.16}$$

and

$$\kappa_y(u) := k_{U^{\leftarrow}(y)}(u). \tag{3.17}$$

A more transparent form of the above equation is the following

$$\frac{1}{u} = \frac{(1 - F)(y + h(y)v(u, y))}{(1 - F)(y)} \tag{3.18}$$

where

$$v(u, y) := h_\gamma(u) + \chi(y)\kappa_y(u).$$

The solution of this equation is given in the final section and leads to the following result. Here again, we use the notation $\eta_\gamma(v) = (1 + \gamma v)^{-1/\gamma}$.

Theorem 3.3 *Assume U satisfies $C_\gamma(a)$ and is continuous. Assume further that for all $u > 0$*

$$\frac{U(ux) - U(x)}{a(x)} - h_\gamma(u) \sim a_2(x)k(u) \quad , \quad x \uparrow \infty$$

where $a_2 \to 0$ belongs to $\mathcal{R}_\rho$ with $\rho \le 0$. Then for $y \to x_*$

$$\frac{1}{\chi(y)}\left\{ \frac{1 - F(y + vh(y))}{1 - F(y)} - \eta_\gamma(v) \right\} \to \psi(v)$$

where $\chi(y) = a_2(U^{\leftarrow}(y))$ and $\psi(v) = \eta_\gamma^{1+\gamma}(v)k\left(\frac{1}{\eta_\gamma(v)}\right)$.

The converse holds too since all steps can be reversed. Similarly to Proposition 3.2, also here a simpler limit function can be obtained when replacing a by $\tilde{a}$.

Proposition 3.4 *Under the assumptions of Theorem 3.3, we have for $\rho < 0$*

$$\lim_{y \to x_*} \frac{1}{\chi(y)}\left\{ \frac{1 - F(y + v\tilde{h}(y))}{1 - F(y)} - \eta_\gamma(v) \right\} = \tilde{c}\eta_\gamma^{1+\gamma}(v)h_{\gamma+\rho}\left(\frac{1}{\eta_\gamma(v)}\right),$$

where $\tilde{h}(y) = \tilde{a}(U^{\leftarrow}(y))$.

3.4 Mathematical Derivations

In this section, we give sketches of the proofs of the results in the previous sections. We first prove the general auxiliary result (3.6) without reference to properties of the underlying distribution or tail quantile function. Then we indicate why (3.8) is valid. We then turn to the results concerning second-order theory.

3.4.1 Proof of (3.6)

We need to rewrite the integrand in such a way that both factors F^{n-k} and $(1 - F)^{k-1}$ can be handled simultaneously. To achieve that, substitute $1 - F(x) := q_n + p_n v$ where the sequences q_n and p_n will be determined in due time. We also write $\bar{q}_n := 1 - q_n$ for convenience. With this substitution, we can write

$$Z_n = \frac{n! p_n}{(k-1)!(n-k)!} \int_{-\frac{q_n}{p_n}}^{\frac{\bar{q}_n}{p_n}} z(\tau_n(v))(q_n + p_n v)^{k-1}(\bar{q}_n - p_n v)^{n-k} \, dv$$

$$:= I_1(q_n, p_n) \int_{-\frac{q_n}{p_n}}^{\frac{\bar{q}_n}{p_n}} z(\tau_n(v)) I_2(q_n, p_n, v) \, dv$$

where we have abbreviated

$$\tau_n(v) := \frac{U\left(\frac{1}{q_n + p_n v}\right) - b_n}{a_n}$$

for convenience. Furthermore, we put

$$I_2(q_n, p_n, v) := \left(1 + \frac{p_n}{q_n} v\right)^{k-1} \left(1 - \frac{p_n}{\bar{q}_n} v\right)^{n-k}$$

and

$$I_1(q_n, p_n) := \frac{n! p_n}{(k-1)!(n-k)!} q_n^{k-1} \bar{q}_n^{n-k}.$$

We first deal with I_2. We take its logarithm and expand to get

$$\log I_2(q_n, p_n, v) = p_n v \left\{ \frac{k-1}{q_n} - \frac{n-k}{\bar{q}_n} \right\} - \frac{v^2}{2} p_n^2 \left\{ \frac{k-1}{q_n^2} + \frac{n-k}{\bar{q}_n^2} \right\} + \cdots.$$

It is now natural to choose q_n in such a way that the first term on the right annihilates. The choice of p_n can then be made to make the coefficient of $-\frac{1}{2}v^2$ equal to 1. An exact solution would lead to $q_n = \frac{k-1}{n-1}$ and $p_n^2 = \frac{(k-1)(n-k)}{(n-1)^3}$. However, the first term still annihilates if we stick to the simpler choice

$$q_n = \frac{k}{n}, \quad p_n^2 = k(n-k)n^{-3}.$$

With this choice, one can then prove that indeed $I_2(q_n, p_n, v) \to e^{-\frac{1}{2}v^2}$. With the help of Stirling's formula $m! \sim m^{m+\frac{1}{2}} e^{-m} \sqrt{2\pi}$, it takes a bit of simple calculus to verify that also $I_1(q_n, p_n) \to (2\pi)^{-\frac{1}{2}}$.

The reason for the above calculations is to show that, once the quantities q_n and p_n are chosen as above, all information on the underlying distribution is contained in the remaining quantity $\tau_n(v)$.

It remains to show that the expression (3.6) is valid. To do that, let ϵ be any positive small quantity. Choose T large enough to have that $\Phi(-T) = 1 - \Phi(T) \leq \epsilon$. That quantity T will need to satisfy some more conditions. If M is a bound for the bounded function $|z|$, then

$$\left| \int_T^{\frac{\bar{q}_n}{p_n}} I_1(q_n, p_n) z_n(v) I_2(q_n, p_n, v) \, dv \right| \leq M \int_T^{\frac{\bar{q}_n}{p_n}} I_1(q_n, p_n) I_2(q_n, p_n, v) \, dv.$$

Turning back to the expression with $1 - F(x) = q_n + p_n v =: s$ as integrating variable, the integral on the right can be rewritten as

$$J_1 := \int_T^{\frac{\bar{q}_n}{p_n}} I_1(q_n, p_n) I_2(q_n, p_n, v) \, dv = \frac{n!}{(k-1)!(n-k)!} \int_{q_n + p_n T}^1 (1-s)^{n-k} s^{k-1} \, ds.$$

However, on the interval $(q_n + p_n T, 1)$, the quantity $((s - q_n)/(p_n T))^2$ is not smaller than 1 and hence

$$J_1 \leq \frac{n!}{(k-1)!(n-k)! p_n^2 T^2} \int_0^1 (s - q_n)^2 (1-s)^{n-k} s^{k-1} \, ds =$$

$$\frac{1}{p_n^2 T^2} \left\{ \frac{k(k+1)}{(n+1)(n+2)} - 2q_n \frac{k}{n+1} + q_n^2 \right\} = O(T^{-2})$$

which can be made smaller than ϵ for T large enough as long as $\limsup \frac{k}{n} < 1$. A similar argument holds for the other limit.

Combining the above estimates, we can write

$$\left| Z_n - \frac{1}{\sqrt{2\pi}} \int_{-\infty}^\infty z(\tau(v)) e^{-\frac{1}{2}v^2} \, dv \right|$$

$$\leq M \int_{-T}^T \left| I_1(q_n, p_n) I_2(q_n, p_n, v) - \frac{1}{\sqrt{2\pi}} e^{-\frac{1}{2}v^2} \right| \, dv + 4M\epsilon.$$

This proves the important auxiliary result.

3.4.2 Proof of (3.8)

Remember that we take $k/n \to \lambda \in (0, 1)$. We return to the crucial quantity $\tau_n(v)$. Rewrite this in the form

$$\tau_n(v) = \frac{U\left(\frac{n}{k} \left(1 - \sqrt{\frac{1-(k/n)}{k/n}} \frac{v}{\sqrt{n}} + o\left(\frac{1}{\sqrt{n}} \right) \right) \right) - U\left(\frac{n}{k} \right)}{a_n \frac{n}{k} \sqrt{\frac{1-(k/n)}{k/n}} \frac{v}{\sqrt{n}}} \left(\frac{n}{k} \sqrt{\frac{1-(k/n)}{k/n}} \frac{v}{\sqrt{n}} \right).$$

Part of this quantity is of the order $-U'(\frac{1}{\lambda}) \frac{1}{\lambda} \sqrt{\frac{1-\lambda}{\lambda}} \frac{v}{\sqrt{n}}$ when the tail quantile function has a derivative at the point $\frac{1}{\lambda}$. If we choose $\sqrt{n} \, a_n = 1$, then under the latter condition

$$\tau_n(v) \to \tau(v) = -U'\left(\frac{1}{\lambda} \right) \frac{1}{\lambda} \sqrt{\frac{1-\lambda}{\lambda}} \, v.$$

Then $\tau(v) = -C_\lambda v$ for some function $C_\lambda > 0$. It follows that for continuous, bounded functions z,

$$E\left\{z\left(\sqrt{n}\left(X_{n-k+1,n} - U\left(\frac{1}{\lambda}\right)\right)\right)\right\} \to \frac{1}{C_\lambda\sqrt{2\pi}} \int_{-\infty}^{\infty} z(y)e^{-\frac{y^2}{2C_\lambda^2}}\, dy.$$

Let us formulate the above result in terms of F.

Assume that $\frac{k}{n} \to \lambda \in (0,1)$ and that F has a density f that is continuous and strictly positive at $U(\lambda^{-1})$. Then

$$\sqrt{\frac{n}{\lambda(1-\lambda)}} f\left(U\left(\frac{1}{\lambda}\right)\right)\left\{X_{n-k+1,n} - U\left(\frac{1}{\lambda}\right)\right\} \xrightarrow{\mathcal{D}} N(0,1)$$

or

$$\sqrt{n}\left\{X_{n-k+1,n} - U\left(\frac{1}{\lambda}\right)\right\} \xrightarrow{\mathcal{D}} N\left(0, \frac{\lambda(1-\lambda)}{f^2\left(U\left(\frac{1}{\lambda}\right)\right)}\right)$$

which is (3.8) as we will now show. We have already shown that $\sqrt{n}(X_{n-k+1,n} - U(\frac{1}{\lambda})) \xrightarrow{\mathcal{D}} N(0,C_\lambda^2)$. However, since F has a derivative f, by the definition of inverse function, the relation $(1-F)(U(y)) = y^{-1}$ is the same as in (2.17). But then

$$C_\lambda = \frac{1}{\lambda}\sqrt{\frac{1-\lambda}{\lambda}} \frac{\lambda^2}{f\left(U\left(\frac{1}{\lambda}\right)\right)} = \frac{\sqrt{\lambda(1-\lambda)}}{f\left(U\left(\frac{1}{\lambda}\right)\right)}$$

which is equivalent to the statement (3.8). Observe that the two boundary cases $\lambda = 0$ and $\lambda = 1$ had to be excluded.

3.4.3 Solution of (3.15)

We follow the approach by Vanroelen (2003). Define

$$W(x) := U(x) - \int_1^x \frac{a(u)}{u}\, du.$$

Then it is easy to derive the following auxiliary equation for the function W:

$$\frac{W(xu) - W(x)}{a_2(x)a(x)} = \frac{1}{a_2(x)}\left[\frac{U(xu) - U(x)}{a(x)} - h_\gamma(u)\right]$$
$$- \int_1^x \frac{1}{a_2(x)}\left[\frac{a(xu)}{a(x)} - u^\gamma\right]\frac{du}{u}.$$

The first part of the right-hand side converges to $k(u)$ in view of our second-order condition on the function U. By the condition on the auxiliary function a_2, also the second part of the right-hand side converges. But then the left-hand side of the above expression converges as well. Automatically, the limit has to be of the form as predicted in section 2.1. The auxiliary function on the left is of regular variation with index $\rho + \gamma$ and hence the limit on the left is of the form $Ah_{\rho+\gamma}(u)$ for some constant A. Solving for $k(u)$ gives the requested expression.

3.4.4 Solution of (3.18)

The equation has to be solved for the quantity u. Two cases arise, depending on the value of γ.

(i) : $\gamma \neq 0$ Substitute $h_\gamma(u) = \frac{1}{\gamma}(u^\gamma - 1)$. Then

$$u^\gamma = (1 + \gamma v) \left\{ 1 - \frac{\gamma \chi(y)\kappa_y(u)}{1 + \gamma v} \right\}.$$

Taking the γ-th root, we can also write this in the form

$$u = \frac{1}{\eta_\gamma(v)} \left\{ 1 - \frac{\gamma \chi(y)\kappa_y(u)}{1 + \gamma v} \right\}^{\frac{1}{\gamma}}.$$

But then we arrive at the following expression for the remainder condition

$$\frac{1 - F(y + vh(y))}{1 - F(y)} - \eta_\gamma(v) = \eta_\gamma(v) \left\{ \left(1 - \frac{\gamma \chi(y)\kappa_y(u)}{1 + \gamma v} \right)^{-\frac{1}{\gamma}} - 1 \right\}.$$

Divide both sides by $\chi(y) \to 0$ when $y \uparrow x_*$. Apply the usual approximation on the right-hand side to see that this side is asymptotically equal to $\eta_\gamma(v)(1 + \gamma v)^{-1}\kappa_y(u(v))$. However, when $y \uparrow x_*$ also $u(v, y) \to \eta_\gamma^{-1}(v)$ so that by the continuity of F and the condition $k_x(u) \to k(u)$ we ultimately find

$$\frac{1}{\chi(y)} \left\{ \frac{1 - F(y + vh(y))}{1 - F(y)} - \eta_\gamma(v) \right\} \to \eta_\gamma^{1+\gamma}(v) k \left(\frac{1}{\eta_\gamma(v)} \right) =: \psi(v).$$

(ii) : $\gamma = 0$ A similar approach applies with the equation

$$\log u + \chi(u)\kappa_y(u) = v(u, y)$$

that has to be solved for u. Following the same path as before, the limit quantity $\psi(v)$ is now equal to $e^{-v}k(e^v)$, which coincides with the previous limit if we straightforwardly put $\gamma = 0$.

4

TAIL ESTIMATION UNDER PARETO-TYPE MODELS

In this chapter, we consider the estimation of the extreme value index, of extreme quantiles and of small exceedance probabilities, in case the distribution is of Pareto-type, that is,

$$\bar{F}(x) = x^{-1/\gamma} \ell_F(x),$$

or equivalently

$$Q(1 - 1/x) = U(x) = x^\gamma \ell_U(x),$$

where ℓ_F and ℓ_U are related s.v. functions as shown in section 2.9.3. We also discuss inferential matters such as point estimators and confidence intervals.

Since the early eighties of the twentieth century, this problem has been studied in great detail in the literature. Hill's estimator (Hill (1975)), which appeared in 1975, continues to be of great importance and constitutes the main subject of this chapter. However, to get a better feeling for the choice of possible estimators, we start out with a few examples of *naive* estimators. What they all have in common is an attempt to avoid the unknown and irrelevant slowly varying part ℓ. We assume from now on that we have a sample of i.i.d. values $\{X_i;\ 1 \le i \le n\}$ from a Pareto-type tail $1 - F$.

Pareto-type tails are systematically used in certain branches of non-life insurance. Also in finance (stock returns) and in telecommunication (file sizes, waiting times), this class is appropriate. In other areas of application of extreme value statistics such as hydrology, the use of Pareto models appears to be much less systematic. However, the estimation problems considered here are typical for extreme value methodology and at the same time the Pareto-type model is more specific and

Statistics of Extremes: Theory and Applications   J. Beirlant, Y. Goegebeur, J. Segers, and J. Teugels
© 2004 John Wiley & Sons, Ltd   ISBN: 0-471-97647-4

simpler to handle. So this chapter has also an instructive purpose; the heavy-tailed distributions are an ideal 'playground' for developing effective methods that are to be extended in the general case $\gamma \in \mathbb{R}$.

4.1 A Naive Approach

Let us try some easy ways to get rid of the function ℓ_U. From Proposition 2.4, we see that for $x \to \infty$,

$$\log U(x) = \gamma \log x + \log \ell_U(x) \sim \gamma \log x .$$

Hence, it looks natural to replace in the above expression the deterministic quantity U by a random quantity whose argument goes to infinity with the sample size. For simplicity, the argument x could be taken to be n or more generally n/k. In the sequel, we set $\hat{U}_n(x) = \hat{Q}_n(1 - 1/x)$ for the natural empirical estimator of U. We then expect to have a probabilistic statement of the type

$$\log \hat{U}_n \left(\frac{n}{k} \right) \sim \gamma \log \frac{n}{k} .$$

However, for any $r \in \{1, 2, \ldots, n\}$, one has

$$\hat{U}_n \left(\frac{n}{n - r} \right) = X_{r,n}$$

and so we expect asymptotically that for $n \to \infty$ that $\log X_{n-k+1,n} \sim \gamma \log(n/k)$. From (3.3), it follows that, replacing $a(n)$ by $\gamma U(n)$, when k is kept fixed,

$$\log \left(\frac{X_{n-k+1,n}}{U(n)} \right) = O_P(1),$$

or

$$\log X_{n-k+1,n} - \gamma \log n - \log \ell_U(n) = O_P(1)$$

from which, with Proposition 2.4, one indeed derives that if F satisfies $(\mathcal{C}_\gamma)$ and $\gamma > 0$

$$\log X_{n-k+1,n} / \log n \overset{P}{\Rightarrow} \gamma.$$

This simple result shows that a single larger order statistic can be used to estimate the extreme value index γ. But there are some serious drawbacks to this naive approach. For example, it looks indeed unsatisfactory to use only one single order statistic in the estimation procedure. Also, what does it mean to keep k fixed? Moreover, from the derivation it follows that the rate of convergence is logarithmically slow.

By basic statistical intuition, we can foresee that an estimator based on more order statistics will be more reliable. One possibility is to consider differences of two different extreme order statistics, such as (see Bacro and Brito (1993)):

$$\frac{\log X_{n-k+1,n} - \log X_{n-2k+1,n}}{\log 2}$$

or generalizations with spacings of different order than k. Using the regular variation of Pareto-type tails, it can easily be seen that this estimator is consistent if $k \to \infty$ and $\frac{n}{k} \to \infty$. It turns out that this statistic improves the consistency rate considerably with respect to the first naive estimator, but still uses only two extreme observations. Hill's estimator will improve on this aspect considerably. But even then, we need to know what large order statistics can be used in the procedure. From the derivations in the previous chapter, we could deduce that, if the sample size tends to ∞, then also k should be allowed to do the same, albeit at a certain rate.

4.2 The Hill Estimator

There are at least four natural ways to introduce this estimator. All of them are inspired by the previous analysis. Moreover, the estimator enjoys a high degree of popularity thanks to some nice theoretical properties but in spite of some serious drawbacks.

4.2.1 Construction

(i) *The quantile view.* The first source of inspiration comes from the quantile plots of the Pareto-type distributions.

  (a) These distributions satisfy

$$\frac{\log Q(1-p)}{-\log p} \to \gamma, \quad \text{as } p \to 0.$$

  From this, it follows that a Pareto quantile plot, that is, an exponential quantile plot based on the log-transformed data, is ultimately linear with slope γ near the largest observations.

  (b) Moreover, the slope of an ultimately linear exponential quantile plot can be estimated by the mean excess values of the type $E_{k,n}$ as discussed in section 1.2.2.

  Combining these two observations leads to the mean excess value of the log-transformed data, known as the *Hill estimator* (Hill (1975)):

$$H_{k,n} = \frac{1}{k} \sum_{j=1}^{k} \log X_{n-j+1,n} - \log X_{n-k,n}. \tag{4.1}$$

An important question can be raised concerning the optimality of the Hill estimator as an estimator of the slope of a quantile-quantile QQ-plot. In fact, the vertical coordinates $\log X_{n-j+1,n}$ are not independent and do not possess the same variance, and hence summarizing the upper part of the Pareto quantile plot $\left(\log = \left(\frac{n+1}{j}\right), \log X_{n-j+1,n}\right)$, $j = 1, \ldots, k$, using a least-squares line $y = \log X_{n-k,n} + \gamma(x - \log((n + 1)/(k + 1)))$ does not seem to be efficient since the classical Gauss-Markov conditions are not met.

Combining the information over a set of possible j-values, we can look for the least-squares straight line that fits best to the points

$$\left\{\left(-\log \frac{j}{n+1}, \log X_{n-j+1,n}\right); j = 1, \ldots, k+1\right\}$$

where we force the straight line to pass through the most left of these points. Such a line has the form

$$y = \log X_{n-k,n} + \gamma\left(-\log \frac{j}{n+1} - \log \frac{n+1}{k+1}\right).$$

A little reflection indicates that it might be wise to give points on the right of the above set variable weights in view of the above-mentioned problem of heteroscedasticity. In order to find the least-squares value of γ, we therefore minimize the quantity

$$\sum_{j=1}^{k} w_{j,k}\left[\log X_{n-j+1,n} - (\log X_{n-k,n} + \gamma \log \frac{k+1}{j})\right]^2$$

where $\{w_{j,k}; j = 1, \ldots, k\}$ are appropriate weights. A simple calculation tells us that the resulting value of γ, say $\hat{\gamma}_k$, is given by

$$\hat{\gamma}_k = \sum_{j=1}^{k} \alpha_{j,k} \log \frac{X_{n-j+1,n}}{X_{n-k,n}}$$

where

$$\alpha_{j,k} = \frac{w_{j,k} \log \frac{k+1}{j}}{\sum_{r=1}^{k} w_{r,k} \left(\log \frac{k+1}{r}\right)^2}.$$

When choosing $\alpha_{j,k} = 1/k$ one arrives at the Hill estimator.

(ii) *The probability view.* The definition of a Pareto-type tail can be rewritten as

$$\frac{1 - F(tx)}{1 - F(t)} \to x^{-1/\gamma} \text{ as } t \to \infty \text{ for any } x > 1.$$

This can be interpreted as

$$P(X/t > x | X > t) \approx x^{-1/\gamma} \text{ for } t \text{ large}, \quad x > 1.$$

Hence, it appears a natural idea to associate a strict Pareto distribution with survival function $x^{-1/\gamma}$ to the distribution of the relative excesses $Y_j = X_i/t$ over a high threshold t conditionally on $X_i > t$, where i is the index of the j-th exceedance in the original sample and $j = 1, \ldots, N_t$. The log-likelihood conditionally on N_t then becomes

$$\log L(Y_1, \ldots, Y_{N_t}) = -N_t \log \gamma - \left(1 + \frac{1}{\gamma}\right) \sum_{j=1}^{N_t} \log Y_j.$$

Since

$$\frac{d \log L}{d\gamma} = -\frac{N_t}{\gamma} + \frac{1}{\gamma^2} \sum_{j=1}^{N_t} \log Y_j,$$

the likelihood equation leads to

$$\hat{\gamma} = \frac{1}{N_t} \sum_{j=1}^{N_t} \log Y_j.$$

Choosing for the threshold t an upper order statistic $X_{n-k,n}$ (so that $N_t = k$), we obtain Hill's estimator again. For t non-random, we get the ratio estimator of Goldie and Smith (1987).

(iii) *Rényi's exponential representation.* There is an alternative way of writing the Hill estimator by introducing the random variables

$$Z_j := j(\log X_{n-j+1,n} - \log X_{n-j,n}) =: jT_j$$

that will play a crucial role later. Through a partial summation, one finds that

$$\sum_{j=1}^{k} Z_j = \sum_{j=1}^{k} jT_j = \sum_{j=1}^{k} \sum_{i=1}^{j} T_j = \sum_{i=1}^{k} \sum_{j=i}^{k} T_j$$

which easily leads to the crucial relation

$$H_{k,n} = \frac{1}{k} \sum_{j=1}^{k} Z_j = \bar{Z}_k.$$

Thanks to a remarkable property about the order statistics of an exponential distribution that was discovered by A. Rényi (see further in section 4.4), it follows that in case of a strict Pareto distribution, the transformed variables Z_j are independent and exponentially distributed:

$$Z_j \overset{\mathcal{D}}{=} \gamma E_j, \quad j = 1, \ldots, k,$$

with $(E_1, E_2, \ldots)$ being standard exponentially distributed. This exponential representation can be interpreted as a generalized linear regression model with $\bar{Z}_k$ the obvious maximum likelihood estimator of γ. Expressing a tail index estimator in terms of spacings between subsequent order statistics follows intuition. For, samples from distributions with heavy tails will be characterized by systematically larger gaps when the index j decreases.

(iv) *Mean excess approach.* Still another alternative derivation is based on the mean excess function of the log-transformed data. If $1 - F \in \mathcal{R}_{-1/\gamma}$ with $\gamma > 0$, then as derived in section 2.6

$$E\{\log X - \log x | X > x\} = \int_x^\infty \frac{1 - F(u)}{1 - F(x)} \frac{du}{u} \to \gamma \text{ as } x \to \infty.$$

Replace the distribution F by its empirical counterpart $\hat{F}_n$, defined in section 1.1, and x by the random sequence $X_{n-k,n}$ that tends to ∞. It is then a pleasant exercise to show that

$$H_{k,n} = \int_{X_{n-k,n}}^\infty \frac{1 - \hat{F}_n(u)}{1 - \hat{F}_n(X_{n-k,n})} \frac{du}{u} .$$

4.2.2 Properties

Mason (1982) showed that $H_{k,n}$ is a consistent estimator for γ (as $k, n \to \infty$, $k/n \to 0$) whatever the slowly varying function ℓ_F (or ℓ_U) may be. This is even true for weakly dependent data (Hsing (1991)) or in case of a linear process (Resnick and Stărică (1995)). Asymptotic normality of $H_{k,n}$ was discussed among others in Hall (1982), Davis and Resnick (1984), Csörgő and Mason (1985), Haeusler and Teugels (1985), Deheuvels *et al.* (1988), Csörgő and Viharos (1998), de Haan and Peng (1998) and de Haan and Resnick (1998). In Drees (1998) and Beirlant *et al.* (2002a), variance and rate optimality of the Hill estimator was derived for large submodels of the Pareto-type model.

However, several problems arise.

(i) For every choice of k, we obtain another estimator for γ. Usually one plots the estimates $H_{k,n}$ against k, yielding the Hill plot: $\{(k, H_{k,n}) : 1 \le k \le n - 1\}$. However, these plots typically are far from being constant, which makes it difficult to use the estimator in practice without further guideline on how to choose the value k. This is illustrated by a simulation from a strict Pareto distribution (Figure 4.1(a)) and from a Burr distribution (Figure 4.1(b)).

Resnick and Stărică (1997) proposed to plot $\{(\log k, H_{k,n}) : 1 \le k \le n - 1\}$, see also Drees *et al.* (2000). While this indeed focuses the graphics on the appropriate area, this procedure does not overcome some of the other problems that are cited next.

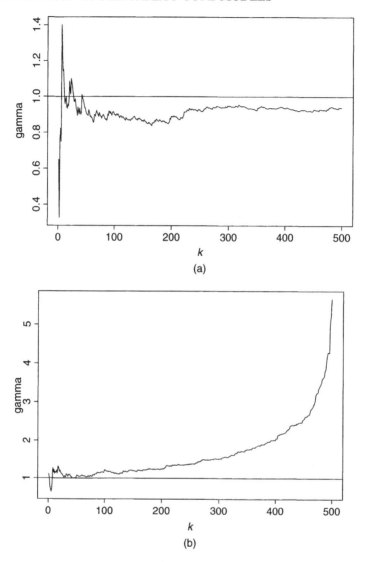

Figure 4.1 $(k, H_{k,n})$ for simulated datasets of size $n = 500$ from (a) the Pa(1) and (b) the Burr(1,1,1) distribution.

(ii) In many instances, a severe bias can appear. This happens when the effect of the slowly varying part in the model disappears slowly in the Pareto quantile plot. Stated differently within the probability view, the assumption that the relative excesses above a certain threshold follow a strict Pareto distribution is sometimes too optimistic. This is illustrated in Figure 4.2 where a Pareto quantile plot from a Burr(1,1,1) is put in contrast with that of

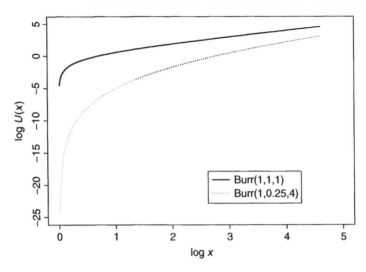

Figure 4.2　$\log U(x)$ as a function of $\log x$ for the Burr(1,1,1) and Burr(1,0.25,4) distributions.

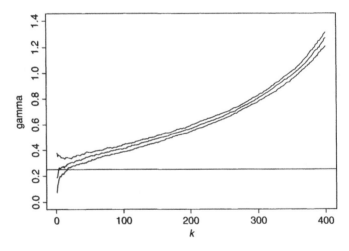

Figure 4.3　Median and quartiles of the Hill estimates $H_{k,n}$ as a function of k, $k = 1, \ldots, 400$, obtained from 100 samples of size 500 from a $|T_4|$ distribution.

a Burr(1,0.25,4) distribution. See also Figure 4.3 for a simulation experiment from a $|T_4|$ distribution with $\gamma = 0.25$ where only for the smallest values of k the median of the Hill estimator touches the correct value.

A large bias leads to poor coverage probabilities of confidence intervals. In many practical cases, systematic over- or underestimation has to be avoided.

For instance, in the valuation of a precious stone deposit, systematic overestimation of the tail of the carat size distribution will lead to over-optimistic predictions.

(iii) The Hill estimator shares a serious defect with many other common estimators that are based on log-transformed data: the estimator is not invariant with respect to shifts of the data. As mentioned by several authors, inadequate use of the Hill estimator in conjunction with a data shift can lead to systematic errors as well. A location-invariant modification of the Hill estimator is proposed in Fraga Alves (2001). To this end, a secondary k-value, denoted by k_0 ($k_0 < k$), is introduced, leading to

$$\hat{\gamma}^{(H)}(k_0, k) = \frac{1}{k_0} \sum_{j=1}^{k_0} \log \frac{X_{n-j+1,n} - X_{n-k,n}}{X_{n-k_0,n} - X_{n-k,n}}.$$

If one lets both $k = k_n$ and $k_0 = k_{0,n}$ tend to infinity with $n \to \infty$, such that $k/n \to 0$ and $k_0/k \to 0$, one can show that $\hat{\gamma}^{(H)}(k_0, k)$ is consistent. An adaptive version of the proposed estimator has been proposed from the best theoretical k_0 given by

$$k_0 \sim [(1 + \gamma)/\sqrt{2\gamma}]^{2/(1+2\gamma)} k^{2\gamma/(1+2\gamma)}$$

departing with some initial estimate $\hat{\gamma}^{(0)}$ in a first step; for instance, obtained by setting $k_0^{(0)} = \lfloor 2k^{2/3} \rfloor$.

4.3 Other Regression Estimators

The Hill estimator has been obtained from the Pareto quantile plot using a quite naive estimator of the slope in the ultimate right end of the quantile plot. Of course, more flexible regression methods on the highest k points of the Pareto quantile plot could be applied. This programme was carried out in detail in Schultze and Steinebach (1996), Kratz and Resnick (1996) and Csörgő and Viharos (1998). We refer to these papers for more mathematical details and confine ourselves here to the derivation of the estimators.

(i) The weighted least-squares fit on the Pareto quantile plot as treated in section 4.2 can be rewritten in the form

$$\hat{\gamma}_k = \frac{\sum_{i=1}^{k} T_i \sum_{j=1}^{i} w_{j,k} \log \frac{k+1}{j}}{\sum_{j=1}^{k} w_{j,k} \log^2 \frac{k+1}{j}}.$$

Setting $K(i/k) = i^{-1} \sum_{j=1}^{i} w_{j,k} \log \frac{k+1}{j}$, this estimator can be approximated by

$$\hat{\gamma}_{K,k}^{+} = \frac{k^{-1} \sum_{i=1}^{k} K(i/k) i \left(\log X_{n-i+1,n} - \log X_{n-i,n} \right)}{k^{-1} \sum_{j=1}^{k} K(j/k)},$$

showing that weighted least-squares estimation leads to the class of *kernel estimators* introduced by Csörgő *et al.* (1985). Here, K denotes a kernel function associating different weights to the different order statistics. However, Csörgő *et al.* (1985) also consider kernel functions with support outside $(0,1]$. Optimal choice of K is possible but is hard to manage in practice. Weighting the spacings Z_i has the advantage that the graphs of the estimates as a function of k are smoother in comparison with, for instance, the Hill estimator where adjacent values of k can lead to quite different estimates.

(ii) The problem of non-smoothness of Hill estimates as a function of k can be solved in another way: simple unconstrained least squares with estimation of a slope γ as well as an intercept, say δ, can already provide more smoothness even without the use of a kernel function. This procedure of fitting lines to (parts of) QQ-plots and especially double-logarithmic plots can be traced back to Zipf from the late 1940s (see Zipf (1949)). Only recently, this procedure has been studied in more depth.

The classical least-squares procedure minimizing

$$\sum_{j=1}^{k} \left(\log X_{n-j+1,n} - \left(\delta + \gamma \log \frac{n+1}{j} \right) \right)^2$$

with respect to δ and γ leads to

$$\hat{\gamma}_{Z,k}^+ = \frac{\frac{1}{k} \sum_{j=1}^{k} \left(\log \frac{n+1}{j} - \frac{1}{k} \sum_{j=1}^{k} \log \frac{n+1}{j} \right) \log X_{n-j+1,n}}{\frac{1}{k} \sum_{j=1}^{k} \log^2 \frac{n+1}{j} - \left(\frac{1}{k} \sum_{j=1}^{k} \log \frac{n+1}{j} \right)^2}$$

$$= \frac{\frac{1}{k} \sum_{j=1}^{k} \left(\log \frac{k+1}{j} - \frac{1}{k} \sum_{j=1}^{k} \log \frac{k+1}{j} \right) \log X_{n-j+1,n}}{\frac{1}{k} \sum_{j=1}^{k} \log^2 \frac{k+1}{j} - \left(\frac{1}{k} \sum_{j=1}^{k} \log \frac{k+1}{j} \right)^2}.$$

This is the estimator proposed in Schultze and Steinebach (1996) and Kratz and Resnick (1996). In Csörgő and Viharos (1998), the asymptotic properties of this estimator are reviewed. These authors also propose a generalization of this estimator that again can be motivated by a weighted least-squares algorithm:

$$\hat{\gamma}_{WLS,k}^+ = \frac{\frac{1}{k} \sum_{j=1}^{k} \left(\int_{(j-1)/k}^{j/k} J(s)ds \right) \log X_{n-j+1,n}}{\frac{1}{k} \sum_{j=1}^{k} \left(\int_{(j-1)/k}^{j/k} J(s)ds \right) \log \frac{k+1}{j}},$$

where J is a non-increasing function defined on $(0,1)$, which integrates to 0. Csörgő and Viharos (1998) propose to use the weight functions J of the type

$$J_\theta(s) := \frac{\theta+1}{\theta} - \frac{(\theta+1)^2}{\theta} s^\theta, \quad s \in [0,1]$$

for some $\theta > 0$.

4.4 A Representation for Log-spacings and Asymptotic Results

In this section, we investigate the most important mathematical properties of the Hill estimator and some selected generalizations as discussed above. In particular, we propose expressions for the asymptotic bias and the asymptotic variance. These results will be helpful later when we discuss the adaptive choice of k. The given results can also assist in providing remedies for some of the problems cited above.

In section 4.2.1 (iii), we derived that Hill's estimator can be written as a simple average of scaled *log*-spacings:

$$H_{k,n} = \frac{1}{k} \sum_{j=1}^{k} Z_j \text{ with } Z_j = j(\log X_{n-j+1,n} - \log X_{n-j,n}).$$

We will now elaborate on these spacings. We continue the discussion as started in 4.2.1 (iii), where we found that in the case of strict Pareto distributions

$$Z_j \overset{\mathcal{D}}{=} \gamma E_j, \quad j = 1, \ldots, k,$$

with $\{E_i; 1 \leq i \leq n\}$ a sample from an exponential distribution with mean 1. In accordance with our conventions, their order statistics are then denoted by

$$E_{1,n} \leq E_{2,n} \leq \cdots \leq E_{n-k+1,n} \leq \cdots \leq E_{n,n}.$$

Double use of the probability integral transform leads to the linking equalities

$$X_{j,n} \overset{\mathcal{D}}{=} U(e^{E_{j,n}}), \quad 1 \leq j \leq n. \tag{4.2}$$

The main reason for using an exponential sample lies in a remarkable property about the order statistics of the latter distribution, discovered by A. Rényi. Indeed,

$$E_{n-j+1,n} - E_{n-k,n} \overset{\mathcal{D}}{=} \sum_{i=j}^{k} \frac{E_i}{i}, \quad 1 \leq j \leq k < n \tag{4.3}$$

where $\{E_i, 1 \leq i \leq n-1\}$ is again an exponential sample with mean 1. From this equation, one can, for example, derive the expectations of the exponential order statistics in that

$$E(E_{n-j,n}) = \sum_{k=j}^{n-1} \frac{1}{k+1} \sim \log\left(\frac{n+1}{j+1}\right)$$

if n is large.

We now combine the above with the second-order properties of the tail quantile function U. From here, we assume that $\log \ell_U$ satisfies $(\mathcal{C}_{-\beta}(b))$ for some $\beta > 0$ and $b \in \mathcal{R}_{-\beta}$. This means that we can write

$$\frac{U(ux)}{U(x)} = u^{\gamma}\left(1 + h_{-\beta}(u)\, b(x) + o(b(x))\right). \tag{4.4}$$

Using the second-order condition (4.4), we expand the distribution of the scaled spacings $Z_j = j(\log X_{n-j+1,n} - \log X_{n-j,n})$, $j = 1, \ldots, k$:

$$
\begin{aligned}
Z_j &= j \log \frac{X_{n-j+1,n}}{X_{n-j,n}} \\[2mm]
&\stackrel{\mathcal{D}}{=} j \log \frac{U\left(e^{E_{n-j+1,n} - E_{n-j,n}}\, e^{E_{n-j,n}}\right)}{U\left(e^{E_{n-j,n}}\right)} \\[2mm]
&\stackrel{\mathcal{D}}{=} j \log \frac{U\left(e^{E_j/j}\, e^{E_{n-j,n}}\right)}{U\left(e^{E_{n-j,n}}\right)} \\[2mm]
&= j\left\{\gamma \log e^{E_j/j} + \log\left[1 + h_{-\beta}\left(e^{E_j/j}\right) b\left(e^{E_{n-j,n}}\right)(1 + o(1))\right]\right\} \\[2mm]
&\approx \gamma E_j + j \log(1 + W_{n,j})
\end{aligned}
$$

where we used (4.3) and the abbreviation

$$W_{n,j} := h_{-\beta}\left(e^{E_j/j}\right) b\left(e^{E_{n-j,n}}\right). \tag{4.5}$$

We therefore get a *stochastic representation* for the spacings

$$Z_j \stackrel{\mathcal{D}}{\approx} \gamma E_j + j \, \log\left(1 + W_{n,j}\right). \tag{4.6}$$

One way of using this result is to replace the log-term on the right by inequalities like

$$\frac{y}{1+y} \leq \log(1+y) \leq y$$

that yield universal, stochastic inequalities for the Hill estimator since $H_{k,n} = \frac{1}{k}\sum_{j=1}^{k} Z_j$. Another possibility is to look at approximations for $\log(1+y)$ for y small. First of all, it is easy to see that for y small, $h_{-\beta}(e^y) = y(1 + o(1))$. Next, we need to gain a bit more insight into the behaviour of the argument of $b(x)$ in (4.5). As long as $j/n \to 0$ when $n \to \infty$, we have $E_{n-j,n}/\log(n/j) \stackrel{P}{\to} 1$ (see section 3.2, case 2, (i)). But then $b\left(e^{E_{n-j,n}}\right) = b\left(\frac{n+1}{j+1}\right)(1 + o_P(1))$. This means that we can approximate $j \log(1 + W_{n,j})$ in distribution by $E_j b\left(\frac{n+1}{j+1}\right)$. Hence, we are lead to the following approximate representation:

$$Z_j \stackrel{\mathcal{D}}{\sim} \left(\gamma + b\left(\frac{n+1}{j+1}\right)\right) E_j, \tag{4.7}$$

or, using the regular variation of b with index $-\beta$,

$$Z_j \overset{\mathcal{D}}{\sim} \left(\gamma + \left(\frac{j}{k+1} \right)^{\beta} b \left(\frac{n+1}{k+1} \right) \right) E_j, \quad j = 1, \ldots, k. \tag{4.8}$$

In the sequel, we use the notation $b_{n,k} = b \left(\frac{n+1}{k+1} \right)$.

The above is just a sketch of the proof of (4.8). In Beirlant *et al.* (2002c), the following result is proven. Similar results can be found in Kaufmann and Reiss (1998) and Drees *et al.* (2000).

Theorem 4.1 *Suppose (4.4) holds. Then there exist random variables $R_{j,n}$ and standard exponential random variables E_j (independent with each n) such that*

$$\sup_{1 \le j \le k} \left| Z_j - \left(\gamma + b_{n,k} \left(\frac{j}{k+1} \right)^{\beta} \right) E_j - R_{j,n} \right| = o_P(b_{n,k}), \tag{4.9}$$

as $k, n \to \infty$ with $k/n \to 0$, where uniformly in $i = 1, \ldots, k$

$$\left| \sum_{j=i}^{k} R_{j,n}/j \right| = o_P \left(b_{n,k} \max \left(\log \left(\frac{k+1}{i} \right), 1 \right) \right).$$

Let us draw some first conclusions concerning the Hill estimator.

(i) The *asymptotic bias* of the Hill estimator can be traced back using the exponential representation. Indeed,

$$ABias(H_{k,n}) \sim b_{n,k} \frac{1}{k} \sum_{j=1}^{k} \left(\frac{j}{k+1} \right)^{\beta}$$

$$\sim \frac{b_{n,k}}{1+\beta}.$$

We notice that the bias will be small only if $b_{n,k}$ is small, which in turn requires k to be small.

(ii) The *asymptotic variance* of the Hill estimator is even easier in that

$$AVar(H_{k,n}) \sim var \left(\frac{\gamma}{k} \sum_{j=1}^{k} E_j \right) \sim \frac{\gamma^2}{k}.$$

Notice that the variance will be small if we take k large.

(iii) Finally, the *asymptotic normality* of the Hill estimator can be expected when $k, n \to \infty$ and $k/n \to 0$. For then, if $\sqrt{k} b_{n,k} \to 0$,

$$\sqrt{k} \left(\frac{H_{k,n}}{\gamma} - 1 \right) \overset{\mathcal{D}}{\to} N(0, 1).$$

This result allows the construction of approximate confidence intervals for γ. At the level $(1 - \alpha)$, this interval is given by

$$\left(\left(1 + \frac{\Phi^{-1}(1 - \alpha/2)}{\sqrt{k}}\right)^{-1} H_{k,n}, \left(1 - \frac{\Phi^{-1}(1 - \alpha/2)}{\sqrt{k}}\right)^{-1} H_{k,n}\right), \quad (4.10)$$

which is an acceptable approach if the bias is not too important, that is, if $\beta \geq 1$. Typically, the condition $\sqrt{k}b_{n,k} \to 0$ severely restricts the range of k-values where the confidence interval works.

We end this section outlining how the above exponential representation result can be used to derive formally the above bias-variance expressions and to deduce asymptotic normality results for kernel type statistics

$$\frac{1}{k}\sum_{j=1}^{k} K\left(\frac{j}{k+1}\right)Z_j$$

as discussed earlier in this chapter. Here, we assume that the kernel K can be written as $K(t) = \frac{1}{t}\int_0^t u(v)dv$, $0 < t < 1$ for some function u defined on $(0, 1)$. Such type of results can be found, for instance, in Csörgő et al. (1985) and Csörgő and Viharos (1998).

Theorem 4.2 *Suppose (4.4) holds. Let $K(t) = \frac{1}{t}\int_0^t u(v)dv$ for some function u satisfying $\left|k\int_{(j-1)/k}^{j/k} u(t)dt\right| \leq f\left(\frac{j}{k+1}\right)$ for some positive continuous function f defined on $(0, 1)$ such that $\int_0^1 \log^+(1/w)f(w)dw < \infty$ and $\int_0^1 |K|^{2+\delta}(w)dw < \infty$ for some $\delta > 0$. Suppose $\sqrt{k}b_{n,k} = O(1)$ as $k, n \to \infty$ with $k/n \to 0$. Then with the same notations as before, we have that*

$$\sqrt{k}\left[\frac{1}{k}\sum_{j=1}^{k} K\left(\frac{j}{k+1}\right)Z_j - \frac{1}{k}\sum_{j=1}^{k} K\left(\frac{j}{k+1}\right)\left(\gamma + b_{n,k}\left(\frac{j}{k+1}\right)^\beta\right)\right]$$

converges in distribution to a $N\left(0, \gamma^2\int_0^1 K^2(u)du\right)$ distribution.

The result follows from the Lindeberg-Feller central limit theorem after showing that

$$\sqrt{k}\left[\sum_{j=1}^{k}\left\{\frac{1}{j}\int_0^{j/k} u(t)dt\right\}Z_j\right.$$

$$\left. -\sum_{j=1}^{k}\left\{\frac{1}{j}\int_0^{j/k} u(t)dt\right\}\left(\gamma + b_{n,k}\left(\frac{j}{k+1}\right)^\beta\right)E_j\right] \overset{P}{\Rightarrow} 0,$$

which follows from $\sqrt{k}b_{n,k} = O(1)$ and the above exponential representation theorem since

$$\sum_{j=1}^{k} \left\{ \frac{1}{j} \int_0^{j/k} u(t)dt \right\} R_{j,n} = \sum_{j=1}^{k} \sum_{i=1}^{j} \left\{ \int_{(i-1)/k}^{i/k} u(t)dt \right\} \frac{R_{j,n}}{j}$$

$$= \sum_{i=1}^{k} \int_{(i-1)/k}^{i/k} u(t)dt \sum_{j=i}^{k} \frac{R_{j,n}}{j},$$

and hence

$$\left| \sum_{j=1}^{k} \left\{ \frac{1}{j} \int_0^{j/k} u(t)dt \right\} R_{j,n} \right| \le \frac{1}{k} \sum_{i=1}^{k} f\left(\frac{i}{k+1}\right) \left| \sum_{j=i}^{k} \frac{R_{j,n}}{j} \right|.$$

From Theorem 4.2, we learn that the asymptotic mean squared error of kernel estimators $\hat{\gamma}_{K,k}^{+}$ equals $k^{-1}\gamma^2 \int_0^1 K^2(u)du + b_{n,k}^2 (\int_0^1 K(u)u^\beta du)^2$. In Csörgő et al. (1985), kernels are derived (with support possibly outside (0, 1]), which minimize $AMSE(\hat{\gamma}_{K,k}^{+})$. In case, $\log \ell_U$ satisfies $(C_{-\beta}(b))$ for some $\beta > 0$ Csörgő et al. (1985) found the kernel

$$K_\beta(t) = \frac{1+\beta}{\beta} \left(\frac{1+2\beta}{2+2\beta}\right)^{1+\beta} \left[\left(\frac{2+2\beta}{1+2\beta}\right)^\beta - t^\beta \right], \quad 0 < t < \frac{2+2\beta}{1+2\beta},$$

to be optimal in the sense that among all kernels satisfying $\int_0^\infty K(u)du = \int_0^\infty K^2(u)du = 1$ the asymptotic mean squared error is minimal for K_β. Note, however, that the optimal kernel depends on β, which makes the application of this approach somewhat difficult. In Csörgő and Viharos (1998), a method for implementing this approach is proposed. The estimation of β will be discussed below.

4.5 Reducing the Bias

In many cases, the Hill estimator overestimates the population value of γ due to slow convergence of $b_{n,k}$ to 0. Some proposals of bias-reduced estimators were recently introduced, for instance, in Peng (1998), Feuerverger and Hall (1999), Beirlant et al. (1999), Gomes et al. (2000) and Gomes and Martins (2002). The last references make use of the exponential representation developed above. Again, one can tackle the problem from the quantile or the probability view.

4.5.1 The quantile view

The representation (4.8) can be considered as a generalized regression model with exponentially distributed responses. For every fixed j, the responses Z_j are approximately exponentially distributed with mean $\gamma + b_{n,k}\left(\frac{j}{k+1}\right)^\beta$. If $b_{n,k} > 0$, then the

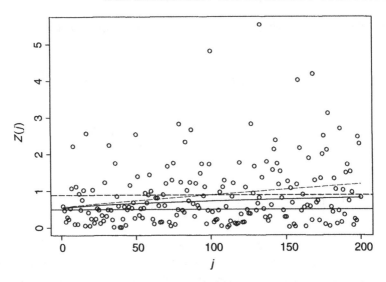

Figure 4.4 Plot of Z_j versus j, $j = 1, \ldots, 200$, for a simulated sample of size $n = 500$ from the Burr(1,1,2) distribution; solid horizontal line: true value of γ; broken horizontal line: $H_{200,500}$; solid curve: $\gamma + b_{n,k} \left(\frac{j}{k+1} \right)^{\beta}$; broken curve: $\hat{\gamma} + \hat{b}_{n,k} \left(\frac{j}{k+1} \right)^{\hat{\beta}}$.

means increase with increasing values of j while the intercept is given by γ. This is illustrated in Figure 4.4 using a simulation from a Burr(1,1,2) distribution of size $n = 500$. We show $k = 200$ points.

Some simple variations of (4.8) were proposed:

$$Z_j \overset{\mathcal{D}}{\sim} \gamma \exp \left(d_{n,k} \left(\frac{j}{k+1} \right)^{\beta} \right) E_j, \quad 1 \leq j \leq k, \tag{4.11}$$

with $d_{n,k} = b_{n,k}/\gamma$, using the approximation

$$1 + d_{n,k} \left(\frac{j}{k+1} \right)^{\beta} \sim \exp \left(d_{n,k} \left(\frac{j}{k+1} \right)^{\beta} \right).$$

Alternatively, changing the generalized linear model (4.8) into a regression model with additive noise (replacing the random factors E_j by their expected values in the bias term), we obtain

$$Z_j \sim \gamma + b_{n,k} \left(\frac{j}{k+1} \right)^{\beta} + \gamma(E_j - 1), \quad 1 \leq j \leq k. \tag{4.12}$$

Joint estimates of γ, $b_{n,k}$ (or $d_{n,k}$) and β can be obtained for each k from (4.8) and (4.11) by maximum likelihood, or from (4.12) by least-squares minimizing

$$\sum_{j=1}^{k} \left(Z_j - \gamma - b_{n,k} \left(\frac{j}{k+1} \right)^{\beta} \right)^2$$

with respect to γ, $b_{n,k}$ and β. We denote the maximum likelihood estimator of γ based on (4.8) by $\hat{\gamma}_{ML}^+$.

In view of the discussion of the properties of the Hill estimator in the preceding section, we will mainly focus on the case $\beta < 1$. Remark, however, that the regression models are not identifiable when β equals 0, for then γ and $b_{n,k}$ together make up the mean response. Necessarily this fact leads to instabilities in case β is close to 0. An algorithm was constructed in Beirlant *et al.* (1999) to search for the maximum likelihood estimate of γ next to estimates $\hat{b}_{n,k}$ and $\hat{\beta}_k$ under (4.8). In order to avoid instabilities in the maximization routines, the restriction $\hat{\beta}_k > 0.5$ was introduced for sample sizes up to $n = 1000$. Simulation experiments indicated that this bound can gradually be relaxed with larger sample sizes, for instance, to $\hat{\beta}_k > 0.25$ for $n = 5000$. Also, to avoid instabilities arising from the optimization procedure ending at a local maximum, some smoothness conditions were added, linking estimates at subsequent values of k: given $\hat{b}_{n,k+1}$ and $\hat{\beta}_{k+1}$, we set $|\hat{b}_{n,k}| \leq 1.1|\hat{b}_{n,k+1}|$ and $\hat{\beta}_k \leq 1.1\hat{\beta}_{k+1}$. A similar program can be carried out on the basis of (4.12) using least squares. The results obtained in this way are very similar to those obtained by maximum likelihood.

The variance of $\hat{\gamma}_{ML}^+$ when $k \to \infty$ and $k/n \to 0$ equals $((1+\beta)/\beta)^4 \gamma^2/k$ in first order (see Beirlant *et al.* (2002c)) showing that the variance of these estimators is much larger than in case of the Hill estimator. The asymptotic bias, however, is zero as long as $\sqrt{k}b_{n,k} = O(1)$, this is in contrast to the Hill estimator, where the asymptotic bias only disappears for relatively small values of k, that is, $\sqrt{k}b_{n,k} \to 0$. The graphs with the resulting estimates as a function of k are much more stable taking away a serious part of the bias of the Hill estimator. Experience suggests that the largest values of k for which both the Hill and the bias-reduced estimator correspond closely provide reasonable estimates for γ. The use of the regression models to choose k adaptively will be further explored in section 4.7. The mean squared errors find their minima at much higher values of k in comparison with the Hill estimator; the respective minima are typically of the same size.

Confidence intervals for γ can now be constructed on the basis of a bias-reduced maximum likelihood estimator with the above-mentioned variance. In contrast to the confidence interval (4.10), this leads to intervals that better approximate the required confidence level $1 - \alpha$. This is illustrated in Figure 4.5 using simulated samples of size $n = 500$ from a Burr(1,0.5,2) distribution, for which $\beta = 0.5$.

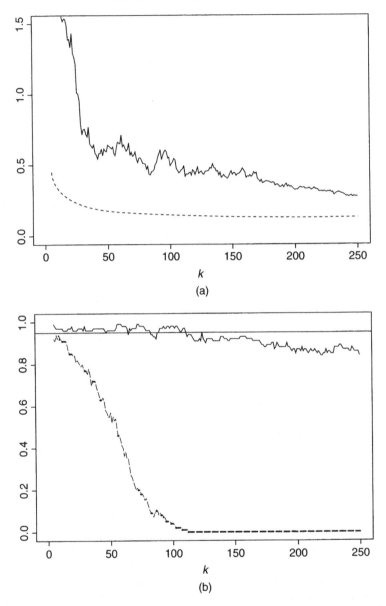

Figure 4.5   (a) Medians of the estimated standard deviations of the Hill estimator (broken line) and the maximum likelihood estimator $\hat{\gamma}^+_{\mathrm{ML}}$ (solid line) as a function of k, $k = 5, \ldots, 250$ and (b) corresponding coverage probabilities of confidence intervals for $k = 5, \ldots, 250$ based on the Hill estimator (broken line) and the maximum likelihood estimator (solid line).

In each of the three regression models considered above, one can also solve for γ and $b_{n,k}$, or γ and $d_{n,k}$, after substituting a consistent estimator $\tilde{\beta} = \tilde{\beta}_{k,n}$ for β. For brevity, we focus on the least-squares estimators based on (4.12), leading to

$$\hat{\gamma}_{LS}^+(\tilde{\beta}) = \bar{Z}_k - \hat{b}_{LS}^+(\tilde{\beta})/(1 + \tilde{\beta})$$

$$\hat{b}_{LS}^+(\tilde{\beta}) = \frac{(1 + \tilde{\beta})^2(1 + 2\tilde{\beta})}{\tilde{\beta}^2} \frac{1}{k} \sum_{j=1}^{k} \left(\left(\frac{j}{k+1} \right)^{\tilde{\beta}} - \frac{1}{1+\tilde{\beta}} \right) Z_j.$$

Here, we approximated $k^{-1} \sum_{j=1}^{k} (j/(k+1))^{\tilde{\beta}}$ by $1/(1 + \tilde{\beta})$ and $k^{-1} \sum_{j=1}^{k} \left((j/(k+1))^{\tilde{\beta}} - 1/(1 + \tilde{\beta}) \right)^2$ by $\tilde{\beta}^2(1 + \tilde{\beta})^{-2}(1 + 2\tilde{\beta})^{-1}$. On the basis of Theorem 4.2, one can show that the asymptotic variance of $\hat{\gamma}_{LS}^+$ equals $k^{-1}\gamma^2((1 + \beta)/\beta)^2$. Here, the increase of the variance in comparison with the Hill estimator is not so large as with $\hat{\gamma}_{ML}^+$, but the question arises concerning an estimator of the second-order parameter β. Drees and Kaufmann (1998) proposed the estimator

$$\tilde{\beta}_{\tilde{k},n,\lambda} = \frac{1}{\log \lambda} \log \frac{H_{\lfloor \lambda^2 \tilde{k} \rfloor, n} - H_{\lfloor \lambda \tilde{k} \rfloor, n}}{H_{\lfloor \lambda \tilde{k} \rfloor, n} - H_{\tilde{k}, n}}$$

for some $\lambda \in (0, 1)$ and with $\tilde{k}$ taken in the range $\sqrt{\tilde{k}}b_{n,\tilde{k}} \to \infty$. An adaptive choice for $\tilde{k}$ in this range is also given. It can also be shown that the estimators of β based on the regression models discussed here share this consistency property as $\sqrt{\tilde{k}}b_{n,\tilde{k}} \to \infty$. For a more elaborate discussion of the estimation of β and several other estimators of β, we refer the reader to Gomes *et al.* (2002), Gomes and Martins (2002) and Fraga Alves *et al.* (2003).

The estimation of β is known to be difficult. Hence, some authors have proposed to set $\beta = 1$ in procedures involving knowledge of β. The resulting estimators yield a compromise between the bias reduction of the estimators involving estimation of β and the smaller variance when using, for instance, the Hill estimator, see, for instance, Gomes and Oliveira (2003). Guillou and Hall (2001) use the estimator of $b_{n,k}$ obtained from (4.12) after setting $\beta = 1$ in the context of adaptive selection rules for the number of extremes k. This will be discussed in section 4.7.

Finally, we mention that $\hat{\gamma}_{ML}^+$ and $\hat{\gamma}_{LS}^+(\tilde{\beta})$, while not shift-invariant in the mathematical sense, are already much more stable under shifts than the Hill estimator. The above-mentioned shift-invariant modification of the Hill estimator proposed by Fraga Alves (2001) also enables for stable plots.

4.5.2 The probability view

Alternatively, Beirlant *et al.* (2004) propose to use a second-order refinement of the probability view. Following the approach discussed in 4.2.1 (ii) where the Hill estimator follows from the approximation of the conditional distribution of the relative excesses $Y_j := X_{n-j+1,n}/X_{n-k,n}$, $j = 1, \ldots, k$, by a strict Pareto distribution, one can argue that the Hill estimator will break down if this approximation is poor. In order to describe the departure of $F_t(x) = P(X/t \leq x | X > t)$ from a strict Pareto distribution, we use the assumption that ℓ_F satisfies (3.14):

$$\frac{1 - F(tx)}{1 - F(t)} = x^{-1/\gamma}(1 + h_{-\tau}(x)B(t) + o(B(t))),   \qquad (4.13)$$

where $\tau > 0$ and B is regularly varying at infinity with index $-\tau$. Condition (4.13) can be rephrased as

$$1 - F_t(x) = x^{-1/\gamma}[1 - B(t)\tau^{-1}(x^{-\tau} - 1) + o(B(t))],$$

as $t \to \infty$. Deleting the error term, this refines the original Pareto approximation to an approximation by a mixture of two Pareto distributions. The idea is now to fit such a perturbed Pareto distribution to the multiplicative excesses Y_j, $j = 1, \ldots, k$, aiming for more accurate estimation of the unknown tail.

Such a perturbed Pareto distribution is then defined by the survival function

$$1 - G(x; \gamma, c, \tau) = (1 - c)x^{-1/\gamma} + cx^{-1/\gamma-\tau}$$

with some $c \in (-1/\tau, 1)$ and $x > 1$. Observe that if $c = 0$, then this mixture coincides with ordinary Pareto distribution.

For $c \downarrow 0$, we can write

$$1 - G(x; \gamma, c, \tau) = \left\{ x[1 + \gamma c(1 - x^{-\tau})] \right\}^{-1/\gamma} + o(c)$$

$$= \left\{ x[(1 + \gamma c) - \gamma c x^{-\tau})] \right\}^{-1/\gamma} + o(c).$$

In practice, it turns out that

$$\bar{G}_{PPD}(x) = x^{-1/\gamma}[(1 + \gamma c) - \gamma c x^{-\tau}]^{-1/\gamma}   \qquad (4.14)$$

fits well by the maximum likelihood method, leading to estimators $\hat{\gamma}_{PPD}^+$, $\hat{c}_{PPD}^+$ and $\hat{\tau}_{PPD}^+$. The likelihood surface can be seen to be rather flat in τ so that the optimization should be handled with care, comparable to the estimation of β in the generalized linear model (4.8).

The perturbed Pareto distribution (4.14) extends the generalized Pareto (GP) distribution, which will be discussed in depth in Chapter 5, in the following way. In statistics of extremes, it is common practice to approximate the distribution of absolute exceedances of a random variable Y above a high-enough threshold u by the generalized Pareto distribution:

$$P(Y - u > y | Y > u) = \left(1 + \frac{\gamma y}{\sigma}\right)^{-\frac{1}{\gamma}},   \qquad y > 0; \ \sigma > 0.   \qquad (4.15)$$

Replacing y by $ux - u$ with $x \geq 1$ transforms (4.15) into a model for relative excesses

$$P(Y/u > x | Y > u) = \left\{ x \left[\frac{\gamma u}{\sigma} - \left(\frac{\gamma u}{\sigma} - 1\right) x^{-1}\right] \right\}^{-\frac{1}{\gamma}},$$

which is clearly (4.14) with $c = u/\sigma - 1/\gamma$ and $\tau = 1$.

4.6 Extreme Quantiles and Small Exceedance Probabilities

In the previous sections on the quantile viewpoint, we fitted a straight line to an ultimately linear part of the Pareto quantile plot. Continuing in this spirit and following the principle for estimating large quantiles and small exceedance probabilities, outlined in Figure 1.2, we are now in the position to estimate extreme quantiles under a Pareto-type model. However, the probability view allows for an alternative interpretation of the available methods.

4.6.1 First-order estimation of quantiles and return periods

We first discuss the simple approach proposed by Weissman (1978) based on the Hill estimator.

We use the Pareto index estimation method based on linear regression of a Pareto quantile plot to derive an estimator for $Q(1 - p)$. *Assuming* that the ultimate linearity of the Pareto quantile plot persists from the largest k observations on (till infinity), that is, assuming that the strict Pareto model persist above this threshold, we can extrapolate along the line with equation

$$y = \log X_{n-k,n} + H_{k,n}\left(x + \log \frac{k+1}{n+1}\right)$$

anchored at the point $\left(-\log \frac{k+1}{n+1}, \log X_{n-k,n}\right)$.

Take $x = -\log p$ to obtain an estimator $\hat{q}^{+}_{k,p}$ of $Q(1 - p)$ given by

$$\hat{q}^{+}_{k,p} = \exp\left(\log X_{n-k,n} + H_{k,n} \log \frac{k+1}{(n+1)p}\right)$$

$$= X_{n-k,n}\left(\frac{k+1}{(n+1)p}\right)^{H_{k,n}}.$$

The asymptotic characteristics of this method can be found through the following expansion: since $Q(1 - p) = p^{-\gamma}\ell_U(1/p)$ and $X_{n-k,n} \overset{\mathcal{D}}{=} U^{-\gamma}_{k+1,n}\ell_U(U^{-1}_{k+1,n})$ where $U_{j,n}$ denote the order statistics from a uniform $(0,1)$ sample, we find that

$$\log \frac{\hat{q}^{+}_{k,p}}{Q(1 - p)} \overset{\mathcal{D}}{=} \log\left[\left(\frac{U_{k+1,n}}{p}\right)^{-\gamma} \frac{\ell_U(U^{-1}_{k+1,n})}{\ell_U(p^{-1})}\left(\frac{k+1}{(n+1)p}\right)^{H_{k,n}}\right]$$

$$= \log\left[\left(\frac{U_{k+1,n}}{(k+1)/(n+1)}\right)^{-\gamma}\left(\frac{k+1}{(n+1)p}\right)^{H_{k,n}-\gamma} \frac{\ell_U(U^{-1}_{k+1,n})}{\ell_U(p^{-1})}\right]$$

$$\overset{\mathcal{D}}{=} \gamma\left(E_{n-k,n} - \log \frac{n+1}{k+1}\right) + (H_{k,n} - \gamma)\log\left(\frac{k+1}{(n+1)p}\right)$$

$$+ \log \frac{\ell_U(U^{-1}_{k+1,n})}{\ell_U(p^{-1})}$$

where we used the same notation as in section 4.4. Under condition (4.4), we can approximate the last term by

$$-b_{n,k}\frac{1-\left(\frac{(n+1)p}{k+1}\right)^{\beta}}{\beta}.$$

Using again $\sqrt{k}\left(E_{n-k,n}-\log\frac{n}{k}\right)\overset{D}{\to}N(0,1)$ as $k,n\to\infty$ and $k/n\to 0$ (see section 3.2, case 2 (ii)), together with the exponential representation of the scaled spacings Z_j, we find the expressions for the asymptotic variance and bias of the Weissman estimator in the log-scale when $p=p_n\to 0$ and $np_n\to c>0$ as $n\to\infty$. We denote the asymptotic expectation by E_∞.

$$E_\infty\left(\log\frac{\hat{q}^+_{k,p}}{Q(1-p)}\right)\sim ABias\,(H_{k,n})\log\left(\frac{k+1}{(n+1)p}\right)-b_{n,k}\frac{1-\left(\frac{(n+1)p}{k+1}\right)^{\beta}}{\beta}$$

$$=\frac{b_{n,k}}{1+\beta}\log\left(\frac{k+1}{(n+1)p}\right)-b_{n,k}\frac{1-\left(\frac{(n+1)p}{k+1}\right)^{\beta}}{\beta},\quad(4.16)$$

$$AVar\,(\log\hat{q}^+_{k,p})\sim\frac{\gamma^2}{k}\left(1+\log^2\left(\frac{k+1}{(n+1)p}\right)\right).\quad(4.17)$$

Furthermore, it can now be shown that when $k,n\to\infty$ and $k/n\to 0$ such that $\sqrt{k}E_\infty\left(\log\frac{\hat{q}^+_{k,p}}{Q(1-p)}\right)\to 0$,

$$\sqrt{k}\left(1+\log^2\left(\frac{k+1}{(n+1)p}\right)\right)^{-1/2}\left(\frac{\hat{q}^+_{k,p}}{Q(1-p)}-1\right)\overset{D}{\to}N(0,\gamma^2).\quad(4.18)$$

An asymptotic confidence interval of level $1-\alpha$ for $Q(1-p)$ when $p_n\to 0$ and $np_n\to c>0$ as $n\to\infty$ is given by

$$\left(\frac{\hat{q}^+_{k,p}}{1+\Phi^{-1}(1-\alpha/2)\frac{H_{k,n}}{\sqrt{k}}\sqrt{1+\log^2\left(\frac{k+1}{(n+1)p}\right)}},\right.$$

$$\left.\frac{\hat{q}^+_{k,p}}{1-\Phi^{-1}(1-\alpha/2)\frac{H_{k,n}}{\sqrt{k}}\sqrt{1+\log^2\left(\frac{k+1}{(n+1)p}\right)}}\right).$$

Alternatively, in order to estimate $P(X>x)$ for x large, one can set the Weissman estimator $\hat{q}^+_{k,p}$ equal to x and solve for p:

$$\hat{p}^+_{k,x}=\left(\frac{k+1}{n+1}\right)\left(\frac{x}{X_{n-k,n}}\right)^{-1/H_{k,n}}.$$

This estimator can also be directly understood from the probability point of view. Indeed, using the approximation $\frac{1-F(ty)}{1-F(t)} = P(X > ty)/P(X > t) \sim y^{-1/\gamma}$ as $t \to \infty$, we find $\hat{p}_{k,x}^+$ when replacing ty by x and, when using $X_{n-k,n}$ as a threshold t, estimating $P(X > t)$ by the empirical estimate $\frac{k+1}{n+1}$.

Again, one can prove asymptotic normality in that

$$\sqrt{k}\left(1 + \gamma^{-2}\log^2\left(\frac{x}{X_{n-k,n}}\right)\right)^{1/2}\left(\frac{\hat{p}_{k,x}^+}{P(X>x)} - 1\right) \xrightarrow{\mathcal{D}} N(0,1). \quad (4.19)$$

This leads to an asymptotic confidence interval of level $1 - \alpha$ for $P(X > x)$:

$$\left(\frac{\hat{p}_{k,x}^+}{1 + \Phi^{-1}(1-\alpha/2)\left(k\left(1 + H_{k,n}^{-2}\log^2\left(\frac{x}{X_{n-k,n}}\right)\right)\right)^{-1/2}},\right.$$

$$\left.\frac{\hat{p}_{k,x}^+}{1 - \Phi^{-1}(1-\alpha/2)\left(k\left(1 + H_{k,n}^{-2}\log^2\left(\frac{x}{X_{n-k,n}}\right)\right)\right)^{-1/2}}\right).$$

4.6.2 Second-order refinements

The quantile view

Using the condition (4.4), one can refine $\hat{q}_{k,p}^+$ exploiting the additional information that is then available concerning the slowly varying function ℓ_U. Using again $X_{n-k,n} \stackrel{\mathcal{D}}{=} U(1/U_{k+1,n})$, we find that

$$\frac{Q(1-p)}{X_{n-k,n}} \stackrel{\mathcal{D}}{=} \frac{p^{-\gamma}}{U_{k+1,n}^{-\gamma}} \frac{\ell_U(1/p)}{\ell_U(1/U_{k+1,n})}$$

$$\sim \left(\frac{U_{k+1,n}}{p}\right)^{\gamma}\exp\left(b(1/U_{k+1,n})\frac{1 - \left(\frac{U_{k+1,n}}{p}\right)^{-\beta}}{\beta}\right)$$

$$\sim \left(\frac{k+1}{(n+1)p}\right)^{\gamma}\exp\left(b_{n,k}\frac{1 - \left(\frac{k+1}{(n+1)p}\right)^{-\beta}}{\beta}\right)$$

where in the last step, we replaced $U_{k+1,n}$ by its expected value $\frac{k+1}{n+1}$. Hence, we arrive at the following estimator for extreme quantiles with $k = 3, \ldots, n-1$:

$$\hat{q}_{k,p}^{(1)} = X_{n-k,n}\left(\frac{k+1}{(n+1)p}\right)^{\hat{\gamma}_{ML}^+}\exp\left(\hat{b}_{n,k}\frac{1 - \left(\frac{k+1}{(n+1)p}\right)^{-\hat{\beta}}}{\hat{\beta}}\right) \quad (4.20)$$

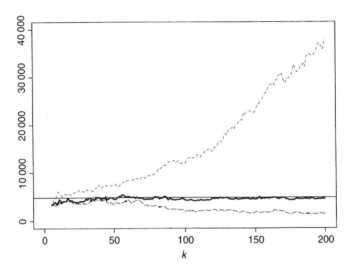

Figure 4.6   Median of $\hat{q}_{k,p}^{(1)}$ (solid line), $\hat{q}_{k,p}^{+}$ (broken-dotted line) and $\hat{q}_{k,p}^{(0)}$ (broken line) with $p = 0.0002$ for 100 simulated samples of size $n = 1000$ from the Burr(1,0.5,2) distribution, $k = 5, \ldots, 200$. The horizontal line indicates the true value of $Q(1 - p)$.

where $\hat{\gamma}_{\mathrm{ML}}^{+}$, $\hat{\beta}$ and $\hat{b}_{k,n}$ denote the maximum likelihood estimators based on (4.8). This estimator was studied in more detail in Matthys and Beirlant (2003). Among others, it was proven that the asymptotic distribution of $\hat{\gamma}_{\mathrm{ML}}^{+}$ and $\hat{q}_{k,p}^{(1)}$ are quite similar. Indeed, compared to (4.18), the asymptotic variance now becomes $\gamma^2 \left(\frac{1+\beta}{\beta} \right)^4$ instead of γ^2 in (4.18). Note that equation (4.20) can also be used to estimate small exceedance probabilities. Indeed, fixing $\hat{q}_{k,p}^{(1)}$ at a high level, (4.20) can be solved numerically for p. The resulting estimator for p will be denoted by $\hat{p}_{k,x}^{(1)}$.

The bias-correcting effect obtained from using $\hat{\gamma}_{\mathrm{ML}}^{+}$ and the factor $\exp\left(\hat{b}_{n,k}[1 - ((k + 1)/(n + 1)p)^{-\hat{\beta}}]/\hat{\beta} \right)$ is illustrated in Figure 4.6 where we show the medians computed over 100 samples of size $n = 1000$ from the Burr(1,0.5,2) distribution and $p = 0.0002$. Next to $\hat{q}_{k,p}^{+}$ and $\hat{q}_{k,p}^{(1)}$, we also show the estimator

$$\hat{q}_{k,p}^{(0)} = X_{n-k,n} \left(\frac{k + 1}{(n + 1)p} \right)^{\hat{\gamma}_{\mathrm{ML}}^{+}}$$

which is in fact $\hat{q}_{k,p}^{+}$ with $H_{k,n}$ replaced by $\hat{\gamma}_{\mathrm{ML}}^{+}$.

The probability view

Following the approach outlined in section 4.5.2 of fitting a perturbed Pareto distribution to the relative excesses Y_j, $j = 1, \ldots, k$, above a threshold $X_{n-k,n}$, results in the following tail estimator

$$\hat{p}_{k,x}^{(2)} = \frac{k+1}{n+1} \bar{G}_{PPD}(x/X_{n-k,n}; \hat{\gamma}_{PPD}^+, \hat{c}_{PPD}^+, \hat{\tau}_{PPD}^+) \tag{4.21}$$

where $\bar{G}_{PPD}$ denotes the survival function of the perturbed Pareto distribution (PPD) (4.14) introduced in section 4.5.2. Fixing $\hat{p}_{k,x}^{(2)}$ at a small value, (4.21) can be solved numerically for x, yielding an extreme quantile estimator. This estimator will be denoted by $\hat{q}_{k,p}^{(2)}$.

An example: the SOA Group Medical Insurance data.

We illustrate the use of the above-introduced estimators for the extreme value index and extreme quantiles with the SOA Group Medical Insurance data. In Figure 4.7(a), we plot $\hat{\gamma}_{ML}^+$ (solid line), $H_{k,n}$ (broken line), $\hat{\gamma}_{Z,k}^+$ (broken-dotted line) and $\hat{\gamma}_{PPD}^+$ (dotted line) for the 1991 claim data against k. This plot indicates a γ estimate around 0.35. Insurance companies typically are interested in an estimate of the claim amount that will be exceeded (on average) only once in, say, 100,000 cases. We illustrate the estimation of extreme quantiles in Figure 4.7(b). In this figure, we plot $\hat{q}_{k,p}^{(1)}$ (solid line), $\hat{q}_{k,p}^+$ (broken line) and $\hat{q}_{k,p}^{(2)}$ (broken-dotted line) for $U(100,000)$ as a function of k.

4.7 Adaptive Selection of the Tail Sample Fraction

We now turn to the estimation of the optimal sample fraction needed to apply a tail index estimator like the Hill estimator. It should be intuitively clear that the estimates of $b_{n,k}$, the parameter that dominates the bias of the Hill estimator as discussed in section 4.4, should be helpful to locate the values of k for which the bias of the Hill estimator is too large, or for which the mean squared error of the estimator is minimal. Several methods have been proposed recently, which we review briefly. See also Hall and Welsh (1985) and Beirlant *et al.* (1996b).

(i) Guillou and Hall (2001) propose to choose $H_{\hat{k},n}$ where $\hat{k}$ is the smallest value of k for which

$$\sqrt{\frac{k}{12}} \frac{|\hat{b}_{LS}^+(-1)|}{H_{k,n}} > c_{crit},$$

where c_{crit} is a critical value such as 1.25 or 1.5.

To understand this standardization, first remark that on the basis of Theorem 4.2, one can show that if $\sqrt{k}b_{n,k} \to c \in \mathbb{R}$, then

$$\sqrt{\frac{k}{12}} \frac{1}{\gamma} \hat{b}_{LS}^+(-1) \overset{D}{\to} N\left(\frac{c\beta\sqrt{3}}{\gamma(2+\beta)(1+\beta)}, 1\right).$$

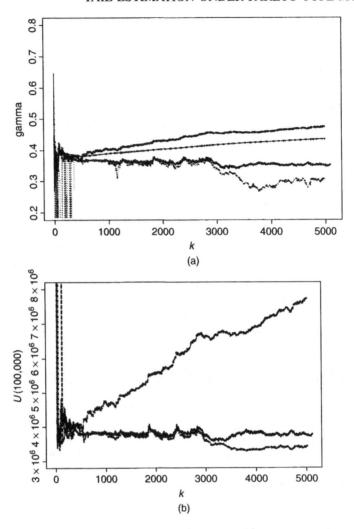

Figure 4.7   SOA Group Medical Insurance data: (a) $\hat{\gamma}^+_{\mathrm{ML}}$ (solid line), $H_{k,n}$ (broken line), $\hat{\gamma}^+_{Z,k}$ (broken-dotted line) and $\hat{\gamma}^+_{PPD}$ (dotted line) as a function of k and (b) $\hat{q}^{(1)}_{k,p}$ (solid line), $\hat{q}^+_{k,p}$ (broken line) and $\hat{q}^{(2)}_{k,p}$ (broken-dotted line) for $U(100,000)$ as a function of k.

So, after appropriate standardization of $\hat{b}^+_{LS}(-1)$, the procedure given in Guillou and Hall (2001) can be considered as an asymptotic test for zero (asymptotic) expectation of $\hat{b}^+_{LS}(-1)$: the bias in the Hill estimator is considered to be too large, and hence the hypothesis of zero bias is rejected, when the asymptotic mean in the limit result appears significantly different from zero.

(ii) An important alternative, popular among statisticians, is to minimize the mean squared error. Then we try to minimize the asymptotic mean squared error of $H_{k,n}$, that is,

$$AMSE(H_{k,n}) = AVar(H_{k,n}) + ABias^2(H_{k,n}) = \frac{\gamma^2}{k} + \left(\frac{b_{n,k}}{1+\beta}\right)^2 \quad (4.22)$$

as derived before. So it appears natural to use the maximum likelihood estimators discussed above and search for the value of $\hat{k}$, which minimizes this estimated mean squared error plot $\{(k, \widehat{AMSE}(H_{k,n})); k = 1, \ldots, n-1\}$. This simple method can of course also be applied to, for instance, the AMSE of Weissman quantile estimators based on the expressions given in (4.16) and (4.17). When estimating $U(100,000)$ in case of the SOA Group Medical Insurance data, we arrive in this way at the value $\hat{k} = 486$ that is to be considered in Figure 4.7(b).

(iii) Let us restrict again to the Hall-class of distributions where the unknown distribution satisfies

$$U(x) = Cx^\gamma \left(1 + Dx^{-\beta}(1 + o(1))\right) \quad (x \to \infty).$$

for some constants $C > 0$, $D \in \mathbb{R}$. Observe that in this case, $b(x) = -\beta D x^\beta$ $(1 + o(1))$ as $x \to \infty$. Then the asymptotic mean squared error of the Hill estimator is minimal for

$$k_{n,opt} \sim (b^2(n))^{-1/(1+2\beta)} \left(\frac{\gamma^2(1+\beta)^2}{2\beta}\right)^{1/(1+2\beta)} \quad (n \to \infty).$$

Here, because of the particular form of b, we obtain

$$k_{n,opt} \sim \left[b^2\left(\frac{n}{k_0}\right)\right]^{-1/(1+2\beta)} k_0^{2\beta/(1+2\beta)} \left(\frac{\gamma^2(1+\beta)^2}{2\beta}\right)^{1/(1+2\beta)} \quad (4.23)$$

for any secondary value $k_0 \in \{1, \ldots, n\}$ with $k_0 = o(n)$. We plug in consistent estimators of b_{n,k_0}, β and γ in this expression as discussed above, all based on the upper k_0 extremes. In this way, we obtain for each value of k_0 an estimator of $k_{n,opt}$.

Then as $k_0, n \to \infty$, $k_0/n \to 0$ and $\frac{\sqrt{k_0} b_{n,k_0}}{\log k_0} \to \infty$, we have that

$$\frac{\hat{k}_{n,k_0}}{k_{n,opt}} \overset{P}{\Rightarrow} 1.$$

Of course, a drawback of this approach is that in practice one needs to identify the k_0-region for which $\sqrt{k_0} b_{n,k_0} \to \infty$ in order to obtain a consistent method. However, graphs of $\log \hat{k}_{n,k_0}$ as a function of k_0 are quite stable,

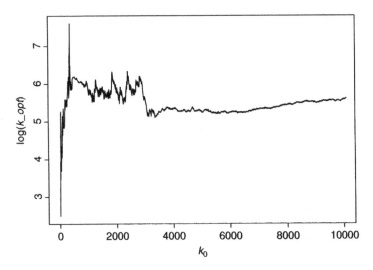

Figure 4.8   SOA Group Medical Insurance data: plot of $\log \hat{k}_{n,k_0}$ versus k_0, $k_0 = 3, \ldots, 10,000$.

except for the k_0-regions corresponding to $\sqrt{k_0} b_{n,k_0} \to 0$. This is illustrated in Figure 4.8 for the SOA Group Medical Insurance data set. The plot of $\log \hat{k}_{n,k_0}$ is stable from $k_0 = 3000$ up to $k_0 = 7000$, indicating a $\log \hat{k}$ value around 5.3. This value corresponds to the endpoint of a stable horizontal area in the Hill plot given in Figure 4.7(b) with height at approximately 0.36.

In order to set up an automatic method, from a practical point of view one can use the median of the first $\lfloor n/2 \rfloor \hat{k}$-values as an overall estimate for $k_{n,opt}$:

$$\hat{k}_{n,med} = \text{median} \left\{ \hat{k}_{n,k_0} : k_0 = 3, \ldots, \left\lfloor \frac{n}{2} \right\rfloor \right\}. \qquad (4.24)$$

(iv) In Hall (1990), a novel resampling technique to estimate the mean squared error of the Hill estimator is proposed. For this purpose, the usual bootstrap does not work properly, especially because it seriously underestimates bias. This problem can be circumvented by taking resamples of smaller size than the original one and linking the bootstrap estimates for the optimal subsample fraction to $k_{n,opt}$ for the full sample. However, in order to establish this link, Hall's method requires that $\beta = 1$, which puts a serious restriction on the tail behaviour of the data. Moreover, an initial estimate is needed to estimate the bias. As pointed out by Gomes and Oliveira (2001), the entire procedure is highly sensitive to the choice of this initial value.

The idea of subsample bootstrapping is taken up in a broader method by Danielsson *et al.* (1997). Instead of bootstrapping the mean squared error of the Hill estimator itself, they use an auxiliary statistic, the mean squared error of which converges at the same rate and which has a known asymptotic mean, independent of the parameters γ and β. Such a statistic is

$$A_{k,n} = H_{k,n}^{(2)} - 2H_{k,n}^2$$

with

$$H_{k,n}^{(2)} = \frac{1}{k} \sum_{j=1}^{k} (\log X_{n-j+1,n} - \log X_{n-k,n})^2.$$

Since both $H_{k,n}^{(2)}/(2H_{k,n})$ and $H_{k,n}$ are consistent estimators for γ, $A_{k,n}$ will converge to 0 for intermediate sequences of k-values as $n \to \infty$. Thus $AMSE(A_{k,n}) = E_\infty(A_{k,n}^2)$, and no initial parameter estimate is needed to calculate the bootstrap counterpart.

Moreover, the k-value that minimizes $AMSE(A_{k,n})$, denoted by $\bar{k}_{n,opt}$, is of the same order in n as $k_{n,opt}$:

$$\frac{\bar{k}_{n,opt}}{k_{n,opt}} \to \left(1 + \frac{1}{\beta}\right)^{\frac{2}{1+2\beta}}, \quad n \to \infty.$$

Unfortunately, the usual bootstrap estimate for $\bar{k}_{n,opt}$ does not converge in probability to the true value; it merely converges in distribution to a random sequence owing to the characteristic balance between variance and squared bias at the optimal threshold. A subsample bootstrap remedies this problem. Taking subsamples of size $n_1 = O(n^{1-\varepsilon})$ for some $0 < \varepsilon < 1$ provides a consistent bootstrap estimate $\hat{\bar{k}}_{n_1,opt}$ for $\bar{k}_{n_1,opt}$.

Further, the ratio of optimal sample and subsample fractions for $A_{k,n}$ is of the order

$$\frac{\bar{k}_{n,opt}}{\bar{k}_{n_1,opt}} \sim \left(\frac{n}{n_1}\right)^{\frac{2\beta}{2\beta+1}}.$$

For $n_1 = O(n^{1-\varepsilon})$, $0 < \varepsilon < 0.5$, this ratio can be estimated through a second subsample bootstrap, now with subsamples of size $n_2 = n_1^2/n$, such that

$$\frac{\bar{k}_{n,opt}}{\bar{k}_{n_1,opt}} \sim \frac{\bar{k}_{n_1,opt}}{\bar{k}_{n_2,opt}}.$$

Combining these results gives

$$k_{n,opt} \sim \frac{(\bar{k}_{n_1,opt})^2}{\bar{k}_{n_2,opt}} \left(1 + \frac{1}{\beta}\right)^{-\frac{2}{2\beta+1}}$$

which leads to the estimator

$$\hat{k}_{n,opt} \sim \frac{(\hat{\bar{k}}_{n_1,opt})^2}{\hat{\bar{k}}_{n_2,opt}} \left(1 + \frac{1}{\hat{\beta}_1}\right)^{-\frac{2}{2\hat{\beta}+1}} \tag{4.25}$$

for $k_{n,opt}$, where

$$\hat{\beta}_1 := \frac{\log \hat{\hat{k}}_{n_1,opt}}{2 \log(\hat{\hat{k}}_{n_1,opt}/n_1)}$$

is a consistent estimator for β.

Under condition $(\mathcal{C}_{-\beta}(b))$ for $\log \ell_U$, it can be shown that the resulting Hill estimator $H_{\hat{k}_{n,opt},n}$ has the same asymptotic efficiency as $H_{k_{n,opt},n}$.

The algorithm for this bootstrap procedure is summarized as follows

(a) Draw B bootstrap subsamples of size $n_1 \in (\sqrt{n}, n)$ from the original sample and determine the value $\hat{\hat{k}}_{n_1,opt}$ that minimizes the bootstrap mean squared error of A_{k,n_1}.

(b) Repeat this for B bootstrap subsamples of size $n_2 = n_1^2/n$ and determine $\hat{\hat{k}}_{n_2,opt}$ where the bootstrap mean squared error of A_{k,n_2} is minimal.

(c) Calculate $\hat{k}_{n,opt}$ from (4.25) and estimate γ by $H_{\hat{k}_{n,opt},n}$.

This procedure considerably extends and improves Hall's original bootstrap method, especially because no preliminary parameter estimate is needed. Only the subsample size n_1 and the number of bootstrap resamples B have to be chosen. In fact, the latter is determined mainly by the available computing time. In simulation studies reported in the literature, the number of resamples ranges from 250 to 5000. As for the subsample size, Danielsson and de Vries (1997) suggest varying n_1 over a grid of values and using a bootstrap diagnostic to select its optimal value adaptively. Gomes and Oliveira (2001), however, found that the method is very robust with respect to the choice of n_1. We also refer to Gomes and Oliveira (2001) for more variations and simulation results on the above bootstrap to choose the optimal sample fraction and for a refined version of Hall's method.

(v) Drees and Kaufmann (1998) present a sequential procedure to select the optimal sample fraction $k_{n,opt}$. From a law of the iterated logarithm, they construct 'stopping times' for the sequence $H_{k,n}$ of Hill estimators that are asymptotically equivalent to a deterministic sequence. An ingenious combination of two such stopping times then attains the same rate of convergence as $k_{n,opt}$. However, the conversion factor to pass from this combination of stopping times to $k_{n,opt}$ involves the unknown parameters γ (which requires an initial estimate $\hat{\gamma}_0$) and β.

We refer to the original paper by Drees and Kaufmann (1998) for the theoretical principles behind this procedure and immediately describe the algorithm with the choices of nuisance parameters as proposed by these authors.

(a) Obtain an initial estimate $\hat{\gamma}_0 := H_{2\sqrt{n},n}$ for γ.

(b) For $r_n = 2.5\hat{\gamma}_0 n^{0.25}$ compute the 'stopping time'

$$\hat{k}_n(r_n) := \min\left\{k \in \{1, \ldots, n-1\} \Big| \max_{1 \le i \le k} \sqrt{i}(H_{i,n} - H_{k,n}) > r_n\right\}.$$

(c) Similarly, compute $\hat{k}_n(r_n^\varepsilon)$ for $\varepsilon = 0.7$.

(d) With a consistent estimator $\hat{\beta}$ for β, calculate

$$\hat{k}_{n,opt} = \left(\frac{\hat{k}_n(r_n^\varepsilon)}{[\hat{k}_n(r_n)]^\varepsilon}\right)^{\frac{1}{1-\varepsilon}} (1 + 2\hat{\beta})^{-\frac{1}{\beta}} (2\hat{\beta}\hat{\gamma}_0)^{\frac{1}{1+2\hat{\beta}}} \qquad (4.26)$$

and estimate γ by $H_{\hat{k}_{n,opt},n}$.

In simulations, it was found that the method mostly performs better if a fixed value β_0 is used for β in (4.26), in particular, for $\hat{\beta} \equiv \beta_0 = 1$.

In Matthys and Beirlant (2000), Beirlant *et al.* (2002c) and Gomes and Oliveira (2001), these adaptive procedures have been compared on the basis of extensive small sample simulations. While both the bootstrap method and the plug-in method tend to give rather variable values for the optimal sample fraction, the results for all four adaptive Hill estimators are well in line. The sequential procedure and the method based on $\hat{k}_{n,med}$ give the best results even when setting $\beta = 1$. The influence of a wrong specification of the parameter β in these methods does not seem to be a major problem. In comparison with other procedures, method (iii) performs best in case of small values of β and even for distributions outside the range of distributions considered by Hall and Welsh (1984), such as the log-gamma distribution. The methods based on the regression models above such as (ii) and (iii) ask most computing effort. The sequential method appears to be the fastest overall.

5

TAIL ESTIMATION FOR ALL DOMAINS OF ATTRACTION

In Chapter 2, we derived the general conditions (C_γ) and (C_γ^*) for a non-degenerate limit distribution of the normalized maximum of a sample of independent and identically distributed random variables to exist:

$$\frac{U(xu) - U(x)}{a(x)} \to \frac{u^\gamma - 1}{\gamma} \text{ for any } u > 0 \text{ as } x \to \infty, \qquad (C_\gamma)$$

for some regularly varying function a with index γ, where $U(x) = Q\left(1 - \frac{1}{x}\right)$, respectively

$$\frac{\bar{F}(t + yb(t))}{\bar{F}(t)} \to (1 + \gamma y)^{-1/\gamma} \text{ for any } y > 0 \text{ as } t \uparrow x_+, \qquad (C_\gamma^*)$$

for some auxiliary function b.

In the preceding chapter, we outlined the extreme value approach for tail estimation in case $\gamma > 0$, that is, when $\bar{F}$ is of Pareto-type. Now, in the present chapter, we discuss statistical tail estimation methods that can serve in all cases, whether the extreme value index (EVI) is positive, negative, or zero. The available methods can be grouped in three sets:

- *the method of block maxima*, inspired by the limit behaviour of the normalized maximum of a random sample,

- *the quantile view* with methods based on (versions of) (C_γ), continuing the line of approach started with Hill's estimator,

---

Statistics of Extremes: Theory and Applications   J. Beirlant, Y. Goegebeur, J. Segers, and J. Teugels
© 2004 John Wiley & Sons, Ltd   ISBN: 0-471-97647-4

- *the probability view, or the peaks over threshold approach (POT)* with methods based on (C^*_γ). Here, one considers the conditional distribution of the excesses over relatively high thresholds t, interpreting $\frac{\bar{F}(t+yb(t))}{\bar{F}(t)}$ as $P\left(\frac{X-t}{b(t)} > y | X > t\right)$.

Next to these approaches, we also briefly mention the possibilities of exponential regression models, generalizing the exponential representations of spacings as considered in section 4.4.

5.1   The Method of Block Maxima

5.1.1   The basic model

In Chapter 2, it was proven that the extreme value distributions are the only possible limiting forms for a normalized maximum of a random sample, at least when a non-degenerate limit exists. On the basis of this result, the EVI can be estimated by fitting the generalized extreme value distribution (GEV)

$$G(x; \sigma, \gamma, \mu) = \begin{cases} \exp\left(-\left(1 + \gamma\frac{x-\mu}{\sigma}\right)^{-\frac{1}{\gamma}}\right), & 1 + \gamma\frac{x-\mu}{\sigma} > 0, \gamma \neq 0, \\ \exp\left(-\exp\left(-\frac{x-\mu}{\sigma}\right)\right), & x \in \mathbb{R}, \gamma = 0, \end{cases} \tag{5.1}$$

with $\sigma > 0$ and $\mu \in \mathbb{R}$ to maxima of subsamples (Gumbel (1958)). This approach is popular in the environmental sciences where the GEV is fitted to, for example, yearly maximal temperatures or yearly maximal river discharges.

5.1.2   Parameter estimation

For notational convenience, we denote the maximum of a sample $X_1, \ldots, X_n$ by Y. Then when a sample $Y_1, \ldots, Y_m$ of independent sample maxima is available, the parameters σ, γ and μ can be estimated in a variety of ways. In Chapter 2, we already discussed the data-analytic method of selecting the γ value that maximizes the correlation coefficient on the GEV quantile plot followed by a least-squares fit to obtain estimates for μ and σ. In this section, we will focus on the maximum likelihood (ML) method and the method of (probability weighted) moments.

The ML method

In case $\gamma \neq 0$, the log-likelihood function for a sample $Y_1, \ldots, Y_m$ of i.i.d. GEV random variables is given by

$$\log L(\sigma, \gamma, \mu) = -m\log\sigma - \left(\frac{1}{\gamma} + 1\right)\sum_{i=1}^{m}\log\left(1 + \gamma\frac{Y_i - \mu}{\sigma}\right)$$

$$-\sum_{i=1}^{m}\left(1 + \gamma\frac{Y_i - \mu}{\sigma}\right)^{-\frac{1}{\gamma}} \tag{5.2}$$

provided $1 + \gamma \frac{Y_i - \mu}{\sigma} > 0$, $i = 1, \ldots, m$. When $\gamma = 0$, the log-likelihood function reduces to

$$\log L(\sigma, 0, \mu) = -m \log \sigma - \sum_{i=1}^{m} \exp\left(-\frac{Y_i - \mu}{\sigma}\right) - \sum_{i=1}^{m} \frac{Y_i - \mu}{\sigma}. \tag{5.3}$$

The ML estimator $(\hat{\sigma}, \hat{\gamma}, \hat{\mu})$ for (σ, γ, μ) is obtained by maximizing (5.2)-(5.3). For computational details, we refer to Prescott and Walden (1980), Prescott and Walden (1983), Hosking (1985) and Macleod (1989). Since the support of G depends on the unknown parameter values, the usual regularity conditions underlying the asymptotic properties of maximum likelihood estimators are not satisfied. This problem is studied in depth in Smith (1985). In case $\gamma > -0.5$, the usual properties of consistency, asymptotic efficiency and asymptotic normality hold. In fact, for $m \to \infty$

$$\sqrt{m}\left((\hat{\sigma}, \hat{\gamma}, \hat{\mu}) - (\sigma, \gamma, \mu)\right) \overset{D}{\to} N(0, V_1) \qquad \gamma > -0.5$$

where V_1 is the inverse of the Fisher information matrix. For more details about the Fisher information matrix, we refer to the Appendix at the end of this chapter. This limit result in principle is valid under the assumption that Y is distributed as a GEV. Remark, however, that the results of Chapter 2 only guarantee that Y is *approximately* GEV.

The method of probability-weighted moments

In general, the probability-weighted moments (PWM) of a random variable Y with distribution function F, introduced by Greenwood et al. (1979), are the quantities

$$M_{p,r,s} = E\{Y^p[F(Y)]^r[1 - F(Y)]^s\} \tag{5.4}$$

for real p, r and s. The specific case of PWM parameter estimation for the GEV is studied extensively in Hosking et al. (1985). In case $\gamma \neq 0$, setting $p = 1, r = 0, 1, 2, \ldots$ and $s = 0$ yields for the GEV

$$M_{1,r,0} = \frac{1}{r+1}\left\{\mu - \frac{\sigma}{\gamma}[1 - (r+1)^\gamma \Gamma(1 - \gamma)]\right\} \qquad \gamma < 1. \tag{5.5}$$

Assume a sample $Y_1, \ldots, Y_m$ of i.i.d. GEV random variables is available. The PWM estimator $(\hat{\sigma}, \hat{\gamma}, \hat{\mu})$ for (σ, γ, μ) is the solution to the following system of equations, obtained from (5.5) with $r = 0, 1, 2$,

$$M_{1,0,0} = \mu - \frac{\sigma}{\gamma}(1 - \Gamma(1 - \gamma)) \tag{5.6}$$

$$2M_{1,1,0} - M_{1,0,0} = \frac{\sigma}{\gamma}\Gamma(1 - \gamma)(2^\gamma - 1) \tag{5.7}$$

$$\frac{3M_{1,2,0} - M_{1,0,0}}{2M_{1,1,0} - M_{1,0,0}} = \frac{3^\gamma - 1}{2^\gamma - 1} \tag{5.8}$$

after replacing $M_{1,r,0}$ by the unbiased estimator (see Landwehr *et al.* (1979))

$$\hat{M}_{1,r,0} = \frac{1}{m} \sum_{j=1}^{m} \left(\prod_{\ell=1}^{r} \frac{(j-\ell)}{(m-\ell)} \right) Y_{j,m}$$

or by the asymptotically equivalent consistent estimator

$$\tilde{M}_{1,r,0} = \frac{1}{m} \sum_{j=1}^{m} \left(\frac{j}{m+1} \right)^{r} Y_{j,m}.$$

Note that to obtain $\hat{\gamma}$, (5.8) has to be solved numerically. Next, (5.7) can be solved for σ, yielding

$$\hat{\sigma} = \frac{\hat{\gamma}(2\hat{M}_{1,1,0} - \hat{M}_{1,0,0})}{\Gamma(1 - \hat{\gamma})(2^{\hat{\gamma}} - 1)}.$$

Finally, given $\hat{\gamma}$ and $\hat{\sigma}$, $\hat{\mu}$ can be obtained from (5.6):

$$\hat{\mu} = \hat{M}_{1,0,0} + \frac{\hat{\sigma}}{\hat{\gamma}} \left(1 - \Gamma(1 - \hat{\gamma}) \right).$$

To derive the limiting distribution of $(\hat{\sigma}, \hat{\gamma}, \hat{\mu})$, we need the limiting behaviour of $(\hat{M}_{1,0,0}, \hat{M}_{1,1,0}, \hat{M}_{1,2,0})$. Define $M = (M_{1,0,0}, M_{1,1,0}, M_{1,2,0})'$ and $\hat{M} = (\hat{M}_{1,0,0}, \hat{M}_{1,1,0}, \hat{M}_{1,2,0})'$. Provided $\gamma < 0.5$, it can be shown that for $m \to \infty$

$$\sqrt{m}(\hat{M} - M) \overset{\mathcal{D}}{\to} N(0, V)$$

where the elements of V are given by

$$v_{r,r} = \left[\frac{\sigma}{\gamma}(r+1)^{\gamma} \right]^{2} \left(\Gamma(1 - 2\gamma)K(r/(r+1)) - \Gamma^{2}(1 - \gamma) \right),$$

$$v_{r,r+1} = \frac{1}{2} \left(\frac{\sigma}{\gamma} \right)^{2} \{ (r+2)^{2\gamma}\Gamma(1 - 2\gamma)K(r/(r+2))$$
$$+ (r+1)^{\gamma} \left[(r+1)^{\gamma} - 2(r+2)^{\gamma} \right] \Gamma^{2}(1 - \gamma) \},$$

$$v_{r,r+s} = \frac{1}{2} \left(\frac{\sigma}{\gamma} \right)^{2} \{ (r+s+1)^{2\gamma}\Gamma(1 - 2\gamma)K(r/(r+s+1))$$
$$- (r+s)^{\gamma}\Gamma(1 - 2\gamma)K((r+1)/(r+s))$$
$$+ 2(r+1)^{\gamma} \left[(r+s)^{\gamma} - (r+s+1)^{\gamma} \right] \Gamma^{2}(1 - \gamma) \} \qquad s \geq 2,$$

and $K(x) = {}_{2}F_{1}(-\gamma, -2\gamma; 1 - \gamma; -x)$, with ${}_{2}F_{1}$ denoting the hypergeometric function.

Table 5.1 ML and PWM estimates for
the Meuse data.

Method	σ	γ	μ
ML	466.468	−0.092	1266.896
PWM	468.358	−0.099	1267.688

Now define $\theta = (\sigma, \gamma, \mu)'$, $\hat{\theta} = (\hat{\sigma}, \hat{\gamma}, \hat{\mu})'$ and write the solution to (5.6), (5.7) and (5.8) as the vector equation $\theta = f(M)$. Further, let G denote the 3×3 matrix with generic elements $g_{i,j} = \partial f_i / \partial M_{1,j,0}$, $i, j = 1, 2, 3$. Application of the delta method yields the limiting distribution of $\hat{\theta}$:

$$\sqrt{m}(\hat{\theta} - \theta) \overset{\mathcal{D}}{\to} N(0, V_2)$$

where $V_2 = GVG'$, as $m \to \infty$ provided $\gamma < 0.5$.

Example 5.1 In Table 5.1, we show the ML and PWM estimates for the parameters (σ, γ, μ) obtained from fitting the GEV to the annual maximum discharges of the Meuse river. Note that the estimates obtained under the two estimation methods agree quite good. The fit of the GEV to these data can be visually assessed by inspecting the GEV quantile plot, introduced in Chapter 2. Figure 5.1 shows the GEV quantile plot obtained with (a) ML and (b) PWM.

We still refer to some other estimation methods for the GEV that have been discussed in literature: best linear unbiased estimation (Balakrishnan and Chan (1992)), Bayes estimation (Lye *et al.* (1993)), method of moments (Christopeit (1994)), and minimum distance estimation (Dietrich and Hüsler (1996)). In Coles and Dixon (1999), it is shown that maximum penalized likelihood estimation improves the small sample properties of a likelihood-based analysis.

5.1.3 Estimation of extreme quantiles

Estimates of extreme quantiles of the GEV can be obtained by inverting the GEV distribution function given by (5.1), yielding

$$q_{Y,p} = \begin{cases} \mu + \frac{\sigma}{\gamma}[(-\log(1-p))^{-\gamma} - 1], & \gamma \neq 0, \\ \mu - \sigma \log(-\log(1-p)), & \gamma = 0, \end{cases} \tag{5.9}$$

and replacing (σ, γ, μ) by either the ML or probability-weighted moments estimates. In case $\gamma < 0$, the right endpoint of the GEV is finite and given by

$$q_{Y,0} = \mu - \frac{\sigma}{\gamma}.$$

The ML estimate of $q_{Y,p}$ can also be obtained directly by a reparametrization such that $q_{Y,p}$ is one of the model parameters, for instance, substituting $q_{Y,p}$ −

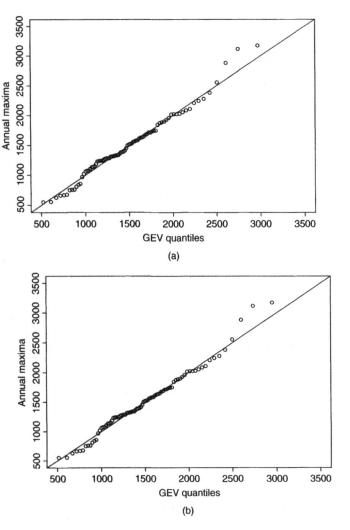

Figure 5.1   GEV QQ-plot of the annual maximum discharges of the Meuse river using (a) ML and (b) PWM estimates.

$\frac{\sigma}{\gamma}[(-\log(1-p))^{-\gamma} - 1]$ for μ. Note that, in case the GEV is used as an approximation to the distribution of the largest observation in a sample, (5.9) yields the quantiles of the maximum distribution. Since $F_{X_{n,n}} = F^n \approx H$, one easily obtains the quantiles of the original X data as

$$q^*_{X,p} = \begin{cases} \mu + \frac{\sigma}{\gamma}[(-\log(1-p)^n)^{-\gamma} - 1], & \gamma \neq 0, \\ \mu - \sigma \log(-\log(1-p)^n), & \gamma = 0, \end{cases}$$

where n is the block length.

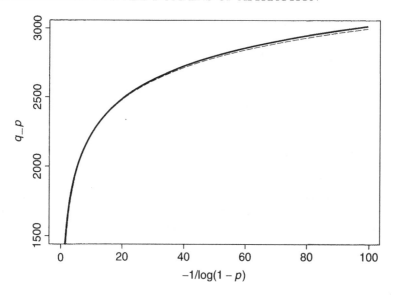

Figure 5.2   GEV-based quantile estimates for the annual maximum discharges of the Meuse river; solid line: based on ML, broken line: based on PWM.

Example 5.1 (continued) In Figure 5.2, we illustrate the estimation of quantiles of the annual maximum discharges of the Meuse river. The solid line (broken line) represents quantile estimates based on the ML (PWM) estimates of the GEV parameters.

5.1.4   Inference: confidence intervals

Confidence intervals and other forms of inference concerning the GEV parameters (σ, γ, μ) follow immediately from the approximate normality of the ML and PWM estimators. For instance, a $100(1 - \alpha)\%$ confidence interval for the tail index γ is given by

$$\hat{\gamma} \pm \Phi^{-1}(1 - \alpha/2)\sqrt{\frac{\hat{v}_{2,2}}{m}}$$

where $\hat{\gamma}$ is the ML or PWM estimate of γ and $\hat{v}_{2,2}$ denotes the second diagonal element of V_1, or V_2, after replacing the unknown parameters by their estimates. Similarly, inference concerning the GEV quantiles can be based on the normal limiting behaviour. Straightforward application of the delta method yields

$$\sqrt{m}(\hat{q}_{Y,p} - q_{Y,p}) \xrightarrow{\mathcal{D}} N(0, \kappa' \tilde{V} \kappa) \text{ as } m \to \infty$$

where $\hat{q}_{Y,p}$ denotes the estimator for $q_{Y,p}$ obtained by plugging the ML or PWM estimators into (5.9), and where $\tilde{V}$ is V_1, or V_2, and

$$
\begin{aligned}
\kappa' &= \left[\frac{\partial q_{Y,p}}{\partial \sigma}, \frac{\partial q_{Y,p}}{\partial \gamma}, \frac{\partial q_{Y,p}}{\partial \mu} \right] \\
&= \left[\frac{1}{\gamma} [(-\log(1-p))^{-\gamma} - 1], \right. \\
&\quad \left. -\frac{\sigma}{\gamma^2} [(-\log(1-p))^{-\gamma} - 1] - \frac{\sigma}{\gamma}(-\log(1-p))^{-\gamma} \log(-\log(1-p)), 1 \right].
\end{aligned}
$$

Inference based on these normal limit results may be misleading as the normal approximation to the true sampling distribution of the respective estimator may be rather poor. In general, better approximations can be obtained by the profile likelihood function. The profile likelihood function (Barndorff-Nielsen and Cox (1994)) of γ is given by

$$
L_p(\gamma) = \max_{\sigma,\mu|\gamma} L(\sigma, \gamma, \mu).
$$

Therefore, the profile likelihood ratio statistic

$$
\Lambda = \frac{L_p(\gamma_0)}{L_p(\hat{\gamma})}
$$

equals the classical likelihood ratio statistic for testing the hypothesis $H_0 : \gamma = \gamma_0$ versus $H_1 : \gamma \neq \gamma_0$, and hence, under H_0, for $m \to \infty$,

$$
-2 \log \Lambda \xrightarrow{\mathcal{D}} \chi_1^2.
$$

The special case of testing $H_0 : \gamma = 0$ (the so-called Gumbel hypothesis) is described in Hosking (1984). Since H_0 will be rejected at significance level α if $-2 \log \Lambda > \chi_1^2(1 - \alpha)$, the profile likelihood–based $100(1 - \alpha)\%$ confidence interval for γ is given by

$$
CI_\gamma = \left\{ \gamma : -2 \log \frac{L_p(\gamma)}{L_p(\hat{\gamma})} \leq \chi_1^2(1 - \alpha) \right\}
$$

or equivalently

$$
CI_\gamma = \left\{ \gamma : \log L_p(\gamma) \geq \log L_p(\hat{\gamma}) - \frac{\chi_1^2(1 - \alpha)}{2} \right\}.
$$

Profile likelihood–based confidence intervals for the other GEV parameters can be constructed in a similar way.

Example 5.1 (continued) The profile likelihood–based 95% confidence intervals for the EVI and the 0.99 quantile of the annual maximum discharges of the Meuse river are given in Figure 5.3(a) and (b) respectively. Note that the 95% confidence interval for γ contains the value 0, so at a significance level of 5%, the hypothesis $H_0 : \gamma = 0$ cannot be rejected. Hence, for practical purposes, the annual maximum discharges can be adequately modelled by the Gumbel distribution.

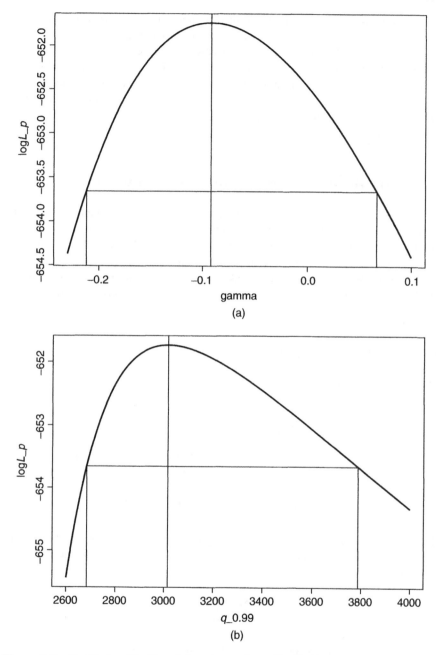

Figure 5.3 Profile log-likelihood function and profile likelihood–based 95% confidence intervals for (a) γ and (b) $q_{0.99}$.

A major weakness with the GEV distribution is that it utilizes only the maximum and thus many data are wasted. Another problem is the determination of an appropriate block size n, especially in case of time series data where the time dependence is to be thinned out by using appropriate independent blocks from which one extracts one maximum; this will be a topic of interest in Chapter 10 on extreme value methods in time series analysis. To lift up the first problem threshold methods and methods based on the k, largest order statistics have been developed. Those are reviewed now.

5.2   Quantile View—Methods Based on $(\mathcal{C}_\gamma)$

Several estimators based on extreme order statistics are available in order to estimate a real-valued EVI, and correspondingly large quantiles and small tail probabilities. These methods rely mainly on the conditions $(\mathcal{C}_\gamma)$ and $(\tilde{\mathcal{C}}_\gamma)$. We discuss here three methods: the estimator proposed by Pickands (1975) and its generalizations, the moment estimator from Dekkers *et al.* (1989) and the estimators based on the generalized quantile plot proposed in Beirlant *et al.* (1996c) and Beirlant *et al.* (2002b).

5.2.1   Pickands estimator

From $(\mathcal{C}_\gamma)$, we obtain

$$\frac{1}{\log 2} \log \left\{ \frac{U(4y) - U(2y)}{U(2y) - U(y)} \right\}$$

$$= \frac{1}{\log 2} \log \left\{ \frac{U(4y) - U(2y)}{a(2y)} \frac{a(2y)}{a(y)} \frac{a(y)}{U(2y) - U(y)} \right\}$$

$$\rightarrow \frac{1}{\log 2} \log\{h_\gamma(2)2^\gamma / h_\gamma(2)\} = \gamma, \qquad y \rightarrow \infty.$$

Treating the limit as an approximate equality for large $y = (n+1)/k$ and replacing $U(x)$ by its empirical version $\hat{U}_n(x) = X_{n-\lceil n/x \rceil+1,n}$ leads to the Pickands (1975) estimator for the EVI

$$\hat{\gamma}_{P,k} = \frac{1}{\log 2} \log \left(\frac{X_{n-\lceil k/4 \rceil+1,n} - X_{n-\lceil k/2 \rceil+1,n}}{X_{n-\lceil k/2 \rceil+1,n} - X_{n-k+1,n}} \right).$$

for $k = 1, \ldots, n$. Pickands original definition uses $4k$ rather k.

The great simplicity of the Pickands estimator $\hat{\gamma}_{P,k}$ is quite appealing but unfortunately offset by its rather large asymptotic variance, equal to $\gamma^2(2^{2\gamma+1} + 1)\{(2^\gamma - 1) \log(2)\}^{-2}$ (Dekkers and de Haan 1989), and its large volatility as a function of k. This motivated the quest for more efficient variants.

A recent proposal by Segers (2004) is the estimator

$$\hat{\gamma}_k(c, \lambda) = \int_0^1 \log(X_{n-\lfloor c\lceil tk\rceil\rfloor,n} - X_{n-\lceil tk\rceil,n}) d\lambda(t)$$

$$= \sum_{j=1}^k \left\{ \lambda\left(\frac{j}{k}\right) - \lambda\left(\frac{j-1}{k}\right) \right\} \log(X_{n-\lfloor cj\rfloor,n} - X_{n-j,n}). \quad (5.10)$$

Here, $0 < c < 1$ while λ is a right-continuous function on $[0, 1]$ such that $\lambda(0) = \lambda(1) = 0$ and $\int_0^1 \lambda(t)t^{-1}dt = 1$. The simplest example is

$$\lambda_v(t) = \begin{cases} 0 & \text{if } 0 \le t < v, \\ 1/\log(1/v) & \text{if } v \le t < 1, \\ 0 & \text{if } t = 1, \end{cases}$$

for some $0 < v < 1$. The estimator $\hat{\gamma}_k(c, \lambda_v)$ is in fact the one proposed by Yun (2002), including as special cases the ones by Pickands (1975) [$c = v = 1/2$], Pereira (1994) and Fraga Alves (1995) [$c = v$], and Yun (2000b) [$1/4 < c < 1$ and $v = (4c)^{-1}$]. A more general example is

$$\lambda_{v,\mu}(t) = \{\mu(t/v) - \mu(t)\}/\log(1/v), \qquad 0 \le t \le 1,$$

where again $0 < v < 1$ and where μ is the distribution function of a probability measure concentrated on $(0, 1]$. The estimator $\hat{\gamma}_k(c, \lambda_{v,\mu})$ can be regarded as a mixture of the estimator $\hat{\gamma}_k(c, \lambda_v)$ over different values of k, encompassing thereby the estimators of Drees (1995) and Falk (1994).

Segers (2004) establishes asymptotic normality of $\hat{\gamma}_k(c, \lambda)$ under the conditions of Theorem 3.1. For fixed $0 < c < 1$ and $\gamma \ne -1/2$, the limiting asymptotic variance, $\sigma^2(\gamma, c, \lambda)$ is minimal for λ equal to $\lambda_{\delta,c}$, where $\delta = |\gamma + 1/2| - 1/2$ and, for $t \in [c^j, c^{j-1})$ (positive integer j),

$$\lambda_{\delta,c}(t) = \begin{cases} (1 - c^{1+\delta})\dfrac{1 - c^{\delta j}}{1 - c^\delta}t, & \text{if } \delta \ne 0, \\ (1 - c^{1+\delta})jt, & \text{if } \delta = 0. \end{cases}$$

In this case,

$$\sigma^2(c, \gamma) = \sigma^2(c, \gamma, \lambda_{c,\gamma}) = \begin{cases} \dfrac{\gamma^2(1 - c^{1+\gamma})^2}{c(1 - c^\gamma)^2}, & \text{for } \gamma > -1/2 \text{ and } \gamma \ne 0, \\ \dfrac{(1 - c)^2}{c(\log c)^2}, & \text{for } \gamma = 0, \\ \gamma^2, & \text{for } \gamma < -1/2. \end{cases}$$

The optimal choice of c is $c \to 1$, in which case

$$\sigma^2(\gamma) = \lim_{c \uparrow 1} \sigma^2(\gamma) = \begin{cases} (1 + \gamma)^2, & \text{for } \gamma > -1/2, \\ \gamma^2, & \text{for } \gamma < -1/2. \end{cases}$$

Clearly, choosing $c = 1$ in (5.10) does not lead to an admissible estimator. This poses no problem in practice, however, as the choice $c = 0.75$ already leads to a relative efficiency of 96%. Observe that for $\gamma > -1/2$, the limiting variance $\sigma^2(\gamma) = (1 + \gamma)^2$ is that of the ML estimator for γ in the GP model (Smith 1987).

The optimal choice for λ depends on γ, which is unknown. The solution is to define $\hat{\gamma} = \hat{\gamma}_k(c, \lambda_{\tilde{\delta},c})$, where $\tilde{\delta} = |\tilde{\gamma} + 1/2| - 1/2$ and $\tilde{\gamma}$ is an arbitrary consistent estimator of γ based on the X_{n-k+i}, $i = 1, \ldots, k$, for instance, $\tilde{\gamma} = \hat{\gamma}_k(c, \lambda_{0,c})$. The asymptotic variance of this two-stage procedure is the same as when we would use γ rather than $\tilde{\gamma}$ (Segers 2004). The estimator is illustrated for the SOA data in Figure 5.5.

5.2.2 The moment estimator

The *moment estimator* has been introduced by Dekkers *et al.* (1989) as a direct generalization of the Hill estimator:

$$M_{k,n} = H_{k,n} + 1 - \frac{1}{2}\left(1 - \frac{H_{k,n}^2}{H_{k,n}^{(2)}}\right)^{-1},$$

where

$$H_{k,n}^{(2)} = \frac{1}{k}\sum_{j=1}^{k}\left(\log X_{n-j+1,n} - \log X_{n-k,n}\right)^2.$$

To understand this estimator, we can proceed as follows: for any $j \in \{1, \ldots, k\}$, we have that

$$\log X_{n-j+1,n} - \log X_{n-k,n} = \log \hat{U}_n\left(\frac{n+1}{j}\right) - \log \hat{U}_n\left(\frac{n+1}{k+1}\right),$$

and hence $\log X_{n-j+1,n} - \log X_{n-k,n}$ can be seen as an estimate of

$$\log U\left(\frac{n+1}{j}\right) - \log U\left(\frac{n+1}{k+1}\right)$$

$$= \log U\left(\left(\frac{n+1}{k+1}\right)\left(\frac{k+1}{j}\right)\right) - \log U\left(\frac{n+1}{k+1}\right).$$

Now, choosing $x = \frac{n+1}{k+1}$ and $u = \frac{k+1}{j}$ in $(\tilde{C}_\gamma)$, then for any $j \in \{1, \ldots, k\}$ as $n/k \to \infty$

$$\log X_{n-j+1,n} - \log X_{n-k,n} \sim \begin{cases} \dfrac{a\left(\frac{n+1}{k+1}\right)}{U\left(\frac{n+1}{k+1}\right)}\log\frac{k+1}{j}, & \text{if } \gamma \geq 0, \\[2em] \dfrac{a\left(\frac{n+1}{k+1}\right)}{U\left(\frac{n+1}{k+1}\right)}\dfrac{\left(\frac{j}{k+1}\right)^{-\gamma}-1}{\gamma}, & \text{if } \gamma < 0. \end{cases}$$

For $k \to \infty$, we have the following limiting results

$$\frac{1}{k} \sum_{j=1}^{k} \log \frac{k+1}{j} \to - \int_0^1 \log u \, du = 1,$$

$$\frac{1}{k} \sum_{j=1}^{k} \left(\log \frac{k+1}{j} \right)^2 \to \int_0^1 (\log u)^2 du = 2,$$

$$\frac{1}{k} \sum_{j=1}^{k} \left\{ \left(\frac{j}{k+1} \right)^{-\gamma} - 1 \right\} \to \int_0^1 (u^{-\gamma} - 1) du = \frac{\gamma}{1-\gamma} \quad (\gamma < 0),$$

$$\frac{1}{k} \sum_{j=1}^{k} \left(\left(\frac{j}{k+1} \right)^{-\gamma} - 1 \right)^2 \to \int_0^1 (u^{-\gamma} - 1)^2 du = \frac{2\gamma^2}{(1-\gamma)(1-2\gamma)} \quad (\gamma < 0).$$

We see therefore that as $k, n \to \infty$ and $k/n \to 0$,

$$\frac{H_{k,n}^2}{H_{k,n}^{(2)}} \overset{P}{\Rightarrow} \begin{cases} \frac{1}{2}, & \text{if } \gamma \geq 0, \\ \frac{1-2\gamma}{2(1-\gamma)}, & \text{if } \gamma < 0. \end{cases}$$

The consistency of the moment estimator now follows since

$$H_{k,n} \overset{P}{\Rightarrow} \begin{cases} \gamma, & \text{if } \gamma \geq 0, \\ 0, & \text{if } \gamma < 0, \end{cases}$$

since in the non Pareto-type case where $\gamma \leq 0$, the slope of the Pareto quantile plot will tend to zero near the higher observations.

5.2.3 Estimators based on the generalized quantile plot

Following (2.15), the function $U(x)e_{\log X}(\log U(x))$ is regularly varying with index γ since indeed also a is a regularly varying function. Therefore,

$$U(x)H(x) := U(x)e_{\log X}(\log U(x)) = x^\gamma \ell_{UH}(x),$$

for some slowly varying function ℓ_{UH}. Hence, as in case of the Pareto quantile plot, when $x \to \infty$

$$\frac{\log \left(U(x)e_{\log X}(\log U(x)) \right)}{\log x} \to \gamma,$$

this is, when plotting $\log \left(U(x)e_{\log X}(\log U(x)) \right)$ versus $\log x$, we obtain an ultimately linear graph with slope γ. In practice, we replace x by $\frac{n+1}{j+1}$ and we estimate $e_{\log X}(\log U(x))$ with the Hill estimator $H_{j,n}$. We obtain that the plot

$$\left(\log \left(\frac{n+1}{j+1} \right), \log(X_{n-j,n} H_{j,n}) \right), \quad j = 1, \ldots, n-1, \tag{5.11}$$

will be ultimately linear with slope γ.

Example 5.2 In Figure 5.4, this is illustrated for the wind-speed data from three cities in the United States, introduced in Chapter 1. These data are the daily fastest-mile speeds measured by anemometers 10 m above the ground. The line structures in the generalized quantile plots are the result of an inherent grouping of the data due to loss of accuracy during the data-collecting process. For the Des Moines daily wind-speed maxima ($n = 5478$), the generalized quantile plot (5.11) clearly shows an increasing behaviour, which reflects a heavy tail for the underlying distribution. The flattening trend in the Grand Rapids dataset ($n = 5478$) suggests a weaker tail with $\gamma = 0$, while for Albuquerque ($n = 6939$) even a negative γ-value, resulting in a distribution with a finite right endpoint, can be expected.

As in the previous chapter, one can now establish an estimation procedure analogous to that induced by the Hill estimator. The slope in the generalized quantile plot is then estimated by

$$\hat{\gamma}_{k,n}^H = k^{-1} \sum_{j=1}^k \log UH_{j,n} - \log UH_{k+1,n},$$

where $UH_{j,n} := X_{n-j,n} H_{j,n}$.

Next to the above-discussed approach based on Hill-type operations on the UH statistics, the slope of the ultimate linear part of the generalized quantile plot can also be estimated by an unconstrained least-squares fit to the k 'last' points on the generalized quantile plot, as proposed by Beirlant *et al.* (2002b). Minimizing

$$\sum_{j=1}^k \left(\log UH_{j,n} - \delta - \gamma \log \frac{n+1}{j+1} \right)^2$$

with respect to δ and γ results in the so-called Zipf estimator:

$$\hat{\gamma}_{k,n}^Z := \frac{\frac{1}{k} \sum_{j=1}^k \left(\log \frac{k+1}{j+1} - \frac{1}{k} \sum_{i=1}^k \log \frac{k+1}{i+1} \right) \log UH_{j,n}}{\frac{1}{k} \sum_{j=1}^k \log^2 \frac{k+1}{j+1} - \left(\frac{1}{k} \sum_{j=1}^k \log \frac{k+1}{j+1} \right)^2}.$$

An interesting property of this estimator is the smoothness of the realizations as a function of k, which alleviates the problem of choosing k to some extent.

Example 5.3 We illustrate the above-introduced quantile-based estimators on the SOA Group Medical Insurance claim data. In Figure 5.5, we plot $\hat{\gamma}_{P,k}$ (solid line), $M_{k,n}$ (broken line), $\hat{\gamma}_{k,n}^H$ (broken-dotted line) and $\hat{\gamma}_{k,n}^Z$ (dotted line) as a function of k. The moment, generalized Hill and Zipf estimator are quite stable when plotted as a function of k and indicate a γ value of around 0.35, a result that is consistent with the estimates obtained in Chapter 4. Also the Pickands estimator indicates a γ estimate of around 0.3 to 0.4 but, compared to the other estimators, shows a much larger variability.

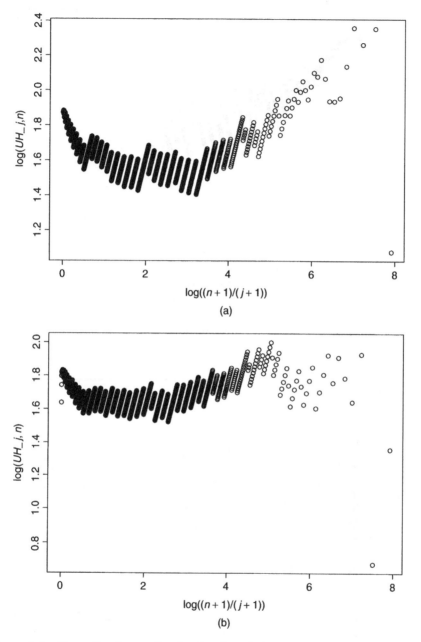

Figure 5.4 Generalized quantile plot for the wind speed data set from (a) Des Moines ($n = 5478$), (b) Grand Rapids ($n = 5478$) and (c) Albuquerque ($n = 6939$).

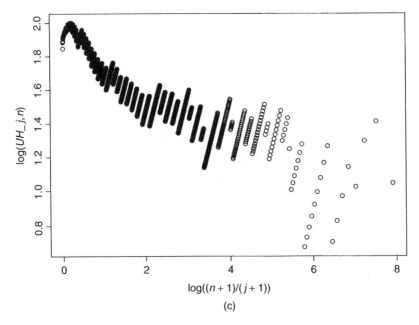

(c)

Figure 5.4   (*continued*)

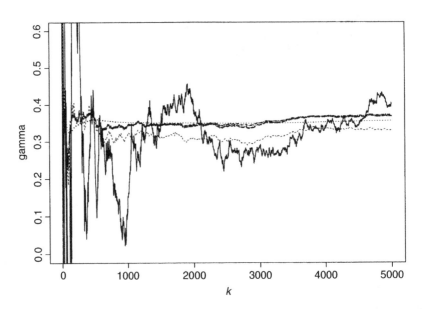

Figure 5.5   SOA Group Medical Insurance data: $\hat{\gamma}_{P,k}$ (solid line), $M_{k,n}$ (broken line), $\hat{\gamma}_{k,n}^{H}$ (broken-dotted line), $\hat{\gamma}_{k,n}^{Z}$ (dotted line) and $\hat{\gamma}_{k}(c, \lambda_{\tilde{\delta},c})$ (broken line, short dashes) as a function of k.

5.3 Tail Probability View—Peaks-Over-Threshold Method

5.3.1 The basic model

The left-hand side in the (C_γ^*) condition can be interpreted as the conditional survival function of the *exceedances* (or *peaks*, or *excesses*) $Y = X - t$ over a threshold t, taken at $yb(t) > 0$:

$$\bar{F}_t(yb(t)) := P\left(Y > yb(t) | Y > 0\right) = P\left(\frac{X - t}{b(t)} > y | X > t\right) = \frac{\bar{F}(t + yb(t))}{\bar{F}(t)}.$$

Hence, from (C_γ^*), it appears a natural statistical procedure to approximate the distribution $\bar{F}_t$ by the distribution given by the right-hand side in (C_γ^*):

$$\bar{F}_t(y) \sim \left(1 + \frac{\gamma y}{b(t)}\right)^{-1/\gamma}. \tag{5.12}$$

Interpreting $b(t)$ in this last expression as a scale parameter σ, we are lead to fit the GP distribution, H, specified by

$$\begin{cases} 1 - \left(1 + \frac{\gamma y}{\sigma}\right)^{-1/\gamma}, & y \in (0, \infty) & \text{if } \gamma > 0, \\ 1 - \exp\left(-\frac{y}{\sigma}\right), & y \in (0, \infty) & \text{if } \gamma = 0, \\ 1 - \left(1 + \frac{\gamma y}{\sigma}\right)^{-1/\gamma}, & y \in (0, -\frac{\sigma}{\gamma}) & \text{if } \gamma < 0, \end{cases} \tag{5.13}$$

to the exceedances over a sufficiently high threshold.

The use of the GP distribution as approximate model for exceedances over high thresholds can also be motivated on the basis of a point process characterization of high-level exceedances. For more details about point processes we refer the reader to section 5.9.2. Let $X_1, \ldots, X_n$ be independent random variables with common distribution function F where F satisfies (C_γ) and consider the two-dimensional point process

$$P_n = \left\{\left(\frac{i}{n + 1}, \frac{X_i - b_n}{a_n}\right); i = 1, \ldots, n\right\},$$

where a_n and b_n normalize $X_{n,n}$ appropriately, as discussed in Chapter 2. It can be shown that on sets that exclude the lower boundary, P_n converges weakly to a two-dimensional Poisson process. The intensity measure Λ of the limiting Poisson process can be immediately derived from the Poisson property. Indeed, since

$$\lim_{n \to \infty} P(\text{no points in } (0, 1) \times (x, \infty)) = \lim_{n \to \infty} P\left(\frac{X_{n,n} - b_n}{a_n} \le x\right)$$

$$= \exp(-(1 + \gamma x)^{-\frac{1}{\gamma}}),$$

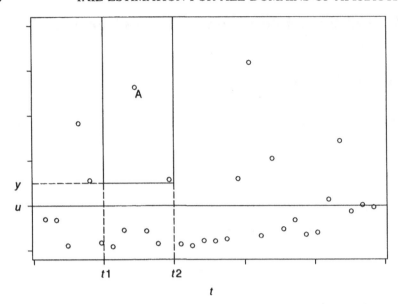

Figure 5.6 Illustration of the point process characterization of high-level exceedances.

we have that for sets $A = (t_1, t_2) \times (x, \infty)$, $t_1 < t_2$,

$$\Lambda(A) = (t_2 - t_1)(1 + \gamma x)^{-\frac{1}{\gamma}}. \tag{5.14}$$

In Figure 5.6, a graphical illustration is provided for this point process interpretation.

Now, for a sufficiently large u

$$P\left(\frac{X_i - b_n}{a_n} > u + x \,\middle|\, \frac{X_i - b_n}{a_n} > u\right) \approx \frac{\Lambda((0, 1) \times (u + x, \infty))}{\Lambda((0, 1) \times (u, \infty))}$$

$$= \left(1 + \frac{\gamma x}{1 + \gamma u}\right)^{-\frac{1}{\gamma}},$$

which is the GP survival function with scale $\sigma(u) = 1 + \gamma u$. For practical purposes, the unknown normalizing constants a_n and b_n can be absorbed in the GEV distribution. So above high thresholds, P_n can be approximated by a two-dimensional Poisson process with intensity measure

$$\Lambda(A) = (t_2 - t_1)\left(1 + \gamma \frac{x - \mu}{\sigma}\right)^{-\frac{1}{\gamma}}.$$

Similarly,

$$\bar{F}_u(x) \approx \left(1 + \frac{\gamma x}{\sigma + \gamma(u - \mu)}\right)^{-\frac{1}{\gamma}}.$$

For a detailed mathematical derivation of these point process results, we refer the interested reader to Leadbetter *et al.* (1983), Falk *et al.* (1994) and Embrechts *et al.* (1997).

5.3.2 Parameter estimation

Given a value of the threshold t and the number of data N_t from the original sample $X_1, \ldots, X_n$ exceeding t, the estimation of the parameters γ and σ can be performed in a variety of ways. We mention the ML method, the method of (probability-weighted) moments and the elemental percentile method (EPM). We denote the absolute exceedances by $Y_j = X_i - t$, provided $X_i > t$, $j = 1, \ldots, N_t$, where i is the index of the j-th exceedance in the original sample. Often, the threshold is taken at one of the sample points, that is, $t = X_{n-k,n}$. In this case, the ordered exceedances are given by $Y_{j,k} = X_{n-k+j,n} - X_{n-k,n}$, $j = 1, \ldots, k$.

The ML method

The log-likelihood function for a sample $Y_1, \ldots, Y_{N_t}$ of i.i.d. GP random variables is given by

$$\log L(\sigma, \gamma) = -N_t \log \sigma - \left(\frac{1}{\gamma} + 1\right) \sum_{i=1}^{N_t} \log\left(1 + \frac{\gamma Y_i}{\sigma}\right)$$

provided $1 + \frac{\gamma Y_i}{\sigma} > 0$, $i = 1, \ldots, N_t$. If $\gamma = 0$, the exponential distribution–based log-likelihood function given by

$$\log L(\sigma, 0) = -N_t \log \sigma - \frac{1}{\sigma} \sum_{i=1}^{N_t} Y_i$$

has to be used. The maximization of $\log L(\sigma, \gamma)$ can be best performed using a reparametrization

$$(\sigma, \gamma) \to (\tau, \gamma) \text{ with } \tau = \frac{\gamma}{\sigma}$$

yielding

$$\log L(\tau, \gamma) = -N_t \log \gamma + N_t \log \tau - \left(\frac{1}{\gamma} + 1\right) \sum_{i=1}^{N_t} \log(1 + \tau Y_i).$$

The ML estimators $\hat{\tau}_{N_t,n}^{ML}$ and $\hat{\gamma}_{N_t,n}^{ML}$ then follow from

$$\frac{1}{\hat{\tau}_{N_t,n}^{ML}} - \left(\frac{1}{\hat{\gamma}_{N_t,n}^{ML}} + 1\right) \frac{1}{N_t} \sum_{i=1}^{N_t} \frac{Y_i}{1 + \hat{\tau}_{N_t,n}^{ML} Y_i} = 0,$$

where

$$\hat{\gamma}_{N_t,n}^{ML} = \frac{1}{N_t} \sum_{i=1}^{N_t} \log \left(1 + \hat{\tau}_{N_t,n}^{ML} Y_i \right).$$

The method of probability-weighted moments

The method of moments (MOM) and the method of probability-weighted moments (PWM) estimators for the GP distribution were introduced by Hosking and Wallis (1987). Both methods share the basic idea that estimators for unknown parameters can be derived from the expressions for the population moments. The r-th moment of the GP distribution exists if $\gamma < 1/r$. Provided that they exist, the mean and the variance of the GP distribution are given by respectively

$$E(Y) = \frac{\sigma}{1 - \gamma}, \tag{5.15}$$

$$\text{var}(Y) = \frac{\sigma^2}{(1 - \gamma)^2 (1 - 2\gamma)}. \tag{5.16}$$

Assume a sample $Y_1, \ldots, Y_{N_t}$ of i.i.d. GP random variables is available. The order statistics associated with $Y_1, \ldots, Y_{N_t}$ are denoted by $Y_{1,N_t} \leq \ldots \leq Y_{N_t,N_t}$. Replacing $E(Y)$ by $\bar{Y} = \sum_{i=1}^{N_t} Y_i / N_t$ and $\text{var}(Y)$ by $S_Y^2 = \sum_{i=1}^{N_t} (Y_i - \bar{Y})^2 / (N_t - 1)$ and solving (5.15)-(5.16) for γ and σ yields the MOM estimators:

$$\hat{\gamma}_{MOM} = \frac{1}{2} \left(1 - \frac{\bar{Y}^2}{S_Y^2} \right),$$

$$\hat{\sigma}_{MOM} = \frac{\bar{Y}}{2} \left(1 + \frac{\bar{Y}^2}{S_Y^2} \right).$$

We now turn to PWM estimation of the GP parameters σ and γ. In case of the GP distribution, it is convenient to consider (5.4) with $p = 1$, $r = 0$ and $s = 0, 1, 2, \ldots$ leading to

$$M_{1,0,s} = \frac{\sigma}{(s + 1)(s + 1 - \gamma)} \qquad \gamma < 1. \tag{5.17}$$

Replacing $M_{1,0,s}$ by its empirical counterpart (as in case of fitting a GEV distribution)

$$\hat{M}_{1,0,s} = \frac{1}{N_t} \sum_{j=1}^{N_t} \left(\prod_{\ell=1}^{s} \frac{(N_t - j - \ell + 1)}{(N_t - \ell)} \right) Y_{j,N_t}$$

or

$$\tilde{M}_{1,0,s} = \frac{1}{N_t} \sum_{j=1}^{N_t} \left(1 - \frac{j}{N_t + 1} \right)^s Y_{j,N_t},$$

and solving (5.17) for $s = 0$ and $s = 1$ with respect to γ and σ yields the PWM estimators

$$\hat{\gamma}_{\text{PWM}} = 2 - \frac{\hat{M}_{1,0,0}}{\hat{M}_{1,0,0} - 2\hat{M}_{1,0,1}},$$

$$\hat{\sigma}_{\text{PWM}} = \frac{2\hat{M}_{1,0,0}\hat{M}_{1,0,1}}{\hat{M}_{1,0,0} - 2\hat{M}_{1,0,1}}.$$

Note that the PWM estimator for γ can be written as a ratio of weighted sums of ordered exceedances. In case $\tilde{M}_{1,0,s}$ is used as an estimator for $M_{1,0,s}$, this then yields

$$\hat{\gamma}_{\text{PWM}} = \frac{\frac{1}{N_t} \sum_{j=1}^{N_t} \left(4\frac{j}{N_t+1} - 3\right) Y_{j,N_t}}{\frac{1}{N_t} \sum_{j=1}^{N_t} \left(2\frac{j}{N_t+1} - 1\right) Y_{j,N_t}}.$$

Application of the MOM and PWM estimators is not without problems. First, in case $\gamma \geq 1$, the MOM and PWM estimators do not exist. Second, the obtained estimates may be inconsistent with the observed data in the sense that in case $\gamma < 0$, some of the observations may fall above the estimate of the right endpoint.

The elemental percentile method

The elemental percentile method (EPM) introduced by Castillo and Hadi (1997) overcomes some of the difficulties associated with the ML method and the method of (probability-weighted) moments. In fact, for this method, there are no restrictions on the value of γ. Here, we will concentrate on the estimation of $\gamma \neq 0$. In case $\gamma = 0$, the parameter σ can be estimated efficiently with the ML method. Assume a sample $Y_1, \ldots, Y_{N_t}$ of i.i.d. GP random variables is available. Consider two distinct order statistics Y_{i,N_t} and Y_{j,N_t}. Equating the GP cumulative distribution function evaluated at these order statistics to the corresponding percentile values gives a system of two equations in two unknowns:

$$1 - (1 + \hat{\tau}_{i,j} Y_{i,N_t})^{-\frac{1}{\hat{\gamma}_{i,j}}} = p_{i,n}, \tag{5.18}$$

$$1 - (1 + \hat{\tau}_{i,j} Y_{j,N_t})^{-\frac{1}{\hat{\gamma}_{i,j}}} = p_{j,n}, \tag{5.19}$$

where, as before $\tau = \gamma/\sigma$ and $p_{i,n} = \frac{i}{n+1}$. Elimination of $\hat{\gamma}_{i,j}$ yields

$$C_j \log(1 + \hat{\tau}_{i,j} Y_{i,N_t}) = C_i \log(1 + \hat{\tau}_{i,j} Y_{j,N_t})$$

where $C_i = -\log(1 - p_{i,n})$, which can be solved numerically for $\hat{\tau}_{i,j}$. Plugging $\hat{\tau}_{i,j}$ into (5.18) (or (5.19)) and solving for $\hat{\gamma}_{i,j}$, we obtain

$$\hat{\gamma}_{i,j} = \frac{\log(1 + \hat{\tau}_{i,j} Y_{i,N_t})}{C_i}$$

and then

$$\hat{\sigma}_{i,j} = \frac{\hat{\gamma}_{i,j}}{\hat{\tau}_{i,j}}.$$

In order to use all available information, $\hat{\gamma}_{i,j}$ and $\hat{\sigma}_{i,j}$ are computed for all distinct pairs of order statistics $Y_{i,N_t} < Y_{j,N_t}$ leading to the final EPM estimators

$$\hat{\gamma}_{EPM} = \text{median}\{\hat{\gamma}_{i,j}; \ i < j\},$$

$$\hat{\sigma}_{EPM} = \text{median}\{\hat{\sigma}_{i,j}; \ i < j\}.$$

In case $i = \frac{N_t}{2}$ and $j = \frac{3N_t}{4}$, it is easy to show that the system of equations (5.18) and (5.19) has a closed-form solution given by

$$\hat{\gamma}_{\lceil \frac{N_t}{2} \rceil, \lceil \frac{3N_t}{4} \rceil} = \frac{1}{\log 2} \log \frac{Y_{\lceil \frac{3N_t}{4} \rceil, N_t} - Y_{\lceil \frac{N_t}{2} \rceil, N_t}}{Y_{\lceil \frac{N_t}{2} \rceil, N_t}}, \tag{5.20}$$

$$\hat{\tau}_{\lceil \frac{N_t}{2} \rceil, \lceil \frac{3N_t}{4} \rceil} = \frac{Y_{\lceil \frac{3N_t}{4} \rceil, N_t} - 2Y_{\lceil \frac{N_t}{2} \rceil, N_t}}{Y_{\lceil \frac{N_t}{2} \rceil, N_t}^2}. \tag{5.21}$$

In fact, (5.20) is the Pickands (1975) estimator for γ as discussed above.

In the above discussion, we always assumed that a sample $Y_1, \ldots, Y_{N_t}$ of i.i.d. GP random variables is available. If the data are not exact GP distributed, one can rely on relation (5.12) and use the GP distribution as an approximation to the conditional distribution of the exceedances. In this case, the GP distribution is fitted to the excesses $Y_j = X_i - t$, in case $X_i > t$, $j = 1, \ldots, N_t$, using one of the above described methods. Note that in the latter case, N_t is random.

Example 5.3 (continued) Applying the POT approach to the SOA Group Medical Insurance data introduced in section 1.3.3 with a threshold of 400,000 USD, we fit the GP distribution to the excesses $y_j = x_i - 400,000$. The ML procedure leads to $\hat{\gamma} = 0.3823$ when $t = 400,000$. The quality of this GP fit to the empirical distribution function of the data Y_i is depicted in Figure 5.7(a). Figure 5.7(b) contains the W-plot of the GP fit to the exceedances over $t = 400,000$. In Table 5.2, we show the ML, MOM, PWM and EPM estimates for the parameters σ and γ obtained from fitting the GP distribution to the excesses over $t = 400,000$.

The choice of the threshold t is very much an open matter and resembles the choice of the value of k in the previous chapter. As in the case of the Hill estimator, a compromise has to be found between high values of t, where the bias of the estimator will be smallest, and low values of t, where the variance will be smallest. In the literature on the POT method, not much attention has been given to this aspect. Davison and Smith (1990) propose to use the mean excess plot. Indeed, the mean excess function of the GP distribution is given by the linear expression

$$e(t) = \frac{\sigma + \gamma t}{1 - \gamma}, \ \text{if} \ \gamma < 1.$$

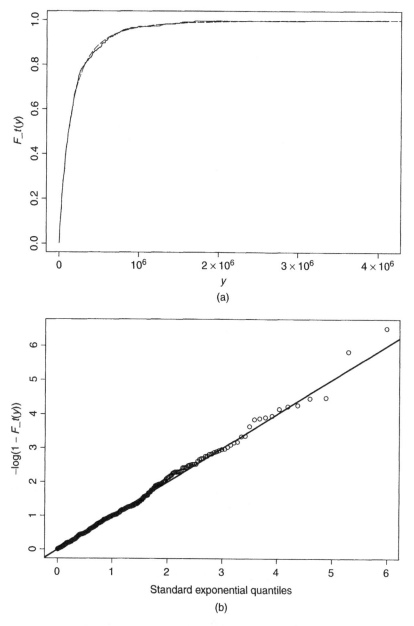

Figure 5.7 SOA Group Medical Insurance data: (a) comparison of the fitted excess distribution with ML estimates for γ and σ (broken line) and the empirical one (solid line) and (b) W-plot for claim sizes exceeding $t = 400,000$.

Table 5.2   ML, MOM, PWM and
EPM estimates for the SOA Group
Medical Insurance data with $t =$
400,000.

Method	σ	γ
ML	142,489	0.3823
MOM	156,841	0.3095
PWM	142,933	0.3707
EPM	139,838	0.4112

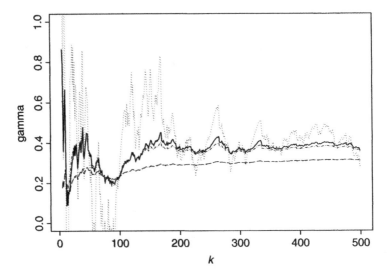

Figure 5.8   SOA Group Medical Insurance data: ML estimate (solid line), MOM estimate (broken line), probability-weighted moments estimate (broken-dotted line) and elemental percentile estimate (dotted line) as a function of k.

Hence, one is lead to the graphical approach choosing $t = X_{n-k,n}$ as the point to the right of which a linear pattern appears in the plot $\{(X_{n-k,n}, E_{k,n}); k = 1, \ldots, n - 1\}$.

The POT method, however, will often lead to stable plots of the estimates $\hat{\gamma}_k$ as a function of k, less volatile than for the case of the Hill plots. An illustration is found in Figure 5.8 concerning the SOA Group Medical Insurance data.

We need to emphasize that, whereas the POT method yields more stable plots for the estimates as a function of the threshold t, the bias can still be quite substantial.

5.4   Estimators Based on an Exponential Regression Model

In this section, we will discuss the approximate representation for log-ratios of spacings of successive order statistics as derived by Matthys and Beirlant (2003). This representation extends the exponential regression models derived in Chapter 4 in a natural way to the general case where $\gamma \in \mathbb{R}$.

Let $U_{j,n}$, $j = 1, \ldots, n$ denote the order statistics of a random sample of size n from the $U(0, 1)$ distribution. Then, for $k = 1, \ldots, n - 1$, $(V_{j,k} := U_{j,n}/U_{k+1,n};$ $j = 1, \ldots, k)$ are jointly distributed as the order statistics of a random sample of size k from the $U(0, 1)$ distribution. As before, E_j, $j = 1, \ldots, n$ denote standard exponential random variables and $E_{j,n}$, $j = 1, \ldots, n$, are the corresponding ascending order statistics. The inverse probability integral transform together with (C_γ) imply that, for $j = 1, \ldots, k$,

$$X_{n-j+1,n} - X_{n-k,n} \stackrel{\mathcal{D}}{=} U(U_{j,n}^{-1}) - U(U_{k+1,n}^{-1})$$

$$\stackrel{\mathcal{D}}{=} U(V_{j,k}^{-1} U_{k+1,n}^{-1}) - U(U_{k+1,n}^{-1})$$

$$\sim a(U_{k+1,n}^{-1}) \frac{V_{j,k}^{-\gamma} - 1}{\gamma},$$

provided $k/n \to 0$. For a log-ratio of spacings of order statistics, we then obtain

$$\log \frac{X_{n-j+1,n} - X_{n-k,n}}{X_{n-j,n} - X_{n-k,n}} \sim \log \frac{V_{j,k}^{-\gamma} - 1}{V_{j+1,k}^{-\gamma} - 1}, \quad j = 1, \ldots, k - 1.$$

Application of the mean value theorem with $E_{j,k}^* \in (E_{k-j,k}, E_{k-j+1,k})$, $V_{j,k}^* = \exp(-E_{j,k}^*)$ and the Rényi representation we have that

$$\log \frac{V_{j,k}^{-\gamma} - 1}{V_{j+1,k}^{-\gamma} - 1} \stackrel{\mathcal{D}}{=} \log \left(\exp(\gamma E_{k-j+1,k}) - 1 \right) - \log \left(\exp(\gamma E_{k-j,k}) - 1 \right)$$

$$= (E_{k-j+1,k} - E_{k-j,k}) \frac{\gamma \exp(\gamma E_{j,k}^*)}{\exp(\gamma E_{j,k}^*) - 1}$$

$$\stackrel{\mathcal{D}}{=} \frac{E_j}{j} \frac{\gamma}{1 - (V_{j,k}^*)^\gamma}.$$

We now replace $V_{j,k}^*$ by $j/(k + 1)$ to obtain the following approximate representation for log-ratios of spacings

$$j \log \frac{X_{n-j+1,n} - X_{n-k,n}}{X_{n-j,n} - X_{n-k,n}} \stackrel{\mathcal{D}}{\approx} \frac{\gamma}{1 - \left(\frac{j}{k+1} \right)^\gamma} E_j, \quad j = 1, \ldots, k - 1, \quad (5.22)$$

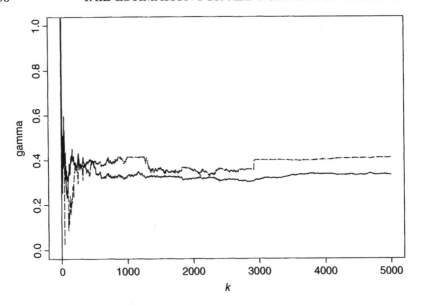

Figure 5.9   SOA Group Medical Insurance data: $\hat{\gamma}_{k,n}^{RMA}$ (solid line) and $\hat{\gamma}_{k,n}^{RMB}$ (broken line) as a function of k.

from which γ can be estimated with the ML method. By construction, the resulting ML estimator, denoted $\hat{\gamma}_{k,n}^{RMA}$, is invariant with respect to a shift and a rescaling of the data. Later on in this chapter, we will refine this estimator by imposing a second-order tail condition on U.

Example 5.3 (continued) In Figure 5.9, we illustrate the use of $\hat{\gamma}_{k,n}^{RMA}$ (solid line) on the SOA Group Medical Insurance claim data set. The exponential regression model approach also indicates a γ estimate of around 0.35, a result that is consistent with the earlier analysis.

5.5   Extreme Tail Probability, Large Quantile and Endpoint Estimation Using Threshold Methods

5.5.1   The quantile view

On the basis of (C_γ), we take $xu = \frac{1}{p}$ and $x = \frac{n+1}{k+1}$, so that $U(xu) = Q(1-p)$ and $\hat{U}(x) = X_{n-k,n}$ lead to the following general form of extreme quantile estimator:

$$\hat{U}\left(\frac{1}{p}\right) = X_{n-k,n} + \hat{a}\left(\frac{n+1}{k+1}\right)\left[\left(\frac{k+1}{(n+1)p}\right)^{\hat{\gamma}} - 1\right]\hat{\gamma}^{-1}, \qquad (5.23)$$

where $\hat{a}\left(\frac{n+1}{k+1}\right)$ and $\hat{\gamma}$ denote estimators of $a\left(\frac{n+1}{k+1}\right)$ and γ respectively. Any one of the estimators of γ discussed above can be used. Concerning $\hat{a}$, it appears natural from (2.15) to consider

$$\hat{a}\left(\frac{n+1}{k+1}\right) = (1 - \hat{\gamma}^-)UH_{k,n} = (1 - \hat{\gamma}^-)X_{n-k,n}H_{k,n}.$$

Alternatively, following Matthys and Beirlant (2003), $a(\frac{n+1}{k+1})$ can also be estimated on the basis of an approximate exponential regression model for spacings of successive order statistics. On the basis of $(\mathcal{C}_\gamma)$, for $k/n \to 0$,

$$X_{n-j+1,n} - X_{n-j,n} \sim a(U_{k+1,n}^{-1})\frac{V_{j,k}^{-\gamma} - V_{j+1,k}^{-\gamma}}{\gamma}, \quad j = 1, \ldots, k-1.$$

Application of the mean value theorem, with the same notation for $E_{j,k}^*$ and $V_{j,k}^*$ as in section 5.4 and using the Rényi representation results in

$$\frac{V_{j,k}^{-\gamma} - V_{j+1,k}^{-\gamma}}{\gamma} \overset{\mathcal{D}}{\cong} \frac{\exp(\gamma E_{k-j+1,k}) - \exp(\gamma E_{k-j,k})}{\gamma}$$

$$= (E_{k-j+1,k} - E_{k-j,k})\exp(\gamma E_{j,k}^*)$$

$$\overset{\mathcal{D}}{\cong} \frac{E_j}{j}\left(V_{j,k}^*\right)^{-\gamma}.$$

Hence, after replacing $V_{j,k}^*$ by $\frac{j}{k+1}$, the following approximate exponential regression model for spacings of successive order statistics is obtained

$$j(X_{n-j+1,n} - X_{n-j,n}) \overset{\mathcal{D}}{\approx} a_{n,k+1}\left(\frac{j}{k+1}\right)^{-\gamma} E_j, \quad j = 1, \ldots, k, \quad (5.24)$$

with $a_{n,k+1} = a(\frac{n+1}{k+1})$. Using straightforward derivations, the log-likelihood function of model (5.24) is maximal at

$$\breve{a}_{n,k+1} = \frac{1}{k}\sum_{j=1}^{k} j(X_{n-j+1,n} - X_{n-j,n})\left(\frac{j}{k+1}\right)^{\gamma}. \quad (5.25)$$

Extreme quantiles can now be estimated using

$$\hat{U}_{k,n}^{\text{RMA}}\left(\frac{1}{p}\right) = X_{n-k,n} + \hat{\breve{a}}_{n,k+1}\frac{\left(\frac{k+1}{p(n+1)}\right)^{\hat{\gamma}_{k,n}^{\text{RMA}}} - 1}{\hat{\gamma}_{k,n}^{\text{RMA}}},$$

where $\hat{\breve{a}}_{n,k+1}$ is as in (5.25) but with γ replaced by $\hat{\gamma}_{k,n}^{\text{RMA}}$.

Concerning the estimation of extreme tail probabilities $P(X > x)$ condition $(\mathcal{C}_\gamma^*)$ similarly leads to setting $U(x) + va(x) =: y$,

$$\hat{\bar{F}}(y) = \frac{k+1}{n+1}\left(1 + \hat{\gamma}\frac{y - X_{n-k,n}}{\hat{a}\left(\frac{n+1}{k+1}\right)}\right)^{-1/\hat{\gamma}}. \quad (5.26)$$

Finally, when $\gamma < 0$, an estimation of x_* is obtained by letting $p \to 0$ in (5.23):

$$\hat{x}_+ = X_{n-k,n} - \frac{1}{\hat{\gamma}} \hat{a} \left(\frac{n+1}{k+1} \right) = X_{n-k,n} - \frac{1-\hat{\gamma}}{\hat{\gamma}} X_{n-k,n} H_{k,n}. \qquad (5.27)$$

5.5.2 The probability view

Extreme quantiles of the GP distribution can be estimated by inverting the GP distribution function given by (5.13), yielding

$$U\left(\frac{1}{p}\right) = \begin{cases} \frac{\sigma}{\gamma}(p^{-\gamma} - 1) & \gamma \neq 0, \\ -\sigma \log p & \gamma = 0, \end{cases} \qquad (5.28)$$

and replacing the unknown parameters by one of the above described estimates. In case $\gamma < 0$, the right endpoint of the GP distribution is finite and can be obtained by letting $p \to 0$ in (5.28):

$$\hat{x}_+ = \frac{\hat{\sigma}}{|\hat{\gamma}|}.$$

If the data are not exact GP distributed, relation (5.12) implies that

$$\bar{F}_t(y) = \frac{\bar{F}(t+y)}{\bar{F}(t)} \sim \left(1 + \frac{\gamma y}{\sigma} \right)^{-1/\gamma},$$

so that with $x = t + y$

$$\bar{F}(x) \sim \bar{F}(t) \left(1 + \frac{\gamma(x-t)}{\sigma} \right)^{-1/\gamma}.$$

Estimating $\bar{F}(t)$ by N_t/n and replacing γ and σ by their respective ML, MOM, PWM or EPM estimates, we obtain that

$$\hat{\bar{F}}(x) = \frac{N_t}{n} \left(1 + \frac{\hat{\gamma}(x-t)}{\hat{\sigma}} \right)^{-1/\hat{\gamma}}. \qquad (5.29)$$

The POT estimator for large quantiles $Q(1-p)$ can now be obtained from inverting the right-hand side in (5.29):

$$\hat{U}\left(\frac{1}{p}\right) = t + \frac{\hat{\sigma}}{\hat{\gamma}} \left(\left(\frac{np}{N_t} \right)^{-\hat{\gamma}} - 1 \right). \qquad (5.30)$$

Note that in case $\gamma < 0$, an estimator for the right endpoint of the support of the distribution $x_* = Q(1)$ is obtained by letting $p \to 0$ in (5.30):

$$\hat{x}_+ = t - \frac{\hat{\sigma}}{\hat{\gamma}}. \qquad (5.31)$$

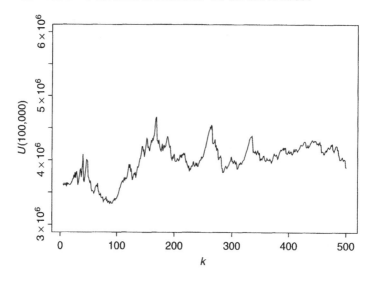

Figure 5.10   SOA Group Medical Insurance data: $\hat{U}(100,000)$ as a function of k.

Note further that in case the parameters of the GP distribution are estimated with the ML method, extreme quantile estimates can be obtained directly by reparametrizing the log-likelihood function in terms of $U(\frac{1}{p})$, for example, setting

$$\sigma = \frac{\gamma(U(\frac{1}{p}) - t)}{\left(\frac{np}{N_t}\right)^{-\gamma} - 1}.$$

Example 5.3 (continued) In Figure 5.10, we illustrate the estimation of extreme quantiles using the POT approach. Figure 5.10 shows the estimates for $U(100,000)$ as a function of k. Here, $\hat{U}$ was obtained by plugging the ML estimates for γ and σ in (5.30). A stable region appears for k from 200 up to 500, leading to an estimate of 4 million.

5.5.3   Inference: confidence intervals

Approximate $100(1 - \alpha)\%$ confidence intervals for the parameters γ and σ of the GP distribution can be constructed on the basis of the asymptotic normality of the ML, MOM and PWM estimators. For instance, a $100(1 - \alpha)\%$ confidence interval for γ is given by

$$\hat{\gamma} \pm \Phi^{-1}(1 - \alpha/2)\sqrt{\frac{\hat{v}_{1,1}}{N_t}}$$

where $\hat{\gamma}$ is either the ML, MOM or PWM estimate for γ and $\hat{v}_{1,1}$ is the first diagonal element of respectively V_1, V_2 or V_3 (for more information on these covariance matrices, we refer to section 5.6) with the unknown parameters replaced by their estimates. Inference about return levels $U(\frac{1}{p})$ can be drawn in a similar way. Straightforward application of the delta method gives

$$\sqrt{N_t}\left(\hat{U}\left(\frac{1}{p}\right) - U\left(\frac{1}{p}\right)\right) \xrightarrow{\mathcal{D}} N(0, \xi'V\xi)$$

where V is either V_1, V_2 or V_3 and

$$\xi' = \left[\frac{\partial U(\frac{1}{p})}{\partial \gamma}, \frac{\partial U(\frac{1}{p})}{\partial \sigma}\right]$$

$$= \left[-\frac{\sigma}{\gamma^2}\left(p^{-\gamma} - 1\right) - \frac{\sigma}{\gamma}p^{-\gamma}\log p, \frac{1}{\gamma}\left(p^{-\gamma} - 1\right)\right],$$

so a $100(1 - \alpha)\%$ confidence interval for $U(\frac{1}{p})$ is given by

$$\hat{U}\left(\frac{1}{p}\right) \pm \Phi^{-1}(1 - \alpha/2)\sqrt{\frac{\xi'V\xi}{N_t}}.$$

Often, better confidence intervals can be constructed on the basis of the profile likelihood ratio test statistic. The profile likelihood function for γ is given by

$$L_p(\gamma) = \max_{\sigma|\gamma} L(\sigma, \gamma).$$

Using similar arguments as in case of the GEV, the $100(1 - \alpha)\%$ profile likelihood confidence interval for the parameter γ can be obtained as

$$CI_\gamma = \left\{\gamma : \log L_p(\gamma) \geq \log L_p(\hat{\gamma}) - \frac{\chi_1^2(1 - \alpha)}{2}\right\}.$$

The special case of testing $H_0 : \gamma = 0$ is described in Marohn (1999).

Example 5.3 (continued) Figure 5.11 illustrates the profile likelihood function–based confidence intervals using the SOA Group Medical Insurance data. In Figure 5.11(a) and (b), we show the profile log-likelihood function of γ and $U(100,000)$ respectively at $k = 200$, together with the 95% confidence interval.

5.6 Asymptotic Results Under $(\mathcal{C}_\gamma)$-$(\mathcal{C}_\gamma^*)$

In order to be able to construct asymptotic confidence intervals or tests concerning γ, we now discuss the most relevant asymptotic results and present some asymptotic comparisons between some of the estimators.

In case of the ML and the probability-weighted moment approach for peaks over thresholds, one can develop asymptotic results *under the assumption that the*

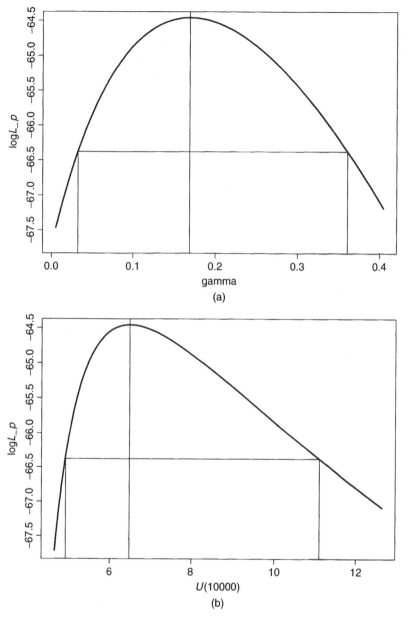

Figure 5.11 SOA Group Medical Insurance data: profile log-likelihood function and profile likelihood–based 95% confidence intervals at $k = 200$ for (a) γ and (b) $U(100,000)$.

excesses exactly follow a GP distribution ; rather than assuming (C_γ) or, equivalently, (C_γ^*) (see also section 5.1 concerning the method of block maxima). This is a restrictive approach that should be validated of course through some goodness-of-fit methods that were developed above. Under this *parametric approach*, the following asymptotic results can be stated concerning the POT methods:

(i) The ML estimators $(\hat{\gamma}_{N_t,n}^{ML}, \hat{\sigma}_{N_t,n}^{ML})$ are asymptotically normal: provided $\gamma > -1/2$, for $N_t \to \infty$

$$\sqrt{N_t}\left((\hat{\gamma}_{N_t,n}^{ML}, \hat{\sigma}_{N_t,n}^{ML}) - (\gamma, \sigma)\right) \xrightarrow{\mathcal{D}} N(0, V_1),$$

where

$$V_1 = (1 + \gamma)\begin{bmatrix} 1 + \gamma & -\sigma \\ -\sigma & 2\sigma^2 \end{bmatrix},$$

while the ML estimator is superefficient when $-1 > \gamma > -1/2$; that is, the ML estimator converges with rate of consistency $N_t^{-\gamma}$.

(ii) The MOM estimators $(\hat{\gamma}_{MOM}, \hat{\sigma}_{MOM})$ satisfy for $N_t \to \infty$

$$\sqrt{N_t}\left((\hat{\gamma}_{MOM}, \hat{\sigma}_{MOM}) - (\gamma, \sigma)\right) \xrightarrow{\mathcal{D}} N(0, V_2),$$

where

$$V_2 = C\begin{bmatrix} (1 - 2\gamma)^2(1 - \gamma + 6\gamma^2) & -\sigma(1 - 2\gamma)(1 - 4\gamma + 12\gamma^2) \\ -\sigma(1 - 2\gamma)(1 - 4\gamma + 12\gamma^2) & 2\sigma^2(1 - 6\gamma + 12\gamma^2) \end{bmatrix},$$

$$C = \frac{(1 - \gamma)^2}{(1 - 2\gamma)(1 - 3\gamma)(1 - 4\gamma)},$$

provided $\gamma < 1/4$.

(iii) The probability-weighted moment estimators $(\hat{\gamma}_{PWM}, \hat{\sigma}_{PWM})$ satisfy, provided $\gamma < 1/2$, as $N_t \to \infty$

$$\sqrt{N_t}((\hat{\gamma}_{PWM}, \hat{\sigma}_{PWM}) - (\gamma, \sigma)) \xrightarrow{\mathcal{D}} N(0, V_3)$$

where

$$V_3 = C\begin{bmatrix} (1 - \gamma)(2 - \gamma)^2(1 - \gamma + 2\gamma^2) & -\sigma(2 - \gamma)(2 - 6\gamma + 7\gamma^2 - 2\gamma^3) \\ -\sigma(2 - \gamma)(2 - 6\gamma + 7\gamma^2 - 2\gamma^3) & \sigma^2(7 - 18\gamma + 11\gamma^2 - 2\gamma^3) \end{bmatrix},$$

$$C = \frac{1}{(1 - 2\gamma)(3 - 2\gamma)}.$$

(iv) Likewise, the asymptotic properties of the initial estimators $\hat{\gamma}_{i,j}$ and $\hat{\sigma}_{i,j}$ in the EPM can be derived. First, consider the limiting behaviour of order statistics (Y_{i,N_t}, Y_{j,N_t}). Let $i = \lfloor N_t p \rfloor$ and $j = \lfloor N_t q \rfloor$, $0 < p < q < 1$, and let Q denote the GP quantile function

$$Q(p) = \frac{\sigma}{\gamma}((1-p)^{-\gamma} - 1).$$

It can be shown that for $N_t \to \infty$

$$\sqrt{N_t}\left((Y_{i,N_t}, Y_{j,N_t}) - (Q(p), Q(q))\right) \overset{\mathcal{D}}{\to} N(0, W)$$

where

$$W = \begin{bmatrix} \sigma^2 p(1-p)^{-2\gamma-1} & \sigma^2 p(1-p)^{-\gamma-1}(1-q)^{-\gamma} \\ \sigma^2 p(1-p)^{-\gamma-1}(1-q)^{-\gamma} & \sigma^2 q(1-q)^{-2\gamma-1} \end{bmatrix}.$$

Remark that these limit results for the marginal distributions were derived in section 3.2, case 2 (iv). Straightforward application of the delta method then yields the limiting behaviour of the initial estimators:

$$\sqrt{N_t}\left((\hat{\gamma}_{i,j}, \hat{\sigma}_{i,j}) - (\gamma, \sigma)\right) \overset{\mathcal{D}}{\to} N(0, CWC')$$

with

$$C = \frac{1}{\eta}\begin{bmatrix} 1 - (1-q)^{-\gamma} & (1-p)^{-\gamma} - 1 \\ -[Q(q) + \sigma\log(1-q)(1-q)^{-\gamma}] & Q(p) + \sigma\log(1-p)(1-p)^{-\gamma} \end{bmatrix}$$

and

$$\eta = \log(1-p)Q(q)(1-p)^{-\gamma} - \log(1-q)Q(p)(1-q)^{-\gamma}.$$

Another more general point of view recognizes that, in general, the excesses are not exactly GP distributed, but that the POT distribution *approaches* a GP distribution for high-enough thresholds as assumed under $(\mathcal{C}_\gamma)$-$(\mathcal{C}_\gamma^*)$. For this, one typically adds an assumption concerning the rate of convergence of the excess distribution to the GP family. This can be found in the second-order theory discussed in Chapter 3. The *semi-parametric point of view* then results in the appearance of an asymptotic bias in the results.

Following the theory of generalized regular variation of second order outlined in de Haan and Stadtmüller (1996), we assume the existence of a positive function a and a second ultimately positive auxiliary function a_2 with $a_2(x) \to 0$ when $x \to \infty$, such that

$$\lim_{x \to \infty} \frac{1}{a_2(x)}\left\{\frac{U(ux) - U(x)}{a(x)} - h_\gamma(u)\right\} = c\int_1^u t^{\gamma-1}h_\rho(t)dt + Ah_{\gamma+\rho}(u). \quad (5.32)$$

In the sequel, we denote the class of generalized second order regularly varying functions U satisfying (5.32) with $GRV_2(\gamma, \rho; a(x), a_2(x); c, A)$.

We restrict the discussion here to the case $\rho < 0$, in which case a clever choice of the auxiliary function a_2 results in a simplification of the limit function in (5.32) with $c = 0$.

In Appendix 5.9.3, we give an overview of possible kinds of GRV_2 functions and the corresponding representations for U and $\log U$ as given in Vanroelen (2003). From this list, it follows that the second-order rate in (5.32) is worse for $\log U$ compared to U when $\rho < \gamma < 0$ and in some cases when $0 < \gamma < -\rho$. In these cases, this will entail asymptotic relative efficiency 0 for estimators based on log-transformed data compared to shift invariant estimators such as the ML estimator or the Pickands type estimators, this is if all these estimators are based on the pertaining optimal number of order statistics.

When $0 < \gamma < -\rho$, this rate problem for $\log U$ arises with the appearance of the constant D in the characterization of U in that case, namely, $U(x) = \ell_+ x^\gamma \{\frac{1}{\gamma} + Dx^{-\gamma} + \frac{A}{\gamma+\rho}a_2(x)(1 + o(1))\}$. When $D = 0$, the original a_2-rate is kept for $\log U$, while it is not when $D \neq 0$, in which case a_2 is replaced by a regularly function with index $-\gamma$. Within the Hall class of Pareto-type distributions (see section 3.3.2), the case $D \neq 0$ occurs when $\beta = \gamma$. This is the case, for instance, for the Fisher F and the GEV distributions. Also remark the special representation in case $\gamma + \rho = 0$, where a slowly varying function L_2 appears, discussed in Appendix 5.9.3. The representations for U, respectively $\log U$, as given in Appendix 5.9.3 can be used to derive the asymptotic mean squared errors (AMSEs) of some well-known estimators, see Appendix 5.9.4. In this, we assume that the slowly varying parts of b and a_2 are asymptotically equivalent to a constant. The optimal values of k, which minimize the different expressions of the AMSEs together with the corresponding minimal AMSE values, are found in Appendix 5.9.5. Matthys and Beirlant (2003) and Segers (2004) contain similar asymptotic results for $\hat{\gamma}_{k,n}^{\text{RMA}}$, $\hat{\gamma}_{k,n}^{\text{RMB}}$ (see section 5.7.1), respectively $\hat{\gamma}_k(c, \lambda)$.

We end this section by specifying the asymptotic distribution of $\hat{U}(1/p)$ as defined in (5.23) with the moment estimator $M_{k,n}$ substituted for $\hat{\gamma}$. Such asymptotic results were first proven in de Haan and Rootzén (1993) and were further explored in Ferreira et al. (2003). Matthys and Beirlant (2003) provide analogous results for $\hat{U}_{k,n}^{\text{RMA}}(1/p)$.

Considering the conditions of Proposition 3.2, Ferreira et al. (2003) defined the function

$$\tilde{a}_2(x) = \begin{cases} a_2(x), & \text{if } \gamma < \rho, \\ x_* - a(x)/U(x), & \text{if } \rho < \gamma \leq 0 \\ & \text{or } 0 < \gamma < -\rho \text{ with } D \neq 0 \\ & \text{or } \gamma = -\rho, \\ \frac{\rho a_2(x)}{\gamma+\rho}, & \text{if } \gamma > -\rho \\ & \text{or } 0 < \gamma < -\rho \text{ with } D = 0, \end{cases}$$

where D is as in the representations for U given in Vanroelen (2003), see also Appendix 5.9.3.

Let $a_n = k/(np_n)$. Then when $U(\infty) > 0$, and $k = k_n \to \infty$ such that $n/k_n \to \infty$, $np_n \to c \geq 0$ (finite), $\sqrt{k}\tilde{a}_2(n/k) \to 0$ and $(\log a_n)/\sqrt{k} \to 0$ as $n \to \infty$, we have that as $\gamma \neq 0$ and $\gamma \neq \rho$

- in case $\gamma > 0$

$$\frac{\gamma\sqrt{k}}{a\left(\frac{n}{k}\right)a_n^\gamma \log a_n}\left(\hat{U}\left(\frac{1}{p_n}\right) - U\left(\frac{1}{p_n}\right)\right) \xrightarrow{\mathcal{D}} N\left(0, (1+\gamma)^2\right),$$

- while in case $\gamma < 0$

$$\frac{\sqrt{k}}{a\left(\frac{n}{k}\right)}\left(\hat{U}\left(\frac{1}{p_n}\right) - U\left(\frac{1}{p_n}\right)\right) \xrightarrow{\mathcal{D}} N\left(0, \frac{(1-\gamma)^2(1-3\gamma+4\gamma^2)}{\gamma^4(1-2\gamma)(1-3\gamma)(1-4\gamma)}\right).$$

In case $\sqrt{k}\tilde{a}_2(n/k) \to \lambda \in \mathbb{R}$ an asymptotic bias appears, see, for instance, Ferreira *et al.* (2003).

5.7 Reducing the Bias

In this section, we show how some of the estimators based on the first-order condition $(\mathcal{C}_\gamma)$ can be refined by taking into account the second-order tail behaviour. This then is analogous to section 4.4. We confine ourselves here to the estimator based on the exponential regression model introduced in section 5.4.

5.7.1 The quantile view

From the discussion in Chapter 2, we have that $F \in \mathcal{D}(G_\gamma)$ implies the existence of a slowly varying function ℓ and a function d with $\pm d \in \mathcal{R}_0$ and $d(x) \to \gamma$ as $x \to \infty$ such that

$$\frac{U(ux) - U(x)}{a(x)} = \frac{1}{d(x)}\left(u^\gamma \frac{\ell(ux)}{\ell(x)} - 1\right). \tag{5.33}$$

Matthys and Beirlant (2003) refined the exponential regression model (5.22) by imposing the second-order condition (3.14) on the function ℓ.

Using the inverse probability integral transform (5.33) and (3.14), one easily obtains for $j = 1, \ldots, k$,

$$X_{n-j+1,n} - X_{n-k,n} \overset{\mathcal{D}}{=} U(U_{k+1,n}^{-1}V_{j,k}^{-1}) - U(U_{k+1,n}^{-1})$$

$$= c_{n,k+1}\left(V_{j,k}^{-\gamma}\frac{\ell(U_{k+1,n}^{-1}V_{j,k}^{-1})}{\ell(U_{k+1,n}^{-1})} - 1\right)$$

$$\sim c_{n,k+1}\left(V_{j,k}^{-\gamma}\exp\left(a_{n,k+1}\frac{V_{j,k}^{-\rho}-1}{\rho}\right) - 1\right),$$

as $k/n \to 0$, where $c_{n,k+1} := a(U_{k+1,n}^{-1})/d(U_{k+1,n}^{-1})$ and $a_{n,k+1} := ca_2(U_{k+1,n}^{-1})$. Taking a log-ratio of spacings results in

$$\log \frac{X_{n-j+1,n} - X_{n-k,n}}{X_{n-j,n} - X_{n-k,n}} \sim \log \frac{V_{j,k}^{-\gamma} \exp\left(a_{n,k+1} \frac{V_{j,k}^{-\rho}-1}{\rho}\right) - 1}{V_{j+1,k}^{-\gamma} \exp\left(a_{n,k+1} \frac{V_{j+1,k}^{-\rho}-1}{\rho}\right) - 1},$$

$$j = 1, \ldots, k-1.$$

We now apply the mean value theorem (with the same notation for $E_{j,k}^*$ and $V_{j,k}^*$ as in section 5.4) and the Rényi representation of standard exponential order statistics to the right-hand side of the above equation:

$$\log \frac{V_{j,k}^{-\gamma} \exp\left(a_{n,k+1} \frac{V_{j,k}^{-\rho}-1}{\rho}\right) - 1}{V_{j+1,k}^{-\gamma} \exp\left(a_{n,k+1} \frac{V_{j+1,k}^{-\rho}-1}{\rho}\right) - 1}$$

$$\overset{\mathcal{D}}{=} \log \frac{\exp\left(\gamma E_{k-j+1,k} + a_{n,k+1} \frac{\exp(\rho E_{k-j+1,k})-1}{\rho}\right) - 1}{\exp\left(\gamma E_{k-j,k} + a_{n,k+1} \frac{\exp(\rho E_{k-j,k})-1}{\rho}\right) - 1}$$

$$= (E_{k-j+1,k} - E_{k-j,k}) \frac{\gamma + a_{n,k+1} \exp(\rho E_{j,k}^*)}{1 - \exp\left(-\gamma E_{j,k}^* + a_{n,k+1} \frac{\exp(\rho E_{j,k}^*)-1}{-\rho}\right)}$$

$$\overset{\mathcal{D}}{=} \frac{E_j}{j} \frac{\gamma + a_{n,k+1} \left(V_{j,k}^*\right)^{-\rho}}{1 - \left(V_{j,k}^*\right)^{\gamma} \exp\left(a_{n,k+1} \frac{\left(V_{j,k}^*\right)^{-\rho}-1}{-\rho}\right)},$$

and hence, after replacing $V_{j,k}^*$ by $\frac{j}{k+1}$, we obtain the following approximate representation for log-ratios of spacings

$$j \log \frac{X_{n-j+1,n} - X_{n-k,n}}{X_{n-j,n} - X_{n-k,n}} \overset{\mathcal{D}}{\approx} \frac{\gamma + a_{n,k+1} \left(\frac{j}{k+1}\right)^{-\rho}}{1 - \left(\frac{j}{k+1}\right)^{\gamma} \exp\left(a_{n,k+1} \frac{\left(\frac{j}{k+1}\right)^{-\rho}-1}{-\rho}\right)} E_j,$$

$$j = 1, \ldots, k-1. \quad (5.34)$$

Note that if $a_{n,k+1} = 0$ model (5.34) reduces to (5.22). The parameters γ, $a_{n,k+1}$ and ρ of model (5.34) can be jointly estimated with the ML method. We denote these ML estimators by $\hat{\gamma}_{k,n}^{\text{RMB}}$, $\hat{a}_{n,k+1}^{\text{RMB}}$ and $\hat{\rho}_{k,n}^{\text{RMB}}$. Since (5.34) is invariant with respect to shifts and rescalings of the data, all estimators based on (5.34) share the same invariance property.

5.7.2   Extreme quantiles and small exceedance probabilities

We continue the discussion of the exponential regression model approach of Matthys and Beirlant (2003). Using (5.33) together with condition (3.14) on ℓ, one easily obtains the following asymptotic representation

$$
U(\tfrac{1}{p}) - X_{n-k,n} \sim c_{n,k+1} \left(\left(\frac{k}{np} \right)^{\gamma} \exp \left(a_{n,k+1} \frac{\left(\frac{k}{np} \right)^{\rho} - 1}{\rho} \right) - 1 \right), \quad (5.35)
$$

where, as before, $c_{n,k+1} := a(U_{k+1,n}^{-1})/d(U_{k+1,n}^{-1})$ and $a_{n,k+1} := ca_2(U_{k+1,n}^{-1})$. Estimating γ, $a_{n,k+1}$ and ρ by respectively $\hat{\gamma}_{k,n}^{\mathrm{RMB}}$, $\hat{a}_{n,k+1}^{\mathrm{RMB}}$ and $\hat{\rho}_{k,n}^{\mathrm{RMB}}$, $c_{n,k+1}$ can be estimated as

$$
\hat{c}_{n,k+1} = \frac{1}{k} \sum_{j=1}^{k} \frac{ j (X_{n-j+1,n} - X_{n-j,n}) \left(\frac{j}{k+1} \right)^{\hat{\gamma}_{k,n}^{\mathrm{RMB}}} }{ \left(\hat{\gamma}_{k,n}^{\mathrm{RMB}} + \hat{a}_{n,k+1}^{\mathrm{RMB}} \left(\frac{j}{k+1} \right)^{-\hat{\rho}_{k,n}^{\mathrm{RMB}}} \right) \exp \left(\hat{a}_{n,k+1}^{\mathrm{RMB}} \frac{ 1 - \left(\frac{j}{k+1} \right)^{-\hat{\rho}_{k,n}^{\mathrm{RMB}}} }{ -\hat{\rho}_{k,n}^{\mathrm{RMB}} } \right) }.
$$

Finally, replacing γ, ρ, $a_{n,k+1}$ and $c_{n,k+1}$ by their respective estimators in (5.35) leads to a bias corrected estimator $\hat{U}_{k,n}^{\mathrm{RMB}}(\tfrac{1}{p})$. As with the refined ML estimator $\hat{\gamma}_{k,n}^{\mathrm{RMB}}$ for γ, $\hat{U}_{k,n}^{\mathrm{RMB}}(\tfrac{1}{p})$ usually succeeds well in reducing the bias of $\hat{U}_{k,n}^{\mathrm{RMA}}(\tfrac{1}{p})$. On the other hand, it has a higher variance and hence is often less optimal in mean squared error (MSE) sense. Note that for a fixed high value of $\hat{U}_{k,n}^{\mathrm{RMB}}(\tfrac{1}{p})$, the above equation can be solved numerically for p, yielding an exceedance probability estimate.

5.8   Adaptive Selection of the Tail Sample Fraction

As we know from the previous chapter, successful practical application of EVI estimators crucially depends on the selection of a good or possibly optimal k-value. We continue the discussion along the lines of section 5.6. In that section, we provided the theoretical optimal k-values for some of the better-known EVI estimators. The optimal k-values clearly depend on the EVI γ and the parameters $b(n)$ and $\tilde{\rho}$ describing the second-order tail behaviour. Replacing these unknown parameters by their respective estimates then yields an estimate for k_{opt}. To this aim, we take a closer look at the regression through the k ultimate points on the generalized quantile plot.

On the basis of condition $(\tilde{C}_{\gamma})$, $F \in \mathcal{D}(G_{\gamma})$ implies $UH \in \mathcal{R}_{\gamma}$ and hence the generalized quantile plot

$$
\left(\log \left(\frac{n+1}{j+1} \right), \log UH_{j,n} \right), \quad j = 1, \ldots, n-1,
$$

will be ultimately linear. Further, the slope of this ultimate linear part approximates γ. Under the Hall (1982) model,

$$U(x) = \begin{cases} Cx^\gamma (1 + Dx^\rho(1 + o(1))) & (x \to \infty) & \text{if } \gamma > 0, \\ x_+ - Cx^\gamma (1 + Dx^\rho(1 + o(1))) & (x \to \infty) & \text{if } \gamma < 0, \end{cases}$$

with $C > 0$ and $D \in \mathbb{R}$, Beirlant *et al.* (2002b) derived the following approximate model:

$$Z_j := (j+1) \log \frac{UH_{j,n}}{UH_{j+1,n}} = \gamma + b(n/k)\left(\frac{j}{k}\right)^{-\tilde{\rho}} + \varepsilon_j, \quad j = 1, \ldots, k,$$

$$(5.36)$$

where ε_j are considered as zero-centred error terms. Ignoring the second term in the right-hand side of (5.36) results in the reduced model

$$(j+1) \log \frac{UH_{j,n}}{UH_{j+1,n}} = \gamma + \varepsilon_j, \quad j = 1, \ldots, k,$$

for which $\hat{\gamma}_{k,n}^H$ is the least-squares estimator. The full model (5.36) can be exploited directly to propose an estimator for γ using a least-squares method, thereby replacing $\tilde{\rho}$ by an estimator $\hat{\tilde{\rho}}$. Beirlant *et al.* (2002c) propose

$$\hat{\tilde{\rho}}_{k,\lambda,n} = -\frac{1}{\log \lambda} \log \frac{\hat{\gamma}_{\lfloor \lambda^2 k \rfloor,n}^H - \hat{\gamma}_{\lfloor \lambda k \rfloor,n}^H}{\hat{\gamma}_{\lfloor \lambda k \rfloor,n}^H - \hat{\gamma}_{k,n}^H}, \quad \lambda \in (0,1),$$

as an appropriate choice, which is a consistent estimator when k is chosen such that $\sqrt{k}b(n/k) \to \infty$. For practical diagnostic purposes it can be sufficient to replace $\tilde{\rho}$ by a canonical choice such as -1. For a given $\hat{\tilde{\rho}}$-value, γ and $b(n/k)$ can be estimated using least squares, resulting in

$$\hat{\gamma}_{\mathrm{LS},k}(\hat{\tilde{\rho}}) = \bar{Z}_k - \hat{b}_{\mathrm{LS},k}(\hat{\tilde{\rho}})/(1 - \hat{\tilde{\rho}}),$$

$$\hat{b}_{\mathrm{LS},k}(\hat{\tilde{\rho}}) = \frac{(1 - \hat{\tilde{\rho}})^2(1 - 2\hat{\tilde{\rho}})}{\hat{\tilde{\rho}}^2} \frac{1}{k} \sum_{j=1}^{k} \left(\left(\frac{j}{k}\right)^{-\hat{\tilde{\rho}}} - \frac{1}{1 - \hat{\tilde{\rho}}}\right) Z_j.$$

These least-squares estimators can now be used to estimate the k_{opt} values as given in section 5.6 in an adaptive way. For brevity, we consider the estimator $\hat{\gamma}_{k,n}^H$ in case $\gamma > 0$. The procedure can, however, be applied without any problem to the other estimators considered in that section. Because of the fact that $a_2 \in \mathcal{R}_\rho$, the AMSE of the simple estimator $\hat{\gamma}_{k,n}^H$ is minimal for

$$k_{opt} \sim \left[b\left(\frac{n}{k_0}\right)\right]^{-2/(1-2\tilde{\rho})} k_0^{-2\tilde{\rho}/(1-2\tilde{\rho})} \left(\frac{(1+\gamma^2)(1-\tilde{\rho})^2}{-2\tilde{\rho}}\right)^{1/(1-2\tilde{\rho})} \quad (5.37)$$

for any secondary value $k_0 \in \{1, \ldots, n-1\}$. Plugging consistent estimators for γ, $b(n/k_0)$ and $\tilde{\rho}$, for instance, the least-squares estimators, into (5.37) yields an estimator for k_{opt}. In this way, for each value of k_0, an estimator of k_{opt} is obtained.

5.9 Appendices

5.9.1 Information matrix for the GEV

Consider a GEV-distributed random variable X. Let $\theta' = (\sigma, \gamma, \mu)$ and denote by g the GEV density function:

$$g(x; \sigma, \gamma, \mu) = \frac{1}{\sigma} \left(1 + \gamma \frac{x - \mu}{\sigma}\right)^{-\frac{1}{\gamma} - 1} \exp\left(-\left(1 + \gamma \frac{x - \mu}{\sigma}\right)^{-\frac{1}{\gamma}}\right).$$

Then the information matrix

$$I(\theta) = -E\left(\frac{\partial^2 \log g(X; \theta)}{\partial \theta \partial \theta'}\right)$$

has as generic elements

$$I_{1,1}(\theta) = \frac{1}{\sigma^2 \gamma^2}\left(1 - 2\Gamma(2 + \gamma) + p\right),$$

$$I_{1,2}(\theta) = -\frac{1}{\sigma \gamma^2}\left(1 - \gamma_* - q + \frac{1 - \Gamma(2 + \gamma)}{\gamma} + \frac{p}{\gamma}\right),$$

$$I_{1,3}(\theta) = -\frac{1}{\sigma^2 \gamma}\left(p - \Gamma(2 + \gamma)\right),$$

$$I_{2,2}(\theta) = \frac{1}{\gamma^2}\left[\frac{\pi^2}{6} + \left(1 - \gamma_* + \frac{1}{\gamma}\right)^2 - \frac{2q}{\gamma} + \frac{p}{\gamma^2}\right],$$

$$I_{2,3}(\theta) = -\frac{1}{\sigma \gamma}\left(q - \frac{p}{\gamma}\right),$$

$$I_{3,3}(\theta) = \frac{p}{\sigma^2},$$

where $\gamma_* = 0.5772157$ is Euler's constant,

$$p = (1 + \gamma)^2 \Gamma(1 + 2\gamma),$$

$$q = \Gamma(2 + \gamma)\left(\psi(1 + \gamma) + \frac{1 + \gamma}{\gamma}\right),$$

with $\psi(x) = d \log \Gamma(x)/dx$.

5.9.2 Point processes

The peaks-over-threshold (POT) method in section 5.3.1 relies on a parametric model for a certain point process. Moreover, point process techniques are useful in inference on multivariate extremes or extremes of time series. For the reader's convenience, we include here a very brief and informal introduction on point processes. For a proper study of the subject, the reader should consult books such as Resnick (1987) or Snyder and Miller (1991), among others.

Let $\{X_i : i \in \mathcal{I}\}$ represent the locations of points, indexed by a set $\mathcal{I}$, occurring randomly in a state space S. A point process N counts the number of points in regions of S:

$$N(A) = \sum_{i \in \mathcal{I}} \mathbf{1}(X_i \in A), \qquad A \subseteq S.$$

The expected number of points in a set A is given by the intensity measure $\Lambda(A) = E[N(A)]$. If the state space S is Euclidean space or a subset thereof and if the intensity measure Λ has a density function $\lambda : S \to [0, \infty)$, that is, if $\Lambda(A) = \int_A \lambda(x)\mathrm{d}x$, then λ is called the *intensity function of the process*.

The most common type of point processes are Poisson processes. A point process N with intensity measure Λ is said to be a Poisson process if the following two conditions are fulfilled: (i) for each set A such that $\Lambda(A) < \infty$ is $N(A)$ a Poisson random variable with mean $\Lambda(A)$; (ii) for all positive integer k and all disjoint sets $A_1, \ldots, A_k$ are the random variables $N(A_1), \ldots, N(A_k)$ independent. A Poisson process on a (subset of) Euclidean space is called *homogenous* if its intensity function λ is constant, $\lambda(x) \equiv \lambda$, and inhomogenous otherwise.

More generally, a marked point process counts for each point X_i a quantity Y_i and has representation

$$N(A) = \sum_{i \in \mathcal{I}} Y_i \mathbf{1}(X_i \in A),$$

the marks $\{Y_i\}_{i \in \mathcal{I}}$ being identically distributed. Observe that a marked point process with all marks equal to unity is simply a point process.

A compound Poisson process is a marked point process for which the points X_i occur according to a Poisson process independently of the marks Y_i, which are themselves independent and identically distributed. We shall denote by $CP(\lambda, \pi)$ a compound Poisson process with intensity function λ and mark distribution π.

A sequence of (marked) point processes N_n on a state space S is said to converge in distribution to a (marked) point process N, notation $N_n \overset{\mathcal{D}}{\to} N$, if for each positive integer k and all sets $A_1, \ldots, A_k$ the vector $(N_n(A_i))_{i=1}^k$ converges in distribution to the vector $(N(A_i))_{i=1}^k$. A typical way to establish convergence of point processes is via convergence of Laplace functionals.

If the intensity function, λ, of a Poisson process N depends on an unknown parameter vector, θ, then we can estimate θ by ML. In order to construct the likelihood, we first have to choose a region A in the sample space such that $\Lambda(A; \theta) < \infty$ for all θ. A crucial property of a Poisson process is now that the points in a region A conditionally on their number $N(A)$ are independent and identically distribution with common density $f(x; \theta) = \lambda(x; \theta)/\Lambda(A; \theta)$. Moreover, $N(A)$ has a Poisson distribution with mean $\Lambda(A; \theta)$. Therefore, if the points falling in A of a realization of N can be enumerated as $x_1, \ldots, x_m$, then the likelihood is

$$L(\theta) = \exp\{-\Lambda(A; \theta)\} \frac{\Lambda(A; \theta)^m}{m!} \prod_{i=1}^{m} \frac{\lambda(x_i; \theta)}{\Lambda(A; \theta)}$$

$$\propto \exp\{-\Lambda(A; \theta)\} \prod_{i=1}^{m} \lambda(x_i; \theta).$$

5.9.3 GRV_2 functions with $\rho < 0$

We restrict to cases where $a_2(x)$ is regularly varying with index $\rho < 0$ and with a slowly varying part being asymptotically equivalent to a constant. Then, without loss of generality, this constant can be set equal to 1.

From Vanroelen (2003), we obtain the following representations of U.

- $0 < -\rho < \gamma$: for $U \in GRV_2(\gamma, \rho; \ell_+ x^\gamma, a_2(x); 0, A)$:

$$U(x) = \ell_+ x^\gamma \left\{ \frac{1}{\gamma} + \frac{A}{\gamma + \rho} a_2(x)(1 + o(1)) \right\},$$

- $\gamma = -\rho$: for $U \in GRV_2(\gamma, -\gamma; \ell_+ x^\gamma, x^{-\gamma} \ell_2(x); 0, A)$ with ℓ_2 some slowly varying function:

$$U(x) = \ell_+ x^\gamma \left\{ \frac{1}{\gamma} + x^{-\gamma} L_2(x) \right\}$$

with $L_2(x) = B + \int_1^x (A + o(1)) \frac{\ell_2(t)}{t} dt + o(\ell_2(t))$ for some constant B,

- $0 < \gamma < -\rho$: for $U \in GRV_2(\gamma, \rho; \ell_+ x^\gamma, a_2(x); 0, A)$:

$$U(x) = \ell_+ x^\gamma \left\{ \frac{1}{\gamma} + D x^{-\gamma} + \frac{A}{\gamma + \rho} a_2(x)(1 + o(1)) \right\},$$

- $\gamma = 0$: for $U \in GRV_2(0, \rho; \ell_+, a_2(x); 0, A)$:

$$U(x) = \ell_+ \log x + D + \frac{A}{\rho} a_2(x)(1 + o(1)),$$

- $\gamma < 0$: for $U \in GRV_2(\gamma, \rho; \ell_+ x^\gamma, a_2(x); 0, A)$:

$$U(x) = U(\infty) - \ell_+ x^\gamma \left\{ \frac{1}{-\gamma} - \frac{A}{\gamma + \rho} a_2(x)(1 + o(1)) \right\},$$

where $\ell_+ > 0$, $A \neq 0$, $D \in \mathbb{R}$.

Concerning $\log U$, the following results are available under these representations:

- If $0 < -\rho < \gamma$ then $\log U \in GRV_2(0, \rho; \gamma, a_2(x); 0, \frac{\rho A}{\gamma + \rho})$;

- If $\gamma = -\rho$ then $\log U \in GRV_2(0, -\gamma; \gamma, x^{-\gamma} L_2(x); 0, -\gamma)$;

- If $0 < \gamma < -\rho$ then $\log U \in GRV_2(0, -\gamma; \gamma, x^{-\gamma}; 0, -\gamma D)$ if $D \neq 0$, and $\log U \in GRV_2(0, \rho; \gamma, a_2(x); 0, \frac{\rho A}{\gamma + \rho})$ if $D = 0$;

- If $\gamma = 0$ then $\log U \in GRV_2(0, 0; \frac{a(x)}{U(x)}, \frac{a(x)}{U(x)}; -1, 0)$;

- If $\gamma < \rho$ then $\log U \in GRV_2(\gamma, \rho; [U(\infty)]^{-1}\ell_+ x^\gamma, a_2(x); 0, A)$;

- If $\rho < \gamma < 0$ then $\log U \in GRV_2(\gamma, \gamma; [U(\infty)]^{-1}\ell_+ x^\gamma, \ell_+ x^\gamma; 0, -\frac{1}{\gamma U(\infty)})$;

- If $\gamma = \rho$ then $\log U \in GRV_2(\gamma, \gamma; [U(\infty)]^{-1}\ell_+ x^\gamma, a_2(x); 0, A - \frac{\ell_+}{\gamma U(\infty)})$.

5.9.4 Asymptotic mean squared errors

In the statement of our results, we will use the following notations:

$$
b(x) = \begin{cases}
\frac{A\rho[\rho+\gamma(1-\rho)]}{(\gamma+\rho)(1-\rho)} a_2(x) & \text{if } 0 < -\rho < \gamma \text{ or if } 0 < \gamma < -\rho \\
& \quad \text{with } D = 0, \\
-\frac{\gamma^3}{(1+\gamma)} x^{-\gamma} L_2(x) & \text{if } \gamma = -\rho, \\
-\frac{\gamma^3 D}{(1+\gamma)} x^{-\gamma} & \text{if } 0 < \gamma < -\rho \text{ with } D \neq 0, \\
\frac{1}{\log^2(x)} & \text{if } \gamma = 0, \\
\frac{A\rho(1-\gamma)}{(1-\gamma-\rho)} a_2(x) & \text{if } \gamma < \rho, \\
-\frac{\gamma}{1-2\gamma} \frac{\ell_+}{U(\infty)} x^\gamma & \text{if } \rho < \gamma < 0, \\
\frac{\gamma}{1-2\gamma}\left[A(1-\gamma) - \frac{\ell_+}{U(\infty)}\right] x^\gamma & \text{if } \gamma = \rho,
\end{cases}
$$

and

$$
\tilde{\rho} = \begin{cases}
-\gamma & \text{if } 0 < \gamma < -\rho \text{ with } D \neq 0, \\
\rho & \text{if } 0 < -\rho \leq \gamma \text{ or if } 0 < \gamma < -\rho \text{ with } D = 0, \\
0 & \text{if } \gamma = 0, \\
\rho & \text{if } \gamma < \rho, \\
\gamma & \text{if } \rho \leq \gamma < 0.
\end{cases}
$$

Below, we derive the AMSEs of the different estimators.

- *For the estimator* $\hat{\gamma}_{k,n}^H$:

$$
AMSE(\hat{\gamma}_{k,n}^H) = \begin{cases}
\frac{1+\gamma^2}{k} + \left(\frac{1}{1-\tilde{\rho}} b\left(\frac{n}{k}\right)\right)^2, & \text{if } \gamma \geq 0, \\
\frac{(1-\gamma)(1+\gamma+2\gamma^2)}{(1-2\gamma)k} + \left(\frac{1}{1-\tilde{\rho}} b\left(\frac{n}{k}\right)\right)^2, & \text{if } \gamma < 0.
\end{cases}
$$

- *For the estimator $\hat{\gamma}_{k,n}^Z$:*

$$
AMSE(\hat{\gamma}_{k,n}^Z) = \begin{cases} \dfrac{2(1+\gamma+\gamma^2)}{k} + \left(\dfrac{1}{(1-\bar{\rho})^2}b\left(\dfrac{n}{k}\right)\right)^2, & \text{if } \gamma \geq 0, \\[3mm] \dfrac{2(1-\gamma)(1+2\gamma+\gamma^2-2\gamma^3)}{(1-2\gamma)(1-\gamma)k} + \left(\dfrac{1}{(1-\bar{\rho})^2}b\left(\dfrac{n}{k}\right)\right)^2, & \text{if } \gamma < 0. \end{cases}
$$

- The AMSE of the moment estimator $M_{k,n}$ can be found from Dekkers *et al.* (1989):

$$
AMSE(M_{k,n}) = \begin{cases} \dfrac{1+\gamma^2}{k} + \left(\dfrac{1}{1-\bar{\rho}}b\left(\dfrac{n}{k}\right)\right)^2, & \text{if } \gamma > 0, \\[3mm] \dfrac{1}{k} + b\left(\dfrac{n}{k}\right), & \text{if } \gamma = 0, \\[3mm] \dfrac{(1-\gamma)^2(1-2\gamma)(6\gamma^2-\gamma+1)}{(1-3\gamma)(1-4\gamma)k} & \text{if } \gamma < \rho, \\[1mm] \quad + \left(\dfrac{1-2\gamma}{1-2\gamma-\bar{\rho}}b\left(\dfrac{n}{k}\right)\right)^2, & \\[3mm] \dfrac{(1-\gamma)^2(1-2\gamma)(6\gamma^2-\gamma+1)}{(1-3\gamma)(1-4\gamma)k} & \text{if } \rho < \gamma < 0, \\[1mm] \quad + \left(\dfrac{1-2\gamma}{\bar{\rho}(1-\bar{\rho})}b\left(\dfrac{n}{k}\right)\right)^2, & \\[3mm] \dfrac{(1-\gamma)^2(1-2\gamma)(6\gamma^2-\gamma+1)}{(1-3\gamma)(1-4\gamma)k} & \text{if } \gamma = \rho. \\[1mm] \quad + \left(\dfrac{(1-2\gamma)}{(1-\gamma)(1-3\gamma)}\dfrac{A(1-\gamma)^2-\frac{2\ell_+}{U(\infty)}}{A(1-\gamma)-\frac{\ell_+}{U(\infty)}}b\left(\dfrac{n}{k}\right)\right)^2, & \end{cases}
$$

- Drees *et al.* (2002) stated the following expressions for the AMSE for the ML estimator based on a generalized Pareto fit:

$$
AMSE\left(\hat{\gamma}_{k,n}^{ML}\right) = \dfrac{(1+\gamma)^2}{k} + \left(\dfrac{\rho(\gamma+1)A}{(1-\rho)(1-\rho+\gamma)}a_2\left(\dfrac{n}{k}\right)\right)^2
$$

$$
\text{if } \gamma > -\tfrac{1}{2}, \rho < 0.
$$

5.9.5 AMSE optimal k-values

Below, the optimal values of k that minimize the different expressions of the AMSEs are given.

- *For the estimator $\hat{\gamma}_{k,n}^H$:*

$$
k_{opt} = \begin{cases} \left(\dfrac{(1+\gamma^2)(1-\bar{\rho})^2}{-2\bar{\rho}}\right)^{\frac{1}{1-2\bar{\rho}}}[b(n)]^{-\frac{2}{1-2\bar{\rho}}}, & \text{if } \gamma > 0, \\[3mm] \tfrac{1}{4}[b(n)]^{-5/2}(1+o(1)), & \text{if } \gamma = 0, \\[3mm] \left(\dfrac{(1+\gamma+2\gamma^2)(1-\bar{\rho})^2(1-\gamma)}{(-2\bar{\rho})(1-2\gamma)}\right)^{\frac{1}{1-2\bar{\rho}}}[b(n)]^{-\frac{2}{1-2\bar{\rho}}}, & \text{if } \gamma < 0, \end{cases}
$$

- *for the estimator* $\hat{\gamma}_{k,n}^Z$:

$$
k_{opt} = \begin{cases}
\left(\dfrac{2(1-\tilde{\rho})^4(1+\gamma+\gamma^2)}{-2\tilde{\rho}}\right)^{\frac{1}{1-2\tilde{\rho}}}[b(n)]^{-\frac{2}{1-2\tilde{\rho}}}, & \text{if } \gamma > 0, \\[3mm]
\frac{1}{2}[b(n)]^{-5/2}(1+o(1)), & \text{if } \gamma = 0, \\[3mm]
\left(\dfrac{2(1-\tilde{\rho})^4(1-\gamma)(1+2\gamma+\gamma^2-2\gamma^3)}{(-2\tilde{\rho})(1-2\gamma)(1-\gamma)}\right)^{\frac{1}{1-2\tilde{\rho}}}[b(n)]^{-\frac{2}{1-2\tilde{\rho}}}, & \text{if } \gamma < 0,
\end{cases}
$$

- *for the estimator* $M_{k,n}$:

$$
k_{opt} = \begin{cases}
\left(\dfrac{(1+\gamma^2)(1-\tilde{\rho})^2}{-2\tilde{\rho}}\right)^{\frac{1}{1-2\tilde{\rho}}}[b(n)]^{-\frac{2}{1-2\tilde{\rho}}}, & \text{if } \gamma > 0, \\[3mm]
\frac{1}{2}[b(n)]^{-3/2}(1+o(1)) & \text{if } \gamma = 0, \\[3mm]
\left(\dfrac{(1-\gamma)^2(1-2\gamma-\tilde{\rho})^2(6\gamma^2-\gamma+1)}{(-2\tilde{\rho})(1-2\gamma)(1-3\gamma)(1-4\gamma)}\right)^{\frac{1}{1-2\tilde{\rho}}}[b(n)]^{-\frac{2}{1-2\tilde{\rho}}}, & \text{if } \gamma < \rho, \\[3mm]
\left(\dfrac{\tilde{\rho}^2(1-\gamma)^4(6\gamma^2-\gamma+1)}{(-2\tilde{\rho})(1-2\gamma)(1-3\gamma)(1-4\gamma)}\right)^{\frac{1}{1-2\tilde{\rho}}}[b(n)]^{-\frac{2}{1-2\tilde{\rho}}}, & \text{if } \rho < \gamma < 0, \\[3mm]
\left(\dfrac{(1-3\gamma)(1-\gamma)^4(6\gamma^2-\gamma+1)}{(-2\tilde{\rho})(1-2\gamma)(1-4\gamma)}\left(\dfrac{A(1-\gamma)-\frac{\ell_+}{U(\infty)}}{A(1-\gamma)^2-\frac{2\ell_+}{U(\infty)}}\right)^2\right)^{\frac{1}{1-2\tilde{\rho}}} \\
\quad [b(n)]^{-\frac{2}{1-2\tilde{\rho}}}, & \text{if } \gamma = \rho,
\end{cases}
$$

- *for the estimator* $\hat{\gamma}_{k,n}^{ML}$:

$$
k_{opt} = \left(\frac{(1-\rho)^2(\gamma-\rho+1)^2}{\rho^2(-2\rho)A^2}\right)^{\frac{1}{1-2\rho}}[a_2(n)]^{-\frac{2}{1-2\rho}} \text{ if } \gamma > -\frac{1}{2}, \rho < 0.
$$

The corresponding minimal AMSE values are then given by

- *for the estimator* $\hat{\gamma}_{k_{opt},n}^H$:

$$
AMSE(\hat{\gamma}_{k_{opt},n}^H) = \begin{cases}
\left(\dfrac{(-2\tilde{\rho})^{\tilde{\rho}}}{(1+\gamma^2)^{\tilde{\rho}}(1-\tilde{\rho})}\right)^{\frac{2}{1-2\tilde{\rho}}}[b(n)]^{\frac{2}{1-2\tilde{\rho}}}(1-2\tilde{\rho}), & \text{if } \gamma > 0, \\[3mm]
b^2(n), & \text{if } \gamma = 0, \\[3mm]
\left(\dfrac{(-2\tilde{\rho})^{\tilde{\rho}}(1-2\gamma)^{\tilde{\rho}}}{(1-\gamma)^{\tilde{\rho}}(1-\tilde{\rho})(1+\gamma+2\gamma^2)^{\tilde{\rho}}}\right)^{\frac{2}{1-2\tilde{\rho}}} & \text{if } \gamma < 0, \\[1mm]
\quad [b(n)]^{\frac{2}{1-2\tilde{\rho}}}(1-2\tilde{\rho}),
\end{cases}
$$

- *for the estimator $\hat{\gamma}^Z_{k_{opt},n}$:*

$$
AMSE(\hat{\gamma}^Z_{k_{opt},n}) = \begin{cases} \left(\dfrac{(-2\tilde{\rho})^{\tilde{\rho}}}{2^{\tilde{\rho}}(1-\tilde{\rho})^2(1+\gamma+\gamma^2)^{\tilde{\rho}}}\right)^{\frac{2}{1-2\tilde{\rho}}} & \text{if } \gamma > 0, \\ [b(n)]^{\frac{2}{1-2\tilde{\rho}}}(1-2\tilde{\rho}), & \\ b^2(n), & \text{if } \gamma = 0, \\ \left(\dfrac{(-2\tilde{\rho})^{\tilde{\rho}}(1-2\gamma)^{\tilde{\rho}}(1-\gamma)^{\tilde{\rho}}}{2^{\tilde{\rho}}(1-\tilde{\rho})^2(1-\gamma)^{\tilde{\rho}}(1+2\gamma+\gamma^2-2\gamma^3)^{\tilde{\rho}}}\right)^{\frac{2}{1-2\tilde{\rho}}} & \text{if } \gamma < 0, \\ [b(n)]^{\frac{2}{1-2\tilde{\rho}}}(1-2\tilde{\rho}), & \end{cases}
$$

- *for the estimator $\hat{\gamma}^M_{k_{opt},n}$:*

$$
AMSE(\hat{\gamma}^M_{k_{opt},n}) =
$$

$$
\begin{cases} \left(\dfrac{(-2\tilde{\rho})^{\tilde{\rho}}}{(1+\gamma^2)^{\tilde{\rho}}(1-\tilde{\rho})}\right)^{\frac{2}{1-2\tilde{\rho}}}[b(n)]^{\frac{2}{1-2\tilde{\rho}}}(1-2\tilde{\rho}), & \text{if } \gamma > 0, \\ b(n), & \text{if } \gamma = 0, \\ \left(\dfrac{(-2\tilde{\rho})^{\tilde{\rho}}(1-\gamma)^{-2\tilde{\rho}}(1-2\gamma)^{1-\tilde{\rho}}(1-3\gamma)^{\tilde{\rho}}(1-4\gamma)^{\tilde{\rho}}}{(1-2\gamma-\tilde{\rho})(6\gamma^2-\gamma+1)^{\tilde{\rho}}}\right)^{\frac{2}{1-2\tilde{\rho}}} & \text{if } \gamma < \rho, \\ [b(n)]^{\frac{2}{1-2\tilde{\rho}}}(1-2\tilde{\rho}), & \\ \left(\dfrac{(-2\tilde{\rho})^{\tilde{\rho}}(1-3\gamma)^{\tilde{\rho}}(1-4\gamma)^{\tilde{\rho}}}{\gamma(1-\gamma)^{1+2\tilde{\rho}}(1-2\gamma)^{\tilde{\rho}-1}(6\gamma^2-\gamma+1)^{\tilde{\rho}}}\right)^{\frac{2}{1-2\tilde{\rho}}} & \text{if } \rho < \gamma < 0, \\ [b(n)]^{\frac{2}{1-2\tilde{\rho}}}(1-2\tilde{\rho}), & \\ \left(\dfrac{(-2\tilde{\rho})^{\tilde{\rho}}(1-3\gamma)^{\tilde{\rho}-1}(1-4\gamma)^{\tilde{\rho}}}{(1-\gamma)^{1+2\tilde{\rho}}(1-2\gamma)^{\tilde{\rho}-1}(6\gamma^2-\gamma+1)^{\tilde{\rho}}}\dfrac{A(1-\gamma)^2-\frac{2\ell_+}{U(\infty)}}{A(1-\gamma)-\frac{\ell_+}{U(\infty)}}\right)^{\frac{2}{1-2\tilde{\rho}}} & \text{if } \gamma = \rho, \\ [b(n)]^{\frac{2}{1-2\tilde{\rho}}}(1-2\tilde{\rho}), & \end{cases}
$$

- *for the estimator $\hat{\gamma}^{ML}_{k_{opt},n}$:*

$$
AMSE(\hat{\gamma}^{ML}_{k_{opt},n}) = \left(\frac{(1+\gamma)^{1-2\rho}(A\rho)(-2\rho)^\rho}{(1-\rho)(1-\rho+\gamma)}\right)^{\frac{2}{1-2\rho}}[a_2(n)]^{\frac{2}{1-2\rho}}(1-2\rho).
$$

6

CASE STUDIES

6.1 The Condroz Data

In this case study, we will concentrate on the Ca content (expressed in mg/100 g of dry soil) of soil samples originating from a particular city (NIS code 61072) in the Condroz region. Although the Ca content is clearly dependent on other factors such as pH level, we ignore this covariate information for the moment and study the univariate properties. Figure 6.1(a) displays a histogram of the Ca measurements of soil samples from this city. When the main interest is in tail modelling, the exponential quantile plot and mean excess plot (which can be considered as a derivative plot of the former) form a good starting point. These plots are given in Figures 6.1(b), (c) and (d). The convex shape of the exponential quantile plot and the increasing behaviour of the mean excess plots in the largest observations give evidence of the HTE nature of the tail of the Ca content distribution.

To assess the fit of a Pareto-type model, a Pareto quantile plot was constructed for these data, given in Figure 6.2. Except for the last seven points, the Pareto quantile plot is linear in the larger observations. The very largest observations that do not follow the ultimate linearity of the Pareto quantile plot are suspect with respect to the Pareto-type model. However, in this analysis, we conditioned on the city but we did not take into account the possible link with other covariates such as pH level. In fact, as can be seen from the Ca versus pH scatterplot given in Figure 6.3, both variables appear to be dependent. Moreover, extreme Ca measurements tend to occur more often at the higher pH levels, indicating the need for a tail analysis conditional on the covariate pH. We will return to this conditioning issue later on in this case study.

As explained in Chapter 2, the ultimate linearity of the Pareto quantile plot can be exploited to construct estimators for the tail index γ. In Figure 6.4(a), we show the results of the maximum likelihood procedure applied to the exponential

Statistics of Extremes: Theory and Applications   J. Beirlant, Y. Goegebeur, J. Segers, and J. Teugels
© 2004 John Wiley & Sons, Ltd   ISBN: 0-471-97647-4

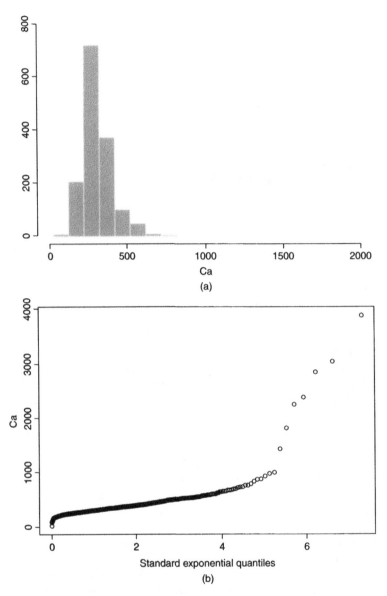

Figure 6.1   (a) Histogram, (b) exponential quantile plot, (c) $e_{k,n}$ versus k and (d) $e_{k,n}$ versus $x_{n-k,n}$ for the Ca measurements.

regression model, $\hat{\gamma}_{ML}^{+}$ (solid line), and the Hill estimates, $H_{k,n}$ (broken line), as a function of k. The maximum likelihood estimates $\hat{\gamma}_{ML}^{+}$ are stable around the value 0.26 and this for k values between 450 and 1500, whereas the Hill estimates show this stability only for k values between 250 and 500. However, the

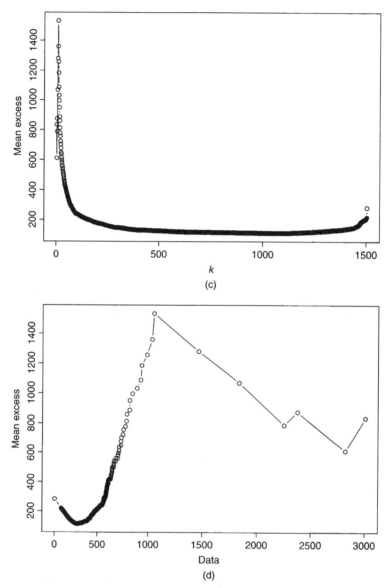

(c)

(d)

maximum likelihood estimator seems to be more sensitive to the seven observations considered as suspect with respect to the Pareto-type model: around $k = 400$, there is an abrupt shift of the optimum found by the ML algorithm. The selection of an optimal k value for the Hill estimator is illustrated in Figure 6.4(b) where we plot the estimated asymptotic mean squared error as a function of k. The minimum is reached at $\hat{k}_{opt} = 402$. The vertical reference line in Figure 6.4(a) and

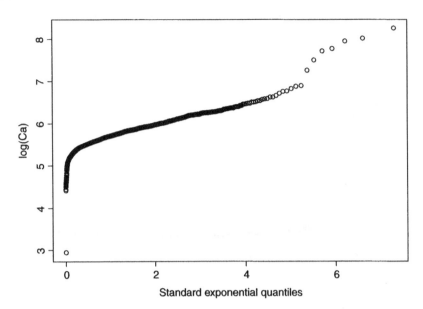

Figure 6.2   Pareto quantile plot of the Ca content measurements.

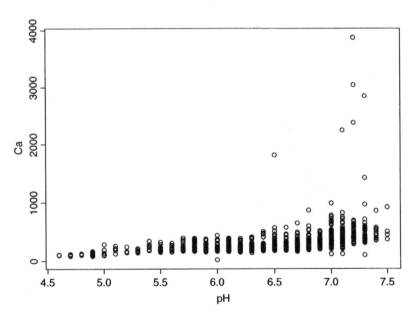

Figure 6.3   Scatterplot of Ca versus pH.

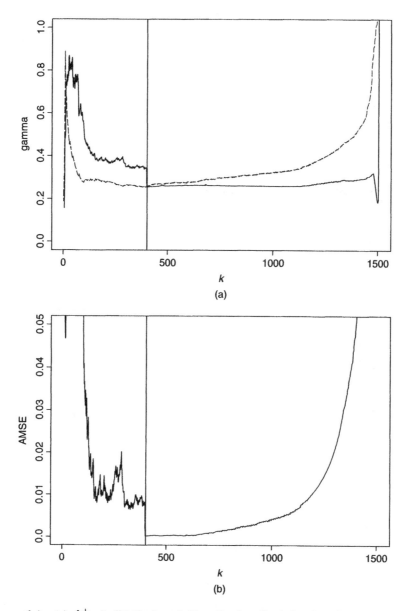

Figure 6.4 (a) $\hat{\gamma}_{\mathrm{ML}}^{+}$ (solid line) and $H_{k,n}$ (broken line) for $k = 5, \ldots, 1504$, (b) $\widehat{AMSE}\,(H_{k,n})$ as a function of k, (c) $\hat{\gamma}_{\mathrm{ML}}^{+}$ (solid line) and $H_{k,n}$ (broken line) and (d) $\widehat{AMSE}\,(H_{k,n})$ as a function of k after removal of the suspect points.

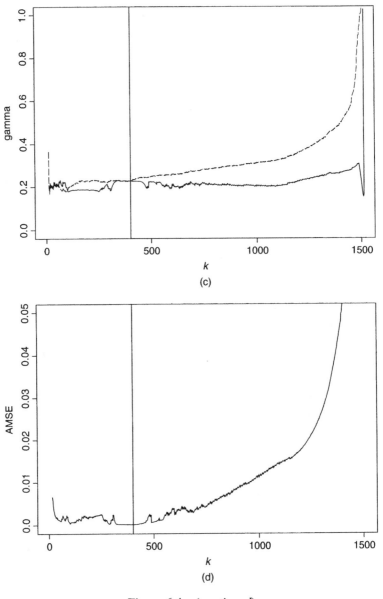

Figure 6.4   (*continued*)

(b) represents this estimated optimal k-value. Note that beyond $\hat{k}_{opt}$, the Hill estimator diverges from the maximum likelihood estimator. In Goegebeur *et al.* (2004), Burr regression models were fitted to these calcium measurements, thereby taking the pH level as covariate. Their analysis identified six points as suspect with respect to the conditional Burr model. Figure 6.4(c) shows $\hat{\gamma}_{ML}^{+}$ and $H_{k,n}$, obtained after deletion of these suspect points as a function of k. The corresponding estimated *AMSE* of $H_{k,n}$ is given in Figure 6.4(d); here the optimum is reached at $k = 391$. Finally, the bias-corrected estimator $\hat{q}_{k,p}^{(1)}$ (solid line) and the Weissman estimator $\hat{q}_{k,p}^{+}$ (broken line) for $Q(0.9995)$ are given as a function of k in Figure 6.5.

We now analyse the data using the extreme value techniques developed for the general case $\gamma \in \mathbb{R}$. Figure 6.6(a) gives the generalized quantile plot ($\log \frac{n+1}{k+1}$, $\log UH_{k,n}$), $k = 1, \ldots, n-1$, for the Ca measurements. The ultimate linear and increasing appearance of the points on this generalized quantile plot gives again evidence in favour of a Pareto-type model. Further, following the discussion given in Chapter 5, the slope of the ultimate linear part of this plot can again be used to construct estimators for γ. The generalized Hill estimator $\hat{\gamma}_{k,n}^{H}$ and the Zipf estimator $\hat{\gamma}_{k,n}^{Z}$, both slope estimators, exploit this ultimate linearity and are plotted in Figure 6.6(b). Note that $\hat{\gamma}_{k,n}^{Z}$ (broken line) gives somewhat higher values for γ than $\hat{\gamma}_{k,n}^{H}$ (solid line). Figure 6.6(b) also shows the moment estimates $M_{k,n}$ (dotted line) and the GP maximum likelihood estimates $\hat{\gamma}_{k,n}^{ML}$ (broken-dotted line). Unlike the plot of $H_{k,n}$ and $\hat{\gamma}_{ML}^{+}$, Figure 6.6(b) does not really show a stable region, making inference about the value for γ more difficult. Finally, the estimation of extreme quantiles on the basis of the GP distribution is illustrated in Figure 6.5, where, next to $\hat{q}_{k,p}^{(1)}$ and $\hat{q}_{k,p}^{+}$, we also show the GP estimate for $Q(0.9995)$ (broken-dotted line).

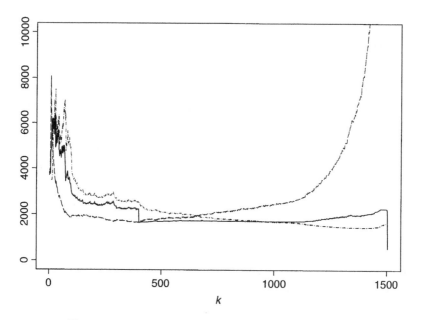

Figure 6.5 $\hat{q}_{k,p}^{(1)}$ (solid line), $\hat{q}_{k,p}^{+}$ (broken line) and $\hat{q}_{POT}$ (broken-dotted line) as a function of k for $p = 0.0005$.

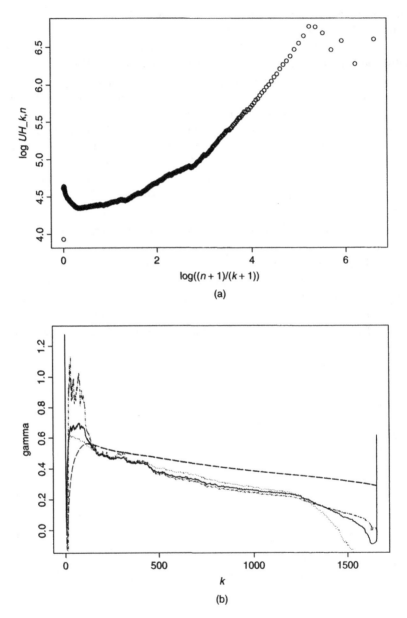

Figure 6.6   (a) Generalized quantile plot of the Ca measurements and (b) $\hat{\gamma}_{k,n}^{H}$ (solid line), $\hat{\gamma}_{k,n}^{Z}$ (broken line), $\hat{\gamma}_{k,n}^{ML}$ (broken-dotted line) and $M_{k,n}$ (dotted line) as a function of (a) k.

We now return to the conditioning issue. Although the extreme value methods especially developed for regression problems will be described extensively in Chapter 7, some straightforward analyses can be performed on the basis of the univariate extreme value methodology discussed so far. First, the fit of a Pareto-type model to the conditional distribution of the dependent variable, given the covariate(s), can be visually assessed by inspection of Pareto quantile plots of the response measurements within narrow bins in the covariate space. Of course, such a procedure based on binning can only be expected to perform well when the conditional distribution of the dependent variable varies smoothly as a function of the covariate(s). This is illustrated in Figure 6.7. Given the discrete nature of the covariate pH, binning is not really necessary here, and all response observations at a particular pH level can be used. The ultimate linearity of the Pareto quantile plots indicates that Pareto-type models provide appropriate fits to the conditional distributions of Calcium, given pH level. Note, however, that the largest observation in Figure 6.7(b) does not follow the linear pattern set by the other large observations. So, even compared to a heavy-tailed model, this point is suspicious and requires special attention.

The tail heaviness of the response distribution conditional on the covariate information can be estimated similarly, using all response observations within a narrow bin in the covariate space. Figure 6.8 shows the different γ estimates as functions of k for the calcium measurements at pH $= 7$, see also the Pareto quantile plot in Figure 6.7(c). Especially note that, compared to the other tail index estimators, $\hat{\gamma}_{\mathrm{ML}}^{+}$ is stable over the whole set of k values.

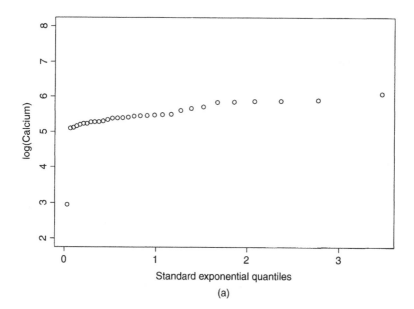

(a)

Figure 6.7   Pareto quantile plots of the calcium measurements at (a) pH $= 6$, (b) pH $= 6.5$ and (c) pH $= 7$.

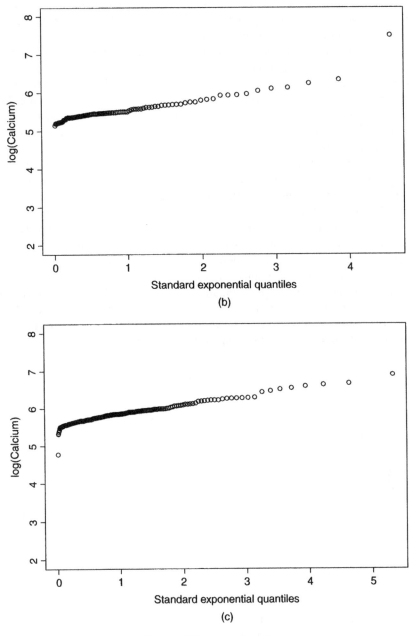

Figure 6.7   (*continued*)

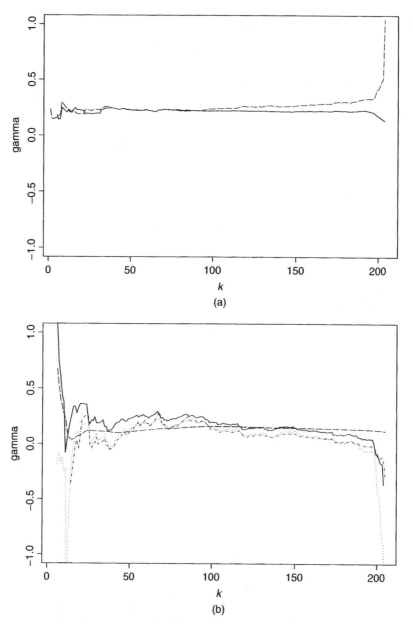

Figure 6.8   Conditional tail index estimation for the Condroz data: (a) $\hat{\gamma}_{\mathrm{ML}}^{+}$ (solid line) and $H_{k,n}$ (broken line) and (b) $\hat{\gamma}_{k,n}^{H}$ (solid line), $\hat{\gamma}_{k,n}^{Z}$ (broken line), $\hat{\gamma}_{k,n}^{\mathrm{ML}}$ (broken-dotted line) and $M_{k,n}$ (dotted line) as a function of k, for $k = 5, \ldots, 203$, using the calcium measurements at pH=7.

6.2 The Secura Belgian Re Data

The Secura Belgian Re data set contains automobile claims from 1988 until 2001, which are at least as large as 1,200,000 Euro. The original claim numbers were corrected, among others, for inflation. This data set contains $n = 371$ observations and is depicted in Figure 6.9. The ultimate goal of this case study is to provide the participating reinsurance companies with an objective statistical analysis in order to assist in the pricing of the unlimited excess-loss layer above an operational priority R. The analysis performed here is based on the methodology described in Beirlant *et al.* (2001).

In an excess-of-loss (XL) reinsurance contract, the reinsurer pays for the claim amount in excess over a given limit. Formally, let X denote the claim size, then, under an XL reinsurance contract with retention level R, the intervention of the reinsurer concerns the random amount $(X - R)_+$. Hence, the net premium $\Pi(R)$ is given by

$$\Pi(R) = E((X - R)_+) = \int_R^{x_*} (1 - F(y))dy.$$

An important ingredient for establishing the net premium is the mean excess function. Indeed, since

$$e(R) = \frac{\int_R^{x_*} (1 - F(y))dy}{1 - F(R)}$$

we have

$$\Pi(R) = e(R)\bar{F}(R).$$

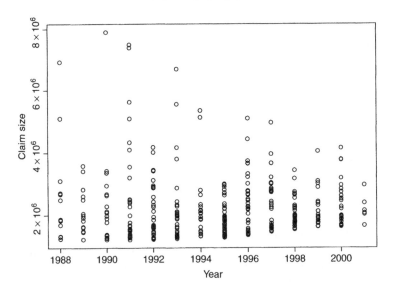

Figure 6.9   Secura data: claim sizes as a function of the year of occurrence.

Special emphasis will be put on the level $R = 5,000,000$ Euro, which is the priority used in practice up to 2003. Remark that only 12 observations are larger than that level. In order to estimate $\Pi(R)$, several possibilities are at our disposal: purely non-parametric methods, semi-parametric methods given by extreme value techniques and fully parametric models where the emphasis lies in trying to model the whole outcome set from 1,200,000 Euro. In contrast to this, extreme value methods will try to fit the tail of the distribution exclusively, from an appropriate (statistical) threshold. Next to the estimation of a net premium, one also needs to estimate the probability for a claim to fall in the layer above R.

6.2.1   The non-parametric approach

Given the importance of the mean excess function for premium calculations, we examine the exponential quantile and mean excess plots first. These are given in Figure 6.10. From the exponential quantile plot, a point of inflection with different slopes to the left and to the right can be detected. This becomes even more apparent in the mean excess plot (Figure 6.10(b) and (c)): behind 2,500,000 the rather horizontal behaviour changes into a positive slope.

Of course, the simplest way to estimate the net premium $\Pi(R)$ is given by

$$\frac{1}{n}\sum_{i=1}^{n}(X_i - R)_+, \qquad\qquad (6.1)$$

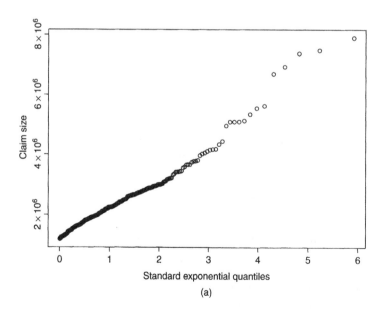

(a)

Figure 6.10   (a) Exponential quantile plot of claim sizes, (b) $e_{k,n}$ versus k and (c) $e_{k,n}$ versus $x_{n-k,n}$.

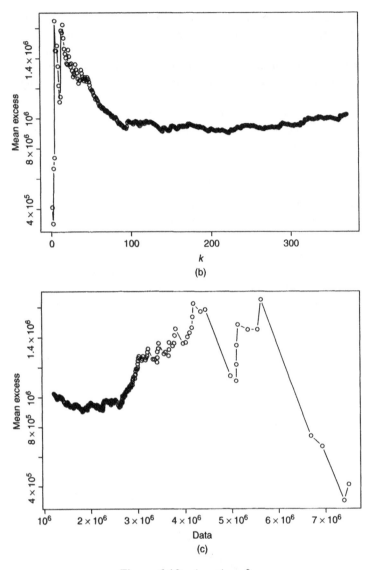

Figure 6.10 (*continued*)

or equivalently, in terms of the mean excess and empirical distribution function

$$\hat{\Pi}(R) = \hat{\bar{F}}_n(R)\hat{e}_n(R).$$

In case R is fixed at one of the sample points, that is, $R = X_{n-k,n}$, the non-parametric estimator is given by

$$\hat{\Pi}(X_{n-k,n}) = \frac{k}{n}E_{k,n}. \tag{6.2}$$

Table 6.1 Non-parametric, Hill and GP-based estimates for $\Pi(R)$.

R	Non-parametric (6.1)	Hill (6.4)	GP (6.5)
3,000,000	161,728.1	163,367.4	166,619.6
3,500,000	108,837.2	108,227.2	111,610.4
4,000,000	74,696.3	75,581.4	79,219.0
4,500,000	53,312.3	55,065.8	58,714.1
5,000,000	35,888.0	41,481,6	45,001.6
7,500,000	-	13,944.5	16,393.3
10,000,000	-	6,434.0	8,087.8

For large R values, this simple non-parametric estimator is of course doubtful because of the small number of observations on which it is effectively constructed. Table 6.1 gives the non-parametric premium estimator (6.1) for some values of R.

6.2.2 Pareto-type modelling

We now further investigate the tail behaviour of the claim size distribution. The Pareto quantile plot given in Figure 6.11 is approximately linear in the largest observations, indicating a good fit of the Pareto model to the tail of the claim size distribution, though at the highest observations, the trend flattens out. Again, the tail index γ can be estimated by measuring the slope of this ultimate linear part. Figure 6.12(a) shows the exponential regression model-based maximum likelihood

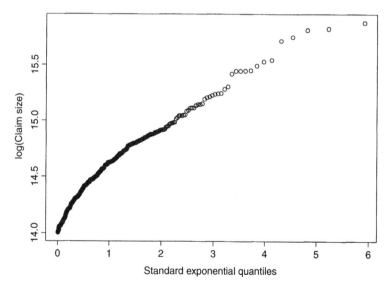

Figure 6.11 Pareto quantile plot.

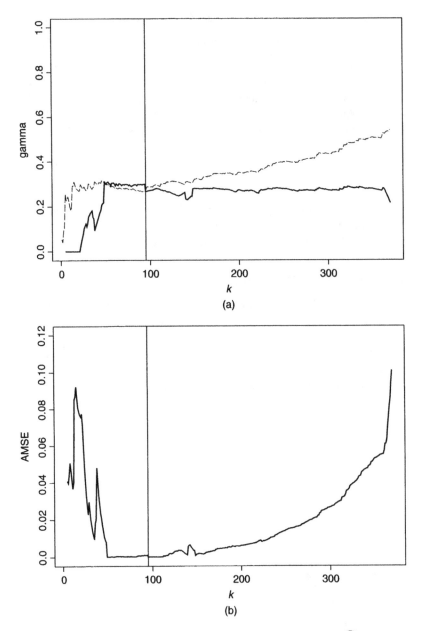

Figure 6.12   (a) $\hat{\gamma}^{+}_{\mathrm{ML}}$ (solid line) and $H_{k,n}$ (broken line), (b) $\widehat{AMSE}(H_{k,n})$ as a function of k with $\hat{k}_{opt} = 95$, (c) Weissman and empirical estimate for $P(X > 5,000,000)$ and (d) estimates for $Q(0.999)$: $\hat{q}^{(1)}_{k,p}$ (solid line) and $\hat{q}^{+}_{k,p}$ (broken line).

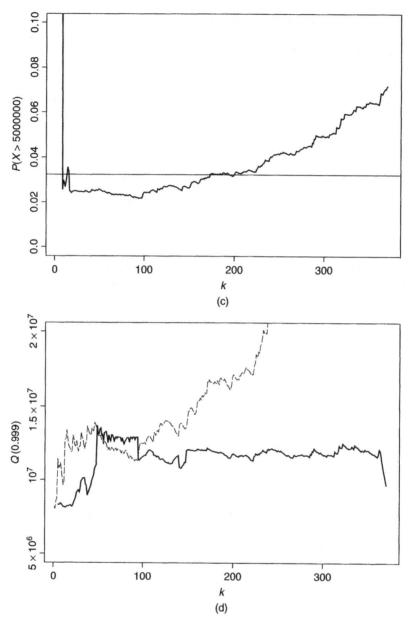

(c)

(d)

estimates $\hat{\gamma}_{\mathrm{ML}}^{+}$ and the Hill estimates $H_{k,n}$ as a function of k. The vertical reference line at $k = 95$ represents the estimated optimal k-value, in the sense of minimum asymptotic mean squared error, for the Hill estimator; see Figure 6.12(b). Note that $\hat{\gamma}_{\mathrm{ML}}^{+}$ and $H_{k,n}$ are almost indistinguishable for k-values between 50 and 100; beyond this interval, the bias of the Hill estimator becomes important while the maximum likelihood estimator remains stable. Next to premium estimation, special attention has to be paid to estimating the probability of exceeding the retention $R = 5,000,000$. The Weissman estimate for $P(X > 5,000,000)$ is given as a function of k in Figure 6.12(c). The horizontal reference line is the empirical exceedance probability, that is, 12/371. Finally, Figure 6.12(d) contains the bias-corrected estimate $\hat{q}_{k,p}^{(1)}$ (solid line) and the Weissman estimate $\hat{q}_{k,p}^{+}$ (broken line) for the 0.999 quantile.

We now turn to XL rating under the Pareto-type model. Recall the basic formula for calculating the net premium of an XL contract is

$$\Pi(R) = e(R)\bar{F}(R)$$

or, after dividing and multiplying the right-hand side by R,

$$\Pi(R) = \frac{e(R)}{R} R\bar{F}(R).$$

For Pareto-type models with $\gamma < 1$, application of Karamata's theorem (Theorem 2.3) gives

$$\frac{e(R)}{R} = \frac{\int_R^\infty u^{-\frac{1}{\gamma}} \ell_F(u)du}{R^{-\frac{1}{\gamma}+1}\ell_F(R)} \rightarrow \frac{1}{\frac{1}{\gamma}-1} \text{ as } R \rightarrow \infty,$$

so that

$$\Pi(R) \sim \frac{1}{\frac{1}{\gamma}-1} R\bar{F}(R), \quad R \rightarrow \infty.$$

When the priority R is situated within the sample, that is, $R = X_{n-k,n}$, the net premium $\Pi(R)$ can be estimated by

$$\hat{\Pi}(R) = \frac{1}{\frac{1}{\tilde{\gamma}_k}-1} X_{n-k,n} \frac{k}{n} \tag{6.3}$$

where $\tilde{\gamma}_k$ denotes an estimator for γ based on k upper order statistics. If R is not fixed at one of the sample points, extreme value formulas can be used to estimate $\bar{F}(R)$. Indeed, for Pareto-type models

$$P\left(\frac{X}{t} > x | X > t\right) \sim x^{-\frac{1}{\gamma}} \text{ as } t \rightarrow \infty$$

so that

$$\bar{F}(tx) \sim \bar{F}(t)x^{-\frac{1}{\gamma}}$$

or, replacing $R = tx$,

$$\bar{F}(R) \sim \bar{F}(t)\left(\frac{R}{t}\right)^{-\frac{1}{\gamma}} \quad R > t.$$

Let $\hat{k}$ denote an appropriate adaptive choice for the number of extreme order statistics and set $t = X_{n-\hat{k},n}$, then $\Pi(R)$ can be estimated as

$$\hat{\Pi}(R) = \frac{1}{\frac{1}{\hat{\gamma}_{\hat{k}}} - 1} R \frac{\hat{k}}{n} \left(\frac{R}{X_{n-\hat{k},n}} \right)^{-\frac{1}{\hat{\gamma}_{\hat{k}}}}. \tag{6.4}$$

In Table 6.1, we illustrate the use of (6.4) for some values of R.

6.2.3   Alternative extreme value methods

In this section, we apply the extreme value methodology developed for the general case $\gamma \in \mathbb{R}$ to this data set. As a first step, we compare several tail index estimators; next we discuss net premium estimation.

First, consider the estimation of the tail parameter γ. In this respect, the generalized quantile plot is a good starting point as the pattern formed by the $UH_{k,n}$ for small k values gives an indication about the tail behaviour. For the Secura data, the generalized quantile plot is given in Figure 6.13(a). The ultimate linear and

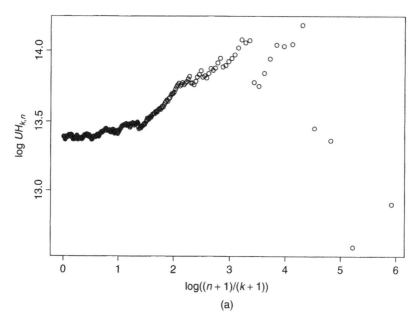

(a)

Figure 6.13   (a) Generalized quantile plot, (b) $\hat{\gamma}_{k,n}^{H}$ (solid line), $\hat{\gamma}_{k,n}^{Z}$ (broken line), $\hat{\gamma}_{k,n}^{ML}$ (broken-dotted line) and $M_{k,n}$ (dotted line) as a function of k, (c) POT (broken line), Weissman (solid line) and empirical estimate for $P(X > 5,000,000)$ and (d) estimates for $Q(0.999)$: POT-based (broken-dotted line), $\hat{q}_{k,p}^{(1)}$ (solid line) and $\hat{q}_{k,p}^{+}$ (broken line).

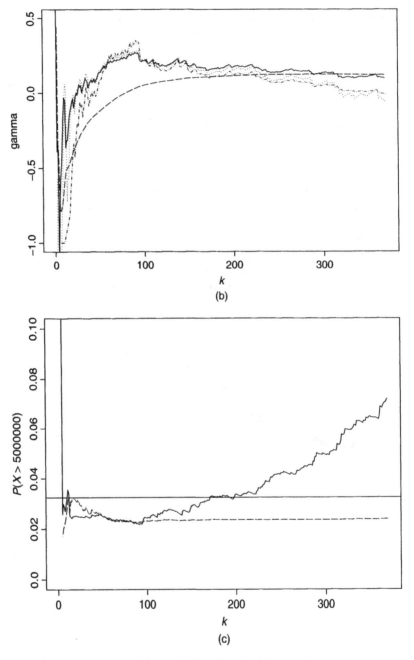

Figure 6.13   (*continued*)

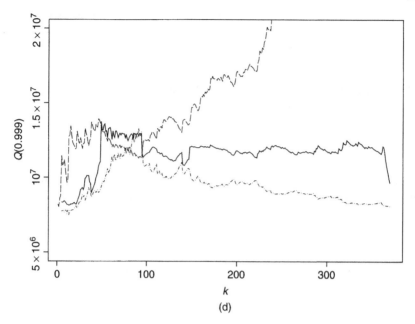

(d)

increasing behaviour indicates a heavy-tailed or Pareto-type model for the claim size distribution, which is in line with the previous analysis. Figure 6.13(b) shows $\hat{\gamma}_{k,n}^{H}$ (solid line), $\hat{\gamma}_{k,n}^{Z}$ (broken line), $\hat{\gamma}_{k,n}^{ML}$ (broken-dotted line) and $M_{k,n}$ as a function of k. The estimation of $P(X > 5, 000, 000)$ is illustrated in Figure 6.13(c) using the Weissman estimates (solid line), the GP approach (broken line) and the empirical exceedance probability (horizontal reference line). Finally, Figure 6.13(d) displays some estimates for $Q(0.999)$.

We now consider net premium calculations on the basis of the GP tail fit. Since

$$\bar{F}(x) \sim \bar{F}(u) \left(1 + \gamma \frac{x - u}{\sigma}\right)^{-\frac{1}{\gamma}}, \quad u \to \infty,$$

we have that for u sufficiently large, provided $\gamma < 1$,

$$e(R) \sim \frac{\sigma}{1 - \gamma} \left(1 + \gamma \frac{R - u}{\sigma}\right), \quad R > u.$$

Setting $u = X_{n-\hat{k},n}$, where $\hat{k}$ denotes again an appropriate choice for the number of upper order statistics, Π can be estimated as

$$\hat{\Pi}(R) = \frac{\hat{k}}{n} \frac{\tilde{\sigma}_{\hat{k}}}{1 - \tilde{\gamma}_{\hat{k}}} \left(1 + \tilde{\gamma}_{\hat{k}} \frac{R - X_{n-\hat{k},n}}{\tilde{\sigma}_{\hat{k}}}\right)^{-\frac{1}{\tilde{\gamma}_{\hat{k}}}+1} \tag{6.5}$$

for some $\tilde{\sigma}_{\hat{k}}$ and $\tilde{\gamma}_{\hat{k}}$. In Table 6.1, we illustrate the use of (6.5) for premium calculations using the Secura data. Note that the net premiums obtained with the three estimators agree quite well. Compared to the estimates based on extreme value methodology, the simple non-parametric estimator clearly does not yield sensible results in case R is outside (or near the end of) the data range.

6.2.4 Mixture modelling of claim sizes

In the previous sections, we discussed XL rating using Pareto-type and GP modelling of the tail of the claim size distributions. This resulted in estimators for $\Pi(R)$ in case R exceeds some sufficiently high threshold $x_{n-\hat{k},n}$. When trying to estimate $\Pi(R)$ for values of R smaller than the threshold $x_{n-\hat{k},n}$, one needs a global statistical model describing the whole range of the possible claim outcomes. The mean excess function given in Figure 6.10(c) suggests a mixture of an exponential and a Pareto distribution:

$$\hat{\bar{F}}_{\text{Exp–Par}}(u) =$$

$$\begin{cases} 1 - \dfrac{n-\hat{k}}{n} \dfrac{1-\exp(-\hat{\lambda}(u-1,200,000))}{1-\exp(-\hat{\lambda}(X_{n-\hat{k},n}-1,200,000))} & \text{if } 1,200,000 < u < X_{n-\hat{k},n}, \\[2ex] \dfrac{\hat{k}}{n}\left(\dfrac{u}{X_{n-\hat{k},n}}\right)^{-1/\hat{\gamma}} & \text{if } u > X_{n-\hat{k},n}, \end{cases}$$

with $\hat{k} = 95$ and $\hat{\lambda} = 1/955,676.55$. In Figure 6.14(a), we plot the empirical distribution function (solid line) together with the fitted Exp-Par mixture model (broken line). As is clear from this plot, the Exp-Par mixture model describes the data quite well. The fit of the Exp-Par mixture model can be further assessed by transforming the data to the Exp(1) framework as follows:

$$E_i = -\log(1 - \hat{\bar{F}}_{\text{Exp–Par}}(X_i)), \ i = 1, \ldots, n, \tag{6.6}$$

followed by a visual inspection of the exponential quantile plot, see Figure 6.14(b). The use of the Exp-Par model for premium computations is illustrated in Table 6.2.

Table 6.2 $\hat{\Pi}(R)$ based on the Exp-Par mixture model.

R	$\hat{\Pi}(R)$
1,250,000	944,217.8
1,500,000	734,371.6
1,750,000	571,314.1
2,000,000	444,275.5
2,250,000	344,965.2
2,500,000	267,000.7

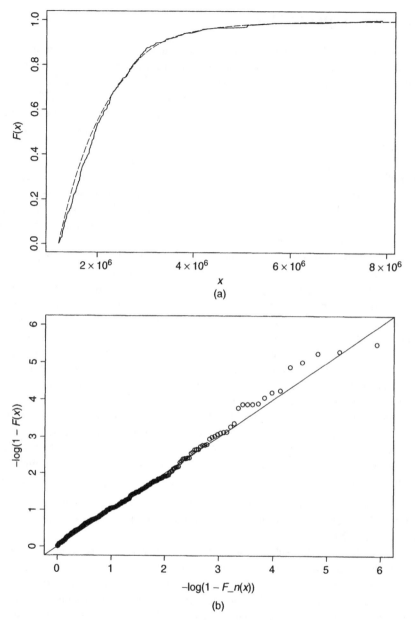

Figure 6.14 (a) Empirical cdf (solid line) and fitted Exp-Par mixture model (broken line), (b) W-plot for the fitted Exp-Par mixture model.

6.3   Earthquake Data

As a third case study, we analyse the earthquake data introduced in Pisarenko and Sornette (2003). This data set is extracted from the Harvard catalog and contains information about the seismic moment (in dyne-cm) of shallow earthquakes (dept < 70 km) over the period from 1977 to 2000. In Pisarenko and Sornette (2003), the tails of the seismic moment distributions for subduction and midocean ridge zones are compared by fitting the GP distribution to seismic moment exceedances over 10^{24} dyne-cm.

The exploratory analysis described in Chapter 1 (Figure 1.17) already indicated for both subduction and midocean ridge zones a *HTE* behaviour of the seismic moment distribution. This is further confirmed by the Pareto quantile plots shown in Figure 6.15. Note, however, that the Pareto quantile plots bend down at the very largest observations, indicating a weaker behaviour of the ultimate tail of the seismic moment distribution.

In Figure 6.16(a) and (b), we show the maximum likelihood estimates $\hat{\gamma}_{ML}^{+}$ (solid line) and the Hill estimates $H_{k,n}$ (broken line) as a function of k for subduction and midocean ridge zones respectively. The subduction zone seismic moment distribution is clearly heavier-tailed than the midocean ridge distribution, a result that is consistent with the analysis performed in Pisarenko and Sornette (2003). Note that only at the very smallest k values $H_{k,n}$ and $\hat{\gamma}_{ML}^{+}$ agree quite well. Beyond these small k values, both estimates tend to increase as a function of k, albeit at a different rate. The selection of an optimal k value for the Hill estimator is illustrated in Figure 6.16(c) and (d) where we plot the estimated asymptotic mean squared errors as a function of k. Imposing the restriction that k should be at least 20, the minimum is reached at $\hat{k}_{opt} = 1157$ for subduction zones and at $\hat{k}_{opt} = 58$ for midocean ridge zones. The vertical reference lines in Figure 6.16 represent these estimated optimal k values. The use of these estimated optimal k values is further illustrated on the Pareto quantile plots given in Figure 6.15 by superimposing the lines through $(\log \frac{n_j+1}{\hat{k}_{opt}^{(j)}+1}, \log x_{n_j-\hat{k}_{opt}^{(j)},n_j}^{(j)})$ with slopes $H_{\hat{k}_{opt}^{(j)},n_j}$, $j = 1, 2$; where $j = 1$ refers to subduction zones and $j = 2$ refers to midocean ridge zones. The horizontal reference lines in Figure 6.15 represent the threshold used in Pisarenko and Sornette (2003).

So far, the data for subduction and midocean ridge zones were considered independently of each other. However, as described in Beirlant and Goegebeur (2004b), combining data originating from several independent data groups may result in improved efficiency. Of course, regression models with dummy explanatory variables describing the groups can be used in combination with classical extreme value models such as the GEV or GP. This regression approach will be further developed in Chapter 7. In this section, we concentrate on a straightforward extension of the exponential regression model for log-spacings of successive order statistics introduced in Chapter 4.

Consider independent and identically distributed positive random variables $X_1^{(j)}, \ldots, X_{n_j}^{(j)}$ with a common distribution function $F_{X^{(j)}}$, $j = 1, \ldots, G$, where

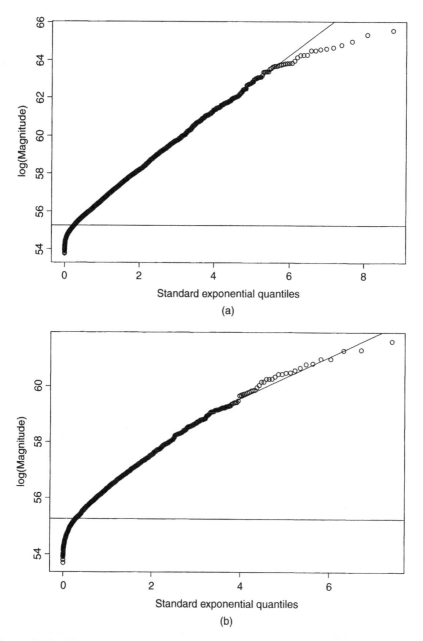

Figure 6.15   Pisarenko and Sornette data: Pareto quantile plot of seismic moment measurements of (a) subduction zones and (b) midocean ridge zones.

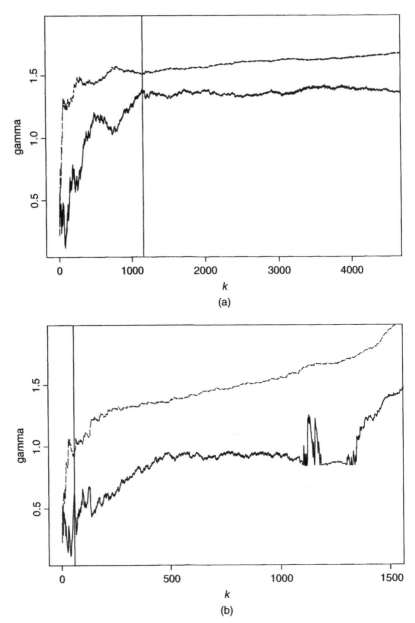

Figure 6.16 Pisarenko and Sornette data: (a) and (b) $\hat{\gamma}_{\mathrm{ML}}^{+}$ (solid line) and $H_{k,n}$ (broken line) as a function of k for subduction and midocean ridge zones respectively and (c) and (d) $\widehat{AMSE}(H_{k,n})$ as a function of k for subduction and midocean ridge zones respectively.

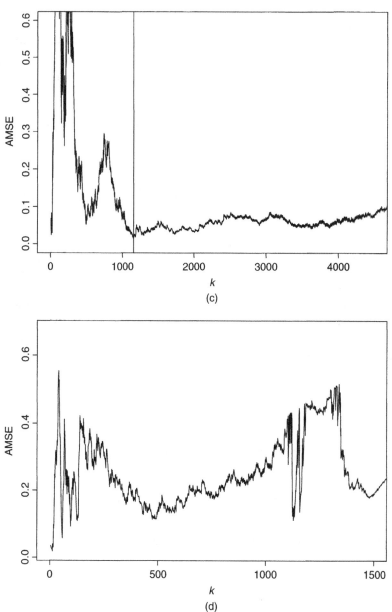

(c)

(d)

G denotes the number of groups. Assume further that the G groups are independent of each other and that the response distributions are of Pareto-type, that is, the tail quantile functions $U_{X^{(j)}}$, $j = 1, \ldots, G$, satisfy

$$U_{X^{(j)}}(x) = x^{\gamma_j} \ell_j(x) \qquad x > 1;\ \gamma_j > 0 \qquad (6.7)$$

where γ_j and ℓ_j denote the extreme value index and the slowly varying function of group j respectively.

As in a classical one-way ANOVA situation, we introduce the parametrization $\gamma_j = \beta_0 + \beta_j$, $j = 1, \ldots, G$, with $\sum_{j=1}^{G} \beta_j = 0$, so that the parameters β_j denote the difference of the extreme value index of group j with respect to the global average overall groups. This transformation will now be combined with the following linear model describing the estimation problem of every γ_j, $j = 1, \ldots, G$.

Under the second order condition (3.14) on the ℓ_j, $j = 1, \ldots, G$, it can be shown as in Beirlant et $al.$ (1999) that the following regression model holds approximately

$$i(\log X^{(j)}_{n_j - i + 1, n_j} - \log X^{(j)}_{n_j - i, n_j}) \approx \left(\gamma_j + b_j \left(\frac{n_j + 1}{k + 1} \right) \left(\frac{i}{k + 1} \right)^{\tau_j} \right) F^{(j)}_i$$
$$i = 1, \ldots, k, \qquad (6.8)$$

with b_j and τ_j denoting the function b and the parameter τ respectively, of group j and the $F^{(j)}_i$, $i = 1, \ldots, k$, independent standard exponential random variables.

The classical way to estimate the parameters γ_j, $j = 1, \ldots, G$ is then given by the Hill (1975) estimates that are obtained as maximum likelihood estimates by omitting the terms $b_j(\frac{n_j+1}{k+1})(\frac{i}{k+1})^{\tau_j}$ in model (6.8) (these terms tend to 0 as $n_j \to \infty$ and $k/n_j \to 0$) leading to a simple average of the scaled log-spacings $i(\log X^{(j)}_{n_j - i + 1, n_j} - \log X^{(j)}_{n_j - i, n_j})$, $i = 1, \ldots, k$, as an estimator of γ_j, and hence

$$\hat{\beta}_0 = \frac{1}{G} \sum_{j=1}^{G} H^{(j)}_{k, n_j} \quad \text{and} \quad \hat{\beta}_j = H^{(j)}_{k, n_j} - \hat{\beta}_0, \quad j = 1, \ldots, G, \qquad (6.9)$$

in which $H^{(j)}_{k, n_j}$ denotes the Hill estimator for group j

$$H^{(j)}_{k, n_j} = \frac{1}{k} \sum_{i=1}^{k} \log X^{(j)}_{n_j - i + 1, n_j} - \log X^{(j)}_{n_j - k, n_j}. \qquad (6.10)$$

Introducing $\Lambda = $ Block-diag$(\gamma_j^2 I_k; \ j = 1, \ldots, G)$ and the $kG \times G$ matrix

$$L = \begin{bmatrix} 1 & 1 & \ldots & 0 \\ 1 & 0 & \ldots & 0 \\ \vdots & \vdots & & \vdots \\ 1 & -1 & \ldots & -1 \end{bmatrix}$$

with 1 denoting a k-vector of ones, we find that the asymptotic covariance matrix of $\hat{\beta}' = (\hat{\beta}_0, \hat{\beta}_1, \ldots, \hat{\beta}_{G-1})$ is given by

$$\text{Acov}(\hat{\beta}) = (L' \Lambda^{-1} L)^{-1}. \qquad (6.11)$$

On the other hand, the main term of the bias of the estimators (when $n_j \to \infty$ and $k/n_j \to 0$) is given by

$$\text{Abias}(\hat{\beta}_0) = \frac{1}{G} \sum_{j=1}^{G} \frac{b_j(\frac{n_j+1}{k+1})}{1+\tau_j}, \tag{6.12}$$

$$\text{Abias}(\hat{\beta}_j) = \frac{b_j(\frac{n_j+1}{k+1})}{1+\tau_j} - \frac{1}{G} \sum_{l=1}^{G} \frac{b_l(\frac{n_j+1}{k+1})}{1+\tau_l} \qquad j = 1, \ldots, G-1. \tag{6.13}$$

Application of the estimators defined by (6.9) and (6.10) involves the selection of the number of extreme order statistics k to be used in the estimation. Remark that we take the tail sample fraction k equal for all groups. If k is chosen too small, the resulting estimators will have a high variance. On the other hand, for larger k values, the estimators will perform quite well with respect to variance but will be affected by a larger bias as observations are used that are not really informative for the tail of $F_{X(j)}$, $j = 1, \ldots, G$. Hence, an appropriate k value should represent a good bias-variance trade-off. Here, we will use the trace of the asymptotic mean squared error (AMSE) matrix as optimality criterion.

Defining the AMSE matrix Ω of $\hat{\beta}$ as

$$\Omega(k) = (L'\Lambda^{-1}L)^{-1} + \kappa\kappa', \tag{6.14}$$

with κ denoting the G-vector containing the asymptotic bias expressions given by (6.12) and (6.13), the optimal number of extremes to be used in the estimation, k_{opt}, is defined as

$$k_{opt} = \arg\min \, \text{tr} \, \Omega(k).$$

Note that $\Omega(k)$ depends on the unknown γ_j, τ_j, $j = 1, \ldots, G$, and $b_j(\frac{n_j+1}{k+1})$, $k = 1, \ldots, n_j - 1$, $j = 1, \ldots, G$, which implies that the optimal k has to be derived from an estimate of $\Omega(k)$. The following algorithm is used to estimate k_{opt} and hence γ_j, $j = 1, \ldots, G$, adaptively:

1. Obtain initial estimates of γ_j, τ_j, $j = 1, \ldots, G$, together with estimates of $b_j(\frac{n_j+1}{k+1})$, $k = 1, \ldots, n_j - 1$, $j = 1, \ldots, G$,

2. for $k = 2, \ldots, \min\{n_j; j = 1, \ldots, G\} - 1$:
   compute tr $\hat{\Omega}(k)$ and let

$$\hat{k}_{opt} = \arg\min \, \text{tr} \, \hat{\Omega}(k),$$

3. repeat step 2 but with the parameter estimates obtained from using a common k and obtain an update of the parameter estimates.

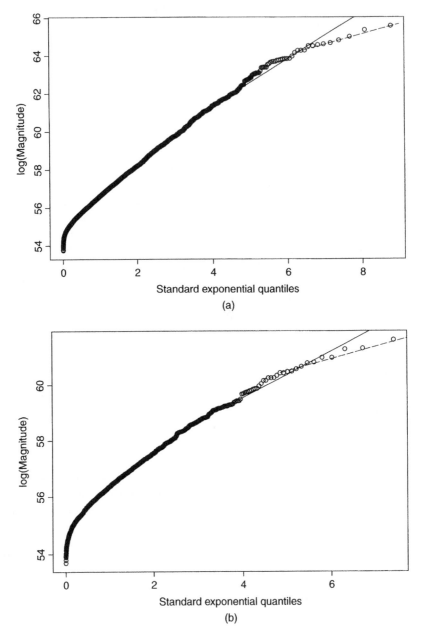

Figure 6.17 Pisarenko and Sornette data: Pareto quantile plots of seismic moments for (a) subduction zones and (b) midocean ridge zones.

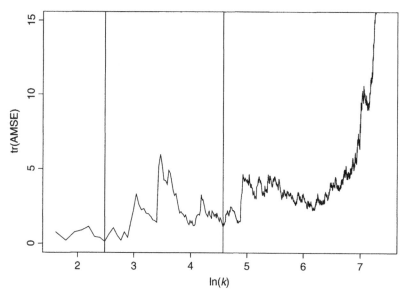

Figure 6.18 Pisarenko and Sornette data: tr $\hat{\Omega}(k)$ as a function of $\log(k)$.

The initial estimates for the unknown parameters (cf. step 1) are obtained by fitting model (6.8) to the k largest observations of each group using a maximum likelihood method (see Beirlant *et al.* (1999)).

Inference about the regression vector β can be drawn using a likelihood ratio test statistic. For k/n_j, $j = 1, \ldots, G$, sufficiently small, the slowly varying nuisance part of (6.8) can be ignored and hence inference can be based on the reduced model $i(\log X^{(j)}_{n_j-i+1,n_j} - \log X^{(j)}_{n_j-i,n_j}) \approx (\beta_0 + \beta_j)F^{(j)}_i$, $i = 1, \ldots, k$, $j = 1, \ldots, G$. As in a 'classical' one-way ANOVA situation, the hypothesis of main interest is $H_0 : \beta_1 = \ldots = \beta_{G-1} = 0$.

We now return to the Pisarenko and Sornette (2003) earthquake data. The procedure described above with $k \geq 20$ yielded $\hat{k}_{opt} = 97$ with $H^{(1)}_{97,6458} = 1.232$ and $H^{(2)}_{97,1665} = 0.821$. In Figure 6.17, we show the Pareto quantile plots of the seismic moments for (a) subduction zones and (b) midocean ridge zones on which we superimposed the lines through $(\log(\frac{n_j+1}{\hat{k}_{opt}+1}), \log x^{(j)}_{n_j-\hat{k}_{opt},n_j})$ with slope $H^{(j)}_{\hat{k}_{opt},n_j}$, $j = 1, 2$ (solid lines). For the hypothesis test of no difference between the tail heaviness of the seismic moment distribution of subduction and midocean ridge zones, a likelihood ratio statistic of 7.92 was obtained, resulting in a rejection of H_0. The GP-based approach described in Pisarenko and Sornette (2001) yielded tail index estimates of 1.51 and 1.02 for subduction and midocean ridge zones respectively, so our results are slightly more conservative. Likewise, these authors found significant differences in the tail heaviness of the seismic moment distributions. As mentioned before, the Pareto quantile plots bend down in the largest observations. Nevertheless, these largest observations still form more or less a straight line pattern. So, also the ultimate tail could be described by a Pareto-type law. This fact is further illustrated in Figure 6.18 where we plot tr $\hat{\Omega}(k)$ as a function of $\log(k)$. Relaxation of the constraint that k should be at least 20 results in the global optimum $\hat{k}_{opt} = 12$ with $H_{12,6458}^{(1)} = 0.541$ and $H_{12,1665}^{(2)} = 0.427$. In Figure 6.17, the resulting optimal fits are plotted with dotted lines. At $\hat{k}_{opt}$, the null hypothesis of no difference in tail behaviour cannot be rejected on the basis of the above-described likelihood ratio test statistic.

7

REGRESSION ANALYSIS

From the discussion in the previous chapters, it became clear that the literature on the estimation of tail characteristics based on an i.i.d. sample is very elaborate. However, a major statistical theme is the description of a variable of primary interest (the dependent variable) in terms of covariates. This regression point of view has been studied much less extensively in extreme value analysis. Further, by using covariate information, data sets originating from different sources may be combined, resulting in opportunities for better point estimates and improved inference. From an extreme value point of view, interest is mainly in estimating conditional tail indices, extreme conditional quantiles and small conditional exceedance probabilities. The available methods together with their references can be grouped in four sets, along

- *the method of block maxima*, fitting the GEV to a sample of maxima, taking one or more of the GEV parameters as a function of the covariates,

- *the quantile view*, extending the exponential regression models for log-spacings of successive order statistics to handle covariate information (Beirlant and Goegebeur (2003)),

- *the probability view*, or POT method, where GP distribution–based regression models are fitted to exceedances over a high threshold (Davison and Smith (1990)),

- *non-parametric estimation procedures* resulting from combining modern smoothing techniques such as maximum penalized likelihood estimation (Green and Silverman (1994)) and local polynomial maximum likelihood estimation (Fan and Gijbels (1996)) with models for extreme values (Davison and Ramesh (2000), Hall and Tajvidi (2000a), Chavez-Demoulin and

Statistics of Extremes: Theory and Applications   J. Beirlant, Y. Goegebeur, J. Segers, and J. Teugels
© 2004 John Wiley & Sons, Ltd   ISBN: 0-471-97647-4

Davison (2001), Pauli and Coles (2001), Chavez-Demoulin and Embrechts (2004), Beirlant and Goegebeur (2004a)).

Before entering into the regression analysis with response distributions in maximal domains of attraction, we recall some facts from classical regression techniques.

7.1 Introduction

The aim of regression analysis is to construct mathematical models that describe or explain relationships that may exist between variables. In general, we are interested in just one variable, the *response or dependent variable*, and we want to study how its distribution depends on a set of variables called the *explanatory or independent variables*. We denote the dependent variable by Y and the vector of covariates by $\mathbf{x}$, that is, $\mathbf{x}' = (x_1, \ldots, x_d)$. The covariates are assumed non-random. Linear regression analysis is one of the oldest and most widely used statistical techniques. The *general linear model* links the dependent variable to the covariates in an approximate linear way:

$$Y = \beta'\mathbf{x} + \varepsilon,$$

where β denotes the vector of regression coefficients, that is, $\beta' = (\beta_1, \ldots, \beta_d)$, and ε is the model error with $\varepsilon \sim N(0, \sigma^2)$, or equivalently

$$Y|\mathbf{x} \sim N(\beta'\mathbf{x}, \sigma^2).$$

Note that the response distribution depends on the covariates through its mean. The general linear model can be extended in different ways. Various non-linear or non-normal regression models have been studied on an individual basis for many years. In 1972, Nelder and Wedderburn (1972) provided a unified and accessible framework for a class of such models, called *generalized linear models* (GLMs). Within this class of generalized linear models, the distribution of the dependent variable is assumed to belong to the one-parameter *exponential family* of distributions, with density function

$$f(y; \theta, \phi) = \exp\left(\frac{\theta y - b(\theta)}{\phi} + c(y, \phi)\right).$$

The parameter θ is the natural parameter of the exponential family and ϕ is a nuisance or scale parameter. Dependence on the covariates is modelled through the mean of the dependent variable using a *link* function g:

$$g(E(Y|\mathbf{x})) = \beta'\mathbf{x},$$

where the link function g is monotone and differentiable. The general linear model is a specific member of this family of generalized linear models with a normal response distribution and identity link function $g(u) = u$.

Note that when dealing with heavy-tailed distributions, the population moments may not be finite, and hence the above-described techniques cannot be used for statistical analysis. Further, from an extreme value point of view, the main interest is in describing conditional tail characteristics such as conditional tail indices, extreme conditional quantiles and small conditional exceedance probabilities rather than modelling conditional means. A straightforward approach to tail analysis in the presence of covariate information consists of modelling one or more of the parameters of a univariate model F, with $F \in \mathcal{D}(G_\gamma)$, as a function of the covariates. In this, parametrizations can be chosen such that the distribution of the response variable depends on the covariates through the extreme value index. Because of its flexibility, the Burr(η, τ, λ) distribution could, for instance, be used to model heavy-tailed data when paying special attention to tail behaviour. Note that for the Burr(η, τ, λ) distribution $\gamma = 1/\lambda\tau$, so in case the main interest is in describing conditional tails λ and/or τ may be taken as a function of the covariates (see Beirlant *et al.* (1998)). This approach results in fully parametric statistical models. These global models are fitted to all available data rather than just to the tail observations and hence do not always provide sufficient flexibility for accurate tail modelling. Therefore, in the subsequent sections, we will focus on regression techniques aimed directly at describing tails of conditional distributions.

7.2 The Method of Block Maxima

7.2.1 Model description

From Chapter 2, we know that the only possible limit distributions for a sequence of normalized maxima are the extreme value distributions. On the basis of this result, the extreme value index can be estimated by fitting the generalized extreme value distribution

$$
G(y; \sigma, \gamma, \mu) = \begin{cases} \exp\left(-\left(1 + \gamma \frac{y-\mu}{\sigma}\right)^{-\frac{1}{\gamma}}\right), & 1 + \gamma \frac{y-\mu}{\sigma} > 0, \gamma \neq 0, \\ \exp\left(-\exp\left(-\frac{y-\mu}{\sigma}\right)\right), & y \in \mathbb{R}, \gamma = 0, \end{cases} \tag{7.1}
$$

with $\mu \in \mathbb{R}$ and $\sigma > 0$ to a sample of maxima. When covariate information is available, it is natural to extend (7.1) to a regression model by taking one or more of its parameters as a function of the covariates. We discuss the estimation problem in its full generality in the sense that the GEV parameters are considered functions of both the covariate vector and the vector of model parameters, that is, $\sigma(\mathbf{x}) = h_1(\mathbf{x}; \beta_1)$, $\gamma(\mathbf{x}) = h_2(\mathbf{x}; \beta_2)$ and $\mu(\mathbf{x}) = h_3(\mathbf{x}; \beta_3)$, with h_1, h_2 and h_3 completely specified functions. In the subsequent discussion, the GEV distribution is referred to as $GEV(\sigma, \gamma, \mu)$.

7.2.2 Maximum likelihood estimation

Consider $Y_1, \ldots, Y_m$ independent random variables and let $\mathbf{x}_i$ represent the covariate vector associated with Y_i such that

$$Y_i | \mathbf{x}_i \sim GEV(\sigma(\mathbf{x}_i), \gamma(\mathbf{x}_i), \mu(\mathbf{x}_i)), \qquad i = 1, \ldots, m.$$

Denoting by β the complete vector of model parameters, that is, $\beta' = (\beta_1', \beta_2', \beta_3')$, the log-likelihood function is simply

$$\log L(\beta) = \sum_{i=1}^{m} \log g(Y_i; \sigma(\mathbf{x}_i), \gamma(\mathbf{x}_i), \mu(\mathbf{x}_i)) \tag{7.2}$$

where g is the GEV density function

$$g(y; \sigma, \gamma, \mu) \tag{7.3}$$

$$= \begin{cases} \frac{1}{\sigma} \left(1 + \gamma \frac{y-\mu}{\sigma}\right)^{-\frac{1}{\gamma}-1} \exp\left(-\left(1 + \gamma \frac{y-\mu}{\sigma}\right)^{-\frac{1}{\gamma}}\right), & 1 + \gamma \frac{y-\mu}{\sigma} > 0, \gamma \neq 0, \\ \frac{1}{\sigma} \exp\left(-\frac{y-\mu}{\sigma}\right) \exp\left(-\exp\left(-\frac{y-\mu}{\sigma}\right)\right), & y \in \mathbb{R}, \gamma = 0. \end{cases}$$

The maximum likelihood estimator $\hat{\beta}$ can be obtained by maximizing (7.2) with respect to β. Approximate asymptotic inference follows in the usual way from the inverse information matrix or the profile likelihood function.

If the data are not exactly GEV distributed but instead we have a sample of maxima at our disposal, then, following condition $(\mathcal{C}_\gamma)$, the GEV can still be used as an approximation to the true maximum distribution. The above-described maximum likelihood procedure is then applied to a sample of maxima. Recall that, in this case, the parameters σ and μ absorb the normalizing constants a_n respectively b_n in the derivation of the limit laws for maxima given in section 2.1. Hence, the parametrization for σ and μ follows immediately from these. In practice, we usually have no knowledge about F and U, and setting up an appropriate GEV parametrization in terms of $\mathbf{x}$ is often difficult. Simulation results, however, indicate that incorrect specifications may lead to unreliable point estimates. One possible solution for this problem is to consider a broad class of distribution functions satisfying $(\mathcal{C}_\gamma)$ followed by the determination of the appropriate μ and σ parametrizations and the resulting limiting form. For instance, ignoring the dependence on the covariates for notational convenience, in case of the Hall class (Hall (1982)) for which

$$U(y) = \begin{cases} Cy^\gamma (1 + Dy^\rho (1 + o(1))), & \gamma > 0, \\ y_+ - Cy^\gamma (1 + Dy^\rho (1 + o(1))), & \gamma < 0, \end{cases} \quad (y \to \infty)$$

with $C > 0$, $D \in \mathbb{R}$ and $\rho < 0$, we can take, for $n \to \infty$,

$$b_n = U(n) \sim \begin{cases} Cn^\gamma, & \gamma > 0, \\ y_+ - Cn^\gamma, & \gamma < 0, \end{cases}$$

$$a_n = |\gamma| Cn^\gamma,$$

and hence, setting $z = b_n + a_n y$,

$$P(Y_{n,n} \leq z) \sim \begin{cases} \exp\left(-\left(\frac{z}{Cn^\gamma}\right)^{-\frac{1}{\gamma}}\right), & \gamma > 0, \\ \exp\left(-\left(\frac{y_+ - z}{Cn^\gamma}\right)^{-\frac{1}{\gamma}}\right), & \gamma < 0. \end{cases} \tag{7.4}$$

Clearly, for this very broad class of distribution functions, the appropriate μ and σ parametrizations can be easily obtained. Of course, application of (7.4) still requires to specify γ and possibly C in terms of the covariates. Note also that since μ and σ depend on n, model (7.4) can accommodate data sets with different subsample sizes.

Example 7.1 The Ca against pH scatterplot shown in Figure 1.19(a) gives an indication that the tail of the conditional Ca distribution may depend on the pH level of soil samples. We analyse these data using the GEV regression approach discussed above. The Pareto quantile plots of the Ca measurements at some fixed pH levels given in Figure 6.7 indicate that Pareto-type models provide appropriate fits to the conditional distributions of Ca given pH. Further, following Goegebeur *et al.* (2004), we model the extreme value index γ in terms of the pH level using an exponential link function. The data set is preprocessed in the sense that observations identified as suspect or incorrect are excluded from the analysis. At each pH level, we compute the maximum Ca value and the number of available Ca observations, see Figure 7.1(a) and (b) respectively. The GEV-Hall model (7.4) with $\gamma(pH) = \exp(\beta_0 + \beta_1 pH)$ is fitted to these 30 subsample maxima using the maximum likelihood method. This results in the point estimates

$$\hat{C} = 96.4173,$$
$$\hat{\beta}_0 = -2.8406,$$
$$\hat{\beta}_1 = 0.2908.$$

The profile likelihood function and the profile likelihood-based 95% confidence interval for β_1 are shown in Figure 7.1(c). The confidence interval for β_1 does not include the value 0, so at a significance level of 5%, the hypothesis $H_0: \beta_1 = 0$ can be rejected. Hence, the tail heaviness of the conditional Ca distribution varies significantly with the pH level of soil samples.

7.2.3 Goodness-of-fit

Having fitted a model to a data set, for instance, using maximum likelihood methods, one should evaluate how well the model describes or explains the available data. This is especially true here since the complicated model was based on the asymptotics of maxima. When dealing with regression models, the goodness-of-fit typically is visually assessed by inspection of various kinds of residual plots. In the present context, the classical residuals $Y_i - \xi_i$, where ξ_i denotes some measure of location, are not very useful as, in general, these are not identically distributed.

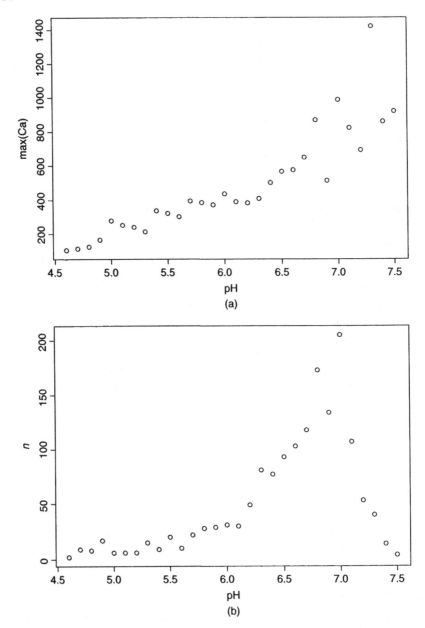

Figure 7.1 Condroz data: (a) Ca-maxima versus pH level, (b) number of observations versus pH level and (c) profile log-likelihood function and profile likelihood-based 95% confidence intervals for β_1.

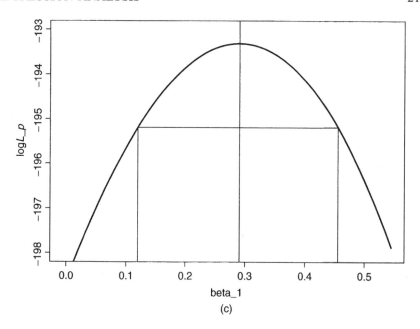

(c)

Therefore, we will look for other quantities or random variables that satisfy the i.i.d. property, and hence can be considered as generalized residuals. Consider

$$Y_i | \mathbf{x}_i \sim GEV(\sigma(\mathbf{x}_i), \gamma(\mathbf{x}_i), \mu(\mathbf{x}_i)), \qquad i = 1, \ldots, m.$$

The transformation

$$R_i = \begin{cases} \frac{1}{\gamma(\mathbf{x}_i)} \log\left(1 + \gamma(\mathbf{x}_i) \frac{Y_i - \mu(\mathbf{x}_i)}{\sigma(\mathbf{x}_i)}\right), & \gamma(\mathbf{x}_i) \neq 0, \\ \frac{Y_i - \mu(\mathbf{x}_i)}{\sigma(\mathbf{x}_i)}, & \gamma(\mathbf{x}_i) = 0, \end{cases} \qquad (7.5)$$

results in a standard Gumbel distributed random variable R_i (Coles (2001)):

$$F_{R_i | \mathbf{x}_i}(r_i)$$

$$= \begin{cases} G\left(\mu(\mathbf{x}_i) + \frac{\sigma(\mathbf{x}_i)}{\gamma(\mathbf{x}_i)}[\exp(\gamma(\mathbf{x}_i)r_i) - 1]; \sigma(\mathbf{x}_i), \gamma(\mathbf{x}_i), \mu(\mathbf{x}_i)\right), & \gamma(\mathbf{x}_i) \neq 0, \\ G(\mu(\mathbf{x}_i) + \sigma(\mathbf{x}_i)r_i; \sigma(\mathbf{x}_i), \gamma(\mathbf{x}_i), \mu(\mathbf{x}_i)), & \gamma(\mathbf{x}_i) = 0, \end{cases}$$

$$= \exp(-\exp(-r_i)). \qquad (7.6)$$

The resulting Gumbel distribution does not any longer depend on the covariates, and hence the random variables $R_1, \ldots, R_m$ are identically distributed. If $Y_1, \ldots, Y_m$ are assumed independent, then $R_1, \ldots, R_m$ are also independent. Hence, analogously to the case of classical linear regression, $R_1, \ldots, R_m$ can be used to construct several kinds of residual plots. Here, we will concentrate on residual quantile plots. The quantile function associated with (7.6) is given by

$$Q(p) = -\log(-\log p), \qquad 0 < p < 1,$$

yielding the Gumbel quantile plot coordinates

$$\left(-\log\left(-\log\frac{i}{m+1}\right), R_{i,m}\right), \qquad i = 1, \ldots, m.$$

In case (7.6) provides an accurate description of the data, we expect the points on the Gumbel quantile plot to be close to the first diagonal.

Alternatively, as discussed in Chapter 1, we can always return to the exponential framework by transforming the data first to the exponential case followed by a subsequent assessment of the exponential quantile fit. To do so, note that

$$G(Y; \sigma(\mathbf{x}), \gamma(\mathbf{x}), \mu(\mathbf{x})) \overset{D}{=} U,$$

where $U \sim U(0, 1)$ and hence

$$-\log(1 - G(Y; \sigma(\mathbf{x}), \gamma(\mathbf{x}), \mu(\mathbf{x}))) \overset{D}{=} E, \qquad (7.7)$$

with $E \sim Exp(1)$. On the basis of this, the fit of the GEV regression model can be assessed by constructing the plot

$$\left(-\log\left(1 - \frac{i}{m+1}\right), -\log\left(1 - U_{i,m}\right)\right), \qquad i = 1, \ldots, m,$$

where $U_i = G(Y_i; \sigma(\mathbf{x}_i), \gamma(\mathbf{x}_i), \mu(\mathbf{x}_i))$ and $U_{1,m} \leq \cdots \leq U_{m,m}$ are the corresponding order statistics, and to inspect the closeness of the points to the first diagonal.

Example 7.1 (continued) We now evaluate how well the GEV-Hall model (7.4) with $\gamma(pH) = \exp(\beta_0 + \beta_1 pH)$ describes the conditional Ca-maxima using the above-introduced quantile plots. In Figure 7.2(a), we show the Gumbel quantile plot of the generalized residuals (7.5). Taking the small sample size and the high variability of the subsample sizes over the pH range into account, we can conclude that the GEV-Hall regression model describes the conditional Ca-maxima distribution quite well. Alternatively, we can evaluate the fit on the basis of the exponential quantile plot of the generalized residuals (7.7), see Figure 7.2(b).

7.2.4 Estimation of extreme conditional quantiles

Extreme conditional quantile estimates can be obtained by inverting the conditional GEV distribution function yielding

$$q_{p,\mathbf{x}} = \begin{cases} \mu(\mathbf{x}) + \frac{\sigma(\mathbf{x})}{\gamma(\mathbf{x})}\left[(-\log(1 - p))^{-\gamma(\mathbf{x})} - 1\right], & \gamma(\mathbf{x}) \neq 0, \\ \mu(\mathbf{x}) - \sigma(\mathbf{x})\log(-\log(1 - p)), & \gamma(\mathbf{x}) = 0, \end{cases} \qquad (7.8)$$

and replacing the unknown parameters by their respective estimates.

Example 7.1 (continued) We continue with the analysis of the conditional Ca-maxima. The estimated conditional 0.95 quantile of the Ca-maxima distribution is

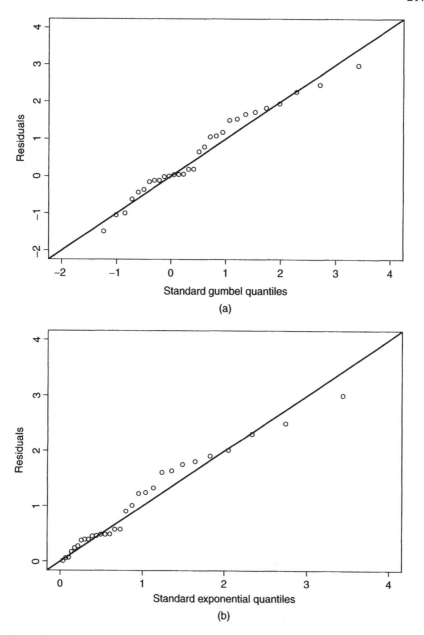

Figure 7.2 Condroz data: (a) Gumbel quantile plot and (b) exponential quantile plot of the generalized residuals.

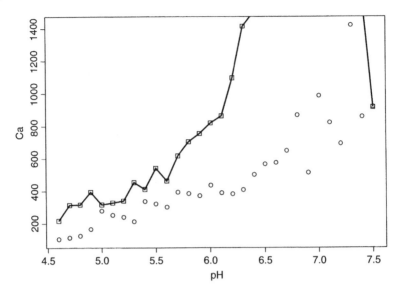

Figure 7.3   Condroz data: conditional Ca-maxima (circles) and estimated conditional 0.95 quantiles (squares).

given as a function of pH in Figure 7.3. Note that one observation exceeds the estimated 0.95 quantile (we expect one or two observations out of 30 above the conditional 0.95 quantile).

Note that in case the GEV regression model is used to approximate the true conditional distribution of the largest value in a sample, (7.8) yields the quantiles of the conditional maximum distribution. The quantiles of the original data can be obtained from

$$q_{p,\mathbf{x}}^* = \begin{cases} \mu(\mathbf{x}) + \frac{\sigma(\mathbf{x})}{\gamma(\mathbf{x})} \left[(-\log(1-p)^n)^{-\gamma(\mathbf{x})} - 1 \right], & \gamma(\mathbf{x}) \neq 0, \\ \mu(\mathbf{x}) - \sigma(\mathbf{x}) \log(-\log(1-p)^n), & \gamma(\mathbf{x}) = 0. \end{cases}$$

7.3   The Quantile View—Methods Based on Exponential Regression Models

7.3.1   Model description

In this section, we extend the exponential regression model for log-spacings of successive order statistics introduced in Chapter 4 to the regression case. This approach has only been worked out in case of Pareto-type response distributions.

Consider $Y_1, \ldots, Y_n$ i.i.d. random variables according to distribution function F, where F is of Pareto-type, that is, the tail quantile function U satisfies

$$U(y) = y^\gamma \ell_U(y), \qquad y > 1; \ \gamma > 0. \tag{7.9}$$

From section 4.4, we know that, when $\log \ell_U$ satisfies $\mathcal{C}_\rho(b)$ for some $\rho < 0$ and $b \in \mathcal{R}_\rho$, log-spacings of successive order statistics can be approximately represented as

$$j \left(\log Y_{n-j+1,n} - \log Y_{n-j,n} \right) \overset{\mathcal{D}}{\sim} \left(\gamma + b_{n,k} \left(\frac{j}{k+1} \right)^{-\rho} \right) E_j,$$

$$j = 1, \ldots, k, \qquad (7.10)$$

where $E_1, \ldots, E_k$ are independent standard exponential random variables and $b_{n,k} = b(\frac{n+1}{k+1})$. When covariate information is available, (7.9) can be extended to a regression model by modelling γ and possibly ℓ_U as a function of the covariates. In this case, the exponential regression model (7.10) cannot be directly applied to the response observations as these are not identically distributed. One possible solution is to transform the response observations into generalized residuals, thereby removing (at least partly) the dependence on the covariates. These generalized residuals then form the basis for applying (7.10).

7.3.2 Maximum likelihood estimation

Consider independent random variables $Y_1, \ldots, Y_n$ with respective associated covariate vectors $\mathbf{x}_1, \ldots, \mathbf{x}_n$ such that the conditional distribution of Y given $\mathbf{x}$ is of Pareto-type, that is, for some $\gamma(\mathbf{x}) > 0$

$$1 - F_{Y|\mathbf{x}}(y) = y^{-1/\gamma(\mathbf{x})} \ell_F(y; \mathbf{x}). \qquad (7.11)$$

As above, we set $\gamma(\mathbf{x}) = h(\mathbf{x}; \beta)$, for some completely specified function h and with β denoting a vector of regression coefficients. Note that aside of the extreme value index γ, $F_{Y|\mathbf{x}}$ may also depend on the covariates through ℓ_F. Since $Y_1, \ldots, Y_n$ are not identically distributed (7.10) cannot be directly applied to the raw input data. However, by transforming the dependent variables, the dependence on the covariates may be at least partly removed. The transformation

$$R = Y^{1/\gamma(x)} \qquad (7.12)$$

standardizes the extreme value index:

$$1 - F_{R|\mathbf{x}}(r) = r^{-1} \ell_F(r^{\gamma(\mathbf{x})}; \mathbf{x}).$$

Next, we restrict the class of distribution functions satisfying (7.11) to the distributions $F_{Y|\mathbf{x}}$ for which

$$\ell_F(r^{\gamma(\mathbf{x})}; \mathbf{x}) = \ell_F(r), \qquad (7.13)$$

or equivalently, to the class of conditional distribution functions for which transformation (7.12) removes the conditioning on $\mathbf{x}$ completely. The random variables $R_1, \ldots, R_n$ obtained by applying (7.12) to $Y_1, \ldots, Y_n$ are now clearly independent (since the Y_i are independent) and identically distributed (γ and ℓ_F no longer depend on **x**). The R_i, $i = 1, \ldots, n$ again form the basis of the statistical analysis. We denote the order statistics associated with $R_1, \ldots, R_n$ by $R_{1,n} \leq \cdots \leq R_{n,n}$.

In case $\log \ell_U$ satisfies $\mathcal{C}_\rho(b)$ for some $\rho < 0$ and $b \in \mathcal{R}_\rho$, using derivations similar to the ones in section 4.4, the following approximate representation for log-spacings of generalized residuals can be proposed

$$Z_j \overset{\mathcal{D}}{\sim} \left(1 + b_{n,k} \left(\frac{j}{k+1}\right)^{-\rho}\right) E_j, \qquad j = 1, \ldots, k, \qquad (7.14)$$

with $Z_j = j(\log R_{n-j+1,n} - \log R_{n-j,n})$, $b_{n,k} = b(\frac{n+1}{k+1})$ and $E_1, \ldots, E_k$ denoting independent standard exponential random variables. The regression coefficients can be estimated jointly with $b_{n,k}$ and ρ using the maximum likelihood method.

The log-likelihood function for $Z_1, \ldots, Z_k$ is given by

$$\log L(\beta, b_{n,k}, \rho) = -\sum_{j=1}^{k} \log\left(1 + b_{n,k}\left(\frac{j}{k+1}\right)^{-\rho}\right)$$
$$-\sum_{j=1}^{k} \frac{Z_j}{1 + b_{n,k}\left(\frac{j}{k+1}\right)^{-\rho}}. \qquad (7.15)$$

Note that the likelihood function depends on the regression coefficients through the ordered residuals and hence is more complicated than in section 4.4. For computational details concerning the numerical maximization of (7.15), we refer to Beirlant and Goegebeur (2003). Inference about the regression coefficients can be drawn using the profile log-likelihood ratio test statistic given by $2(\log L_p(\hat{\beta}_{(0)}) - \log L_p(\beta^*_{(0)}))$ with $\log L_p(\beta_{(0)})$ denoting the profile log-likelihood function of some subset $\beta_{(0)}$ of β. This statistic equals the classical likelihood ratio statistic for testing the hypothesis $H_0: \beta_{(0)} = \beta^*_{(0)}$. As discussed in Beirlant and Goegebeur (2003), the classical χ^2 approximation to the null distribution of the test statistic is inappropriate. We therefore propose to simulate the reference distribution by using a parametric bootstrap procedure (Efron and Tibshirani (1993)). Bootstrap samples are generated from a strict Pareto distribution with parameters $\beta^*_{(0)}$ and the maximum likelihood estimates of the remaining regression coefficients given $\beta^*_{(0)}$.

Example 7.2 We illustrate the proposed procedure with the diamond data introduced in section 1.3.5. In a first attempt, trying to fit regression models over the whole range of the variable size, the application of model (7.11) with $Y = value$ and $\gamma(size) = \exp(\beta_0 + \beta_1 \, size)$ does not provide an appropriate fit: the Pareto quantile plot of the residuals $R = Y^{\exp(-\beta_1 \, size)}$ becomes horizontal for the largest observations, that is, $\exp(\beta_0) = 0$ (cf infra). Rather, the extreme value index is found to vary polynomially with size. The scatterplot of value versus $\log(size)$ is given in Figure 7.4(a). In Figure 7.4(b), we show the profile log-likelihood function of β_1 for $k = 200, 250, 300, 350, 400$. These profile likelihood functions clearly indicate a β_1 estimate of approximately 0.3. Finally, Figure 7.4(c) and (d) contain the maximum likelihood estimates of respectively β_0 and β_1 as a function of k.

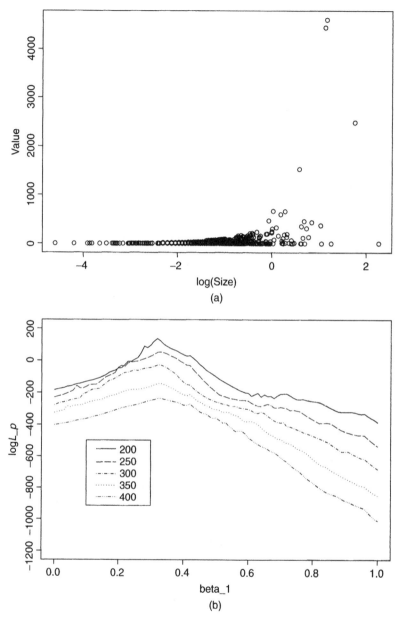

Figure 7.4   Diamond data: (a) value against log($size$) scatterplot, (b) profile log-likelihood function of β_1 for $k = 200, 250, 300, 350, 400$, (c) $\hat{\beta}_0$ as a function of k and (d) $\hat{\beta}_1$ as a function of k.

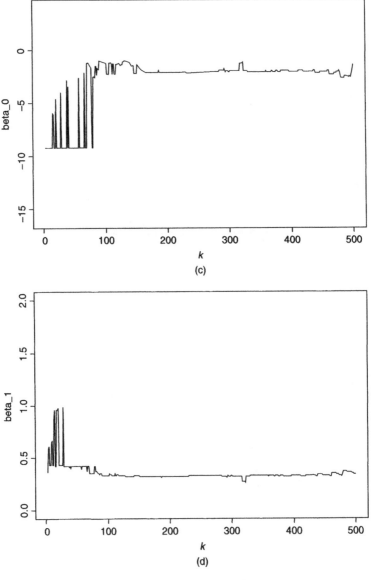

Figure 7.4   (*continued*)

7.3.3   Goodness-of-fit

The fit of the conditional Pareto-type model given by (7.11) can be assessed in several ways. Under (7.11) and (7.13), $R_1, \ldots, R_n$ form an i.i.d. sample from a Pareto-type model with $\gamma = 1$ and hence can be used to construct a Pareto quantile plot. Alternatively, following (7.14)

$$\tilde{R}_j = \frac{Z_j}{1 + b_{n,k}\left(\frac{j}{k+1}\right)^{-\rho}}, \qquad j = 1, \ldots, k,$$

are approximately distributed as an i.i.d. sample from the standard exponential distribution and can be used to construct an exponential quantile plot.

Example 7.2 (continued) Figure 7.5(a) shows the Pareto quantile plot of the generalized residuals $\log r_i = size_i^{-\beta_1} \log value_i$, $i = 1, \ldots, 1914$. The residuals are computed using the β_1 estimate at $k = 200$. The Pareto quantile plot is clearly linear in the largest observations, indicating a good fit of a Pareto-type model to the residual distribution. In a similar way, we also constructed a Pareto quantile plot of the residuals $\log r_i = \exp(-\beta_1 size_i) \log value_i$, see Figure 7.5(b). The ultimate horizontal appearance indicates that the residual distribution cannot be adequately described by a Pareto-type model.

7.3.4 Estimation of extreme conditional quantiles

In case of an i.i.d. sample from a Pareto-type distribution, $Y_1, \ldots, Y_n$, extreme quantiles can be estimated by extrapolation along a line through the anchor point $(\log \frac{n+1}{k+1}, \log Y_{n-k,n})$ with slope $\hat{\gamma}_k$ on the Pareto quantile plot, resulting in the estimator (see e.g. Weissman (1978))

$$Q_{Y,k}(p) = Y_{n-k,n}\left(\frac{k+1}{(n+1)p}\right)^{\hat{\gamma}_k}, \qquad k = 1, \ldots, n-1, \qquad (7.16)$$

where $\hat{\gamma}_k$ denotes an estimator for γ based on the k largest order statistics. When covariate information is available (7.16) cannot be applied directly to the raw data since the observations are not longer identically distributed. In this situation, the observations will be first transformed to i.i.d. data using (7.12). Next, (7.16) will be used in the extrapolation step, yielding an estimator of an extreme quantile of the residual distribution. Finally, the quantile estimator of the generalized residuals will be transformed back to the original observations by inverting (7.12). This results in the following estimator for the $(1 - p)$-th quantile of $F_{Y|\mathbf{x}}$:

$$Q_{Y,k}(p; \mathbf{x}) = \left(\hat{R}_{n-k,n}\frac{k+1}{(n+1)p}\right)^{\hat{\gamma}_k(\mathbf{x})}, \qquad k = d+1, \ldots, n-1,$$

with $\hat{\gamma}_k(\mathbf{x})$ denoting the estimator for $\gamma(\mathbf{x})$ obtained by using the k largest order statistics and $\hat{R}_{n-k,n}$ representing the $(k + 1)$-th largest order statistic of the generalized residuals obtained by using $\hat{\gamma}_k(\mathbf{x})$ in (7.12).

Example 7.2 (continued) In Figure 7.6, we show the value versus $\log(size)$ scatterplot with the estimated conditional 0.99 quantile obtained at $k = 102$ superimposed. The k value used to compute the 0.99 quantiles is selected so as to

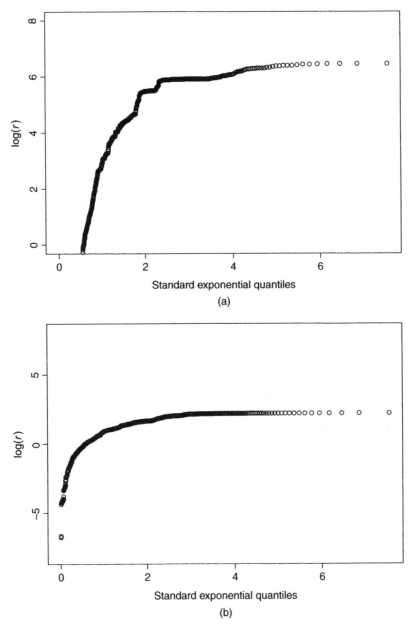

Figure 7.5   Diamond data: (a) Pareto quantile plot of the generalized residuals for the regression model with $\log(size)$ as explanatory variable, (b) Pareto quantile plot of the generalized residuals for the regression model with $size$ as explanatory variable (in (a) and (b), the generalized residuals are computed using the β_1-estimates at $k = 200$).

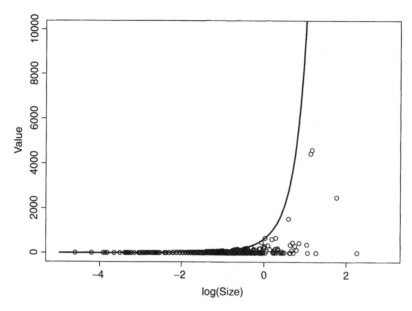

Figure 7.6 Diamond data: scatterplot of value versus $\log(size)$ with $Q_{Y,102}(0.01; \log(size))$ superimposed.

minimize

$$\left| \frac{1}{1914} \sum_{i=1}^{1914} I\left(value_i \geq Q_{Y,k}(0.01; \log(size_i))\right) - 0.01 \right|$$

with respect to k.

7.4 The Tail Probability View—Peaks Over Threshold (POT) Method

7.4.1 Model description

Consider a random variable Z with distribution function F satisfying $F \in \mathcal{D}(G_\gamma)$. Following (C_γ^*), the conditional distribution of $Y = Z - u$ given $Z > u$ can be well approximated by the GP distribution at least for threshold values that are sufficiently large. On the basis of this, it is natural to model exceedances over a high-specified threshold by the GP distribution with distribution function

$$H(y; \sigma, \gamma) = \begin{cases} 1 - \left(1 + \gamma \frac{y}{\sigma}\right)^{-\frac{1}{\gamma}}, & 1 + \gamma \frac{y}{\sigma} > 0, \gamma \neq 0, \\ 1 - \exp\left(-\frac{y}{\sigma}\right), & y > 0, \gamma = 0, \end{cases}$$

where $\sigma > 0$ is the scale parameter of the GP family. Similar to the approach followed with the GEV, we extend the GP distribution to a regression model by taking σ and/or γ as a function of the covariate vector and the vectors of regression coefficients, that is, $\sigma(\mathbf{x}) = h_1(\mathbf{x}; \beta_1)$ and $\gamma(\mathbf{x}) = h_2(\mathbf{x}; \beta_2)$, see, for instance, Davison and Smith (1990).

7.4.2 Maximum likelihood estimation

Let $Y_1, \ldots, Y_n$ be independent random variables and let $\mathbf{x}_i$ denote the covariate vector associated with Y_i such that

$$Y_i | \mathbf{x}_i \sim GP(\sigma(\mathbf{x}_i), \gamma(\mathbf{x}_i)), \qquad i = 1, \ldots, n.$$

Again we denote the complete parameter vector with β, so $\beta' = (\beta_1', \beta_2')$. The log-likelihood function is then

$$\log L(\beta) = \sum_{i=1}^{n} \log h(Y_i; \sigma(\mathbf{x}_i), \gamma(\mathbf{x}_i)), \qquad (7.17)$$

where h is the GP density function:

$$h(y; \sigma, \gamma) = \begin{cases} \frac{1}{\sigma}\left(1 + \gamma \frac{y}{\sigma}\right)^{-\frac{1}{\gamma}-1}, & 1 + \gamma \frac{y}{\sigma} > 0, \gamma \neq 0, \\ \frac{1}{\sigma} \exp\left(-\frac{y}{\sigma}\right), & y > 0, \gamma = 0. \end{cases}$$

The maximum likelihood estimator $\hat{\beta}$ can be obtained by maximizing (7.17) with respect to β. Approximate inference about the regression coefficients can be drawn on the basis of the limiting normal distribution of the maximum likelihood estimator or using the profile likelihood approach. The profile log-likelihood function is usually not quadratic in small and moderate samples and provides a better basis for confidence intervals than the observed expected information (see Davison and Smith (1990)).

We now turn to the case where the data are not exactly GP distributed. Consider a random variable Z with associated covariate vector $\mathbf{x}$ such that the conditional distribution of Z given $\mathbf{x}$, $F_{Z|\mathbf{x}}$, is in the max-domain of attraction of the GEV, $F_{Z|\mathbf{x}} \in \mathcal{D}(G_{\gamma(\mathbf{x})})$. Here, the notation $\gamma(\mathbf{x})$ stresses the possible dependence of the tail index on the covariates. On the basis of (C_γ^*), the GP distribution can be used to approximate the conditional distribution of $Z - u_{\mathbf{x}}$ given $Z > u_{\mathbf{x}}$ where $u_{\mathbf{x}}$ denotes a sufficiently high threshold. Given independent random variables $Z_1, \ldots, Z_n$ and associated covariate vectors $\mathbf{x}_1, \ldots, \mathbf{x}_n$, the above-described maximum likelihood procedure is then applied to the exceedances $Y_j = Z_i - u_{\mathbf{x}_i}$, provided $Z_i > u_{\mathbf{x}_i}$, $j = 1, \ldots, N_{u_{\mathbf{x}}}$, where i is the index of the j-th exceedance in the original sample and $N_{u_{\mathbf{x}}}$ denotes the number of exceedances over the threshold 'function' $u_{\mathbf{x}}$. Of course, the covariate vectors need to be re-indexed in an analogous way. Similar to the i.i.d. case, applying the GP approach involves the selection of an appropriate threshold $u_\mathbf{x}$. In the regression case, the specification of a threshold gets even more difficult since, in principle, the threshold can depend on the covariates in order to take the relative extremity of the observations into account (see also Davison and Smith (1990) and Coles (2001)). Up to now, solutions seem to be more ad hoc and depending on the data set at hand. One possibility to a scientifically better-founded approach is to proceed as follows. Let $\gamma(\mathbf{x})$ and $\sigma(\mathbf{x})$ be as defined above and $u_\mathbf{x} = u(\mathbf{x}; \theta)$ denotes the threshold function, depending on both the covariates and a vector of regression coefficients θ, where $\theta' = (\theta_1, \ldots, \theta_d)$. The following mixed-integer programming formulation allows to estimate β_1, β_2 and θ for the GP regression model such that exactly k observations fall above $u_\mathbf{x}$:

$$\max_{\beta_1, \beta_2, \theta, \delta} \sum_{i=1}^{n} \left\{ -\log \sigma(\mathbf{x}_i) - \left(\frac{1}{\gamma(\mathbf{x}_i)} + 1 \right) \log \left(1 + \gamma(\mathbf{x}_i) \frac{Z_i - u_{\mathbf{x}_i}}{\sigma(\mathbf{x}_i)} \right) \right\} (1 - \delta_i)$$

subject to

$$\beta_1, \beta_2, \theta \in \mathbb{R}^d,$$
$$Z_i + M\delta_i \geq u_{\mathbf{x}_i}, \qquad i = 1, \ldots, n,$$
$$\sum_{i=1}^{n} (1 - \delta_i) = k,$$
$$\delta_i \in \{0, 1\}, \qquad i = 1, \ldots, n,$$

with M a big number. From a computational point of view, however, this approach is very difficult. Alternatively, the Koenker and Bassett (1978) quantile regression methodology may be used to obtain a covariate dependent threshold. Suppose the conditional quantile function $Q(p; \mathbf{x})$ associated with $F_{Z|\mathbf{x}}$ can be modelled by a completely specified function $u(\mathbf{x}; \theta_p)$, that is, $Q(p; \mathbf{x}) = u(\mathbf{x}; \theta_p)$. The p-th ($0 < p < 1$) quantile regression estimator $\hat{\theta}_p$ of θ_p is then defined as a solution to the following optimization problem

$$\min_{\theta_p} \sum_{i=1}^{n} \left(p(Z_i - u(\mathbf{x}_i; \theta_p))^+ + (1 - p)(Z_i - u(\mathbf{x}_i; \theta_p))^- \right),$$

with $x^+ = \max(0, x)$ and $x^- = \max(0, -x)$. When working with the GP distribution, the threshold can be set at a particular regression quantile, that is, $u_\mathbf{x} = u(\mathbf{x}; \theta_p)$. The estimated conditional quantile function is then used to compute exceedances that are in turn plugged into the maximum likelihood estimation. This procedure may be performed for $p = \frac{n-k}{n+1}$, $k = d + 1, \ldots, n - 1$ and, similar to the i.i.d. case, the point estimates plotted as a function of k.

Example 7.1 (continued) We illustrate the GP regression modelling of conditional exceedances with the Condroz data. A GP regression model with $\gamma(pH) = \exp(\beta_0 + \beta_1 pH)$ is fitted to the Ca exceedances over both a constant threshold and a covariate dependent threshold. The constant threshold is taken as the $(k + 1)$th largest observation on the dependent variable, $k = 5, \ldots, n - 1$. This is illustrated in Figure 7.7(a) for $k = 20$. Alternatively, we used the Koenker and Bassett (1978)

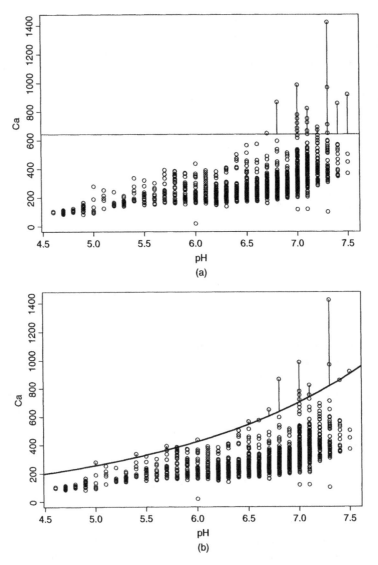

Figure 7.7   Condroz data: exceedances over (a) the 20th largest response observation and (b) regression quantile 0.9877 (corresponding to $k = 20$).

quantile regression methodology to obtain a covariate dependent threshold. Here
we set

$$u_{pH} = \exp(\theta_{0,p} + \theta_{1,p} pH),$$

where $\theta_{0,p}$ and $\theta_{1,p}$ denote the pth regression quantile, $p = (n - k)/(n + 1)$,
$k = 5, \ldots, n - 1$, see Figure 7.7(b). Note that the covariate dependent threshold

yields exceedances over the whole range of the covariate, whereas for the constant threshold, exceedances exclusively occur at the higher pH levels. In Figure 7.8(a), (b) and (c), we show the maximum likelihood estimates of respectively σ, β_0 and β_1 for the above-described GP regression model fitted to exceedances over a constant threshold (broken line) and the quantile regression threshold (solid line). Finally, the profile log-likelihood function of β_1 using $k = 500$ exceedances over the covariate dependent threshold and the corresponding 95% confidence interval are shown in Figure 7.8(d). The 95% interval does not contain the value $\beta_1 = 0$, so the hypothesis $H_0:\beta_1 = 0$ can be rejected at the 5% significance level.

7.4.3 Goodness-of-fit

Similar to the discussion in section 7.2.3, we focus on the use of residual quantile plots to assess the fit of a GP regression model. Consider

$$Y_i|\mathbf{x}_i \sim GP(\sigma(\mathbf{x}_i), \gamma(\mathbf{x}_i)), \qquad i = 1, \ldots, n. \tag{7.18}$$

Since the exponential distribution is a special member of the GP family, it is natural to apply a transformation to the exponential case. The transformation (Coles (2001))

$$R_i = \begin{cases} \frac{1}{\gamma(\mathbf{x}_i)} \log\left(1 + \gamma(\mathbf{x}_i)\frac{Y_i}{\sigma(\mathbf{x}_i)}\right), & \gamma(\mathbf{x}_i) \neq 0, \\ \frac{Y_i}{\sigma(\mathbf{x}_i)}, & \gamma(\mathbf{x}_i) = 0. \end{cases} \tag{7.19}$$

results in a standard exponential random variable:

$$F_{R_i|\mathbf{x}_i}(r_i) = \begin{cases} H\left(\frac{\sigma(\mathbf{x}_i)}{\gamma(\mathbf{x}_i)}[\exp(\gamma(\mathbf{x}_i)r_i) - 1]; \sigma(\mathbf{x}_i), \gamma(\mathbf{x}_i)\right), & \gamma(\mathbf{x}_i) \neq 0, \\ H(\sigma(\mathbf{x}_i)r_i; \sigma(\mathbf{x}_i), \gamma(\mathbf{x}_i)), & \gamma(\mathbf{x}_i) = 0, \end{cases}$$
$$= 1 - \exp(-r_i).$$

If $Y_1, \ldots, Y_n$ are independent, then $R_1, \ldots, R_n$ are i.i.d. random variables, and hence can be used to validate model (7.18), for instance, using an exponential quantile plot

$$\left(-\log\left(1 - \frac{i}{n+1}\right), R_{i,n}\right), \qquad i = 1, \ldots, n.$$

When regression model (7.18) indeed gives an accurate description of the data, we expect the points on the exponential quantile plot to scatter around the first diagonal.

Example 7.1 (continued) We evaluate the fit of the GP regression model with $\gamma(pH) = \exp(\beta_0 + \beta_1 pH)$ to the 500 Ca exceedances of regression quantile 0.6667 using an exponential quantile plot. This plot is shown in Figure 7.9. The ordered residuals scatter quite well around the first diagonal, giving evidence of a good fit of the GP regression model to the Ca exceedances.

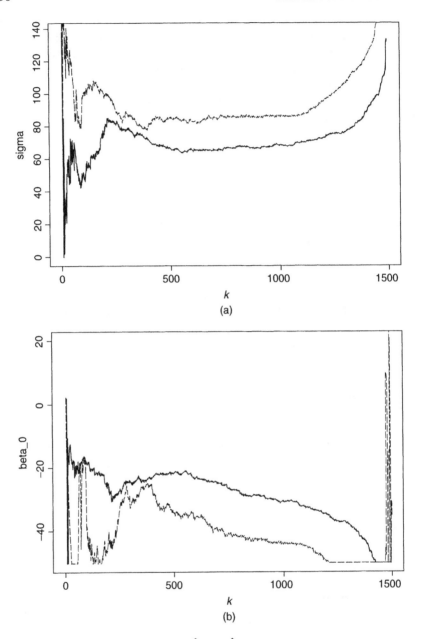

Figure 7.8   Condroz data: (a) $\hat{\sigma}$, (b) $\hat{\beta}_0$, (c) $\hat{\beta}_1$ as a function of k for GP regression model with constant threshold (broken line) and covariate dependent threshold (solid line) and (d) profile log-likelihood function and profile likelihood-based 95% confidence interval of β_1 at $k = 500$ (GP regression model with covariate dependent threshold).

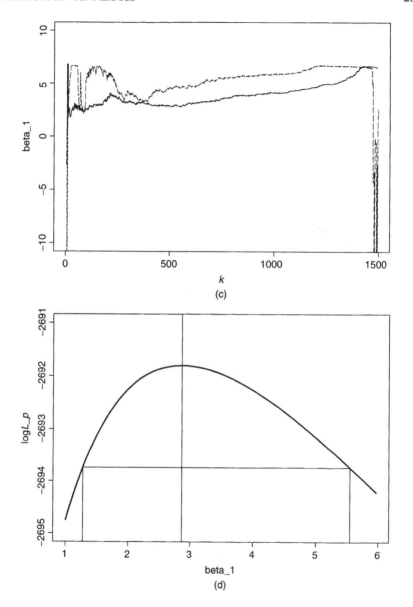

(c)

(d)

7.4.4   Estimation of extreme conditional quantiles

Estimates of extreme conditional quantiles can be obtained from the GP quantile function

$$
U\left(\frac{1}{p};\mathbf{x}\right) = \begin{cases} \frac{\sigma(\mathbf{x})}{\gamma(\mathbf{x})}\left(p^{-\gamma(\mathbf{x})} - 1\right), & \gamma(\mathbf{x}) \neq 0, \\ -\sigma(\mathbf{x})\log p, & \gamma(\mathbf{x}) = 0, \end{cases}
$$

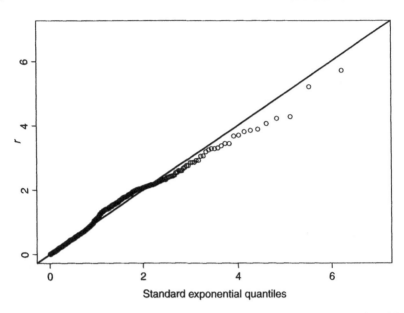

Figure 7.9   Condroz data: exponential quantile plot of the generalized residuals (7.19) at $k = 500$.

after replacing the unknown parameter functions by their respective maximum likelihood estimates. In case the GP distribution was used as an approximation to the conditional distribution of $Z - u_{\mathbf{x}}$ given $Z > u_{\mathbf{x}}$, then, on the basis of (C_{γ}^{*}), setting $z := u_{\mathbf{x}} + y$, for $u_{\mathbf{x}} \to z_{*}$

$$\bar{F}_{Z|\mathbf{x}}(z) \sim \begin{cases} \bar{F}_{Z|\mathbf{x}}(u_{\mathbf{x}}) \left(1 + \frac{\gamma(\mathbf{x})(z-u_{\mathbf{x}})}{\sigma(\mathbf{x})}\right)^{-\frac{1}{\gamma(\mathbf{x})}}, & \gamma(\mathbf{x}) \neq 0, \\ \bar{F}_{Z|\mathbf{x}}(u_{\mathbf{x}}) \exp\left(-\frac{z-u_{\mathbf{x}}}{\sigma(\mathbf{x})}\right), & \gamma(\mathbf{x}) = 0. \end{cases} \tag{7.20}$$

Solving (7.20) for z and replacing the unknown quantities by their respective point estimates yields

$$\hat{U}^{*}\left(\frac{1}{p}; \mathbf{x}\right) = \begin{cases} u_{\mathbf{x}} + \frac{\hat{\sigma}(\mathbf{x})}{\hat{\gamma}(\mathbf{x})}\left(\left(\frac{p}{\hat{F}_{Z|\mathbf{x}}(u_{\mathbf{x}})}\right)^{-\hat{\gamma}(\mathbf{x})} - 1\right), & \gamma(\mathbf{x}) \neq 0, \\ u_{\mathbf{x}} - \hat{\sigma}(\mathbf{x}) \log \frac{p}{\hat{F}_{Z|\mathbf{x}}(u_{\mathbf{x}})}, & \hat{\gamma}(\mathbf{x}) = 0. \end{cases}$$

If the covariate dependent threshold is set at a non-extreme regression quantile obtained with, for instance, the quantile regression methodology of Koenker and Bassett (1978), that is, $u_{\mathbf{x}} = \hat{U}^{*}(\frac{n}{k}; \mathbf{x})$, then the above expression reduces to

$$\hat{U}^{*}\left(\frac{1}{p}; \mathbf{x}\right) = \begin{cases} u_{\mathbf{x}} + \frac{\hat{\sigma}(\mathbf{x})}{\hat{\gamma}(\mathbf{x})}\left(\left(\frac{np}{k}\right)^{-\hat{\gamma}(\mathbf{x})} - 1\right), & \gamma(\mathbf{x}) \neq 0, \\ u_{\mathbf{x}} - \hat{\sigma}(\mathbf{x}) \log \frac{np}{k}, & \hat{\gamma}(\mathbf{x}) = 0. \end{cases} \tag{7.21}$$

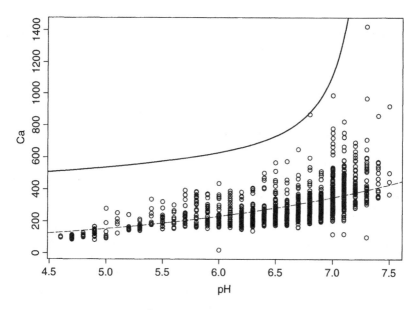

Figure 7.10 Condroz data: $\hat{U}^*(1000; pH)$ at $k = 500$ (solid line) and regression quantile 0.6667 (broken line) as a function of pH.

Example 7.1 (continued) In Figure 7.10, we show the Ca versus pH scatterplot together with $\hat{U}^*(1000; pH)$ (solid line) at $k = 500$. The broken line represents the threshold that is here set to regression quantile 0.6667.

7.5 Non-parametric Estimation

The methods considered so far all require the specification of a functional form for the model parameters. In practice, this often turns out to be a hard job. Moreover, completely parametric models are often smoother than a visual inspection of the data would suggest, and their lack of flexibility can lead to models with large numbers of parameters still providing poor fits. As an alternative to the parametric models discussed in the previous sections, the approach taken here is non-parametric, that is, we let the data themselves describe the functional relationship for the model parameters. In this section, we focus on modern smoothing techniques such as maximum penalized likelihood estimation (Green and Silverman (1994)) and local polynomial maximum likelihood estimation (Fan and Gijbels (1996)) and combine these with the GP distribution as approximate model for exceedances over high thresholds. Although we restrict the discussion to the GP modelling of exceedances, the non-parametric procedures may be combined with the GEV equally well. In this respect, some relevant references are Davison and Ramesh (2000), Hall and Tajvidi (2000a) and Pauli and Coles (2001).

Consider a random variable Z with associated covariate x such that $F_{Z|x} \in \mathcal{D}(G_{\gamma(x)})$. For simplicity of notation, we restrict the discussion to the single covariate case. Following $(\mathcal{C}_{\gamma}^*)$, the conditional distribution of $Y = Z - u_x$ given $Z > u_x$ can be well approximated by the GP distribution, at least when the threshold u_x is set sufficiently high.

7.5.1 Maximum penalized likelihood estimation

Given independent observations $Z_1, \ldots, Z_n$ and associated covariates $x_1 < \cdots < x_n$, we fit a GP regression model to the exceedances $Y_j = Z_j - u_{x_j}$, provided $Z_j > u_{x_j}$, $j = 1, \ldots, N_{u_x}$. Thereby, we do not impose a particular functional form that describes how the parameters σ and γ depend on the covariate. Note that the covariate is re-indexed in the sense that x_i denotes the x observation associated with exceedance Y_i. Take $\sigma_i = \exp(s(x_i))$ and $\gamma_i = t(x_i)$ with s and t unknown functions. The purpose of the notation is to stress that the σ_i and γ_i are parameters whose similarity over the covariate space is determined by the smoothness of the continuous functions s and t, which are assumed to be twice differentiable over $[x_1, x_n]$. The penalized log-likelihood function is defined as

$$
\Pi(s, t) = \sum_{i=1}^{N_{u_x}} \log g(Y_i; \exp(s(x_i)), t(x_i)) - \frac{1}{2}\lambda_1 \int (s''(x))^2 \, \mathrm{d}x
$$
$$
- \frac{1}{2}\lambda_2 \int (t''(x))^2 dx, \tag{7.22}
$$

where g is the GP density function. The penalized log-likelihood function is a difference of two terms. The first term is the classical log-likelihood function for the GP distribution, the second term is a penalty function whose magnitude reflects the integrated roughness of the functions s and t. The amount of smoothing is determined by the parameters λ_1 and λ_2. For small λ_1 and λ_2, the (over parametrized) log-likelihood dominates Π, leading to estimates that follow the data closely. Increasing λ_1 and λ_2 results in larger penalties and hence produces smoother fits. Computing the maximum penalized likelihood estimates involves the maximization of Π over the entire functional space of s and t. However, using the fundamental theorems concerning natural cubic splines given in sections 2.1 and 2.2 of Green and Silverman (1994), it can be shown that the maximization of (7.22) is equivalent to the maximization of a finite dimensional system corresponding to the σ_i and γ_i, $i = 1, \ldots, N_{u_x}$, followed by a cubic spline fit to construct the complete s and t curves.

Using the notation of Green and Silverman (1994), define band matrices Q and R as follows. Let $h_i = x_{i+1} - x_i$, $i = 1, \ldots, N_{u_x} - 1$. Define Q as a $N_{u_x} \times (N_{u_x} - 2)$ matrix with elements $q_{i,j}$, $i = 1, \ldots, N_{u_x}$, $j = 2, \ldots, N_{u_x} - 1$, given by

$$q_{i,j} = \begin{cases} h_{j-1}^{-1}, & \text{if } i = j - 1, \\ -h_{j-1}^{-1} - h_j^{-1}, & \text{if } i = j, \\ h_j^{-1}, & \text{if } i = j + 1, \\ 0, & \text{otherwise}, \end{cases}$$

and R as a $(N_{u_x} - 2) \times (N_{u_x} - 2)$ symmetric matrix with elements $r_{i,j}$, $i, j = 2, \ldots, N_{u_x-1}$, given by

$$r_{i,j} = \begin{cases} \frac{1}{3}(h_{i-1} + h_i), & \text{if } i = j, \\ \frac{1}{6}h_i, & \text{ifs } i = j - 1 \text{ or } i = j + 1, \\ 0, & \text{otherwise}. \end{cases}$$

Finally, define $K = QR^{-1}Q'$. Maximization of (7.22) with respect to s and t is equivalent to the maximization with respect to $\mathbf{s}$ and $\mathbf{t}$ of

$$\Pi(\mathbf{s}, \mathbf{t}) = \sum_{i=1}^{N_{u_x}} \log g(Y_i; \exp(s(x_i)), t(x_i)) - \frac{1}{2}\lambda_1 \mathbf{s}' K \mathbf{s} - \frac{1}{2}\lambda_2 \mathbf{t}' K \mathbf{t},$$

where $\mathbf{s}' = (s(x_1), \ldots, s(x_{N_{u_x}}))$ and $\mathbf{t}' = (t(x_1), \ldots, t(x_{N_{u_x}}))$, followed by a cubic spline fit to link the estimates together. The precision of the maximum penalized likelihood estimators $\hat{\mathbf{s}}$ and $\hat{\mathbf{t}}$ can be assessed by the bootstrap (Chavez-Demoulin (1999), Chavez-Demoulin and Davison (2001)) or on the basis of a Bayesian interpretation of the penalized likelihood function (Wahba (1978), Green and Silverman (1994), Pauli and Coles (2001)).

Example 7.1 (continued) In section 7.4, the Condroz data were analysed by fitting the regression model $GP(\sigma, \exp(\beta_0 + \beta_1 pH))$ to the conditional Ca exceedances. Here, maximum penalized likelihood estimation will be used to obtain a nonparametric estimate of $\gamma(pH)$. Similar to the analysis in section 7.4, we take a constant scale parameter σ. The maximum penalized likelihood estimates $\hat{\sigma}$ and $\hat{\mathbf{t}}$ are obtained by maximizing

$$\Pi(\sigma, \mathbf{t}) = \sum_{i=1}^{N_{u_{pH}}} \log g(Y_i; \sigma, t(pH_i)) - \frac{1}{2}\lambda \mathbf{t}' K \mathbf{t}$$

with respect to σ and $\mathbf{t}$. Note that maximum penalized likelihood estimation easily accommodates fully parametric specifications for some of the model parameters. In Figure 7.11(a), we show the 50 exceedances over regression quantile 0.9673. Figure 7.11(b) contains the maximum penalized likelihood estimates of $\gamma(pH)$ for three different values of λ together with the estimate obtained from fitting the parametric $GP(\sigma, \exp(\beta_0 + \beta_1 pH))$ regression model. The corresponding results for the 200 exceedances over regression quantile 0.866 are shown in Figure 7.11(c) and (d). Note how increasing the parameter λ, and hence increasing the penalty assigned to roughness, leads to smoother γ estimates.

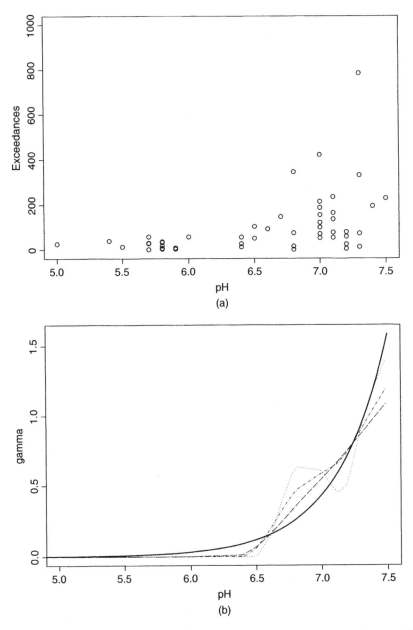

Figure 7.11   Condroz data: (a) exceedances over regression quantile 0.9673
($k = 50$) versus pH and (b) maximum penalized likelihood estimates of $\gamma(pH)$
for $\lambda = 0.1$ (broken line), $\lambda = 0.01$ (broken-dotted line) and $\lambda = 0.001$ (dotted
line) together with the estimate obtained with the parametric regression model
$GP(\sigma, \exp(\beta_0 + \beta_1 pH))$ (solid line). Figures (c) and (d) show the results obtained
with a threshold taken as regression quantile 0.866 ($k = 200$).

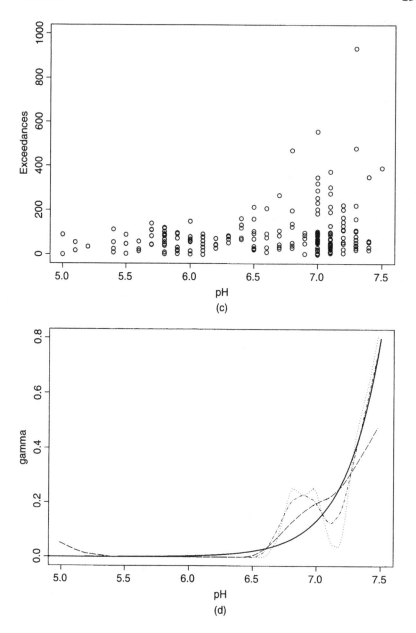

(c)

(d)

Similar to the parametric GP regression modelling, the fit of the maximum penalized likelihood estimate can be assessed by a visual inspection of the exponential quantile plot of the generalized residuals (7.19). Non-parametric estimates for extreme conditional quantiles can be obtained from (7.21), thereby replacing $\sigma(x)$ and $\gamma(x)$ by their maximum penalized likelihood estimates.

Example 7.1 (continued) We evaluate the fit of the maximum penalized likelihood estimation with $\lambda = 0.1$ at $k = 200$ by means of an exponential quantile plot of the generalized residuals (7.19), see Figure 7.12(a). The residuals scatter quite well around the first diagonal, indicating an appropriate fit of the Ca exceedances by the maximum penalized likelihood estimates. In Figure 7.12(b), we show the Ca versus pH scatterplot together with $\hat{U}^*(1000; pH)$ at $k = 200$. Here, $\hat{U}^*(1000; pH)$ is obtained by plugging the maximum penalized likelihood estimates obtained with $\lambda = 0.1$ into (7.21).

7.5.2 Local polynomial maximum likelihood estimation

Alternatively, the parameter functions σ and γ can be estimated by repeated local fits of the GP distribution. Consider independent random variables $Z_1, \ldots, Z_n$ and associated covariate observations $x_1, \ldots, x_n$. Suppose we are interested in estimating σ and γ at x^*. Fix a high *local* threshold u_{x^*} and compute the exceedances $Y_j = Z_i - u_{x^*}$, provided $Z_i > u_{x^*}$, $j = 1, \ldots, N_{u_{x^*}}$, where i is the index of the j-th exceedance in the original sample and $N_{u_{x^*}}$ denotes the number of exceedances over the threshold u_{x^*}. Re-index the covariates in an appropriate way such that x_i denotes the covariate observation associated with exceedance Y_i. Let h denote a bandwidth parameter. Since σ and γ are unknown, we approximate them by polynomials centred at x^*. Indeed, assuming σ and γ are p_1 and p_2 respectively, times differentiable we have, for $|x_i - x^*| \le h$,

$$\sigma(x_i) = \sum_{j=0}^{p_1} \beta_{1j}(x_i - x^*)^j + o(h^{p_1}),$$

$$\gamma(x_i) = \sum_{j=0}^{p_2} \beta_{2j}(x_i - x^*)^j + o(h^{p_2}),$$

where $\beta_{1j} = \frac{1}{j!} \frac{\partial^j \sigma(x_i)}{\partial x_i^j}\Big|_{x_i=x^*}$, $j = 0, \ldots, p_1$, and $\beta_{2j} = \frac{1}{j!} \frac{\partial^j \gamma(x_i)}{\partial x_i^j}\Big|_{x_i=x^*}$, $j = 0, \ldots, p_2$. The coefficients of these approximations can be estimated by local maximum likelihood fits of the GP distribution. Thereby, the contribution of the observations to the log-likelihood is governed by a kernel function K, where K is such that observations close to x^* receive more weight. Further, K is assumed to be a symmetric density function on $[-1, 1]$ and h rescales K as $K_h(x) = K(x/h)/h$. Clearly, h determines the amount of smoothing. The local polynomial maximum likelihood estimator $(\hat{\beta}'_1, \hat{\beta}'_2) = (\hat{\beta}_{10}, \ldots, \hat{\beta}_{1p_1}, \hat{\beta}_{20}, \ldots, \hat{\beta}_{2p_2})$ is the maximizer of the kernel weighted log-likelihood function

$$L_{N_{u_{x^*}}}(\beta_1, \beta_2)$$

$$= \frac{1}{N_{u_{x^*}}} \sum_{i=1}^{N_{u_{x^*}}} \log g\left(Y_i; \sum_{j=0}^{p_1} \beta_{1j}(x_i - x^*)^j, \sum_{j=0}^{p_2} \beta_{2j}(x_i - x^*)^j\right) K_h(x_i - x^*)$$

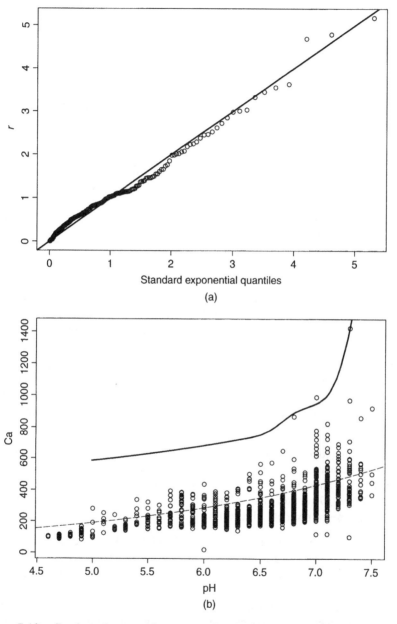

Figure 7.12   Condroz data: maximum penalized likelihood estimation with $\lambda = 0.1$ at $k = 200$: (a) exponential quantile plot of the generalized residuals (7.19) and (b) $\hat{U}^*(1000; pH)$ (solid line) and regression quantile 0.866 (broken line) as a function of pH.

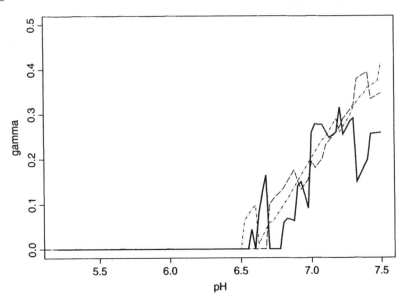

Figure 7.13   Condroz data: local polynomial maximum likelihood estimates of $\gamma(pH)$ obtained with a normal kernel function, $p_1 = 0$, $p_2 = 1$, $k = 75$ and $h = 0.3$ (solid line), $h = 0.5$ (broken line), $h = 0.7$ (broken-dotted line).

with respect to $(\beta_1', \beta_2') = (\beta_{10}, \ldots, \beta_{1p_1}, \beta_{20}, \ldots, \beta_{2p_2})$, where g denotes the GP density. Note that local polynomial fitting provides estimates of $\gamma(x^*)$ and $\sigma(x^*)$ and their derivatives up to order p_1 respectively p_2. Beirlant and Goegebeur (2004a) proved consistency and asymptotic normality of the local polynomial maximum likelihood estimator in case $\gamma(x) > 0$.

Example 7.1 (continued) In Figure 7.13, we show the local polynomial maximum likelihood estimates of $\gamma(pH)$ obtained with the above-described procedure with a normal kernel function, $p_1 = 0$, $p_2 = 1$ and $h = 0.3$ (solid line), $h = 0.5$ (broken line) and $h = 0.7$ (broken-dotted line). In this analysis, we set the local threshold at the 76th largest response observation within each window, so $k = 75$.

Consistent with this local approach, Hall and Tajvidi (2000a) proposed to use local quantile plots as a basis for the goodness-of-fit evaluation. Consider a window centred at x^* with length $2h$ and let $(Y_1', x_1'), \ldots, (Y_k', x_k')$ denote the observations (Y_i, x_i) for which $x_i \in [x^* - h, x^* + h]$. Given a local polynomial fit, we transform $(Y_1', x_1'), \ldots, (Y_k', x_k')$ into generalized residuals (7.19), thereby replacing the unknown parameter functions by their polynomial approximation, and use these to construct an exponential quantile plot. Non-parametric estimates of extreme quantiles of $F_{Z|x^*}$ can be obtained from

$$\hat{U}^* \left(\frac{1}{p}; x^* \right) = u_{x^*} + \frac{\hat{\sigma}(x^*)}{\hat{\gamma}(x^*)} \left[\left(\frac{n^* p}{k} \right)^{-\hat{\gamma}(x^*)} - 1 \right]$$

with n^* the number of observations in $[x^* - h, x^* + h]$, k the number of exceedances receiving positive weight and $\hat{\sigma}(x^*)$ and $\hat{\gamma}(x^*)$ denoting the local polynomial maximum likelihood estimates for respectively $\sigma(x^*)$ and $\gamma(x^*)$.

Example 7.1 (continued) We evaluate the local polynomial fit of the GP distribution with $h = 0.5$ and $k = 75$ at $pH^* = 7$ using a local exponential quantile plot of the generalized residuals, see Figure 7.14(a). In Figure 7.14(b), we show the threshold (broken line), which is set here at the 76th largest response observation within each window of length $2h$, and $\hat{U}^*(1000; pH)$ obtained with $h = 0.5$ and $k = 75$ as a function of pH.

7.6 Case Study

Insurance companies often use reinsurance contracts to safeguard themselves against portfolio contaminations caused by extreme claims. In an excess-of-loss reinsurance contract, the reinsurer pays for the claim amount in excess of a given retention. For the reinsurer, accurate description of the upper tail of the claim size distribution is of crucial importance for competitive price setting. In this process, taking covariate information into account allows to differentiate premiums according to the risks involved.

In this section, we illustrate how parametric and non-parametric GP modelling of conditional exceedances may help in describing tails of conditional claim-size distributions. Consider the AoN Re Belgium fire portfolio data introduced in section 1.3.3. In Figure 7.15(a), we show the claim size versus $\log(sum\ insured)$ $(\log(SI))$ scatterplot of claims generated by the office buildings portfolio. Given this point cloud with some really large claims for $7 < \log(SI) < 10$, we propose to use the covariate dependent threshold

$$\log(u_{SI}) = \theta_{0,p} + \theta_{1,p} \log(SI) + \theta_{2,p} \log^2(SI)$$

where $\theta_{0,p}$, $\theta_{1,p}$ and $\theta_{2,p}$ $(0 < p < 1)$ are estimated using the quantile regression methodology of Koenker and Bassett (1978). Figure 7.15(b) contains the regression quantiles $p = 0.9116$ (60 exceedances) and $p = 0.7875$ (150 exceedances). The exceedances over these regression quantiles are shown in Figure 7.15(c) and (d).

A $GP(\sigma(SI), \gamma(SI))$ regression model with $\log(\sigma(SI)) = \beta_{1,0} + \beta_{1,1} \log(SI) + \beta_{1,2} \log^2(SI)$ and $\log(\gamma(SI)) = \beta_{2,0} + \beta_{2,1} \log(SI) + \beta_{2,2} \log^2(SI)$ is fitted to both sets of exceedances. In Table 7.1 and Table 7.2, we show the resulting parameter estimates together with the value of the log-likelihood function for the full model and some reduced models. The reduced models are obtained by sequentially removing non-significant parameters. Significance is decided upon by performing a classical likelihood ratio test. For instance, at $k = 60$, the likelihood ratio test statistic for testing the hypothesis $H_0 : \beta_{1,2} = 0$ equals $2(143.8726 - 143.8677) = 0.0098$. This value is smaller than the critical value $\chi_1^2(0.95) = 3.8415$ and hence H_0 cannot be rejected at significance level $\alpha = 0.05$. In this way, removing non-significant

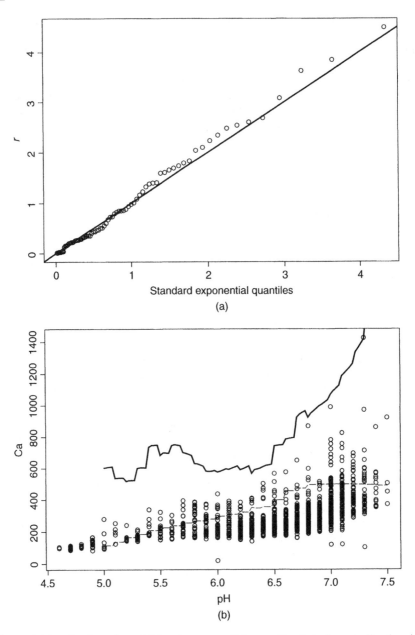

Figure 7.14   Condroz data: (a) local exponential quantile plot of generalized residuals at $pH^* = 7$ and (b) $\hat{U}^*(1000; pH)$ (solid line) and threshold u_{pH} (broken
line) as a function of pH.

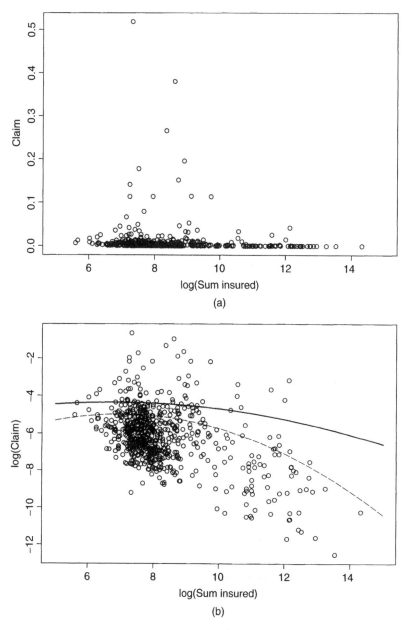

Figure 7.15   AoN Re Belgium data: (a) scatterplot of claim size versus $\log(SI)$, (b) $\log(claim)$ versus $\log(SI)$ with regression quantile 0.7875 (broken line) and regression quantile 0.9116 (solid line) superimposed, (c) exceedances over regression quantile 0.9116 ($k = 60$) and (d) exceedances over regression quantile 0.7875 ($k = 150$).

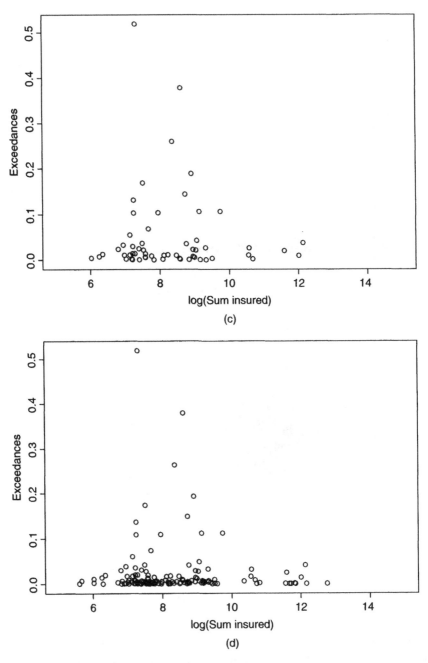

Figure 7.15 (*continued*)

Table 7.1 AoN Re Belgium data: GP modelling at $k = 60$.

Model	$\hat{\beta}_{1,0}$	$\hat{\beta}_{1,1}$	$\hat{\beta}_{1,2}$	$\hat{\beta}_{2,0}$	$\hat{\beta}_{2,1}$	$\hat{\beta}_{2,2}$	$\log L$
I	−4.4069	−0.0564	0.0080	−35.7814	8.7077	−0.5238	143.8726
II	−5.0653	0.0927	0	−35.9648	8.7633	−0.5280	143.8677
III	−4.2476	0	0	−35.7165	8.6223	−0.5150	143.5624
IV	−4.2806	0	0	0.3177	−0.0479	0	141.3641

Table 7.2 AoN Re Belgium data: GP modelling at $k = 150$.

Model	$\hat{\beta}_{1,0}$	$\hat{\beta}_{1,1}$	$\hat{\beta}_{1,2}$	$\hat{\beta}_{2,0}$	$\hat{\beta}_{2,1}$	$\hat{\beta}_{2,2}$	$\log L$
I	−13.0059	1.9757	−0.1262	−1.8591	0.3018	−0.0078	513.6545
II	−13.3194	2.0426	−0.1297	−1.1827	0.1538	0	513.6464
III	−14.4144	2.2034	−0.1330	0.1181	0	0	512.6212
IV	−4.1386	−0.1666	0	0.1091	0	0	509.9120

parameters one by one, we finally obtain a GP regression model with $\log(\sigma(SI)) = \beta_{1,0}$ and $\log(\gamma(SI)) = \beta_{2,0} + \beta_{2,1}\log(SI) + \beta_{2,2}\log^2(SI)$ (model III) at $k = 60$ and with $\log(\sigma(SI)) = \beta_{1,0} + \beta_{1,1}\log(SI) + \beta_{1,2}\log^2(SI)$ and $\log(\gamma(SI)) = \beta_{2,0}$ (model III) at $k = 150$. The final parameter functions are shown in Figure 7.16. At $k = 150$, the tail dependence on the covariate SI is modelled through the scale parameter of the GP distribution while at $k = 60$, that is, deeper in the conditional tails, tail dependence goes through the extreme value index. We evaluate the fit of both GP regression models based on an exponential quantile plot of the generalized residuals (7.19), see Figure 7.17. Both plots show residuals that scatter quite well around the first diagonal indicating a reasonable fit of the respective GP regression models.

Generally, the estimation of the extreme value index is not a goal on its own and is often performed as a kind of in-between step when ultimate interest is in extreme conditional quantiles or small conditional exceedance probabilities. Likewise, the primary interest of a reinsurer will not focus on the extreme value index estimates but rather on the claim level that will be exceeded only once in, say, 1000 claims, thereby taking into account the possible influence of covariate information. In Figure 7.18, we show the claim size versus SI scatterplot with the estimated 0.995 conditional quantile at $k = 60$ (solid line) and $k = 150$ (broken line) superimposed. At $k = 60$, the extreme value index was found to vary significantly with SI. This is reflected here in extreme conditional quantile estimates that follow the data better than the estimates obtained at $k = 150$.

As a final step, we analyse the AoN Re Belgium claim data in a non-parametric way. We restrict the non-parametric analysis to the 60 exceedances over regression quantile 0.9116 and fit a $GP(\sigma, \gamma(SI))$ regression model using maximum penalized likelihood estimation. Figure 7.19 contains the maximum penalized likelihood estimates of $\gamma(SI)$ for $\lambda = 0.1$ (solid line), $\lambda = 0.05$ (broken-dotted line),

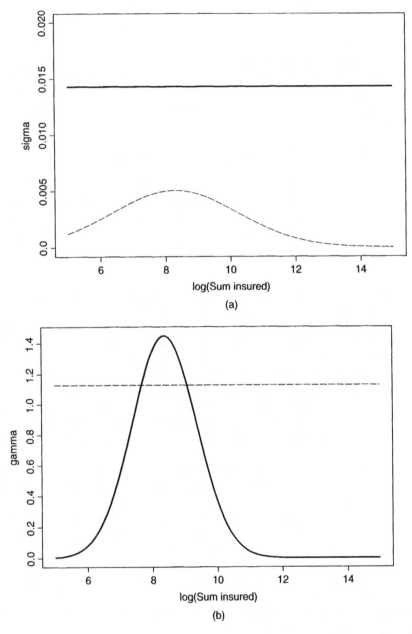

Figure 7.16   AoN Re Belgium data: (a) $\sigma(SI)$ and (b) $\gamma(SI)$ for $k = 60$ (solid line) and $k = 150$ (broken line).

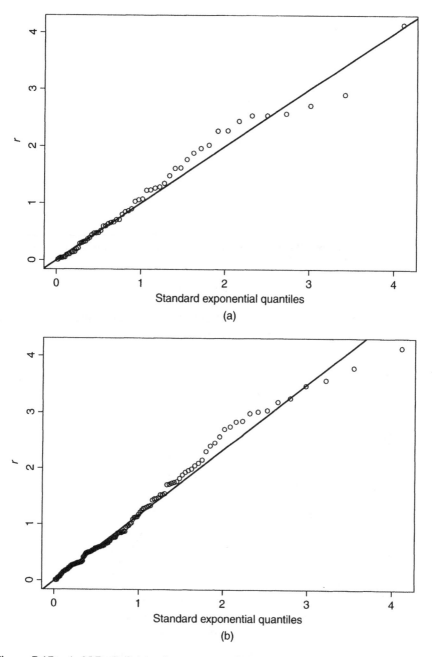

Figure 7.17   AoN Re Belgium data: exponential quantile plot of generalized residuals at (a) $k = 60$ and (b) $k = 150$.

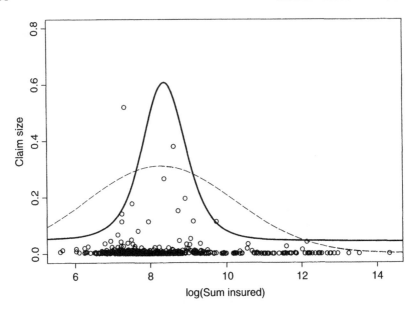

Figure 7.18   AoN Re Belgium data: $\hat{U}^*(200; SI)$ at $k = 60$ (solid line) and $k = 150$ (broken line) as a function of $\log(SI)$.

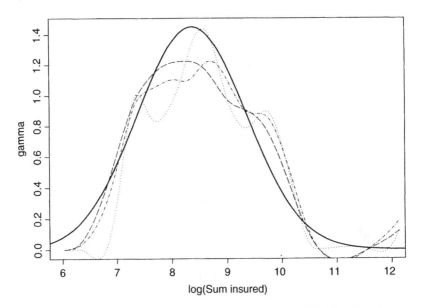

Figure 7.19   AoN Re Belgium data: maximum penalized likelihood estimates of $\gamma(SI)$ obtained at $k = 60$ for $\lambda = 0.1$ (broken line), $\lambda = 0.05$ (broken-dotted line) and $\lambda = 0.01$ (dotted line) together with the estimate obtained with the parametric regression model $GP(\sigma, \exp(\beta_0 + \beta_1 \log(SI) + \beta_2 \log^2(SI)))$ (solid line).

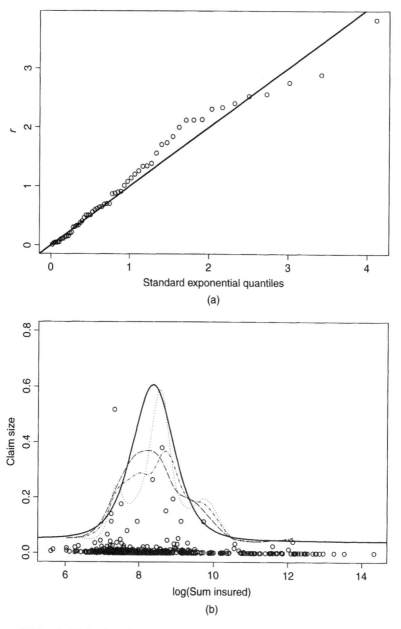

Figure 7.20   AoN Re Belgium data: maximum penalized likelihood estimation
at $k = 60$: (a) exponential quantile plot of the generalized residuals (7.19) with
$\lambda = 0.05$ and (b) $\hat{U}^*(200; SI)$ for $\lambda = 0.1$ (broken line), $\lambda = 0.05$ (broken-dotted
line), $\lambda = 0.01$ (dotted line) and parametric estimate (solid line).

$\lambda = 0.01$ (dotted line) and the parametric estimate obtained before (solid line). Remark that increasing λ, and hence giving more weight to the roughness penalty, produces smoother fits. Besides providing an alternative estimation procedure here the penalized likelihood estimates strongly confirm the previously performed completely parametric analysis. We evaluate the fit of the non-parametric estimate on the basis of the exponential quantile plot of the generalized residuals (7.19), see Figure 7.20(a). Non-parametric estimates of $U^*(200; SI)$ can be obtained from (7.21), thereby replacing the unknown parameters by the maximum penalized likelihood estimates. In Figure 7.20(b), we show the claim size versus $\log(SI)$ scatterplot with $\hat{U}^*(200; SI)$ for the three values of λ considered before superimposed. Again, the non-parametric analysis confirms the previously obtained completely parametric results.

8

MULTIVARIATE EXTREME VALUE THEORY

8.1 Introduction

Many problems involving extreme events are inherently multivariate. Gumbel and Goldstein (1964) already investigate the maximum annual discharges of the Ocmulgee River in Georgia at two different stations located upstream and downstream. Coles and Tawn (1996a) and Schlather and Tawn (2003) undertake a spatial analysis of daily rainfall extremes in south-west England in the context of risk assessment for hydrological structures such as reservoirs, river flood networks and drainage systems. De Haan and de Ronde (1998) and de Haan and Sinha (1999) estimate the probability that a storm will cause a certain sea-dike near the town of Petten, the Netherlands, to collapse because of a dangerous combination of sea level and wave height. In a financial context, Stărică (1999) analyses the occurrence of joint extreme returns in pairs of exchange rates of various European currencies (in the pre-euro era) versus the US dollar, while Longin and Solnik (2001) investigate the dependence between international equity markets in periods of high volatility. Surprisingly, multivariate techniques also come into play in the analysis of univariate time series, for instance, in the construction in Smith *et al.* (1997) of a Markov model for the extremes of a series of daily minimum temperatures recorded at Wooster, Ohio. These and many other examples demonstrate the need for statistical methods for analysing extremes of multivariate data.

Already at a first attempt of imagining how a statistical methodology for multivariate extremes could look like, we stumble upon a fundamental difficulty: what exactly makes a multivariate observation 'extreme'? Is it sufficient that just a single coordinate attains an exceptional value, or should it be extreme in all dimensions

Statistics of Extremes: Theory and Applications   J. Beirlant, Y. Goegebeur, J. Segers, and J. Teugels
© 2004 John Wiley & Sons, Ltd   ISBN: 0-471-97647-4

simultaneously? More technically, what meaning to attach in a multivariate setting to concepts such as order statistics, sample maximum, tail quantiles, threshold
exceedances, which are all so useful in univariate extreme value statistics? The
answers to these questions may depend on the situation at hand.

A fundamentally new issue that arises when there is more than one variable
is that of *dependence*. How do extremes in one variable relate to those in another
one? What are the possible dependence structures? And how to estimate them? As
in the univariate case, one of the aims of the exercise is to extrapolate outside the
range of the data. When more than one variable is involved, we can only hope to
reliably do so if we take proper account of the possibility of extremes in several
coordinates to occur jointly.

The study of multivariate extremes, then, splits apart into two components: the
marginal distributions and the dependence structure. This distinction is reflected in
both theory and practice. Typically, first, the margins are dealt with, and second,
after a transformation standardizing the margins to a common scale, the dependence. The first step merely involves the univariate techniques developed in the
previous chapters. The second step, however, is new.

We will discover that the class of possible limiting dependence structures cannot
be captured in a finite-dimensional parametric family. This is a major setback in
comparison to the univariate case, where we could rely on parametric techniques
based on the GEV or GP distributions. This time, we will either have to shift to
non-parametric techniques or construct sensible parametric models.

There exists a great variety of equivalent descriptions of extreme value dependence structures, and although each of them has its own merits, this multitude of
sometimes seemingly unconnected approaches may cause confusion and hamper
the flow from theory to practice. It is one of the aims of this text to fit the pieces
together and give the reader a panoramic view of the state of the art. Some new
insights and results form a pleasant by-product of this unification exercise.

The text on multivariate extremes is divided into two chapters. In the present
chapter, we explore the probability theory on extremes of a sample of independent,
identically distributed random vectors. This forms the necessary preparation for
the statistical methodology in the next chapter. All in all, the material is vast,
and a complete coverage of the literature would have filled a book by itself. The
interested reader may find further reading in, for instance, Galambos (1978, 1987),
Resnick (1987), Kotz and Nadarajah (2000), Coles (2001), Drees (2001), Reiss
and Thomas (2001), Fougères (2004), and of course the many papers cited in the
next two chapters.

The outline of the present chapter is as follows. In the remainder of this introduction, we formulate the multivariate version of the domain-of-attraction problem,
which, as in the univariate case, is a convenient starting point. In section 8.2, we
study multivariate extreme value distributions, focusing mainly on their dependence structure, while section 8.3 describes their domains of attraction. Some
additional topics are briefly touched upon in section 8.4. Section 8.5 summarizes
the essential things to know before attacking the statistical issues in Chapter 9.

Finally, the appendix (section 8.6) includes, amongst others, a directory of formulas connecting the various equivalent descriptions of multivariate extreme value distributions.

The multivariate domain-of-attraction problem

The road from univariate to multivariate extreme value theory is immediately confronted with an obstacle: there is no obvious way to order multivariate observations. Barnett (1976) considers not less than four different categories of order relations for multivariate data, each being of potential use. The most useful order relation in multivariate extreme value theory is a special case of what is called *marginal ordering*: for d-dimensional vectors $x = (x_1, \ldots, x_d)$ and $y = (y_1, \ldots, y_d)$, the relation $x \leq y$ is defined as $x_j \leq y_j$ for all $j = 1, \ldots, d$. Unlike in one dimension, not every two vectors can be ordered in this way—imagine two bivariate vectors, one at the upper-left of the other. The component-wise maximum of x and y, defined as

$$x \vee y := (x_1 \vee y_1, \ldots, x_d \vee y_d),$$

is in general different from both x and y.

Consider a sample of d-dimensional observations, $X_i = (X_{i,1}, \ldots, X_{i,d})$ for $i = 1, \ldots, n$. The sample maximum, M_n, is now defined as the vector of component-wise maxima, that is, the components of $M_n = \bigvee_{i=1}^{n} X_i$ are given by

$$M_{n,j} = \bigvee_{i=1}^{n} X_{i,j}, \qquad j = 1, \ldots, d.$$

Observe that the sample maximum need not be a sample point; in this sense, the definition might seem artificial. Still, from its study a rich theory emanates that leads to a broad set of statistical tools for analysing extremes of multivariate data.

Of course, we could just as well study the component-wise minimum rather than the maximum. But clearly, just as in the univariate case, results for one of the two can be immediately transferred to the other through the relation

$$\bigwedge_{i=1}^{n} X_i = -\bigvee_{i=1}^{n} (-X_i).$$

Therefore, we can, without loss of generality, focus on maxima alone. Notations will be greatly simplified if we adopt the following convention: unless mentioned otherwise, all operations and order relations between vectors are understood to be component-wise. Observe that we have already employed this convention in the definitions above of '$\leq$' and '$\vee$' for vectors.

The distribution function of the component-wise maximum, M_n, of an independent sample $X_1, \ldots, X_n$ from a distribution function F is given by

$$P[M_n \leq x] = P[X_1 \leq x, \ldots, X_n \leq x] = F^n(x), \qquad x \in \mathbb{R}^d.$$

Like in the univariate case, we will need to normalize M_n in some way in order to get a non-trivial limit distribution as the sample size tends to infinity. The domain-of-attraction problem then reads as follows: find sequences of vectors, $(a_n)_n$ and $(b_n)_n$, where $a_n > 0 = (0, \ldots, 0)$, such that $a_n^{-1}(M_n - b_n)$ converges in distribution to a non-degenerate limit, that is, such that there exists a d-variate distribution function G with non-degenerate margins such that

$$F^n(a_n x + b_n) \overset{\mathcal{D}}{\to} G(x), \qquad n \to \infty. \tag{8.1}$$

If (8.1) holds, we say that F is in the *(max-)domain of attraction* of G, notation $F \in D(G)$. Moreover, G is called a *(multivariate) extreme value distribution function*.

The study of equation (8.1) then splits into two parts: (i) characterize the class of extreme value distribution functions, and (ii) for a given extreme value distribution function, characterize its domain of attraction. We will take up these parts separately in the next two sections.

Before we start off, there is a simple but crucial observation to be made. Let F_j and G_j denote the jth marginal distribution functions of F and G respectively. Recall that, by assumption, G_j is non-degenerate. Since a sequence of random vectors can only converge in distribution if the corresponding marginal sequences do, we obtain for $j = 1, \ldots, d$,

$$F_j^n(a_{n,j} x_j + b_{n,j}) \overset{\mathcal{D}}{\to} G_j(x_j), \qquad n \to \infty.$$

Therefore, each G_j is by itself a univariate extreme value distribution function and F_j is in its domain of attraction. This has been extensively studied in Chapter 2. In the present chapter, then, we can focus on the dependence structures of F and G.

A final remark: since the marginal distributions of G are continuous, G itself is continuous, so that the convergence in (8.1) holds not only in distribution but also for every $x \in [-\infty, \infty]$—even uniformly.

8.2 Multivariate Extreme Value Distributions

Unlike the univariate case, multivariate extreme value distributions cannot be represented as a parametric family indexed by a finite-dimensional parameter vector. The reason is that the class of dependence structures is too large. Instead, the family of multivariate extreme value distributions is indexed, for instance, by a class of convex of functions, or, in another description, by a class of finite measures.

8.2.1 Max-stability and max-infinite divisibility

Let us start from equation (8.1). From the theory on univariate extremes in Chapter 2, we know that for positive integer k, there exist vectors $\alpha_k > 0$ and β_k such that $a_n^{-1} a_{nk} \to \alpha_k$ and $a_n^{-1}(b_{nk} - b_n) \to \beta_k$ as $n \to \infty$. But since, for positive integer k and $x \in \mathbb{R}^d$, also $F^{nk}(a_{nk}x + b_{nk}) \to G(x)$ as $n \to \infty$ as well as $a_{nk}x + b_{nk} = a_n\{a_n^{-1}a_{nk}x + a_n^{-1}(b_{nk} - b_n)\} + b_n$, we obtain

$$G^k(\alpha_k x + \beta_k) = G(x), \qquad x \in \mathbb{R}^d. \tag{8.2}$$

A d-variate distribution function G such that for every positive integer k we can find vectors $\alpha_k > 0$ and β_k such that (8.2) holds is called *max-stable*. The meaning is the same as in the univariate case: If $Y, Y_1, Y_2, \ldots$ are independent random vectors with distribution function G, then

$$\alpha_k^{-1}\left(\bigvee_{i=1}^k Y_i - \beta_k\right) \stackrel{\mathcal{D}}{=} Y, \qquad k = 1, 2, \ldots$$

Clearly, a max-stable distribution function is in its own domain of attraction; in particular, it must be an extreme value distribution function. This argument together with the previous paragraph shows that the classes of extreme value and max-stable distribution functions actually coincide.

A consequence of (8.2) is that $G^{1/k}$ is a distribution function for every positive integer k, that is, G is *max-infinitely divisible* (Balkema and Resnick 1977). In particular, there exists a measure, μ, on $[-\infty, \infty)$, such that

$$G(x) = \exp\{-\mu([-\infty, \infty) \setminus [-\infty, x])\}, \qquad x \in [-\infty, \infty], \tag{8.3}$$

whence the name *exponent measure*.

The exponent measure, μ, is in general not unique, and in the future we will always use the following particular choice. For $j = 1, \ldots, d$, let q_j be the lower end-point of the jth margin, G_j, of G, that is, $q_j = \inf\{x \in \mathbb{R} : G_j(x) > 0\}$. Define $q = (q_1, \ldots, q_d)$. Then, as $G(x) = 0$ for $x \not\geq q$, there exists an exponent measure μ that is concentrated on $[q, \infty) \setminus \{q\}$. Moreover, there is only one such exponent measure.

8.2.2 Exponent measure

Reduction to standard Fréchet margins

To study the dependence structure of a max-stable distribution, it is convenient to standardize the margins so that they are all the same. The precise choice of marginal distribution itself is not so important. Still, a particularly useful choice is that of standard Fréchet margins, as in that case the exponent measure must satisfy a useful homogeneity property. Connections with other choices for the margins employed in the literature are discussed in section 8.2.6.

Let $G_j^{\leftarrow}$ denote the quantile function of G_j, the jth margin of the max-stable distribution function G, that is, $G_j^{\leftarrow}(p) = x$ if and only if $G_j(x) = p$, where $0 < p < 1$. Observe that if, in the usual parametrization,

$$G_j(x_j) = \exp\left\{-\left(1 + \gamma_j \frac{x_j - \mu_j}{\sigma_j}\right)_+^{-1/\gamma_j}\right\}, \qquad x_j \in \mathbb{R}^d, \tag{8.4}$$

for some $\gamma_j \in \mathbb{R}$, $\mu_j \in \mathbb{R}$ and $\sigma_j > 0$, then

$$G_j^{\leftarrow}(e^{-1/z_j}) = \mu_j + \sigma_j \frac{z_j^{\gamma_j} - 1}{\gamma_j}, \qquad 0 < z_j < \infty,$$

with appropriate interpretations if $\gamma_j = 0$.

Let Y be a random vector with distribution function G, and let G_* be the distribution function of $(-1/\log G_1(Y_1), \ldots, -1/\log G_d(Y_d))$, that is,

$$G_*(z) = G\{G_1^{\leftarrow}(e^{-1/z_1}), \ldots, G_d^{\leftarrow}(e^{-1/z_d})\}, \qquad z \in (0, \infty). \tag{8.5}$$

Then the margins of G_* are standard Fréchet, as $P[-1/\log G_j(Y_j) \le z] = e^{-1/z}$ for $0 < z < \infty$. Conversely,

$$G(x) = G_*\{-1/\log G_1(x_1), \ldots, -1/\log G_d(x_d)\}, \qquad x \in \mathbb{R}^d, \tag{8.6}$$

where, taking appropriate limits, $-1/\log(0) := 0$ and $-1/\log(1) := \infty$.

Not only does G_* have max-stable margins, it is itself max-stable as well: Since $G_j^k(\alpha_{k,j} x_j + \beta_{k,j}) = G_j(x_j)$ for every $j = 1, \ldots, d$, $k = 1, 2, \ldots$, and $x_j \in \mathbb{R}$, it follows that

$$G_*^k(kz) = G_*(z), \qquad z \in \mathbb{R}^d; \ k = 1, 2, \ldots$$

In particular, $G_*^k(kz) = G_*^m(mz)$ for arbitrary positive integer k and m, and thus $G_*^r(rz) = G_*(z)$ for all positive rational r. By continuity,

$$G_*^s(sz) = G_*(z), \qquad z \in \mathbb{R}^d; \ 0 < s < \infty. \tag{8.7}$$

An extreme value distribution (function) with standard Fréchet margins is sometimes called *simple*.

Exponent measure

Let μ_* be an exponent measure of the simple extreme value distribution function G_*. Without loss of generality, we can assume that μ_* is concentrated on $[0, \infty) \setminus \{0\}$, so that

$$V_*(z) := -\log G_*(z) = \mu_*([0, \infty) \setminus [0, z]), \qquad z \in [0, \infty]. \tag{8.8}$$

Observe that $V_*(z) = \infty$ as soon as $z_j = 0$ for some $j = 1, \ldots, d$. Also, since the margins of G_* are standard Fréchet,

$$V_*(\infty, \ldots, \infty, z_j, \infty, \infty) = \mu_*(\{x \in [0, \infty) : x_j > z_j\}) = z_j^{-1} \tag{8.9}$$

for all $j = 1, \ldots, d$ and $0 < z_j < \infty$.

The exponent measures μ and μ_* of G and G_* are related in the following way. For $x \in [q, \infty]$ and $z \in [0, \infty]$ related by $z_j = -1/\log G_j(x_j)$,

$$\mu([q, \infty) \setminus [q, x]) = -\log G(x)$$

$$= -\log G_*(z) = \mu_*([0, \infty) \setminus [0, z]). \tag{8.10}$$

Equation (8.7) now implies

$$s\mu_*(s([0, \infty) \setminus [0, z])) = \mu_*([0, \infty) \setminus [0, z]), \qquad z \in [0, \infty); \; 0 < s < \infty.$$

By a measure-theoretic argument, this homogeneity relation actually holds for all Borel subsets of $[0, \infty) \setminus \{0\}$, that is,

$$\mu_*(s \cdot) = s^{-1}\mu_*(\cdot), \qquad 0 < s < \infty. \tag{8.11}$$

Stable tail dependence function

The *stable tail dependence function* is defined by

$$\begin{aligned} l(v) &= V_*(1/v_1, \ldots, 1/v_d) \\ &= \mu_*\big([0, \infty] \setminus [0, (1/v_1, \ldots, 1/v_d)]\big), \qquad v \in [0, \infty] \end{aligned} \tag{8.12}$$

(Huang 1992). In terms of the original max-stable distribution function G, it is given by

$$l(v) = -\log G\{G_1^{\leftarrow}(e^{-v_1}), \ldots, G_d^{\leftarrow}(e^{-v_d})\}, \qquad v \in [0, \infty]. \tag{8.13}$$

Conversely, we can reconstruct a max-stable distribution function G from its margins G_j and its stable tail dependence function l through

$$-\log G(x) = l\{-\log G_1(x_1), \ldots, -\log G_d(x_d)\}, \qquad x \in \mathbb{R}^d. \tag{8.14}$$

By (8.9) and (8.11), a stable tail dependence function l has the following properties:

(L1) $l(s \cdot) = sl(\cdot)$ for $0 < s < \infty$;

(L2) $l(e_j) = 1$ for $j = 1, \ldots, d$, where e_j is the jth unit vector in $\mathbb{R}^d$;

(L3) $v_1 \vee \cdots \vee v_d \leq l(v) \leq v_1 + \cdots + v_d$ for $v \in [0, \infty)$.

The upper and lower bounds in (L3) are itself valid stable tail dependence functions: the lower bound, $l(v) = v_1 \vee \cdots \vee v_d$, corresponds to complete dependence, $G(x) = G_1(x_1) \wedge \cdots \wedge G_d(x_d)$, whereas the upper bound, $l(v) = v_1 + \cdots + v_d$, corresponds to independence, $G(x) = G_1(x_1) \cdots G_d(x_d)$. Moreover, from (8.23) below it follows that

(L4) l is convex, that is, $l\{\lambda v + (1 - \lambda)w\} \leq \lambda l(v) + (1 - \lambda)l(w)$ for $\lambda \in [0, 1]$.

Note that, except for the bivariate case, properties (L1) to (L4) do not characterize the class of stable tail dependence functions, that is, a function l that satisfies (L1) to (L4) is not necessarily a stable tail dependence function. As a counter-example in the trivariate case, put $l(v_1, v_2, v_3) = (v_1 + v_2) \vee (v_2 + v_3) \vee (v_3 + v_1)$. Properties (L1) to (L4) are clearly fulfilled. Still, l cannot be a stable tail dependence function, because $l(1, 1, 0) = l(1, 0, 1) = l(0, 1, 1) = 2$ would imply pairwise independence and thus, as we will see in section 8.2.4, full independence, in contradiction with $l(1, 1, 1) = 2 \neq 3$.

8.2.3 Spectral measure

The homogeneity property (8.11) of the exponent measure μ_* yields a versatile representation in terms of (pseudo-)polar coordinates. We start from two arbitrary norms, $\|\cdot\|_1$ and $\|\cdot\|_2$, on $\mathbb{R}^d$. Typical choices include the L_p-norms $\|x\| = (|x_1|^p + \cdots + |x_d|^p)^{1/p}$ for $1 \leq p < \infty$ or the max-norm $\|x\| = \max(|x_1|, \ldots, |x_d|)$, see below. Let $\mathbb{S}_2 = \{\omega \in \mathbb{R}^d : \|\omega\|_2 = 1\}$ be the unit sphere with respect to the norm $\|\cdot\|_2$. Define the mapping T from $\mathbb{R}^d \setminus \{0\}$ to $(0, \infty) \times \mathbb{S}_2$ by

$$T(z) = (r, \omega), \qquad \text{where } r = \|z\|_1 \text{ and } \omega = z/\|z\|_2, \qquad (8.15)$$

that is, r is the radial part and ω the angular part of z. Observe that T is one-to-one and onto, because $T(z) = (r, \omega)$ if and only if $z = r\omega/\|\omega\|_1 = T^{-1}(r, \omega)$.

Now define a measure, S, on $\Xi = \mathbb{S}_2 \cap [0, \infty)$ by

$$S(B) = \mu_*\big(\{z \in [0, \infty) : \|z\|_1 \geq 1, \, z/\|z\|_2 \in B\}\big) \qquad (8.16)$$

for Borel subsets B of Ξ. The measure S is called the *spectral measure*. It is determined uniquely by the exponent measure μ_* and the chosen norms by (8.16) and (8.19) below. The homogeneity of μ_* expressed in (8.11) implies

$$\mu_*\big(\{z \in [0, \infty) : \|z\|_1 \geq r, \, z/\|z\|_2 \in B\}\big) = r^{-1} S(B)$$

for $0 < r < \infty$ and Borel subsets B of Ξ. The interpretation is that in polar coordinates (r, ω), the exponent measure μ_* factors into a product of two measures, one in the radial coordinate that is always equal to $r^{-2}dr$, and one in the angular coordinate, equal to the spectral measure S. This property is usually written as

$$\mu_* \circ T^{-1}(dr, d\omega) = r^{-2} dr \, S(d\omega), \qquad (8.17)$$

which is called the *spectral decomposition* of the exponent measure. It is essentially due to de Haan and Resnick (1977), who considered the special case where the two norms are equal to the Euclidean norm.

The spectral decomposition (8.17) can be used to calculate the integral of a real-valued function g on $[0, \infty) \setminus \{0\}$ with respect to μ_* by

$$\int_{[0,\infty)\setminus\{0\}} g(z)\mu_*(dz) = \int_\Xi \int_0^\infty g(r\omega/\|\omega\|_1)r^{-2}dr \, S(d\omega)$$

$$= \int_\Xi \int_0^\infty g(r\omega)r^{-2}dr \, \|\omega\|_1^{-1} S(d\omega). \qquad (8.18)$$

Conversely, for a real-valued, S-integrable function f on Ξ, we have

$$\int_\Xi f(\omega)S(d\omega) = \int_{\|z\|_1 \geq 1} f(z/\|z\|_2)\mu_*(dz). \qquad (8.19)$$

Combining (8.8) and (8.18), we can write $V_* = -\log G_*$ in terms of the spectral measure S:

$$V_*(z) = \int_{[0,\infty)\setminus\{0\}} 1\left(\bigvee_{j=1}^{d} \frac{y_j}{z_j} > 1\right) \mu_*(dy)$$

$$= \int_{\Xi} \bigvee_{j=1}^{d} \left(\frac{\omega_j}{\|\omega\|_1} \frac{1}{z_j}\right) S(d\omega), \qquad z \in [0, \infty]. \tag{8.20}$$

The requirement that the margins of G_* are standard Fréchet is equivalent to

$$\int_{\Xi} \frac{\omega_j}{\|\omega\|_1} S(d\omega) = 1, \qquad j = 1, \ldots, d. \tag{8.21}$$

Conversely, any positive measure S on Ξ satisfying (8.21) is the spectral measure of the d-variate extreme value distribution $G_* = \exp(-V_*)$ given by (8.20). In terms of the original max-stable distribution function G, we find, combining (8.6) and (8.20),

$$\log G(x) = \int_{\Xi} \bigwedge_{j=1}^{d} \left\{\frac{\omega_j}{\|\omega\|_1} \log G_j(x_j)\right\} S(d\omega), \qquad x \in \mathbb{R}^d, \tag{8.22}$$

with the convention $\log(0) = -\infty$. In case the two norms are equal, then the previous formulas simplify slightly as $\|\omega\|_1 = 1$ for $\omega \in \Xi$. Finally, combining (8.12) with (8.20) yields

$$l(v) = \int_{\Xi} \bigvee_{j=1}^{d} \left(\frac{\omega_j}{\|\omega\|_1} v_j\right) S(d\omega), \qquad v \in [0, \infty]. \tag{8.23}$$

A useful consequence of (8.23) is that the stable tail dependence function l is convex.

Independence and complete dependence

Two interesting special cases are those of independence and complete dependence. Let G be a multivariate extreme value distribution with spectral measure S as in (8.22). Let e_j denote the jth unit vector in $\mathbb{R}^d$, that is, the jth coordinate of e_j is one and all other coordinates are zero. Then

$$G(x) = \prod_{j=1}^{d} G_j(x_j), \qquad x \in \mathbb{R}^d,$$

that is, the margins of G are *independent*, if and only if S consists of point masses of size $\|e_j\|_1$ at the points $e_j/\|e_j\|_2$, that is, if

$$\int_{\Xi} f(\omega) S(d\omega) = \sum_{j=1}^{d} \|e_j\|_1 f(e_j/\|e_j\|_2),$$

for any real-valued, S-integrable function f on Ξ. On the other hand, let the point $\omega_0 = (\omega_0, \ldots, \omega_0)$ be the intersection of Ξ and the line $\{x \in \mathbb{R}^d : x_1 = \ldots = x_d\}$. Then

$$G(x) = \bigwedge_{j=1}^{d} G_j(x_j), \qquad x \in \mathbb{R}^d,$$

that is, the margins of G are *completely dependent*, if and only if S collapses to a single point mass of size $\|\omega_0\|_1/\omega_0$ at the point ω_0, that is, if

$$\int_{\Xi} f(\omega) S(d\omega) = \frac{\|\omega_0\|_1}{\omega_0} f(\omega_0),$$

for any real-valued, S-integrable function f on Ξ.

Special cases

We specialize the spectral decomposition (8.20) for a number of choices of the two norms, $\| \cdot \|_1$ and $\| \cdot \|_2$.

Sum-norm. The most popular choice for the two norms $\| \cdot \|_1$ and $\| \cdot \|_2$ is the *sum-norm*, $\|x\| = |x_1| + \cdots + |x_d|$. In that case, the spectral measure S is typically denoted by H, and the space it is defined on, Ξ, is equal to the *unit simplex*,

$$S_d = \{\omega \in [0, \infty) : \omega_1 + \cdots + \omega_d = 1\}.$$

Representations (8.23) and (8.22) become

$$l(v) = \int_{S_d} \bigvee_{j=1}^{d} (\omega_j v_j) H(d\omega), \qquad v \in [0, \infty], \tag{8.24}$$

$$\log G(x) = \int_{S_d} \bigwedge_{j=1}^{d} \{\omega_j \log G_j(x_j)\} H(d\omega), \qquad x \in \mathbb{R}^d, \tag{8.25}$$

and the requirement (8.21) on H reads

$$\int_{S_d} \omega_j H(d\omega) = 1, \qquad j = 1, \ldots, d. \tag{8.26}$$

In particular, the total mass of H is always $H(S_d) = d$. By (8.16), the measure H can obtained from the exponent measure μ_* through

$$H(B) = \mu_*(\{z \in [0, \infty) : z_1 + \cdots + z_d \geq 1, (z_1 + \cdots + z_d)^{-1} z \in B\}), \tag{8.27}$$

for Borel subsets B of S_d. Independence occurs if and only if H consists of unit point masses at the vertices $e_1, \ldots, e_d$ of the simplex S_d, while complete dependence occurs if and only if H consists of a single point mass of size d at the centre point $(1/d, \ldots, 1/d)$. Representation (8.25) already appears, without proof, in Galambos (1978).

In the bivariate case, $d = 2$, the unit simplex S_2 is usually identified with the unit interval $[0, 1]$ by identifying $(\omega, 1 - \omega)$ with ω. The spectral measure H is then defined on $[0, 1]$ and is given by

$$H([0, \omega]) = \mu_*\big(\{(z_1, z_2) \in [0, \infty)^2 : z_1 + z_2 \geq 1, z_1/(z_1 + z_2) \leq \omega\}\big), \quad (8.28)$$

for $\omega \in [0, 1]$. The constraints on H are

$$\int_{[0,1]} \omega H(d\omega) = 1 = \int_{[0,1]} (1 - \omega) H(d\omega). \quad (8.29)$$

The stable tail dependence function is given by

$$l(v_1, v_2) = \int_{[0,1]} (\omega v_1) \vee \{(1 - \omega)v_2\} H(d\omega), \qquad (v_1, v_2) \in [0, \infty]^2. \quad (8.30)$$

Euclidean norm.   Mainly in the bivariate case, other choices for the two norms have been considered as well. If, as originally in de Haan and Resnick (1977), both norms are equal to the Euclidean norm, $\|(x_1, x_2)\| = (|x_1|^2 + |x_2|^2)^{1/2}$, then by (8.22) and (8.23), identifying $\omega = (\cos\theta, \sin\theta)$ with $\theta \in [0, \pi/2]$,

$$l(v_1, v_2) = \int_{[0,\pi/2]} \{\cos(\theta)v_1\} \vee \{\sin(\theta)v_2\} S(d\theta),$$

$$\log G(x_1, x_2) = \int_{[0,\pi/2]} \{\cos(\theta) \log G_1(x_1)\} \wedge \{\sin(\theta) \log G_2(x_2)\} S(d\theta),$$

where S is a finite measure on $[0, \pi/2]$ such that

$$\int_{[0,\pi/2]} \cos(\theta) S(d\theta) = 1 = \int_{[0,\pi/2]} \sin(\theta) S(d\theta),$$

see (8.21). By (8.16), the spectral measure S and the exponent measure μ_* are related through

$$S([0, \theta]) = \mu_*\big(\{(z_1, z_2) \in [0, \infty)^2 : z_1^2 + z_2^2 \geq 1, z_2/z_1 \leq \tan(\theta)\}\big), \quad (8.31)$$

for $\theta \in [0, \pi/2]$. Independence occurs if and only if S puts unit point masses at the end-points 0 and $\pi/2$. On the other hand, complete dependence occurs if and only if S puts a single point mass of size $\sqrt{2}$ at the mid-point $\pi/4$.

Max-norm and Euclidean norm.   Einmahl et al. (1997) set the first norm equal to the max-norm, $\|(x_1, x_2)\|_1 = \max(|x_1|, |x_2|)$ and the second norm to the Euclidean norm, $\|(x_1, x_2)\|_2 = (|x_1|^2 + |x_2|^2)^{1/2}$. Identifying again $\omega = (\cos\theta, \sin\theta)$ with $\theta \in [0, \pi/2]$, we find from (8.23)

$$l(v_1, v_2) = \int_{[0,\pi/2]} \{\cot(\theta \vee \pi/4)v_1\} \vee \{\tan(\theta \wedge \pi/4)v_2\} S(d\theta) \quad (8.32)$$

as well as from (8.22)

$$\log G(x_1, x_2)$$
$$= \int_{[0,\pi/2]} \{\cot(\theta \vee \pi/4) \log G_1(x_1)\} \wedge \{\tan(\theta \wedge \pi/4) \log G_2(x_2)\} S(d\theta),$$

where S is a finite measure on $[0, \pi/2]$ such that

$$\int_{[0,\pi/2]} \cot(\theta \vee \pi/4) S(d\theta) = 1 = \int_{[0,\pi/2]} \tan(\theta \wedge \pi/4) S(d\theta),$$

see (8.21). The relation between the spectral measure S and the exponent measure μ_* is that

$$S([0, \theta]) = \mu_*\big(\{(z_1, z_2) \in [0, \infty)^2 : z_1 \vee z_2 \geq 1, \ z_2/z_1 \leq \tan(\theta)\}\big), \qquad (8.33)$$

for $\theta \in [0, \pi/2]$. Independence occurs if and only if S puts unit point masses at the end-points 0 and $\pi/2$. On the other hand, complete dependence occurs if and only if S degenerates to a unit point mass at the mid-point $\pi/4$.

The spectral measure S considered in (8.32) is often connected to an alternative exponent measure directly related to l. Let ψ denote the transformation of $[0, \infty]^2$ into itself given by $\psi(v_1, v_2) = (1/v_1, 1/v_2)$. Consider the measure $v = \mu_* \circ \psi$ on the space $(0, \infty]^2 \setminus \{(\infty, \infty)\}$. By (8.8) and (8.12),

$$l(v_1, v_2) = v\big((0, \infty]^2 \setminus ([v_1, \infty] \times [v_2, \infty])\big), \qquad (v_1, v_2) \in [0, \infty]^2.$$

The measure v is sometimes called the *exponent measure* as well. By (8.11), it satisfies the homogeneity property

$$v(s \cdot) = s v(\cdot), \qquad 0 < s < \infty.$$

The spectral measure S of (8.32) can be found from v through

$$S([0, \theta]) = v\big(\{(v_1, v_2) \in (0, \infty]^2 : v_1 \wedge v_2 \leq 1, \ v_1/v_2 \leq \tan(\theta)\}\big),$$

for $\theta \in [0, \pi/2]$. Independence occurs if and only if v is concentrated on the lines through infinity $\{(v_1, \infty) : 0 < v_1 < \infty\}$ and $\{(\infty, v_2) : 0 < v_2 < \infty\}$, whereas complete dependence occurs if and only if v is concentrated on the diagonal.

Spectral densities

Consider the spectral measure H of (8.27) on the unit simplex $S_d = \{\omega \in [0, \infty) : \omega_1 + \cdots + \omega_d = 1\}$ for $d \geq 2$. If G_* is absolutely continuous, then we may reconstruct the densities of H from the derivatives of the function $V_* = -\log G_*$. We say "densities" and not "density" because, in general, H may have a density on the interior of S_d and also on each of the lower-dimensional subspaces of S_d.

More specifically, the unit simplex S_d is partitioned in a natural way in so-called *faces*, with dimensions ranging from 0 (the d vertices) up to $d - 1$ (the interior of S_d). In particular, for a non-empty subset a of $\{1, \ldots, d\}$, define

$$S_{d,a} = \{\boldsymbol{\omega} \in S_d : \omega_j > 0 \text{ if } j \in a; \omega_j = 0 \text{ if } j \notin a\}.$$

For instance, if $d = 2$, we have

$$S_{d,\{1\}} = \{(1, 0)\},$$

$$S_{d,\{2\}} = \{(0, 1)\},$$

$$S_{d,\{1,2\}} = \{(\omega, 1 - \omega) : 0 < \omega < 1\}.$$

In $d = 3$ dimensions, we obtain three vertices, three edges, and the interior of the triangle. For general d, the sets $S_{d,a}$ partition S_d into $2^d - 1$ subsets.

Now let us consider the restriction of the spectral measure H to the face $S_{d,a}$. First, if a is a singleton, $\{j\}$, then $S_{d,a}$ is just the vertex $\{e_j\}$, the jth unit vector in $\mathbb{R}^d$. Even if G_* is absolutely continuous, the spectral measure H may still assign positive mass to these vertices; for instance, when the margins of G_* are independent, $H(\{e_j\}) = 1$ for all $j = 1, \ldots, d$. Let us denote this mass by $h_a = h_a(e_j)$, to be thought of as the density of H with respect to the unit point mass at e_j.

Next, let a be a subset of $\{1, \ldots, d\}$ with $|a|$, the number of elements of a, at least two. Clearly, the number of free variables in $S_{d,a}$ is $k = |a| - 1$. Now, assume that the spectral measure H has a density h_a on $S_{d,a}$, the latter set being identified with the open region Δ_k of $\mathbb{R}^k$ defined by

$$\Delta_k = \{\boldsymbol{u} \in (0, \infty)^k : u_1 + \cdots + u_k < 1\}.$$

More precisely, integrals over $S_{d,a}$ with respect to H may be calculated through

$$\int_{S_{d,a}} f(\boldsymbol{\omega}) H(d\boldsymbol{\omega}) = \int_{\Delta_k} f\{I_a(\boldsymbol{u})\} h_a(\boldsymbol{u}) du_1 \cdots du_k;$$

here I_a is the map identifying $\boldsymbol{u}$ in Δ_k with $I_a(\boldsymbol{u}) = \boldsymbol{\omega}$ in $S_{d,a}$, that is, if $a = \{j_1, \ldots, j_{k+1}\}$, then $\omega_{j_i} = u_i$ $(i = 1, \ldots, k)$, $\omega_{j_{k+1}} = 1 - (u_1 + \cdots + u_k)$, and $\omega_j = 0$ for $j \notin a$.

Coles and Tawn (1991) found a way to compute the spectral densities h_a from the partial derivatives of V_*. For $a = \{j_1, \ldots, j_m\} \subset \{1, \ldots, d\}$ and $(z_j)_{j \in a}$ such that $0 < z_j < \infty$, we have

$$\lim_{\substack{z_j \to 0 \\ j \notin a}} \frac{\partial^m V_*}{\partial z_{j_1} \cdots \partial z_{j_m}}(\boldsymbol{z})$$

$$= -r^{-(m+1)} h_a\left(\frac{z_{j_1}}{r}, \ldots, \frac{z_{j_{m-1}}}{r}\right) \qquad \text{where } r = \sum_{j \in a} z_j. \qquad (8.34)$$

A new proof of (8.34) is given in section 8.6.1.

It is useful to spell out (8.34) explicitly in the bivariate case and to rewrite it in terms of the stable tail dependence function, $l(v_1, v_2) = V_*(1/v_1, 1/v_2)$. As usual, we identify $(\omega, 1 - \omega)$ in S_2 with ω in $[0, 1]$. The point masses of H on 0 and 1 are

$$H(\{0\}) = -\lim_{z_1 \to 0} \frac{\partial V_*}{\partial z_2}(z_1, z_2) = \lim_{v_1 \to \infty} \frac{\partial l}{\partial v_2}(v_1, v_2),$$

$$H(\{1\}) = -\lim_{z_2 \to 0} \frac{\partial V_*}{\partial z_1}(z_1, z_2) = \lim_{v_2 \to \infty} \frac{\partial l}{\partial v_1}(v_1, v_2),$$

(8.35)

while its density on the interior of the unit interval is, for $0 < \omega < 1$,

$$h(\omega) = -\frac{\partial^2 V_*}{\partial z_1 \partial z_2}(\omega, 1 - \omega)$$

$$= -\{\omega(1 - \omega)\}^{-1} \frac{\partial^2 l}{\partial v_1 \partial v_2}(1 - \omega, \omega). \tag{8.36}$$

Example 8.1 The bivariate asymmetric logistic model (Tawn 1988a) is given by its stable tail dependence function

$$l(v_1, v_2) = (1 - \psi_1)v_1 + (1 - \psi_2)v_2 + \{(\psi_1 v_1)^{1/\alpha} + (\psi_2 v_2)^{1/\alpha}\}^\alpha.$$

Here $0 < \alpha \leq 1$ and $0 \leq \psi_j \leq 1$ for $j = 1, 2$. Computing the partial derivatives of l and applying (8.35) and (8.36), we find $H(\{0\}) = 1 - \psi_2$, $H(\{1\}) = 1 - \psi_1$, and

$$h(\omega) = (\alpha^{-1} - 1)(\psi_1 \psi_2)^{1/\alpha}\{\omega(1 - \omega)\}^{1/\alpha - 2}[\{\psi_1(1 - \omega)\}^{1/\alpha} + (\psi_2\omega)^{1/\alpha}]^{\alpha - 2}$$

(8.37)

for $0 < \omega < 1$.

Change of norms

The spectral measure S of an exponent measure μ_* depends on the choice of norms $\| \cdot \|_i$ $(i = 1, 2)$. This choice is basically a matter of convenience, and in the literature, different authors use different norms, see above. Still, the transition from one choice of norms to another only involves a simple formula.

Let $\| \cdot \|_i$ and $\| \cdot \|_i'$ $(i = 1, 2)$ be four norms on $\mathbb{R}^d$ and let S and S' be the spectral measures of the exponent measure μ_* w.r.t. $\| \cdot \|_i$ $(i = 1, 2)$ and $\| \cdot \|_i'$ $(i = 1, 2)$, respectively. The supports of S and S' are $\Xi = \{\omega \geq 0 : \|\omega\|_2 = 1\}$ and $\Xi' = \{\omega' \geq 0 : \|\omega'\|_2' = 1\}$. Then for a real-valued, S-integrable function f on Ξ, we have by (8.18) and (8.19)

$$\int_\Xi f(\omega)S(d\omega) = \int_{\Xi'} f\left(\frac{\omega'}{\|\omega'\|_2}\right) \frac{\|\omega'\|_1}{\|\omega'\|_1'} S'(d\omega').$$

In particular, for a Borel subset B of Ξ, we have

$$S(B) = \int_{\Xi'} \mathbf{1}\left(\frac{\omega'}{\|\omega'\|_2} \in B\right) \frac{\|\omega'\|_1}{\|\omega'\|_1'} S'(d\omega'). \tag{8.38}$$

Spectral functions

Alternatively, a simple multivariate distribution function G_* is max-stable if and only if there exist non-negative, integrable functions $f_1, \ldots, f_d$ on $[0, 1]$ satisfying $\int_0^1 f_j(t) dt = 1$ such that

$$-\log G_*(z) = \int_0^1 \bigvee_{j=1}^d \frac{f_j(t)}{z_j} dt, \qquad z \in [0, \infty] \tag{8.39}$$

(de Haan 1984). That this defines a distribution function follows from a special point-process construction described in the same paper; see also section 9.2.1, where this point-process construction will be used to motivate parametric models for multivariate extreme value distributions. Clearly, G_* has standard Fréchet margins and is max-stable.

To show that a representation (8.39) is always possible, let G_* be a simple multivariate extreme value distribution with spectral measure S for some equal choice of the two norms. Let $Q(\cdot) = S(\cdot)/S(\Xi)$, which is a probability measure on Ξ. By a multivariate extension of the probability integral transform, there exist non-negative functions $g_1, \ldots, g_d$ on $[0, 1]$ such that, with U uniformly distributed on $(0, 1)$, the distribution of the random vector $(g_1(U), \ldots, g_d(U))$ is Q (Skorohod 1956). Then for $z \in [0, \infty]$,

$$-\log G_*(z) = S(\Xi) \int_{S_d} \bigvee_{j=1}^d \frac{\omega_j}{z_j} Q(d\omega) = S(\Xi) \int_0^1 \bigvee_{j=1}^d \frac{g_j(t)}{z_j} dt.$$

Defining $f_j = S(\Xi) g_j$, we obtain (8.39).

In terms of the original max-stable distribution function G, we find, combining (8.6) and (8.39),

$$\log G(x) = \int_0^1 \bigwedge_{j=1}^d \{ f_j(t) \log G_j(x_j) \} dt, \qquad x \in \mathbb{R}^d. \tag{8.40}$$

Observe that the *spectral functions* f_j in (8.40) are not unique. In particular, independence arises as soon as the supports of the spectral functions are disjoint, while total dependence arises as soon as all spectral functions are equal.

8.2.4 Properties of max-stable distributions

The fact that a max-stable distribution function G is linked to its margins $G_1, \ldots, G_d$ by means of a spectral measure H as in (8.22) has large repercussions on its dependence structure.

Positive association

A multivariate extreme value distribution G, in the terminology of Lehmann (1966), is necessarily *positively quadrant dependent*, that is,

$$G(x) \geq G_1(x_1) \cdots G_d(x_d), \qquad x \in \mathbb{R}^d, \tag{8.41}$$

a property originally noted in the bivariate case by Sibuya (1960), and in the general case by Tiago de Oliveira (1962/1963). In particular, a random variable Y with distribution function G has $\mathrm{cov}[f_i(Y_i), f_j(Y_j)] \geq 0$ for any $1 \leq i, j \leq d$ and any pair of non-decreasing functions f_i and f_j such that the relevant expectations exist. Relation (8.41) follows from (8.14) and the fact that $l(v) \leq v_1 + \cdots + v_d$. Observe also that (8.41) implies that $G(x) > 0$ as soon as $G_j(x_j) > 0$ for all $j = 1, \ldots, d$.

Multivariate extreme value distributions satisfy even stronger concepts of positive dependence. Marshall and Olkin (1983) show that they are *associated* (Esary *et al* 1967) in the sense that $\mathrm{cov}[\xi(Y), \eta(Y)] \geq 0$ for every pair of non-decreasing functions ξ and η on $\mathbb{R}^d$ for which the relevant expectations exist; see also Resnick (1987) for an alternative proof. Bivariate extreme value distributions are shown to satisfy a property called *total positivity of order two* by Kimeldorf and Sampson (1987) and to be monotone regression dependent (Lehmann 1966) by Garralda Guillem (2000).

Independence and complete dependence

Next we turn to characterizations of the two extreme cases of independence and complete dependence. We start with the case of independence. Takahashi (1994) showed that $G(x) = G_1(x_1) \cdots G_d(x_d)$ *for all* $x \in \mathbb{R}^d$ *if and only if there exists* $y \in \mathbb{R}^d$ *with* $0 < G_j(y_j) < 1$ *for all* $j = 1, \ldots, d$ *such that* $G(y) = G_1(y_1) \cdots G_d(y_d)$.

The 'if'-part may be proved as follows. Denoting $v_j = -\log G_j(y_j)$, we must have $\int_{S_d} \{\sum_{j=1}^d (\omega_j v_j) - \bigvee_{j=1}^d (\omega_j v_j)\} H(d\omega) = 0$. Since the integrand is non-negative, the H-measure of the set where it is positive must be zero. But then, since $0 < v_j < \infty$ for all $j = 1, \ldots, d$, the set $\{\omega \in S_d : \exists 1 \leq i < j \leq d : \omega_i > 0, \omega_j > 0\}$ must have H-measure zero. Consequently, H is concentrated on the complement of the set above, which is equal to $\{e_1, \ldots, e_d\}$. Restriction (8.26) forces $H(\{e_j\}) = 1$ for all j, which by (8.25) implies independence.

A closely related characterization of independence, going back to Berman (1961), is in terms of the bivariate margins G_{ij} for $1 \leq i < j \leq d$, that is, the bivariate distribution functions of the pairs (Y_i, Y_j), where Y is a random vector with distribution function G. We have $G(x) = G_1(x_1) \cdots G_d(x_d)$ *for all* $x \in \mathbb{R}^d$ *if and only if there exists* $y \in \mathbb{R}^d$ *with* $0 < G_j(y_j) < 1$ *for all* $j = 1, \ldots, d$ *such that* $G_{ij}(y_i, y_j) = G_i(y_i)G_j(y_j)$ *for all* $1 \leq i < j \leq d$. In words, pairwise independence implies total independence. The proof is similar as the one of the characterization above.

On the other extreme is the case of complete dependence. Takahashi (1994) noted that $G(x) = G_1(x_1) \wedge \cdots \wedge G_d(x_d)$ *for all* $x \in \mathbb{R}^d$ *if and only if there exists*

$y \in \mathbb{R}^d$ with $0 < G_1(y_1) = \cdots = G_d(y_d) < 1$ such that $G(y) = G_1(y_1)$. The 'if'-part can be easily proven as follows. Denoting $v = -\log G_1(y_1) \in (0, \infty)$, we have by (8.26) and (8.25), for $i = 1, \ldots, d$,

$$\int_{S_d} \left(\bigvee_{j=1}^{d} \omega_j - \omega_i \right) H(d\omega) = \frac{1}{v} \left\{ \int_{S_d} \bigvee_{j=1}^{d} (\omega_j v) H(d\omega) - v \int_{S_d} \omega_i H(d\omega) \right\} = 0.$$

Since the integrand on the left is non-negative, the H-measure of the set $\{\omega \in S_d : \omega_1 \vee \cdots \vee \omega_d > \omega_i\}$ must be zero for all $i = 1, \ldots, d$. Hence H must be concentrated on the mid-point $(1/d, \ldots, 1/d)$. Restriction (8.26) then forces $H(\{(1/d, \ldots, 1/d)\}) = d$, which by (8.25) implies complete dependence.

Closure property

Finally, we mention the following closure properties of the class of max-stable distributions. If G is a max-stable distribution function with spectral measure S as in (8.22), then for all $0 < \beta < \infty$, the function G^β is a max-stable distribution function as well, its spectral measure being again S. More generally, if $G_1, \ldots, G_m$ are d-variate max-stable distribution functions such that for each $j = 1, \ldots, d$ the marginal distribution functions $G_{i,j}$ are the same for all $i = 1, \ldots, m$, then for all non-negative $\beta_1, \ldots, \beta_m$ such that $\beta_1 + \cdots + \beta_m > 0$, the distribution function

$$G = G_1^{\beta_1} \cdots G_m^{\beta_m} \tag{8.42}$$

is max-stable as well (Gumbel 1962), its spectral measure being $S = w_1 S_1 + \cdots + w_m S_m$, where $w_i = \beta_i / (\beta_1 + \cdots + \beta_m)$ and where S_i is the spectral measure of G_i. In particular, any convex combination of stable tail dependence functions is again a stable tail dependence function.

8.2.5 Bivariate case

Let G be a bivariate extreme value distribution function with margins G_1 and G_2. Apart from the spectral measure H or the stable tail dependence function l, various alternative ways to describe the dependence structure of G have been proposed in the literature.

Pickands dependence function

Quite popular is *Pickands dependence function*

$$A(t) = l(1 - t, t), \qquad t \in [0, 1] \tag{8.43}$$

(Pickands 1981). Equation (8.43) is Pickands original definition. Later authors, including Pickands himself, sometimes define Pickands dependence function as $l(t, 1 - t) = A(1 - t)$ for $t \in [0, 1]$. Pickands dependence function can be viewed as the restriction of the stable tail dependence function to the unit simplex. In a higher-dimensional setting, the restriction of the stable tail dependence function to the unit simplex is sometimes called *Pickands dependence function* as well.

The function A completely determines the stable tail dependence function l, as

$$l(v_1, v_2) = (v_1 + v_2)A\left(\frac{v_2}{v_1 + v_2}\right), \tag{8.44}$$

for $0 \leq v_j < \infty$ ($j = 1, 2$) such that $v_1 + v_2 > 0$. In particular, a bivariate max-stable distribution G is determined by its margins, G_1 and G_2, and its Pickands dependence function, A, through

$$G(y_1, y_2) = \exp\left[\log\{G_1(y_1)G_2(y_2)\}A\left(\frac{\log\{G_2(y_2)\}}{\log\{G_1(y_1)G_2(y_2)\}}\right)\right], \tag{8.45}$$

for $(y_1, y_2) \in \mathbb{R}^2$.

By (L3) and (L4), a Pickands dependence function A satisfies the following two properties:

(A1) $(1 - t) \vee t \leq A(t) \leq 1$ for $t \in [0, 1]$;

(A2) A is convex.

In (A1), the lower bound, $A(t) = (1 - t) \vee t$, corresponds to complete dependence, $G(x_1, x_2) = G_1(x_1) \wedge G_2(x_2)$, whereas the upper bound, $A(t) = 1$, corresponds to independence, $G(x_1, x_2) = G_1(x_1, x_2)$.

We can connect Pickands dependence function to the spectral measure H of μ_* with respect to the sum-norm, see (8.28). Combining (8.30) with (8.43), we find

$$A(t) = \int_{[0,1]} \{\omega(1 - t)\} \vee \{(1 - \omega)t\}H(d\omega) \tag{8.46}$$

$$= t\int_{[0,t]} (1 - \omega)H(d\omega) + (1 - t)\int_{(t,1]} \omega H(d\omega)$$

Now by (8.29),

$$\int_{(t,1]} \omega H(d\omega) = H((t, 1]) - \int_{(t,1]} (1 - \omega)H(d\omega)$$

$$= \{2 - H([0, t])\} - \left\{1 - \int_{[0,t]} (1 - \omega)H(d\omega)\right\}$$

$$= 1 - H([0, t]) + \int_{[0,t]} (1 - \omega)H(d\omega),$$

so that

$$A(t) = \int_{[0,t]} (1 - \omega)H(d\omega) + (1 - t)\{1 - H([0, t])\}.$$

Moreover, since

$$\int_{[0,t]} (1 - \omega)H(d\omega) = \int_{[0,t]} \int_{\omega}^{1} d\,u\,H(d\omega)$$

$$= \int_{0}^{1} \int_{[0,u\wedge t]} H(d\omega)\,du$$

$$= \int_{0}^{t} H([0,u])du + (1-t)H([0,t]),$$

we obtain the convenient formula

$$A(t) = 1 - t + \int_{0}^{t} H([0,\omega])d\omega, \qquad t \in [0,1].$$

In particular, H can be computed from A through

$$H([0,\omega]) = \begin{cases} 1 + A'(\omega) & \text{if } \omega \in [0,1), \\ 2 & \text{if } \omega = 1, \end{cases} \tag{8.47}$$

where A' is the right-hand derivative of A. The point masses of H at 0 and 1 are given by

$$H(\{0\}) = 1 + A'(0), \qquad H(\{1\}) = 1 - A'(1), \tag{8.48}$$

with $A'(1) = \sup_{0 \le t < 1} A'(t)$. If A' is absolutely continuous, then H is absolutely continuous on the interior of the unit interval with density $h = A''$. Incidentally, equation (8.47) shows that any real-valued function A defined on [0, 1] that satisfies the properties (A1)–(A2) is necessarily a Pickands dependence function, with the spectral measure H given by (8.47).

Pickands dependence function A can also be linked to the spectral measure S for a general choice of the two norms $\| \cdot \|_i$ ($i = 1, 2$). Combining (8.23) and (8.43),

$$A(t) = \int_{\Xi} \left\{ (1-t)\frac{\omega_1}{\|(\omega_1,\omega_2)\|_1} \right\} \vee \left\{ t\frac{\omega_2}{\|(\omega_1,\omega_2)\|_1} \right\} S(d(\omega_1,\omega_2)). \tag{8.49}$$

Retrieving S from A is more difficult. First, we need to find H in terms of A using (8.47), and second, we need to compute S in terms of H using (8.38), which specializes to

$$S(B) = \int_{[0,1]} \mathbf{1}\left\{ \frac{(\omega, 1-\omega)}{\|(\omega, 1-\omega)\|_2} \in B \right\} \|(\omega, 1-\omega)\|_1 H(d\omega)$$

for Borel subsets B of Ξ.

Huang's level sets. In some sense dual to Pickands dependence function, A, are the level sets of l,

$$Q_c = \{(v_1, v_2) \in [0, \infty)^2 : l(v_1, v_2) = c\}, \qquad 0 < c < \infty,$$

first studied by Huang (1992); see also de Haan and de Ronde (1998). Clearly,

$$Q_c = \{(r\omega, r(1 - \omega)) : 0 \le \omega \le 1, r = c/A(1 - \omega)\}, \qquad 0 < c < \infty.$$

The set Q_c is the graph of a non-increasing, concave function through the points $(0, c)$ and $(c, 0)$, and $Q_c = cQ_1$. From Q_c, we can reconstruct A and hence l. Independence occurs if $Q_c = \{(v_1, v_2) \in [0, \infty)^2 : v_1 + v_2 = c\}$, and complete dependence occurs if $Q_c = \{(v_1, v_2) \in [0, \infty)^2 : v_1 \vee v_2 = c\}$.

Some history

The oldest descriptions of bivariate extreme value distributions date back to Tiago de Oliveira (1958), Sibuya (1960) and Geffroy (1958/59); see also the early review by Gumbel (1962). Each of these authors introduced a different function to characterize the dependence structure. However, the representation discovered by Pickands (1981) turned out to be far more convenient than its predecessors and reduced the popularity of the latter to virtually zero. Still, Obrenetov (1991) studied multivariate extensions of the dependence functions of Tiago de Oliveira and Sibuya.

Tiago de Oliveira (1958, 1962) obtained the representation

$$G(x_1, x_2) = \exp\left[\log\{G_1(x_2)G_2(x_2)\}k\left(\log\left\{\frac{\log G_1(x_1)}{\log G_2(x_2)}\right\}\right)\right].$$

The dependence function k is related to Pickands dependence function A by

$$k(x) = A\left(\frac{1}{e^x + 1}\right), \qquad x \in \mathbb{R}. \tag{8.50}$$

Since $(1 + t)k(\log t) = l(t, 1)$ for $0 < t < \infty$, necessary and sufficient conditions for a function k on $\mathbb{R}$ to be a Tiago de Oliveira dependence function are (i) $t \vee 1 \le (1 + t)k(\log t) \le t + 1$ and (ii) $(1 + t)k(\log t)$ is a convex function of t.

Next, Geffroy (1958/59) considered the representation

$$G(x_1, x_2) = \exp\left[\left\{1 + \varphi\left(\frac{\log G_1(x_1)}{\log G_2(x_2)}\right)\right\}\log G_2(x_2)\right].$$

In terms of the stable tail dependence function l, we have

$$\varphi(t) = l(t, 1) - 1, \qquad 0 < t < \infty.$$

Necessary and sufficient conditions for a function φ on $(0, \infty)$ to be a Geffroy dependence function are (i) $0 \vee (t - 1) \le \chi(t) \le t$ and (ii) χ is convex.

Finally, Sibuya (1960) introduced the representation

$$G(x_1, x_2) = \exp\left[\left\{1 + \chi\left(\frac{\log G_2(x_2)}{\log G_1(x_1)}\right)\right\} \log\{G_1(x_1)\} + \log\{G_2(x_2)\}\right].$$

In terms of the stable tail dependence function, l, we have

$$\chi(t) = l(1, t) - (1 + t), \qquad 0 < t < \infty.$$

Hence, necessary and sufficient conditions for a function χ on $(0, \infty)$ to be a Sibuya dependence function are (i) $-(t \wedge 1) \leq \chi(t) \leq 0$ and (ii) χ is convex.

8.2.6 Other choices for the margins

Reductions to other margins than standard Fréchet have been considered in the literature as well, other popular distributions being the exponential, extreme value Weibull, Gumbel, or uniform distribution. Although of course the choice of marginal distribution essentially makes no difference, some properties or characterizations are most naturally seen for one particular choice. Also, different choices sometimes motivate different statistical methods.

Exponential margins

One such choice, by Pickands (1981), is the standard exponential distribution, which is a univariate extreme value distribution for minima rather than for maxima. Let the random vector Y have the extreme value distribution function G. Then $(-\log G_1(Y_1), \ldots, -\log G_d(Y_d))$ has a multivariate extreme value distribution for minima with standard exponential margins, $P[-\log G_j(Y_j) \leq v] = 1 - e^{-v}$ for $v \geq 0$. Its joint survivor function is given by

$$P[-\log G_1(Y_1) > v_1, \ldots, -\log G_d(Y_d) > v_d] = \exp\{-l(v)\} \tag{8.51}$$

for $v \in [0, \infty]$. In the bivariate case,

$$P[-\log G_1(Y_1) > v_1, -\log G_2(Y_2) > v_2] = \exp\left\{-(v_1 + v_2)A\left(\frac{v_2}{v_1 + v_2}\right)\right\},$$

for $(v_1, v_2) \in [0, \infty]^2$.

Extreme value Weibull margins

Rather than exponential margins, Falk et al. (1994) prefer extreme value Weibull or reversed exponential margins. For $w \in [-\infty, 0]$, we have

$$P[\log G_1(Y_1) \leq w_1, \ldots, \log G_d(Y_d) \leq w_d] = \exp\{-l(-w)\},$$

with marginal distribution $P[\log G_j(Y_j) \leq w] = e^w$ for $w \leq 0$. In the bivariate case,

$$P[\log G_1(Y_1) \leq w_1, \log G_2(Y_2) \leq w_2] = \exp\left\{(w_1 + w_2)A\left(\frac{w_2}{w_1 + w_2}\right)\right\},$$

for $(w_1, w_2) \in [-\infty, 0]^2$.

Gumbel margins

In the early days of multivariate extreme value theory, it was customary to standardize to Gumbel margins, probably by the influence of the classical monograph by Gumbel (1958). Recall that the Gumbel distribution function is defined by $\Lambda(x) = \exp(-e^{-x})$ for $x \in \mathbb{R}$. If G is a multivariate extreme value distribution function, then the distribution function of the random vector $(-\log\{-\log G_1(Y_1)\}, \ldots, -\log\{-\log G_d(Y_d)\})$ is, with slight abuse of notation, given by

$$\Lambda(x) = \exp\{-l(e^{-x_1}, \ldots, e^{-x_d})\}, \qquad x \in \mathbb{R}^d,$$

which is a multivariate extreme value distribution function with Gumbel margins. In the bivariate case, we find

$$\Lambda(x_1, x_2) = \exp\left\{-(e^{-x_1} + e^{-x_2})k(x_2 - x_1)\right\}, \qquad (x_1, x_1) \in \mathbb{R}^2,$$

where k is Tiago de Oliveira's dependence function (8.50).

Uniform margins

A popular way to describe the dependence structure of a multivariate distribution function is through its *copula*. In general, for any multivariate distribution function F with margins $F_1, \ldots, F_d$, there exists a distribution function C_F with uniform margins on $(0, 1)$ such that

$$F(x) = C_F\{F_1(x_1), \ldots, F_d(x_d)\}, \qquad x \in \mathbb{R}^d$$

(Sklar 1959). Such a C_F is called a copula for F. If the margins, F_j, are continuous, then the copula, C_F, is unique and is given by

$$C_F(u) = F\{F_1^{\leftarrow}(u_1), \ldots, F_d^{\leftarrow}(u_d)\}, \qquad u \in [0, 1]^d,$$

so C_F is the distribution function of the random vector $(F_1(X_1), \ldots, F_d(X_d))$, where X is a random vector with distribution function F. Here $F_j^{\leftarrow}$ denotes the quantile function of F_j, defined by $F_j^{\leftarrow}(p) = \inf\{x \in \mathbb{R} : F(x) \geq p\}$.

The copula of a multivariate extreme value distribution G is the distribution function of the random vector $(G_1(Y_1), \ldots, G_d(Y_d))$ and is given by

$$C_G(u) = \exp[-l\{-\log(u_1), \ldots, -\log(u_d)\}], \qquad u \in [0, 1]^d, \qquad (8.52)$$

see (8.14). Such a copula necessarily satisfies the stability property

$$C_G^s(\mathbf{u}) = C_G(u_1^s, \ldots, u_d^s), \qquad \mathbf{u} \in [0, 1]^d. \tag{8.53}$$

Conversely, any copula that satisfies (8.53) is the copula of a multivariate extreme value distribution. A bivariate extreme value copula can be written in terms of Pickands dependence function as

$$C_G(u, v) = \exp\left[\log(uv) A\left\{\frac{\log(v)}{\log(uv)}\right\}\right], \qquad (u, v) \in [0, 1]^2, \tag{8.54}$$

see (8.44) and (8.52).

8.2.7 Summary measures for extremal dependence

The dependence structure of a max-stable distribution can be described in various ways: the preceding paragraphs featured exponent measures, spectral measures, the stable tail dependence function, Pickands dependence function, the copula, and so on. These quantities are infinite-dimensional objects and therefore not always easy to handle. A possible solution consists of choosing a finite-dimensional but hopefully large enough sub-class of dependence structures, that is, restricting attention to a parametric model (section 9.2). An alternative solution is to summarize the main properties of the dependence structure in a number of well-chosen coefficients that give a rough but representative picture of the full dependence structure.

Extremal coefficients

Let G be a max-stable distribution function with margins $G_1, \ldots, G_d$, spectral measure S with respect to two norms $\|\cdot\|_i$ $(i = 1, 2)$ on $\mathbb{R}^d$, and stable tail dependence function l. For a non-empty subset V of $\{1, \ldots, d\}$, let $\mathbf{e}_V$ be the d-dimensional vector of which the jth coordinate is one or zero according to $j \in V$ or $j \notin V$. For such V, the coefficients

$$\theta_V = l(\mathbf{e}_V) = \int_{\Xi} \bigvee_{j \in V} (\omega_j / \|\boldsymbol{\omega}\|_1) \, S(d\boldsymbol{\omega}) \tag{8.55}$$

satisfy

$$P[Y_j \leq G_j^{\leftarrow}(p), \forall j \in V] = p^{\theta_V}, \qquad 0 < p < 1,$$

where $\mathbf{Y}$ is a random vector with distribution function G (Coles 1993; Smith 1991). In particular, stronger dependence corresponds to smaller extremal coefficients θ_V. Clearly, $\theta_\emptyset = 0$ and $\theta_{\{j\}} = 1$ for all $j = 1, \ldots, d$, so that the only relevant coefficients θ_V are those for which V has at least two elements.

Hence, in the bivariate case, the only non-trivial coefficient is

$$\theta = \theta_{\{1,2\}} = l(1, 1) = 2A(1/2), \tag{8.56}$$

where A is Pickands dependence function (8.43). The coefficient θ must lie in the interval $[1, 2]$, and satisfies

$$P[Y_1 \le G_1^{\leftarrow}(p), \; Y_2 \le G_2^{\leftarrow}(p)] = p^{\theta}, \qquad 0 < p < 1.$$

In view of the conditions (A1)–(A2) on A, it is clear that θ strongly restricts the shape of A. In particular, independence occurs if and only if $\theta = 2$, while complete dependence occurs if and only if $\theta = 1$.

In the d-dimensional case, we have $1 \le \theta_V \le |V|$ for non-empty $V \subset \{1, \ldots, d\}$, where $|V|$ denotes the number of elements of V. The upper and lower bounds correspond to independence and complete dependence, respectively. Conversely, $\theta_{\{1,\ldots,d\}} = d$ implies independence, whereas $\theta_{\{1,\ldots,d\}} = 1$ implies complete dependence, as follows from the characterizations due to Takahashi (1994) given earlier. Schlather and Tawn (2002, 2003) give necessary and sufficient conditions on a collection of numbers θ_V indexed by the non-empty subsets V of $\{1, \ldots, d\}$ to be the extremal coefficients of a multivariate extreme value distribution.

Other summary measures for bivariate dependence

Two popular distribution-free measures of dependence between the components of a bivariate random vector are *Kendall's tau* and *Spearman's rho*. Applied to a bivariate max-stable distribution, they can also be used as useful summaries of the dependence structure.

Let F be a bivariate distribution function, and let (X_1, Y_1) and (X_2, Y_2) be independent random vectors with distribution function F. Kendall's tau is defined by

$$\tau = P[(X_1 - X_2)(Y_1 - Y_2) > 0] - P[(X_1 - X_2)(Y_1 - Y_2) < 0], \qquad (8.57)$$

that is, the difference between the probabilities of concordance and discordance. If the margins, F_X and F_Y, of F are continuous, and if C_F is the (necessarily unique) copula function of F, then τ is given by

$$\tau = 4E[C_F(U, V)] - 1,$$

where $(U, V) = (F_X(X), F_Y(Y))$ has distribution function C (Nelsen 1999). Next, Spearman's rho is defined as the Pearson correlation coefficient of (U, V), that is,

$$\rho_S = \text{corr}(U, V) = 12E[UV] - 3.$$

Tiago de Oliveira (1980) already gave expressions for Kendall's tau and Spearman's rho of a bivariate extreme value distribution G in terms of his dependence function k, see (8.50). These expressions were rediscovered later, but then in terms of Pickands dependence function A, see (8.54). Let $A'(t)$ be the right-hand derivative of A in $t \in [0, 1)$; also let $A'(1) = \sup_{0 \le t < 1} A'(t)$. Then Kendall's tau is given by

$$\tau = \int_0^1 \frac{t(1-t)}{A(t)} \, dA'(t)$$

$$= 1 - \int_0^1 \left\{ 1 + (1-t)\frac{A'(t)}{A(t)} \right\} \left\{ 1 - t\frac{A'(t)}{A(t)} \right\} dt,$$

(Ghoudi et al 1998; Hürlimann 2003), and Spearman's rho by

$$\rho_S = \mathrm{corr}[G_1(Y_1), G_2(Y_2)] = 12 \int_0^1 \frac{1}{\{1 + A(t)\}^2} \, dt - 3$$

(Hürlimann 2003); see also the unpublished 1995 Université Laval doctoral dissertation by A Khoudraji. For both τ and ρ_S, the extreme cases 0 and 1 correspond to independence and complete dependence, respectively. Hürlimann (2003) also shows that for bivariate extreme value copulas,

$$-1 + (1 + 3\tau)^{1/2} \le \rho_S \le \min\left(\frac{3}{2}\tau, 2\tau - \tau^2\right),$$

thereby proving a special case of a conjecture of Hutchinson and Lai (1990).

To conclude, dependence measures for bivariate extreme value distributions can also be obtained by studying the correlation of the two components of the random vector for a particular choice of marginal distributions. In all cases, they can be expressed in terms of Pickands dependence function, A. First, the reduction to uniform margins on $(0, 1)$ leads to Spearman's rho, ρ_S. Next, choosing Gumbel margins, Tiago de Oliveira (1980) obtains

$$\mathrm{corr}[-\log\{-\log G_1(Y_1)\}, -\log\{-\log G_2(Y_2)\}] = -\frac{6}{\pi^2} \int_0^1 \frac{\log A(t)}{t(1-t)} dt. \quad (8.58)$$

Finally, Tawn (1988a), choosing standard exponential margins, mentions

$$\mathrm{corr}[-\log G_1(Y_1), -\log G_2(Y_2)] = \int_0^1 \frac{1}{A^2(t)} dt - 1. \quad (8.59)$$

For all correlation coefficients, the two extreme cases 0 and 1 correspond to independence and complete dependence, respectively.

8.3   The Domain of Attraction

Consider again the domain-of-attraction equation

$$\lim_{n \to \infty} F^n(a_n x + b_n) = G(x), \qquad x \in [-\infty, \infty], \quad (8.60)$$

where $a_n \in (0, \infty)$ and $b_n \in \mathbb{R}^d$. In section 8.2, we have focused on the right-hand side of this equation, that is, on the class of multivariate extreme value distributions. In this section, then, we will consider the left-hand side of equation (8.60). More precisely, we will formulate a range of equivalent descriptions of the domain of attraction, $D(G)$, of an extreme value distribution function G. Of particular interest will be the connection between the dependence structure at extreme levels of a distribution function F in $D(G)$ and the various equivalent descriptions of the dependence structure of G. The reinforcement of (8.60) to density convergence is briefly mentioned in section 8.4.

The domain-of-attraction conditions form the groundwork for the statistical threshold methods in section 9.4. The conditions are always phrased as limit relations, which, taken as approximate equalities, generate approximations of F over certain regions of its support in terms of G. These approximations then serve as a tool to devise statistical models and corresponding inference methods. For proper understanding, we will denote the approximations by the symbol "$\approx$" —not to be confused, by the way, with the symbol "$\sim$", which has the precise meaning $a(t) \sim b(t)$ if and only if $a(t)/b(t) \to 1$ as t tends to its limit value, typically 0 or ∞.

8.3.1   General conditions

The domain-of-attraction condition as stated in (8.60) is not very convenient to work with. In itself, it does not tell us much about the distribution of a random vector X with distribution function F given that X is in some sense extreme. To obtain that kind of information, we have to manipulate (8.60) carefully.

Tail function

Writing $F^n = [1 - n^{-1}\{n(1 - F)\}]^n$ and using the fact that $(1 - n^{-1}x_n)^n \to e^{-x} \in [0, 1]$ if and only if $x_n \to x \in [0, \infty]$ as $n \to \infty$, we find that (8.60) holds if and only if

$$\lim_{n \to \infty} n\{1 - F(a_n x + b_n)\} = -\log G(x), \qquad x \in [-\infty, \infty], \qquad (8.61)$$

with the usual convention $-\log(0) = \infty$. By max-stability (8.2), we may rewrite the previous equation as

$$1 - F(a_n x + b_n) \sim -\log G(\alpha_n x + \beta_n)$$
$$\sim 1 - G(\alpha_n x + \beta_n), \qquad n \to \infty, \qquad (8.62)$$

for x such that $0 < G(x) < 1$.

Relation (8.62) may be used as a starting point for statistical inference on $F(x)$ in x-regions for which each $F_j(x_j)$ is sufficiently close to one. Let u be such that $F_j(u_j)$ is close to one for every $j = 1, \ldots, d$. Equation (8.62) suggests the approximation

$$F(x) \approx G(\alpha_n a_n^{-1} x - \alpha_n a_n^{-1} b_n + \beta_n) =: \tilde{G}(x), \qquad x \geq u.$$

Since G and $\tilde{G}$ differ only in scale and location, $\tilde{G}$ is an extreme value distribution as well, with the same stable tail dependence function, l, and the same extreme value indices, γ_j. Hence

$$F(x) \approx \exp\{-l(v)\}, \qquad x \geq u, \tag{8.63}$$

where, for $j = 1, \ldots, d$ and $x_j \geq u_j$,

$$v_j = -\log \tilde{G}_j(x_j)$$

$$= \left(1 + \gamma_j \frac{x - \tilde{\mu}_j}{\tilde{\sigma}_j}\right)_+^{-1/\gamma_j} = \lambda_j \left(1 + \gamma_j \frac{x - u_j}{\sigma_j}\right)_+^{-1/\gamma_j} \tag{8.64}$$

with $\lambda_j = -\log \tilde{G}_j(u_j)$ and $\sigma_j = \tilde{\sigma}_j + \gamma_j(u_i - \tilde{\mu}_i)$. Together, equations (8.63) and (8.64) form a semi-parametric model for F in the region $[u, \infty)$. If we also assume a parametric model for l, we end up with a fully parametric model. This is the basis for the so-called censored-likelihood approach of Ledford and Tawn (1996), see section 9.4.2.

Multivariate-threshold exceedances

Like in the univariate case, the domain-of-attraction condition (8.60) can be cast in terms of exceedances over a high threshold. The event $\{X \not\leq b_n\}$ is called an exceedance over the (multivariate) threshold b_n. It entails that there is at least one coordinate variable X_j that exceeds the corresponding threshold $b_{n,j}$, although the precise coordinate where this happens remains unspecified. Conditionally on the exceedance $X \not\leq b_n$, the vector $a_n^{-1}(X - b_n)$ is the vector of (scaled) excesses; observe that some coordinates of the excess vector may be negative, although under the conditioning event, at least one coordinate must be positive.

We are interested in the asymptotic distribution of the excess vector $a_n^{-1}(X - b_n)$ conditionally on $X \not\leq b_n$. Without loss of generality, assume that $0 < G(0) < 1$. For x such that $G(x) > 0$, we obtain after some calculation that

$$P[a_n^{-1}(X - b_n) \leq x \mid X \not\leq b_n] \rightarrow \frac{1}{-\log G(0)} \log \left\{\frac{G(x)}{G(x \wedge 0)}\right\}, \qquad n \rightarrow \infty.$$

Now let $q = (q_1, \ldots, q_d)$ with q_j the lower end-point of G_j, the jth margin of G. Then the limit relation above implies

$$P[a_n^{-1}(X - b_n) \vee q \in \cdot \mid X \not\leq b_n] \overset{\mathcal{D}}{\rightarrow} P[W \in \cdot], \qquad n \rightarrow \infty, \tag{8.65}$$

where W is a random vector with distribution function

$$P[W \leq x] = \frac{1}{-\log G(0)} \log \left\{\frac{G(x)}{G(x \wedge 0)}\right\}, \qquad x > q. \tag{8.66}$$

Observe that $W_j \geq q_j$ and $\bigvee_{j=1}^{d} W_j > 0$ with probability one, although $P[W_j \leq 0] = \lim_{n \to \infty} P[X_j \leq b_{n,j} \mid X \not\leq b_n]$ may be positive. In view of (8.65), we may call the distribution of W a multivariate Generalized Pareto (GP) distribution. Up to our knowledge, it has not been studied before. It seems likely that (8.65) and (8.66) can form the basis of new statistical procedures modelling multivariate-threshold excesses.

From (8.65), we can also derive the asymptotic distribution of the excess vector given that there is a threshold exceedance in a specific coordinate: for $j = 1, \ldots, d$, we have

$$P[a_n^{-1}(X - b_n) \vee q \in \cdot \mid X_j > b_{n,j}] \overset{\mathcal{D}}{\to} P[W \in \cdot \mid W_j > 0] \qquad (8.67)$$

as $n \to \infty$. Observe that the distribution of W_j given $W_j > 0$ is univariate GP.

A closely related definition of multivariate GP distributions appears in Tajvidi (1996). For a multivariate-threshold exceedance $\{X \not\leq b_n\}$, he suggested to set every coordinate of the excess vector where the threshold is not exceeded equal to zero. In the notation of (8.65), this gives

$$P[a_n^{-1}(X - b_n) \vee 0 \in \cdot \mid X \not\leq b_n] \overset{\mathcal{D}}{\to} P[W \vee 0 \in \cdot], \qquad n \to \infty, \qquad (8.68)$$

the distribution of the limiting random vector $W \vee 0$ being given by

$$P[W \vee 0 \leq x] = \frac{1}{-\log G(0)} \log \left\{ \frac{G(x)}{G(0)} \right\}, \qquad x \geq 0.$$

Observe that in two dimensions or more, the margins of this vector can be zero with positive probability.

In the bivariate case, yet another definition of multivariate GP distributions is proposed by Kaufmann and Reiss (1995): for a bivariate extreme value distribution function G with stable tail dependence function l as in (8.14), define

$$H(x_1, x_2) = \{1 + \log G(x_1, x_2)\}_+ = [1 - l\{-\log G_1(x_1), -\log G_2(x_2)\}]_+.$$

This H is a bivariate distribution function with translated GP margins $H_i(x_i) = \{1 + \log G_i(x_i)\}_+$ and copula $C_H(u_1, u_2) = \{1 - l(1 - u_1, 1 - u_2)\}_+$. In three or more dimensions, however, the formula $H(x) = \{1 + \log G(x)\}_+$ does not, in general, lead to a valid distribution function, a counter-example being the case where the margins of G are independent.

Equal margins

In case all margins of F are equal to, say, F_1, the previous reformulations of the domain-of-attraction condition (8.60) can be simplified somewhat. Let $x_* = \sup\{x \in \mathbb{R} : F_1(x) < 1\}$ be the right end-point of F_1. By Pickands (1975), $F_1 \in D(G_1)$ for some univariate extreme value distribution function G_1 with

$0 < G_1(0) < 1$ if and only if there exists a positive function $\sigma(u)$ defined on $u < x_*$ such that $[1 - F_1\{u + \sigma(u)x\}]/\{1 - F_1(u)\} \to -\log G_1(x)$ as $u \uparrow x_*$, see Chapter 2. In that case, the normalizing constants may be taken equal to $a_n = \sigma(b_n)$ and $b_n = F_1^{\leftarrow}(1 - 1/n)$.

Now let G be a d-variate extreme value distribution function with all margins equal to G_1. Denote the lower end-point of G_1 by q; observe that $q < 0$ because of our assumption $G(0) > 0$. Using monotonicity and continuity, we can then show that (8.61) and hence (8.60) is equivalent to

$$\frac{1 - F\{u + \sigma(u)x_1, \ldots, u + \sigma(u)x_d\}}{1 - F_1(u)} \to -\log G(x), \qquad u \uparrow x_* \qquad (8.69)$$

for all x such that $x_j > q$ for all $j = 1, \ldots, d$. The latter criterion is a reformulation of results obtained by Marshall and Olkin (1983), who considered the more general case that the extreme value indices of the margins of F all have the same sign. For absolutely continuous distributions, Yun (1997) gives sufficient conditions for (8.69) in terms of convergence of certain conditional densities. We will come back to this in section 10.4 when studying the extremes of Markov chains.

When the margins are equal, criteria involving exceedances over multivariate thresholds get simpler as well. With W as in (8.66), equation (8.65) is equivalent to

$$P\left[\left(\frac{X_j - u}{\sigma(u)} \vee q\right)_{i=1}^d \in \cdot \left| \bigvee_{i=1}^d X_i > u\right] \xrightarrow{\mathcal{D}} P[W \in \cdot], \qquad u \uparrow x_*,$$

and equation (8.67) to

$$P\left[\left(\frac{X_i - u}{\sigma(u)} \vee q\right)_{i=1}^d \in \cdot \left| X_j > u\right] \xrightarrow{\mathcal{D}} P[W \in \cdot \mid W_j > 0], \qquad u \uparrow x_*,$$

for all $j = 1, \ldots, d$. The latter formulation is used in Segers (2003a) to study the extremes of univariate stationary time series.

Exponent measure

Condition (8.61) has an interesting interpretation in terms of exponent measures. Recall from (8.3) that G has an exponent measure, μ, concentrated on $[q, \infty) \setminus \{q\}$, given by $\mu([q, \infty) \setminus [q, x]) = -\log G(x)$ for $x \geq q$, where q_j is the lower end-point of G_j, the jth margin of G. Observe that, by (8.41), $-\log G(x)$ is finite if $x > q$ and infinite otherwise, so that $\mu(B)$ is finite for Borel sets B of $[q, \infty)$ bounded away from q. Also, define the measures μ_n on $[q, \infty) \setminus \{q\}$ by

$$\mu_n(\cdot) = nP[X_{1,n} \in \cdot], \qquad \text{where } X_{i,n} = a_n^{-1}(X_i - b_n) \vee q. \qquad (8.70)$$

Since $\mu_n([q, \infty) \setminus [q, x]) = n\{1 - F(a_n x + b_n)\}$ for $x \in [q, \infty]$, equation (8.61) may now be written in terms of the measures μ_n and μ as $\mu_n(B) \to \mu(B)$ as $n \to \infty$ for every set $B = [q, \infty) \setminus [q, x]$ with $x \geq q$. Since both μ_n and μ put zero mass on $[q, \infty] \setminus [q, \infty)$, a measure-theoretic argument now yields that (8.61) and hence (8.60) is equivalent to

$$\mu_n \text{ converges vaguely to } \mu, \text{ notation } \mu_n \overset{v}{\to} \mu, \text{ on } [q, \infty] \setminus \{q\}, \qquad (8.71)$$

to be interpreted as $\mu_n(B) \to \mu(B)$ as $n \to \infty$ for every Borel set B in $[q, \infty] \setminus \{q\}$ with compact closure and such that $\mu(\partial B) = 0$, where ∂B denotes the topological boundary of B. Observe that $B \subset [q, \infty] \setminus \{q\}$ has compact closure if and only if there exists $x > q$ such that $B \subset [q, \infty] \setminus [q, x]$. For more information on vague convergence of measures, see Resnick (1987) or Kallenberg (1983).

Point processes

Consider the following point processes on $[0, \infty) \times [q, \infty)$:

$$N_n(\cdot) = \sum_{i=1}^{\infty} \mathbf{1}\{(i/n, X_{i,n}) \in \cdot\},$$

with $X_{i,n}$ as in (8.70). See section 5.9.2 for a short introduction on point processes. Recall from (8.71) that the domain-of-attraction condition (8.60) is equivalent to $\mu_n \overset{v}{\to} \mu$. By Proposition 3.21 of Resnick (1987), this is in turn equivalent to

$$N_n \overset{D}{\to} \text{ Poisson process with mean measure } dt \, \mu(dx). \qquad (8.72)$$

A particular consequence, useful for statistical inference, is the following convergence of point processes on $[q, \infty)$:

$$\sum_{i=1}^{n} \mathbf{1}(X_{i,n} \in \cdot) \overset{D}{\to} \text{ Poisson process with mean measure } \mu. \qquad (8.73)$$

Discrete versus continuous index

In the previous equations, the integer variable n can be replaced by a continuous variable t tending to infinity. For instance, with $\lfloor t \rfloor$ denoting the integer part of the real number t, equations (8.60), (8.61) and (8.71) can be extended to

$$\lim_{t \to \infty} F^t(a_{\lfloor t \rfloor} x + b_{\lfloor t \rfloor}) = G(x), \qquad (8.74)$$

$$\lim_{t \to \infty} t\{1 - F(a_{\lfloor t \rfloor} x + b_{\lfloor t \rfloor})\} = -\log G(x), \qquad (8.75)$$

$$\mu_t(\cdot) = t P[X_{1, \lfloor t \rfloor} \in \cdot] \overset{v}{\to} \mu(\cdot), \qquad t \to \infty, \qquad (8.76)$$

the argument being that $t/\lfloor t \rfloor \to 1$ as $t \to \infty$.

8.3.2 Convergence of the dependence structure

When studying multivariate extremes, it is often convenient to separate the marginal distributions from the dependence structure. The fact that we are allowed to do so follows from the property that weak convergence of multivariate distribution functions is equivalent to weak convergence of (i) the marginal distribution functions and (ii) the copula functions, provided the margins of the limit distribution are continuous; see, for example, Deheuvels (1984).

So let $F_1, \ldots, F_d$ be the margins of F and assume that for every $j = 1, \ldots, d$ there exist real sequences $(a_{n,j})_n$ and $(b_{n,j})_n$ with $a_{n,j} > 0$ and an extreme value distribution function G_j such that

$$F_j^n(a_{n,j}x_j + b_{n,j}) \to G_j(x_j), \qquad n \to \infty. \tag{8.77}$$

Then which extra condition is needed on F in order to have (8.60) for some multivariate extreme value distribution function G with margins $G_1, \ldots, G_d$? By the property in the previous paragraph, what is needed is convergence of the dependence structure, to be specified next.

For convenience, we will assume that all the margins F_j are continuous. This has the particular advantage that each $F_j(X_j)$ is uniformly distributed on $(0, 1)$; here $(X_1, \ldots, X_d)$ denotes a random vector with distribution function F. Also, for $0 \leq u_j \leq 1$, the four events $\{X_j \leq F_j^\leftarrow(u_j)\}$, $\{X_j < F_j^\leftarrow(u_j)\}$, $\{F_j(X_j) \leq u_j\}$, and $\{F_j(X_j) < u_j\}$ only differ on an event of probability zero and hence can be interchanged freely. Finally, the copula of F is unique and given by

$$C_F(u) = F\{F_1^\leftarrow(u_1), \ldots, F_d^\leftarrow(u_d)\}, \qquad u \in [0, 1]^d, \tag{8.78}$$

that is, C_F is the distribution function of $(F_1(X_1), \ldots, F_d(X_d))$.

Copula convergence

Let $X_1, X_2, \ldots$ be a sequence of independent random vectors with distribution function F. The copula of F^n, the distribution function of the sample maximum, $M_n = X_1 \vee \cdots \vee X_n$, is

$$C_{F^n}(u) = F^n\{(F_1^n)^\leftarrow(u_1), \ldots, (F_d^n)^\leftarrow(u_d)\}$$
$$= F^n\{F_1^\leftarrow(u_1^{1/n}), \ldots, F_d^\leftarrow(u_d^{1/n})\} = C_F^n(u_1^{1/n}, \ldots, u_d^{1/n}),$$

for $u \in [0, 1]^d$.

Now let G be an extreme value distribution function with margins G_j and copula C_G. We obtain that $F \in D(G)$ if and only if (8.77) together with

$$\lim_{n \to \infty} C_F^n(u_1^{1/n}, \ldots, u_d^{1/n}) = C_G(u), \qquad u \in [0, 1]^d. \tag{8.79}$$

Since the limit copula, C_G, is continuous, the above convergence holds uniformly in $u \in [0, 1]^d$. Hence, in (8.79), we can replace the discrete variable n by the continuous variable t:

$$\lim_{t\to\infty} C_F^t(u_1^{1/t}, \ldots, u_t^{1/t}) = C_G(u), \qquad u \in [0, 1]^d. \tag{8.80}$$

By the stability relation (8.53) of C_G, we obtain from (8.80) the approximation $C_F(u) \approx C_G(u)$ for u such that all u_j are sufficiently close to one. Writing C_G in terms of the stable tail dependence function l as in (8.52) and substituting $F_j(x_j)$ for u_j yields the approximation

$$F(x) \approx \exp[-l\{-\log F_1(x_1), \ldots, -\log F_d(x_d)\}], \tag{8.81}$$

for x such that all $F_j(x_j)$ are close to unity.

Reduction to standard Fréchet or standard Pareto margins

Alternatively, we can transform the random vector X in such a way that its margins become standard Fréchet: define the random vector X_* with distribution function F_* by

$$
\begin{aligned}
X_{*j} &= -1/\log F_j(X_j), \qquad j = 1, \ldots, d, \\
F_*(z) &= F\{F_1^\leftarrow(e^{-1/z_1}), \ldots, F_d^\leftarrow(e^{-1/z_d})\},
\end{aligned}
\tag{8.82}
$$

where $0 < z_j < \infty$ for $j = 1, \ldots, d$. Conversely, F can be obtained from F_* and its margins F_j through $F(x) = F_*\{-1/\log F_1(x_1), \ldots, -1/\log F_d(x_d)\}$.

The margins of F_* are all standard Fréchet, while its copula is the same as the copula of F. Since the standard Fréchet distribution is in its own domain of attraction, copula convergence as in (8.79) is equivalent to $F_* \in D(G_*)$, that is,

$$\lim_{t\to\infty} F_*^t(tz) = G_*(z), \tag{8.83}$$

where G_* is obtained from G after a transformation to standard Fréchet margins as in (8.5). Alternative formulations of (8.83) are

$$\lim_{t\to\infty} t\{1 - F_*(tz)\} = -\log G_*(z), \qquad z \in [0, \infty], \tag{8.84}$$

as well as

$$
\begin{aligned}
1 - F_*(tz) &\sim -\log G_*(tz) \\
&\sim 1 - G_*(tz), \qquad 0 < z < \infty; \, t \to \infty.
\end{aligned}
\tag{8.85}
$$

Taking (8.85) as an approximation for large t leads again to the approximation (8.81).

With $e = (1, \ldots, 1) \in \mathbb{R}^d$, equation (8.84) implies

$$\lim_{t\to\infty} \frac{1 - F_*(tz)}{1 - F_*(te)} = \frac{-\log G_*(z)}{-\log G_*(e)}, \qquad z \in [0, \infty], \tag{8.86}$$

that is, $1 - F_*$ is *multivariate regularly varying* on the cone $(0, \infty)$ (Resnick 1987); see also section 8.4. Conversely, it is not hard to see that (8.86) also implies $F_* \in D(G_*)$.

Finally, we could equally well have transformed to standard Pareto rather than to standard Fréchet margins, that is, (8.79) is still equivalent with each of (8.83), (8.84), (8.85), or (8.86) if, rather than (8.82), we would have put

$$X_{*j} = 1/\{1 - F_1(X_j)\}, \qquad j = 1, \ldots, d,$$
$$F_*(z) = F\{F_1^{\leftarrow}(1 - 1/z_1), \ldots, F_d^{\leftarrow}(1 - 1/z_d)\}, \tag{8.87}$$

for $1 < z_j < \infty$, $j = 1, \ldots, d$.

Tail dependence function convergence

Closely related to the copula of F is its *tail dependence function*

$$D_F(u) = 1 - F\{F_1^{\leftarrow}(1 - u_1), \ldots, F_d^{\leftarrow}(1 - u_d)\}. \tag{8.88}$$

Observe that $D_F(u) = 1 - C_F(1 - u_1, \ldots, 1 - u_d) = P[\bigcup_{j=1}^d \{F_j(X_j) > 1 - u_j\}]$ and $1 - F(x) = D_F\{1 - F_1(x_1), \ldots, 1 - F_d(x_d)\}$. Using (8.84) and $D_F(u) = 1 - F_*(1/u_1, \ldots, 1/u_d)$ with F_* as in (8.87) (Pareto margins), we find that (8.79) is equivalent to

$$\lim_{s \downarrow 0} s^{-1} D_F(sv) = l(v), \qquad v \geq 0. \tag{8.89}$$

A few equivalent formulations of (8.89) are

$$l(v) = \lim_{s \downarrow 0} s^{-1}\{1 - C_F(1 - sv_1, \ldots, 1 - sv_d)\} \tag{8.90}$$

$$= \lim_{s \downarrow 0} s^{-1} P[\exists j = 1, \ldots, d : F_j(X_j) > 1 - sv_j]$$

$$= \lim_{t \to \infty} t P[\bigvee_{j=1}^d [v_j/\{1 - F_j(X_j)\}] > t]$$

for $v \geq 0$. Since the convergence in the previous equations is locally uniform in $v \in [0, \infty)$, we may replace $1 - sv_j$ by any function of the form $1 - sv_j + o(s)$ as $s \downarrow 0$, for instance, $(1 - s)^{v_j}$ or e^{-sv_j}. In the bivariate case, a necessary and sufficient condition is

$$\lim_{s \downarrow 0} s^{-1}[1 - C_F\{1 - s(1 - t), 1 - st\}] = A(t), \qquad t \in [0, 1], \tag{8.91}$$

where $A(t) = l(1 - t, t)$ is the Pickands dependence function of G_*. A useful consequence is

$$\lim_{s \downarrow 0} s^{-1}\{1 - C_F(1 - s, 1 - s)\} = 2A(1/2) = \theta, \tag{8.92}$$

the extremal coefficient of (8.56).

Equation (8.89) and its reformulations point the way to non-parametric estimation of l from observations from F (section 9.4.1). Moreover, since $sl(v) = l(sv)$ for $s > 0$, equation (8.89) has the interesting interpretation

$$1 - F(x) \approx l\{1 - F_1(x_1), \ldots, 1 - F_d(x_d)\} \tag{8.93}$$

provided all $1 - F_j(x_j)$ are sufficiently small. Approximation (8.93) is in fact a first-order expansion of the one in (8.81). However, (8.81) is preferable over (8.93) as the latter approximation undervalues the probability of joint exceedances in different margins: for instance, if $d = 2$ and $l(v_1, v_2) = v_1 + v_2$ (independence), then $P[X_1 > x_1, X_2 > x_2] \approx P[X_1 > x_1]P[X_2 > x_2]$ under (8.81) while $P[X_1 > x_1, X_2 > x_2] \approx 0$ under (8.93). Also, in three or more dimensions, the right-hand side of (8.93) does in general not define a valid distribution.

Exponent and spectral measure

Let X_* and F_* be as in (8.82) or (8.87). The condition $F_* \in D(G_*)$ can also be linked to the exponent measure μ_* and the spectral measure S of G_*, see (8.8) and (8.16). First, by (8.76), $F_* \in D(G_*)$ is equivalent to

$$\mu_{*t}(\cdot) = tP[t^{-1}X_* \in \cdot] \overset{v}{\to} \mu_*(\cdot) \tag{8.94}$$

on $[0, \infty] \setminus \{0\}$. Taking (8.94) as an approximation for large t leads to a recipe for a non-parametric estimation of the exponent measure μ_* in section 9.4.1.

Second, let T be the transformation to pseudo-polar coordinates as in (8.15) determined by two norms $\| \cdot \|_1$ and $\| \cdot \|_2$ on $\mathbb{R}^d$. Applying T to $t^{-1}X_*$ in (8.94) and using (8.17), we find that $F_* \in D(G_*)$ is equivalent to

$$tP[(t^{-1}\|X_*\|_1, X_*/\|X_*\|_2) \in \cdot] \overset{v}{\to} r^{-2}dr\,S(d\omega), \qquad t \to \infty, \tag{8.95}$$

on $(0, \infty] \times \Xi$. Equation (8.95), and hence $F_* \in D(G_*)$, is equivalent to

$$tP[\|X_*\|_1 > t, X_*/\|X_*\|_2 \in \cdot] \overset{v}{\to} S(\cdot) \tag{8.96}$$

on Ξ, which, in turn, is equivalent to

$$\left. \begin{array}{c} P[\|X_*\|_1 > t] \sim t^{-1}S(\Xi) \\[2mm] P[X_*/\|X_*\|_2 \in \cdot \mid \|X_*\|_1 > t] \overset{\mathcal{D}}{\to} S(\cdot)/S(\Xi) \end{array} \right\} \quad t \to \infty \tag{8.97}$$

(de Haan 1985). Equations (8.96) and (8.97) give an interpretation of the spectral measure S in terms of the distribution of the angular component of X_* in the region where its radial component is large. As for the exponent measure, interpreting limits as approximations for large t points the way to non-parametric estimators of S in section 9.4.1.

Point processes

In terms of point processes, we have, by (8.73), $F_* \in D(G_*)$ if and only if

$$\sum_{i=1}^{n} \mathbf{1}\left(n^{-1}X_{*i} \in \cdot\right) \xrightarrow{\mathcal{D}} \text{Poisson process with mean measure } \mu_*, \qquad (8.98)$$

where the X_{i*} are independent copies of X_* (de Haan 1985). This point-process characterization can be used for likelihood-based statistical inference on the spectral measure S in the context of a parametric model, see section 9.4.2.

Asymptotic independence and complete dependence

The two boundary cases within the class of dependence structures of multivariate extreme value distributions are those of independence and complete dependence. Although the latter is merely of academic importance, the former is highly relevant in practice as many multivariate distributions, including the non-degenerate multivariate normal, lie in the domain of attraction of a multivariate extreme value distribution with independent margins, a result dating back to Sibuya (1960); see also Example 9.3. Because of this, section 9.5 is devoted to more refined models in case of asymptotic independence. Here, we restrict ourselves to some characterizations of the domains of attraction of the two cases.

Asymptotic independence. A multivariate distribution function F with copula C_F is called *asymptotically independent* if C_F satisfies (8.79) with the independent copula as limit, that is, $C_G(u) = u_1 \cdots u_d$ for $u \in [0, 1]^d$. In terms of the tail dependence function $D = D_F$ defined in (8.88), asymptotic independence can be written as

$$\lim_{s \downarrow 0} s^{-1} D(sv) = v_1 + \cdots + v_d, \qquad v \in [0, \infty), \qquad (8.99)$$

see (8.89). If additionally each marginal distribution F_j of F is in the domain attraction of an extreme value distribution G_j, then F is in the domain of attraction of the extreme value distribution G given by $G(x) = G_1(x_1) \cdots G_d(x_d)$.

Berman (1961) already showed that a random vector $(X_1, \ldots, X_d)$ is asymptotically independent as soon as all pairs (X_i, X_j) with $i \neq j$ are asymptotically independent. Let D_{ij} be the bivariate tail dependence function of the pair (X_i, X_j); observe that $D_{ij}(v_i, v_j) = D(v)$ where the kth coordinate of v is v_k if $k \in \{i, j\}$ and zero otherwise. Elementary Bonferroni inequalities give

$$u_1 + \cdots + u_d \geq D(u) \geq u_1 + \cdots + u_d - \sum_{1 \leq i < j \leq d} \{u_i + u_j - D_{ij}(u_i, u_j)\}.$$

Hence $s^{-1} D_{ij}(sv_i, sv_j) \to v_i + v_j$ as $s \downarrow 0$ for all $i \neq j$ and all $(v_i, v_j) \in [0, \infty)^2$ indeed implies asymptotic independence.

Observe that the pair (X_i, X_j) is asymptotically independent if

$$\lim_{s\downarrow 0} s^{-1} P[F_i(X_i) > 1 - sv_i, F_j(X_j) > 1 - sv_j] = 0, \qquad (v_i, v_j) \in [0, \infty)^2.$$

By monotonicity, it is sufficient to have the stated convergence for a *single* $(v_i, v_j) \in (0, \infty)^2$; in particular, the pair (X_i, X_j) is asymptotically independent if

$$\exists (v_i, v_j) \in (0, \infty)^2 : \lim_{s\downarrow 0} P[F_i(X_i) > 1 - sv_i \mid F_j(X_j) > 1 - sv_j] = 0.$$

Typically, this result is stated with $v_i = 1 = v_j$. In conjunction with the previous paragraph, we obtain that the random vector $(X_1, \ldots, X_d)$ is asymptotically independent if

$$\lim_{s\downarrow 0} P[F_i(X_i) > 1 - s \mid F_j(X_j) > 1 - s] = 0, \qquad 1 \le i < j \le d. \qquad (8.100)$$

In terms of the copula C_{ij} of the pair (X_i, X_j), that is, $C_{ij}(u_i, u_j) = P[F_i(X_i) \le u_i, F_j(X_j) \le u_j]$, asymptotic independence can be written as

$$C_{ij}(1 - s, 1 - s) = 1 - 2s + o(s), \qquad s \downarrow 0; \ 1 \le i < j \le d.$$

Takahashi (1994) also showed that asymptotic independence arises as soon as

$$\exists v \in (0, \infty) : \lim_{s\downarrow 0} s^{-1} D(sv) = v_1 + \cdots + v_d. \qquad (8.101)$$

Necessity of (8.101) follows from (8.99). But (8.101) is also sufficient: From the inequalities

$$s^{-1} D(sv) \le s^{-1} D_{ij}(sv_i, sv_j) + \sum_{\substack{k=1,\ldots,d \\ k \neq i,j}} v_k \le \sum_{k=1}^{d} v_k, \qquad 1 \le i < j \le d,$$

it follows that (8.101) implies $s^{-1} D_{ij}(sv_i, sv_j) \to v_i + v_j$ as $s \downarrow 0$ for all $1 \le i < j \le d$, whence indeed (pairwise) asymptotic independence.

Asymptotic complete dependence. In some sense opposite to the case of asymptotic independence, a multivariate distribution function F with copula C_F is called *asymptotically completely dependent* if C_F satisfies (8.79) with the completely dependent copula as limit, that is, $C_G(u) = u_1 \wedge \cdots \wedge u_d$ for $u \in [0, 1]^d$. In terms of the tail dependence function $D = D_F$ defined in (8.88), asymptotic complete dependence can be written as

$$\lim_{s\downarrow 0} s^{-1} D(sv) = v_1 \vee \cdots \vee v_b, \qquad v \in [0, \infty),$$

see (8.89). If additionally each marginal distribution F_j of F is in the domain attraction of an extreme value distribution G_j, then F is in the domain of attraction of the extreme value distribution G given by $G(x) = G_1(x_1) \wedge \cdots \wedge G_d(x_d)$.

Takahashi (1994) showed that asymptotic complete dependence arises as soon as

$$\exists 0 < w < \infty : \lim_{s \downarrow 0} s^{-1} D(sw, \cdots, sw) = w. \tag{8.102}$$

To see that the above condition is indeed sufficient, take $v \in [0, \infty) \setminus \{0\}$ and set $v = v_1 \vee \cdots \vee v_d > 0$. Then

$$v \leq s^{-1} D(sv) \leq s^{-1} D(sv, \cdots, sv)$$
$$= (v/w)(sv/w)^{-1} D\{(sv/w)w, \ldots, (sv/w)w\}.$$

Since the right-hand side converges to v, we obtain indeed $s^{-1} D(sv) \to v$ as $s \downarrow 0$.

Also, pairwise asymptotic complete dependence implies asymptotic complete dependence: The pairwise case entails $s^{-1} P[F_j(X_j) > 1 - s \geq F_i(X_i)] \to 0$ as $s \downarrow 0$ for all $1 \leq i < j \leq d$ and thus

$$1 \leq s^{-1} D(s, \ldots, s)$$
$$\leq s^{-1} P[F_1(X_1) > 1 - s] + \sum_{j=2}^{d} P[F_j(X_j) > 1 - s \geq F_1(X_1)] \to 1,$$

which by (8.102) forces asymptotic complete dependence.

8.4 Additional Topics

We collect some topics that did not find their way into the main part of the text.

Multivariate regular variation

A rather popular condition implying that a distribution is in the domain of attraction of a multivariate extreme value distribution is multivariate regular variation. We have already encountered it in (8.86) as a necessary and sufficient condition for the dependence structure of a distribution to be in the domain of attraction of an extreme value dependence structure. More generally, let F be a d-variate distribution function with support $[0, \infty)$. Put $e = (1, \ldots, 1) \in \mathbb{R}^d$. We say that F is regularly varying on $(0, \infty)$ if there exists a function $\lambda : (0, \infty) \to (0, \infty)$ such that

$$\lim_{t \to \infty} \frac{1 - F(tx)}{1 - F(te)} = \lambda(x), \qquad x \in (0, \infty).$$

It follows that there exists a measure ν on $[0, \infty) \setminus \{0\}$ such that $\lambda(x) = \nu([0, \infty) \setminus [0, x])$ for all $x > 0$. Observe that (8.86) says that F_* is regularly varying on $(0, \infty)$ with limit measure $\nu(\cdot) = \mu_*(\cdot)/\mu_*([0, \infty) \setminus [0, e])$.

Most properties we discovered for μ_* extend also to ν. For instance, there must exist $0 < \alpha < \infty$ such that $\nu(t \cdot) = t^{-\alpha}\nu(\cdot)$ for all $0 < t < \infty$. For t_n such that $1 - F(t_n e) \sim n^{-1}$ as $n \to \infty$, we get $F^n(t_n x) \to \exp\{-\lambda(x)\}$, an extreme value distribution with Fréchet margins. Also, ν admits a spectral decomposition of the same kind as we found for μ_* in section 8.2.3. As in (8.97), the normalized spectral measure can be interpreted as the limiting distribution of the angular component of a random vector with distribution function F given that its radial component is large. A detailed account of multivariate regular variation can be found in Resnick (1987) and Mikosch (2004); see also Bingham et al. (1987). Far-stretching generalizations are developed in the monograph by Meerschaert and Scheffler (2001).

Now suppose that F is an absolutely continuous d-variate distribution function with density f supported on $[0, \infty)$. Sufficient conditions in terms of f for F to be regularly varying on $(0, \infty)$ are stated in de Haan and Resnick (1987). These are useful as most multivariate models are defined in terms of their densities rather than their distribution functions. Typical examples where the conditions can be applied are the (restriction to $[0, \infty)$ of the) multivariate t-distribution and F-distribution. In combination with (8.86), the conditions can serve as a tool to prove that the dependence structure of some absolutely continuous distribution is in the domain of attraction of an extreme value dependence structure.

Special classes of distributions

For certain non-parametric classes of distributions, the domain-of-attraction conditions in section 8.3 can be worked out explicitly. For instance, Hult and Lindskog (2002) study the multivariate extremes of elliptical distributions, focusing in particular on the limiting spectral measure.

Alternatively, Capéraà et al (2000) study the class of bivariate copulas given by

$$C(u, v) = \phi^{-1}\left[\{\phi(u) + \phi(v)\}A\left\{\frac{\phi(v)}{\phi(u) + \phi(v)}\right\}\right], \qquad (u, v) \in [0, 1]^2.$$

Here A is a Pickands dependence function, $\phi : (0, 1] \to [0, \infty)$ is convex and decreasing and verifies $\phi(1) = 0$, the function ϕ^{-1} is the inverse function of ϕ, and we employed the conventions $\phi(0) = \lim_{u \downarrow 0} \phi(u)$ and $\phi^{-1}(s) = 0$ if $s \geq \phi(0)$. The class unifies the families of bivariate extreme value copulas ($\phi = -\log$) and Archimedean copulas by Genest and MacKay (1986) ($A = 1$), whence the name Archimax copulas. Within the class, it is easy to construct non-trivial examples of copulas in the domain of attraction of any given bivariate extreme value copula.

Other extreme-related quantities

Rather than the coordinate-wise maximum or the exceedances over a high multivariate threshold, other quantities related to the extremes of a multivariate sequence have been studied in the literature as well.

Cheng *et al.* (1995), for instance, study multivariate intermediate order statistics, defined as follows. Let X_i, $i = 1, \ldots, n$ be independent, identically distributed d-dimensional random vectors. For $j = 1, \ldots, d$, let $X_{(1),j} \leq \cdots \leq X_{(n),j}$ be the ascending order statistics corresponding to the observations $X_{1,j}, \ldots, X_{n,j}$. For every $j = 1, \ldots, d$, let $(k_{n,j})_n$ be an intermediate sequence of positive integers, that is, $k_{n,j} \to \infty$ and $k_{n,j}/n \to 0$ as $n \to \infty$. Suppose also that all $k_{n,j}$ grow at the same rate. Cheng *et al.* (1995) then find the asymptotic distribution as $n \to \infty$ of the sequence of vectors $X_{(k_n),n} = (X_{(k_{n,1}),n}, \ldots, X_{(k_{n,d}),n})$.

Records can also be studied in the multivariate case, although a natural definition of multivariate records is not obvious because of the lack of a natural ordering for multivariate observations. The principle of marginal ordering suggests the following definition: X_n is a record in the sequence $X_1, \ldots, X_n$ if $X_n > \bigvee_{i=1}^{n-1} X_i$, that is, if there is a record simultaneously in all coordinates. The asymptotic distribution of the sequence of such records is the topic of Goldie and Resnick (1995) and the references therein. Alternatively, in the context of Gaussian processes, Habach (1997) defines X_n to be a record as soon as there is a record in one of the coordinates, that is, if $X_{n,j} > \bigvee_{i=1}^{n-1} X_{i,j}$ for some $j = 1, \ldots, d$.

A concept that is inherently multivariate is that of concomitants or induced order statistics. For instance, let (X_{i1}, X_{i2}), $i = 1, \ldots, n$, be a sample of bivariate random pairs and let $X_{(1),1} \leq \cdots \leq X_{(n),1}$ be the ascending order statistics in the first coordinate. Then the value of the second coordinate of the pair of which the first coordinate is equal to $X_{(i),1}$ is called the concomitant of that order statistic and is denoted by $X_{[i],2}$. For example, $X_{[n],2}$ is the second coordinate of the pair with the largest first coordinate. The distribution of concomitants of extreme order statistics is investigated in David (1994) and Nagaraja and David (1994). Ledford and Tawn (1998) focus on the concomitant of the largest order statistic in case the marginal bivariate survivor function is bivariate regularly varying, see section 9.5. In particular, they give an asymptotic expansion for the tail function of that concomitant and find the asymptotic probability that the pair of coordinate-wise maxima is an actual observation, that is, $P[X_{[n],2} = X_{(n),2}]$.

Rates of convergence

Recall from Chapters 4 and 5 that in one dimension, because of slow convergence in the domain-of-attraction condition, estimators of the tail of a distribution sometimes suffer from a substantial bias. A similar problem may arise in higher dimensions, an extra issue being the rate of convergence of the dependence structure.

Omey and Rachev (1991) and Falk and Reiss (2002) investigate the rate of convergence of the copula of the sample maximum to the limiting extreme value copula (8.79) with respect to the uniform metric, whereas de Haan and Peng (1997) employ the stronger total variation metric. Alternatively, Kaufmann and Reiss (1995) consider the rate of convergence of certain point processes of exceedances to the limiting Poisson process, corollaries being rates of convergence for the joint distributions of upper order statistics, although their error term appears to be sub-optimal. Finally, Nadarajah (2000) gives asymptotic expansions for the convergence of spectral densities in (8.96).

More general settings than i.i.d. sequences

Up to now, we always started from a sequence of independent, identically distributed random vectors. This setting can be generalized in a number of ways.

A first possibility is to drop the assumption of stationarity. For instance, Hüsler (1989b), building on work by Gerritse (1986), characterizes the class of limit distributions of normalized maxima of sequences of independent, non-identically distributed random vectors and states a number of properties of the dependence structure of the possible limit laws. Moreover, Hüsler (1989a) gives conditions under which the extremes of a general non-stationary, possibly dependent sequence of random vectors have the same asymptotic distribution as the corresponding sequence with independent random vectors.

Alternatively, one can drop the assumption of independence. Hsing (1989) and Hüsler (1990) examine the asymptotic distribution of normalized maxima of sequences of general stationary sequences of random vectors; see also section 10.5. The asymptotic distribution of point processes of exceedances and vectors of extreme order statistics for multivariate stationary normal sequences is the topic of Wiśniewski (1996).

Finally, interesting results can also be obtained for a triangular array $\{X_{in} : n = 1, 2, \ldots; i = 1, \ldots, n\}$ of independent d-dimensional random vectors. Hüsler and Reiss (1989) consider the case where every row $X_{1n}, \ldots, X_{nn}$ consists of centred, unit-variance normal random vectors with correlation matrix ρ_n depending on n. For instance, in the bivariate case, they find that if $(1 - \rho_n) \log(n) \to \lambda^2 \in [0, \infty]$ as $n \to \infty$ then the suitably normalized maximum $M_n = \bigvee_{i=1}^{n} X_{in}$ converges weakly to a parametric family of multivariate extreme value distributions with dependence structure depending on λ, see section 9.2. More general triangular arrays are considered in Hüsler (1994).

8.5 Summary

For the reader's convenience, we provide a summary of the essential facts to be remembered from the theory of multivariate extremes.

We work in d-dimensional space. The distribution functions G with non-degenerate margins that can arise as the limit in $\lim_{n \to \infty} F^n(a_n x + b_n) = G(x)$, where F is a d-variate distribution function and a_n and b_n are arbitrary vectors, the entries of a_n being positive, are called *multivariate extreme value distribution functions*. We say that F is in the *(max-)domain of attraction* of G. The interpretation is that G is the limit distribution of the properly normalized component-wise maximum of an independent sample from F as the sample size tends to infinity. The class of extreme value distributions coincides with that of max-stable distributions.

The margins, G_j, of a max-stable distribution function G are univariate extreme value distribution functions themselves. They have the corresponding margins, F_j, of F in their respective domains of attraction. In order to study the dependence structure of G, we may, without loss of generality, standardize the margins of G to the standard Fréchet distribution by $G_*(z) = G\{G_1^{\leftarrow}(e^{-1/z_1}), \ldots, G_d^{\leftarrow}(e^{-1/z_d})\}$ for $z \in [0, \infty]$.

The function $V_* = -\log G_*$ satisfies the homogeneity relation $sV_*(sz) = V_*(z)$ for $0 < s < \infty$. Moreover, there exists a measure, μ_*, on $[0, \infty) \setminus \{0\}$, the *exponent measure*, such that $V_*(z) = \mu_*(\{x \geq 0 : x \not\leq z\})$. The exponent measure μ_* inherits a similar homogeneity property from V_*.

In polar coordinates, the measure μ_* factorizes as a product measure in the radial and angular components. More specifically, identifying z with (r, ω), where $r = z_1 + \cdots + z_d$ is the "radius" and $\omega = (z_1/r, \ldots, z_d/r)$ is the "angle", we have $\mu(dz) = r^{-2}dr\,H(d\omega)$. Here, the *spectral measure* H is a finite measure on the unit simplex, $S_d = \{\omega \geq 0 : \omega_1 + \cdots + \omega_d = 1\}$. The only requirement on a positive measure H on S_d to be the spectral measure of an extreme value distribution is that $\int_{S_d} \omega_j H(d\omega) = 1$ for all $j = 1, \ldots, d$. Alternative definitions of the spectral measure are possible, starting from a different choice of the radial and angular components.

The *stable tail dependence function* l is given by $l(v) = V_*(1/v_1, \ldots, 1/v_d)$ for $0 \leq v < \infty$. It satisfies the homogeneity relation $l(sv) = sl(v)$ for $0 < s < \infty$ and is connected to the extreme value distribution G and the spectral measure H through

$$G(x) = \exp[-l\{-\log G_1(x_1), \ldots, -\log G_d(x_d)\}],$$

$$l(v) = \int_{S_d} \bigvee_{j=1}^{d} (\omega_j v_j) H(d\omega).$$

The partial derivatives of l or V_* can be used to compute the densities of the spectral measure H on the $2^d - 1$ faces of the unit simplex S_d.

The two extreme cases for the dependence structure of an extreme value distribution are those of *independence* and *complete dependence*. In general, the dependence structure lies between these cases. In particular, extreme value distributions always exhibit *positive association*. In case of independence, the spectral measure H consists of unit point masses at each of the d vertices of the unit simplex S_d and the stable tail dependence function is given by $l(v) = v_1 + \cdots + v_d$. Independence arises as soon as all pairs are independent. In case of complete dependence, H reduces to a single point mass of size d at the centre-point of S_d, and $l(v) = v_1 \vee \cdots \vee v_d$.

In two dimensions, all information on the dependence structure is contained in *Pickands dependence function* $A(t) = l(1 - t, t)$, where $t \in [0, 1]$. It is convex and satisfies $t \vee (1 - t) \leq A(t) \leq 1$. These are the only restrictions for a function A to be the Pickands dependence function of a bivariate extreme value distribution. The lower and upper boundaries on A correspond to complete dependence and independence, respectively. Identifying the unit simplex S_2 with the unit interval $[0, 1]$, we can easily obtain from A the point masses of H on 0 and 1 and its density h on $(0, 1)$.

For a given extreme value distribution function G, we formulate various equivalent conditions for a distribution function F to lie in its domain of attraction. A number of these conditions are in terms of the limit distribution of *excesses over a high multivariate threshold*. Equally useful is to tear the domain-of-attraction condition apart into two pieces: first, the margins of F must lie in the (univariate) domain-of-attraction of the corresponding margins of G, and second, informally stated, the dependence structure of F must lie in the domain of attraction of the dependence structure of G.

A particularly useful interpretation of the domain-of-attraction condition is the approximation

$$F(x) \approx \exp[-l\{-\log F_1(x_1), \ldots, -\log F_d(x_d)\}]$$

for x such that $1 - F_j(x_j)$ is small for all $j = 1, \ldots, d$, and with l the stable tail dependence function of the limiting extreme value distribution. Combined with a Generalized Pareto model for the marginal tails, this leads to (semi-)parametric models for F in the regions of the form $[u, \infty)$ for high multivariate thresholds u. A related, slightly less accurate approximation is $1 - F(x) \approx l\{1 - F_1(x_1), \ldots, 1 - F_d(x_d)\}$. Alternatively, we find a condition in terms of convergence of certain point processes to a non-homogeneous Poisson process with intensity measure μ_*.

8.6 Appendix

8.6.1 Computing spectral densities

We give a proof of (8.34) expressing the densities of the spectral measure H on the faces of the unit simplex S_d in terms of the derivatives of $V_* = -\log G_*$. Let $z > 0$. By the inclusion-exclusion formula,

$$V_*(z) = \mu\big(\{x \geq 0 : x_j > z_j \text{ for some } j = 1, \ldots, d\}\big)$$

$$= \sum_{\emptyset \neq b \subset \{1, \ldots, d\}} (-1)^{|b|-1} \mu\big(\{x \geq 0 : x_j > z_j \text{ for all } j \in b\}\big).$$

Now let $a = \{j_1, \ldots, j_m\}$ be a non-empty subset of $\{1, \ldots, d\}$, and let D_a be the differential operator $\partial^m / (\partial z_{j_1} \cdots \partial z_{j_m})$. Applying D_a to both sides of the previous equation, we only retain those terms for which b contains a, that is,

$$D_a V_*(z)$$

$$= \sum_{a \subset b \subset \{1, \ldots, d\}} (-1)^{|b|-1} D_a \mu\big(\{x \geq 0 : x_j > z_j \text{ for all } j \in b\}\big).$$

Denote $a^c = \{1, \ldots, d\} \setminus a$. We can split the sum above in two parts: the term corresponding to $b = a$, and the terms corresponding to $b = a \cup b'$ with $\emptyset \neq b' \subset a^c$. We get

$$D_a V_*(z)$$
$$= (-1)^{|a|-1} D_a \mu(\{x \geq \mathbf{0} : x_j > z_j \text{ for all } j \in a\})$$
$$+ (-1)^{|a|} D_a \sum_{\emptyset \neq b' \subset a^c} (-1)^{|b'|-1} \mu(\{x \geq \mathbf{0} : x_j > z_j \text{ for all } j \in a \cup b'\}).$$

Applying the inclusion-exclusion formula again, we get

$$D_a V_*(z)$$
$$= (-1)^{|a|-1} D_a \mu(\{x \geq \mathbf{0} : x_j > z_j \text{ for all } j \in a\})$$
$$+ (-1)^{|a|} D_a \mu(\{x \geq \mathbf{0} : x_j > z_j \text{ for all } j \in a \text{ and some } j \in a^c\})$$
$$= (-1)^{|a|-1} D_a \mu(\{x \geq \mathbf{0} : x_j > z_j \text{ for all } j \in a; x_j \leq z_j \text{ for all } j \in a^c\}).$$

Now if we let $z_j \to 0$ for all $j \in a^c$, we get

$$\lim_{\substack{z_j \to 0 \\ j \notin a}} D_a V_*(z)$$
$$= (-1)^{|a|-1} D_a \mu(\{x \geq \mathbf{0} : x_j > z_j \text{ for all } j \in a; x_j = 0 \text{ for all } j \in a^c\}).$$

Let h_a be the density of H on $S_{d,a}$. Using the spectral decomposition (8.17) and the multivariate change-of-variables formula, we get

$$\mu(\{x \geq \mathbf{0} : x_j > z_j \text{ for all } j \in a; x_j = 0 \text{ for all } j \in a^c\})$$
$$= \int_{z_{j_1}}^{\infty} \cdots \int_{z_{j_m}}^{\infty} h_a \left(\frac{x_1}{\sum x_i}, \ldots, \frac{x_{m-1}}{\sum x_i} \right) \left(\sum x_i \right)^{-(|a|+1)} dx_1 \cdots dx_m.$$

Apply the operator D_a on both sides of this equation to get (8.34).

8.6.2 Representations of extreme value distributions

Let G be a d-variate extreme value distribution function with margins G_j for $j = 1, \ldots, d$. We have seen equivalent descriptions of the dependence structure of G in terms of, amongst others, the simple max-stable distribution function $G_* = \exp(-V_*)$, the exponent measure μ_*, the stable tail dependence function l, the spectral measure S w.r.t. two norms $\| \cdot \|_i$ ($i = 1, 2$) on $\mathbb{R}^d$, the copula C_G, and, in the bivariate case, Pickands dependence function A. For easy reference, we collect here the connections between these various descriptions.

Formulas for G

$$
\begin{aligned}
G(x) &= G_*\{-1/\log G_1(x_1), \ldots, -1/\log G_d(x_d)\} \\
&= \exp\left\{-\mu_*\big([0, \infty] \setminus [0, (-1/\log G_1(x_1), \ldots, -1/\log G_d(x_d))]\big)\right\} \\
&= \exp[-l\{-\log G_1(x_1), \ldots, -\log G_d(x_d)\}] \\
&= \exp\left[\int_\Xi \bigwedge_{j=1}^d \left\{\frac{\omega_j}{\|\omega\|_1} \log G_j(x_j)\right\} S(d\omega)\right] \\
&= C_G\{G_1(x_1), \ldots, G_d(x_d)\}
\end{aligned}
$$

Formulas for G_*

$$
\begin{aligned}
G_*(z) &= G\{G_1^\leftarrow(e^{-1/z_1}), \ldots, G_d^\leftarrow(e^{-1/z_d})\} \\
&= \exp\{-\mu_*([0, \infty] \setminus [0, z])\} \\
&= \exp\{-l(1/z_1, \ldots, 1/z_d)\} \\
&= \exp\left\{-\int_\Xi \bigvee_{j=1}^d \left(\frac{\omega_j}{\|\omega\|_1} \frac{1}{z_j}\right) S(d\omega)\right\} \\
&= C_G(e^{-1/z_1}, \ldots, e^{-1/z_d})
\end{aligned}
$$

Formulas for l

$$
\begin{aligned}
l(v) &= -\log G\{G_1^\leftarrow(e^{-v_1}), \ldots, G_d^\leftarrow(e^{-v_d})\} \\
&= -\log G_*(1/v_1, \ldots, 1/v_d) \\
&= \mu_*\big([0, \infty] \setminus [0, (1/v_1, \ldots, 1/v_d)]\big) \\
&= \int_\Xi \bigvee_{j=1}^d \left(\frac{\omega_j}{\|\omega\|_1} v_j\right) S(d\omega) \\
&= -\log C_G(e^{-v_1}, \ldots, e^{-v_d})
\end{aligned}
$$

Formulas for S

$$
\begin{aligned}
S(B) &= \mu_*\big(\{z \in [0, \infty) : \|z\|_1 \geq 1, \, z/\|z\|_2 \in B\}\big) \\
&= \int_{\Xi'} \mathbf{1}\left(\frac{\omega'}{\|\omega'\|_2} \in B\right) \frac{\|\omega'\|_1}{\|\omega'\|_1'} S'(d\omega')
\end{aligned}
$$

Formulas for C_G

$$
\begin{aligned}
C_G(\boldsymbol{u}) &= G\{G_1^{\leftarrow}(u_1), \ldots, G_d^{\leftarrow}(u_d)\} \\
&= G_*(-1/\log u_1, \ldots, -1/\log u_d) \\
&= \exp\{-\mu_*([\boldsymbol{0}, \infty] \setminus [\boldsymbol{0}, (-1/\log u_1, \ldots, -1/\log u_d)])\} \\
&= \exp\{-l(-\log u_1, \ldots, -\log u_d)\} \\
&= \exp\left\{\int_\Xi \bigwedge_{j=1}^d \left(\frac{\omega_j}{\|\boldsymbol{\omega}\|_1} \log u_j\right) S(\mathrm{d}\boldsymbol{\omega})\right\}
\end{aligned}
$$

Bivariate case: Formulas in terms of A

$$
G(x_1, x_2) = \exp\left[\log\{G_1(x_1)G_2(x_2)\}A\left(\frac{\log\{G_2(x_2)\}}{\log\{G_1(x_1)G_2(x_2)\}}\right)\right]
$$

$$
G_*(z_1, z_2) = \exp\left\{-\left(\frac{1}{z_1} + \frac{1}{z_2}\right)A\left(\frac{z_1}{z_1 + z_2}\right)\right\}
$$

$$
l(v_1, v_2) = (v_1 + v_2)A\left(\frac{v_2}{v_1 + v_2}\right)
$$

$$
H([0, \omega]) = \mu_*(\{(z_1, z_2) \in [0, \infty)^2 : z_1 + z_2 \geq 1, \, z_1/(z_1 + z_2) \leq \omega\})
$$

$$
= \begin{cases} 1 + A'(\omega) & \text{if } \omega \in [0, 1) \\ 2 & \text{if } \omega = 1 \end{cases}
$$

$$
S(B) = \int_{[0,1]} \mathbf{1}\left\{\frac{(\omega, 1 - \omega)}{\|(\omega, 1 - \omega)\|_2} \in B\right\} \|(\omega, 1 - \omega)\|_1 \mathrm{d}H([0, \omega])
$$

$$
C_G(u_1, u_2) = \exp\left[\log(u_1 u_2)A\left\{\frac{\log(u_2)}{\log(u_1 u_2)}\right\}\right]
$$

Formulas for A

$$
\begin{aligned}
A(t) &= -\log G[G_1^{\leftarrow}\{\mathrm{e}^{-(1-t)}\}, G_2^{\leftarrow}(\mathrm{e}^{-t})] \\
&= -\log G_*\{(1 - t)^{-1}, t^{-1}\} \\
&= \mu_*([0, \infty]^2 \setminus [0, ((1 - t)^{-1}, t^{-1})]) \\
&= l(1 - t, t) \\
&= \int_\Xi \left\{\frac{\omega_1}{\|(\omega_1, \omega_2)\|_1}(1 - t)\right\} \vee \left\{\frac{\omega_2}{\|(\omega_1, \omega_2)\|_1}t\right\} S(\mathrm{d}(\omega_1, \omega_2)) \\
&= -\log C_G\{\mathrm{e}^{-(1-t)}, \mathrm{e}^{-t}\}
\end{aligned}
$$

9

STATISTICS OF MULTIVARIATE EXTREMES

co-authored by Björn Vandewalle

9.1 Introduction

Given a sample of multivariate observations, assumed to be generated by independent and identically distributed random vectors, how to estimate the tail of the underlying multivariate distribution? In particular, how to estimate with good relative accuracy the probability of an event in a region of the sample space with none or only very few of the observations? As for statistics of univariate extremes, this calls for generally applicable models based on which it is justified to extrapolate outside of the sample region. If the interest is in the occurrence of joint extremes in several coordinates, then proper modelling of the marginal distributions should be complemented by a correct assessment of the dependence structure at extreme levels.

A successful class of models and inference techniques is based on the multivariate extreme value distributions, studied extensively in Chapter 8. The argument in favour of these distributions is summarized by the property that, as in the univariate case, the tail of a distribution in the domain of attraction of an extreme value distribution can be approximated by the tail of that extreme value distribution itself.

As the class of multivariate extreme value distributions does not admit a finite-dimensional parametrization, a quite popular approach is to perform inference within a well-chosen parametric sub-model. A number of such models have been shown in various case studies to be particularly successful in combining analytical tractability with practical applicability. Of course, new situations may ask for new

Statistics of Extremes: Theory and Applications   J. Beirlant, Y. Goegebeur, J. Segers, and J. Teugels
© 2004 John Wiley & Sons, Ltd   ISBN: 0-471-97647-4

models, so it is useful to have tools to construct parametric families of multivariate extreme value distributions. All this is treated in section 9.2.

As for the univariate case, historically the first statistical methods for multivariate extremes follow the annual maximum approach (Gumbel and Goldstein 1964). The approach consists of partitioning a sample of multivariate observations into blocks, each typically corresponding to one year of observations, and fitting a multivariate extreme value distribution to the sample of component-wise block maxima. The crucial point here is to estimate the multivariate extreme value dependence structure or copula. Section 9.3 describes both parametric and non-parametric techniques to do this.

Reducing the sample to a single observation per year disregards the possibility of a given year to witness several relevant events. More efficient is to use all data that are in some sense large, for instance all observations for which at least one coordinate exceeds a high threshold, which may differ according to the coordinate. The modelling assumption, motivated by the domain-of-attraction conditions of section 8.3, is that the dependence structure of the underlying distribution may at extreme levels be approximated by a max-stable dependence structure. Again, the choice is between parametric inference within a subclass or general non-parametric techniques, most of which are motivated by the spectral decomposition of a multivariate extreme value distribution.

Both the annual maximum approach and the threshold approach are founded in the paradigm of multivariate extreme value distributions, motivated by the theory in Chapter 8. The resulting models are therefore restricted to either perfect independence or asymptotic dependence. Neither of these may be satisfactory for cases of asymptotic independence with positive or negative association at penultimate thresholds, such as, for instance, the bivariate normal with positive or negative correlation. This calls for more refined models for the joint survivor function of a random vector in case of asymptotic independence, and these are presented in section 9.5.

We conclude the chapter with a number of additional topics in section 9.6 and a summary in section 9.7.

Loss-ALAE data

We will illustrate the methods in this chapter on the Loss-ALAE data set comprising 1500 liability claims in an insurance set-up, see Figure 1.15 in section 1.3.3. Each claim consists of a loss or indemnity payment and an Allocated Loss Adjustment Expense. ALAEs can be seen as additional costs for the insurance company, such as lawyers' fees and investigation expenses resulting from individual claim settlements. The scatterplot of the two variables in Figure 9.1(a) suggests a strong relationship between losses and other expenses at intermediate levels, as confirmed by the value of the correlation coefficient, 0.4.

Starting from (Loss,ALAE) observations (x_{i1}, x_{i2}), $i = 1, \ldots, n$, we can obtain an informal, margin-free picture of dependence by transforming the data to have uniform $(0, 1)$ marginal distributions using the (modified) empirical marginal distribution

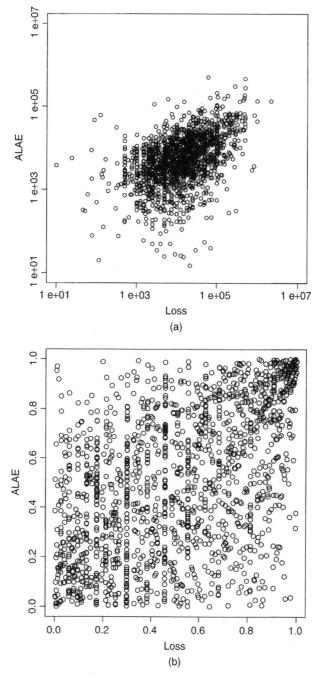

Figure 9.1 Scatterplot of ALAE versus Loss: (a) original data (log-scale), (b) data transformed to uniform (0, 1) margins.

functions. For $i = 1, \ldots, n$ and $j = 1, 2$, define

$$u_{ij} = \frac{1}{n+1} \sum_{k=1}^{n} \mathbf{1}(x_{kj} \leq x_{ij}). \tag{9.1}$$

If we consider the (x_{i1}, x_{i2}) as being realizations of independent random variables with common distribution function F, then the (u_{i1}, u_{i2}) can be interpreted as realizations from the copula, C, of F. The scatterplot of the (u_{i1}, u_{i2}) in Figure 9.1(b) suggests some dependence between losses and ALAEs at levels where both are high.

The case studies in this chapter were partially performed with the R package 'evd' by Alec Stephenson, freely available from cran.r-project.org, including some routines by Chris Ferro, and with routines written by Björn Vandewalle.

9.2   Parametric Models

Recall from section 8.2 that the family of d-variate extreme value distributions is indexed by a positive measure on the unit simplex S_d satisfying a number of moment restrictions. In particular, unlike the univariate case, the family does not admit a finite-dimensional parametrization. As a consequence, we lose the comfort of parametric likelihood machinery, which guarantees efficient estimation, easy assessment of estimation uncertainty, hypothesis testing and inclusion of covariate information. This is a major setback.

In order to be able to still enjoy the mentioned features, one can, rather than to work in the general class, postulate a parametric subfamily. Of course, there is a price to pay: sacrificing generality comes at the risk of model mis-specification. A good balance then must be struck between model flexibility and analytical tractability.

This raises the issue of model construction and model choice. No model can be expected to work well in all situations. New data may require new models. However, because of the constraints that a dependence structure must fulfil in order to be an extreme value dependence structure, it is not straightforward to generate valid parametric families, let alone useful ones. In section 9.2.1, we list a number of tools for generating multivariate extreme value models. An overview of the most popular models is given in section 9.2.2.

9.2.1   Model construction methods

Max-stable processes

Loosely speaking, max-stable processes are stochastic processes of which all finite-dimensional distributions are multivariate extreme value distributions. They can be viewed as infinite-dimensional generalizations of extreme value distributions. A spectral representation for such processes by de Haan (1984) was turned into a versatile tool by Smith (1991) for a construction method for multivariate extreme value distributions that allows certain characteristics of a physical process under study to be incorporated into the model.

Consider the following situation. A certain system V is affected by a collection of shock events of possible different sizes and different types. Event number i has size $0 < r_i < \infty$ and is of type s_i in some classification space S. The impact caused to an element v of the system V by an event of size r and type s is equal to $rf(s, v)$, where $f : S \times V \to (0, \infty)$ is the so-called event profile function. The aggregate impact Z_v to an element v caused by all events i is equal to the maximal impact by each of the events, that is, $Z_v = \max_i\{r_i f(s_i, v)\}$.

Now we make the following assumption: (i) the events (r_i, s_i) form a Poisson process on the space $(0, \infty) \times S$ with intensity measure $r^{-2}dr\, v(ds)$ for some measure v on S, and (ii) for all $v \in V$ we have $\int_S f(s, v)v(ds) = 1$. The measure v is called the *frequency measure*, as it describes the relative frequency with which events of certain types occur. The second assumption is just a normalization.

Under this assumption, we can find the distribution of the vector $(Z_v : v \in V_0)$, where $V_0 = \{v_1, \dots, v_d\} \in V$. For $0 < z_v < \infty$, we have

$$P[Z_v \leq z_v, \forall v \in V_0]$$
$$= P[r_i f(s_i, v) \leq z_v, \forall i, \forall v \in V_0]$$
$$= P\left[r_i \max_{v \in V_0}\{z_v^{-1} f(s_i, v)\} \leq 1, \forall i\right]$$
$$= \exp\left(-\int_S \int_0^\infty \mathbf{1}\left[r \max_{v \in V_0}\{z_v^{-1} f(s, v)\} > 1\right] r^{-2}dr\, v(ds)\right)$$
$$= \exp\left[-\int_S \max_{v \in V_0}\{z_v^{-1} f(s, v)\}v(ds)\right]. \tag{9.2}$$

Because of the normalization assumption, we have $P[Z_v \leq z_v] = \exp(-1/z_v)$, that is, the marginal distributions are standard-Fréchet. Moreover, the distribution of the vector $(Z_v : v \in V_0)$ satisfies the max-stability relation (8.7). All in all, we conclude that $(Z_v : v \in V_0)$ has a multivariate extreme value distribution with standard-Fréchet margins.

Examples of this construction are the Gaussian model for spatial extremes of rain storms in Smith (1991) and Coles and Tawn (1996a), and the directional model for extreme wind speeds in Coles and Tawn (1994); see also section 9.2.2. The extremal coefficients of the process $(Z_v : v \in V)$ in the sense of (8.55) are given by $\theta_{V_0} = \int_S \max_{v \in V_0} f(s, v)v(ds)$.

Spectral densities

Recall from Chapter 8 that one of the representations of the dependence structure of a multivariate extreme value distribution was in terms of a so-called spectral measure H, which in the bivariate case is the positive measure on $[0, 1]$ given by (8.28). Hence, assuming that H is absolutely continuous, we may construct models for H by modelling its density h. However, we must take care that the constraint (8.29) is fulfilled, that is, we need $\int_0^1 \omega h(\omega)d\omega = 1$ and $\int_0^1 (1 - \omega)h(\omega)d\omega = 1$.

Coles and Tawn (1991) describe a way to modify an arbitrary non-negative function h^* on $(0, 1)$ that does not satisfy the constraints to a function h that does. Let

$$m_1 = \int_0^1 u h^*(u) \mathrm{d}u, \qquad m_2 = \int_0^1 (1 - u) h^*(u) \mathrm{d}u. \qquad (9.3)$$

Without loss of generality, assume $m_1 > 0$ and $m_2 > 0$. From Pickands (1981), we know that

$$P[X_1 > x_1, X_2 > x_2] = \exp\left(-\int_0^1 [(ux_1) \vee \{(1 - u)x_2\}] h^*(u) \mathrm{d}u\right)$$

is the joint survivor function of a bivariate min-stable distribution with exponential margins with expectations $E[X_j] = 1/m_j$ for $j = 1, 2$. Hence, writing $P[m_1 X_1 > v_1, m_2 X_2 > v_2] = \exp\{-l(v_1, v_2)\}$, we find that

$$l(v_1, v_2) = \int_0^1 \left\{\left(\frac{u}{m_1} v_1\right) \vee \left(\frac{1 - u}{m_2} v_2\right)\right\} h^*(u) \mathrm{d}u$$

is a stable tail dependence function, see also (8.51). Change variables $u = m_1 \omega / \{m_1 \omega + m_2 (1 - \omega)\}$ to find $l(v_1, v_2) = \int_0^1 [(\omega v_1) \vee \{(1 - \omega) v_2\}] h(\omega) \mathrm{d}\omega$, where

$$h(\omega) = \frac{m_1 m_2}{\{m_1 \omega + m_2 (1 - \omega)\}^3} h^* \left\{\frac{m_1 \omega}{m_1 \omega + m_2 (1 - \omega)}\right\} \qquad (9.4)$$

for $0 < \omega < 1$. This h then must be the density of a spectral measure H. One can also verify directly that the constraints are satisfied.

The procedure can be extended to accommodate for spectral measures H with point masses at 0 or 1. More importantly, Coles and Tawn (1991) generalize the argument to higher dimensions. For a non-negative function h^* on the unit simplex S_d such that $m_j = \int_{S_d} \omega_j h^*(\boldsymbol{\omega}) \lambda(\mathrm{d}\boldsymbol{\omega})$ is positive and finite [where $\lambda(\mathrm{d}\boldsymbol{\omega}) = \mathrm{d}\omega_1 \cdots \mathrm{d}\omega_{d-1}$ denotes the $(d - 1)$-dimensional Lebesgue measure on S_d],

$$h(\boldsymbol{\omega}) = \frac{m_1 \ldots m_d}{(m_1 \omega_1 + \cdots + m_d \omega_d)^{d+1}} h^*(u_1, \ldots, u_d),$$

$$\text{where} \quad u_j = \frac{m_j \omega_j}{m_1 \omega_1 + \cdots + m_d \omega_d} \qquad (9.5)$$

defines the density of a measure H concentrated on the interior of S_d and satisfying (8.26).

Order restrictions

Sometimes, the variables that we want to model satisfy certain order restrictions. For instance, if M_1 and M_2 denote the maxima of respectively the hourly and two-hourly aggregated rainfall amounts during a certain period at a certain location, then necessarily $M_1 \leq M_2 \leq 2M_1$. Nadarajah et al. (1998) propose ways to construct models for bivariate extreme value distributions that can incorporate such restrictions.

For simplicity, we restrict attention to the case where the margins are standard-Fréchet. Let G_* be a bivariate extreme value distribution with standard-Fréchet margins, and let the random pair (Z_1, Z_2) have distribution function G_*. Let H be the spectral measure of G_* as in (8.28), that is

$$P[Z_1 \leq z_1, Z_2 \leq z_2] = G_*(z_1, z_2) = \exp\left\{ \int_{[0,1]} \frac{\omega}{z_1} \vee \frac{1-\omega}{z_2} H(d\omega) \right\}.$$

For $0 < m < \infty$, we have

$$P[Z_2 < mZ_1] = \lim_{\delta \downarrow 0} \sum_{k=0}^{\infty} P[k\delta < Z_1 \leq (k+1)\delta, Z_2 \leq mk\delta]$$

$$= \lim_{\delta \downarrow 0} \delta \sum_{k=0}^{\infty} \frac{G_*\{(k+1)\delta, mk\delta\} - G_*(k\delta, mk\delta)}{\delta}$$

$$= \int_0^{\infty} \lim_{\delta \downarrow 0} \frac{G_*(z+\delta, mz) - G_*(z, mz)}{\delta} dz$$

$$= \int_0^{\infty} G_*(z, mz) z^{-2} dz \int_{(1/(m+1), 1]} \omega H(d\omega).$$

Hence $P[Z_2 \geq mZ_1] = 1$ provided $H((1/(m+1), 1]) = 0$, that is, the spectral measure is concentrated on $[0, 1/(m+1)]$. Similarly, $P[Z_1 \geq mZ_2] = 1$ provided $H([0, m/(m+1))) = 0$, that is, H is concentrated on $[m/(m+1), 1]$. The requirements (8.29) force $m \leq 1$.

All in all, we can implement order restrictions on Z_1 and Z_2 by letting the spectral measure H be concentrated on a subinterval $[a, b]$ of $[0, 1]$, where $0 \leq a \leq 1/2 \leq b \leq 1$. Observe that if $a = 1/2$ or $b = 1/2$, then in view of (8.29), H must be concentrated on $1/2$, corresponding to complete dependence. So assume $a < 1/2 < b$.

Nadarajah et al. (1998) describe a method to construct such H starting from an initial measure H^* with density h^* and satisfying (8.29). For

$$0 \leq \gamma_a \leq \frac{2b-1}{b-a}, \qquad 0 \leq \gamma_b \leq \frac{1-2a}{b-a},$$

define H by its point masses on a and b and its density h on (a, b) through $H(\{a\}) = \gamma_a$, $H(\{b\}) = \gamma_b$, and, for $a < \omega < b$,

$$h(\omega) = \frac{(b-a)(\alpha\beta)^2}{\{\alpha(\omega-a) + \beta(\omega-b)\}^3} h^* \left\{ \frac{\alpha(w-a)}{\alpha(w-a) + \beta(b-w)} \right\},$$

where $\alpha = 2b - 1 - (b-a)\gamma_a$ and $\beta = 1 - 2a - (b-a)\gamma_b$. Then H satisfies (8.29) and is concentrated on $[a, b]$, as desired. If γ_a and γ_b are equal to their respective upper boundaries, then $h = 0$, so that H merely consists of atoms at $\{a\}$ and $\{b\}$, which is the so-called natural model, already introduced by Tiago de Oliveira (1980, 1989b).

9.2.2   Some parametric models

Logistic model and variations

Basic form.   In its simplest form, the logistic model has stable tail dependence function

$$l(v_1, v_2) = (v_1^{1/\alpha} + v_2^{1/\alpha})^\alpha, \qquad v_j \geq 0, \tag{9.6}$$

with parameter $0 < \alpha \leq 1$. Introduced by Gumbel (1960a,b), it is the oldest parametric family of bivariate extreme value dependence structures. Because of its simplicity, it is still the most popular one. From (8.35), we can compute easily that the corresponding spectral measure H does not have point masses on 0 or 1, while by (8.36), its spectral density h on $(0, 1)$ is

$$h(\omega) = \frac{1 - \alpha}{\alpha} \{\omega(1 - \omega)\}^{1/\alpha - 2} \{(1 - \omega)^{1/\alpha} + \omega^{1/\alpha}\}^{\alpha - 2}.$$

The parameter α measures the strength of dependence between the two coordinates. In particular, independence and complete dependence correspond to $\alpha = 1$ and $\alpha \downarrow 0$, respectively. An interesting interpretation of the parameter α is given in Ledford and Tawn (1998): they show that in a random sample from a bivariate extreme value distribution with logistic dependence structure, the probability that the maximum values in both coordinates occur at the same pair of observations converges to $1 - \alpha$ as the sample size tends to infinity. Further, Kendall's tau (8.57) is given by $\tau = 1 - \alpha$ (Oakes and Manatunga 1992), whereas the correlation between the two coordinates after transformation to Gumbel or exponential margins as in (8.58) or (8.59) is equal to $1 - \alpha^2$ or $\alpha \Gamma^2(\alpha) \{\Gamma(2\alpha)\}^{-1} - 1$ respectively (Tawn 1988a; Tiago de Oliveira 1980). The extremal coefficient in (8.56) is $l(1, 1) = 2^\alpha$. All in all, the strength of dependence increases as α decreases.

Asymmetric logistic model.   The logistic model has the disadvantage that it is symmetric in the two variables. An asymmetric extension, proposed by Tawn (1988a), is

$$l(v_1, v_2) = (1 - \psi_1)v_1 + (1 - \psi_2)v_2 + \{(\psi_1 v_1)^{1/\alpha} + (\psi_2 v_2)^{1/\alpha}\}^\alpha \tag{9.7}$$

for $v_j \geq 0$, with parameters $0 < \alpha \leq 1$ and $0 \leq \psi_j \leq 1$ for $j = 1, 2$. For $\psi_1 = \psi_2$, we obtain a mixture of independence and the logistic model; in particular, for $\psi_1 = \psi_2 = 1$, the model reduces to the logistic model (9.6). Independence arises as soon as $\alpha = 1$ or $\psi_1 = 0$ or $\psi_2 = 0$. If $\alpha < 1$, the corresponding spectral measure H has point masses $H(\{0\}) = 1 - \psi_2$ and $H(\{1\}) = 1 - \psi_1$, while the spectral density h is given by (8.37). Figure 9.2(a) shows the Pickands dependence function $A(t) = l(1 - t, t)$ for a number of parameter values.

For $\alpha \downarrow 0$, we get the non-differentiable model

$$l(v_1, v_2) = \max\{(1 - \psi_1)v_1 + v_2, v_1 + (1 - \psi_2)v_2\}. \tag{9.8}$$

Its spectral measure H is concentrated on three points: $H(\{0\}) = 1 - \psi_2$, $H(\{\psi_1/(\psi_1 + \psi_2)\}) = \psi_1 + \psi_2$, and $H(\{1\}) = 1 - \psi_1$. If $\psi_1 = \psi_2$ in (9.8), we get a bivariate model discovered by Marshall and Olkin (1967) in the context of survival analysis and recognized as an extreme value dependence structure by Tiago de Oliveira (1971), who called it the *Gumbel* model. Also, choosing $\psi_1 = 1$ or $\psi_2 = 1$ in (9.8) yields the so-called *bi-extremal* model (Tiago de Oliveira 1969, 1974). Complete dependence arises if $\psi_1 = \psi_2 = 1$.

Bilogistic model. In the asymmetric logistic model, the spectral measure H of (8.30) may put non-negative mass on the boundary points 0 and 1, which complicates likelihood inference in certain point-process models for high-threshold exceedances, see section 9.4.2. Therefore, starting from the representation (8.39) of a bivariate extreme value dependence structure in terms of spectral functions, Joe *et al.* (1992) propose the model

$$l(v_1, v_2) = \int_0^1 \max\{(1 - \alpha)t^{-\alpha}v_1, (1 - \beta)(1 - t)^{-\beta}v_2\}dt, \qquad (9.9)$$

where $0 \le \alpha < 1$ and $0 \le \beta < 1$. The model is another asymmetric extension of the logistic model (9.6) to which, it reduces if $\alpha = \beta$. The parameter $(\alpha + \beta)/2$ may be thought of as measuring the strength of dependence, while $\alpha - \beta$ measures the amount of asymmetry. From (8.35) we find that the spectral measure H does not put any mass on 0 or 1, whereas (8.36) only leads to an implicit formula for its density h on $(0, 1)$ in terms of the root of a certain equation.

Tajvidi's generalized symmetric logistic model. Tajvidi (1996) proposes the following extension of the bivariate symmetric logistic model (9.6): for $v_j \ge 0$,

$$l(v_1, v_2) = \{v_1^{2/\alpha} + 2(1 + \psi)v_1^{1/\alpha}v_2^{1/\alpha} + v_2^{2/\alpha}\}^{\alpha/2}$$

where $0 < \alpha \le 1$ and $-1 < \psi \le 2(\alpha^{-1} - 1)$. The model seems to have an identifiability problem as it reduces to (9.6) with shape parameter α for $\psi = 0$ and to (9.6) with shape parameter $\alpha/2$ for $\psi \downarrow -1$. Complete dependence arises as soon as $\alpha \downarrow 0$, while independence occurs as $\alpha = 1$ and $\psi = 0$.

Multivariate extensions. With the aim of constructing spatial models for environmental extremes, Tawn (1990) proposes the following generalization of the asymmetric logistic model (9.7) to an arbitrary number, d, of dimensions. Let C_d be the collection of non-empty subsets c of $\{1, \ldots, d\}$. The multivariate asymmetric logistic model is defined by

$$l(v) = \sum_{c \in C_d} \left\{ \sum_{j \in c} (\psi_{c,j}v_j)^{1/\alpha_c} \right\}^{\alpha_c} \qquad (9.10)$$

for $v \in [0, \infty)$; here $0 < \alpha_c \le 1$, $\psi_{c,j} \ge 0$, and $\sum_{c \ni j} \psi_{c,j} = 1$ for $j = 1, \ldots, d$. If $\alpha_c \downarrow 0$ for all $c \in C_d$, we get a model originally due to Marshall and Olkin (1967),

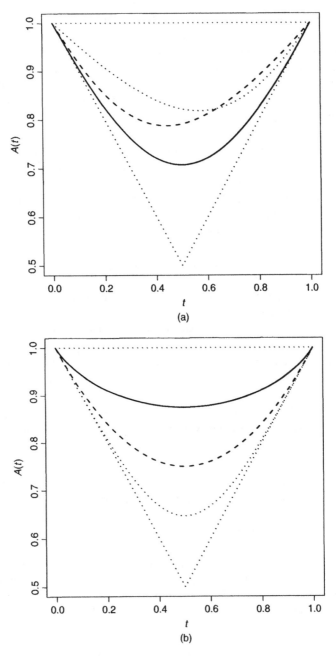

Figure 9.2 Pickands dependence functions: (a) asymmetric logistic (9.7) with
(α, ψ_1, ψ_2) equal to $(0.5, 1, 1)$ (——), $(0.5, 0.6, 0.9)$ (- - - -), $(0.5, 0.8, 0.5)$
($\cdots\cdots$); (b) negative logistic (9.13) with $\psi_1 = \psi_2 = 1$ and α equal to -2 (——),
-1 (- - - -), -0.5 ($\cdots\cdots$).

while the model arising when the $\psi_{c,j}$ do not depend on j is already studied by McFadden (1978). Smith *et al.* (1990) apply a certain tri-variate sub-model to sea-level data on the coast of England.

The spectral densities corresponding to (9.10) can be computed from (8.34) and are given explicitly in Coles and Tawn (1991). Simulation methods for the multivariate logistic model are developed in Stephenson (2003) and the references therein.

A simple special case of (9.10) is

$$l(\boldsymbol{v}) = (v_1^{1/\alpha} + \cdots + v_d^{1/\alpha})^\alpha, \tag{9.11}$$

the multivariate symmetric logistic distribution. Genest and Rivest (1989) characterize the copulas corresponding to (9.11) as the only extreme value copulas that are also Archimedean copulas.

Tawn (1990) also mentions an extension of (9.10) in a kind of nested structure involving a hierarchy of levels, thereby generalizing models studied in McFadden (1978). A special tri-variate case applied in Coles and Tawn (1991) to oceanographic data is the nested logistic model,

$$l(v_1, v_2, v_3) = \{(v_1^{1/\alpha} + v_2^{1/\alpha})^{\alpha/\beta} + v_3^{1/\beta}\}^\beta \tag{9.12}$$

where $0 < \alpha \le \beta \le 1$, featuring bivariate symmetric logistic dependence with parameter α for the first two coordinates and with parameter β for the other pairs of coordinates. Proceeding in a recursive manner from (9.12) to higher dimensions leads to a model described by Joe (1994).

Yet another multivariate extension of the logistic model is the time series logistic model by Coles and Tawn (1991). The idea is to start from a first-order Markov process $X_1, \ldots, X_d$ for which the bivariate dependence structures of the pairs (X_j, X_{j+1}) fall in the (differentiable) domain of attraction of the bivariate symmetric logistic model (9.6). Then, by Markov dependence, actually the joint dependence structure of the vector $(X_1, \ldots, X_d)$ lies in the domain of attraction of a d-variate dependence structure, coined the time series logistic model. In section 10.4, this is implicitly used to model the extremes of time series with Markov structure.

Negative logistic models and extensions. The negative logistic model introduced by Joe (1990) is quite similar in form to the logistic model. In its asymmetric version, the bivariate negative logistic model is defined by

$$l(v_1, v_2) = v_1 + v_2 - \{(\psi_1 v_1)^{1/\alpha} + (\psi_2 v_2)^{1/\alpha}\}^\alpha, \tag{9.13}$$

where $-\infty \le \alpha \le 0$ and $0 \le \psi_j \le 1$ for $j = 1, 2$. Independence arises as soon as $\alpha \to -\infty$ or $\psi_1 = 0$ or $\psi_2 = 0$. If $\alpha \to 0$, we rediscover the non-differentiable model (9.8). The model is symmetric for $\psi_1 = \psi_2$. Figure 9.2(b) shows the Pickands dependence function $A(t) = l(1 - t, t)$ for $\psi_1 = \psi_2 = 1$ and a number of values for α.

We can, as usual, compute the spectral measure H from (8.35) and (8.36): we find $H(\{0\}) = 1 - \psi_2$, $H(\{1\}) = 1 - \psi_1$, and spectral density

$$h(\omega) = (1 - \alpha^{-1})(\psi_1\psi_2)^{1/\alpha}\{\omega(1 - \omega)\}^{1/\alpha - 2}[\{\psi_1(1 - \omega)\}^{1/\alpha} + (\psi_2\omega)^{1/\alpha}]^{\alpha - 2}$$

for $0 < \omega < 1$.

In the same way as the bilogistic model is an asymmetric extension of the bivariate symmetric logistic model, the *negative bilogistic model* by Coles and Tawn (1994) is an extension of the bivariate symmetric negative logistic model. The stable tail dependence function is again (9.9), but now with parameter ranges $-\infty < \alpha < 0$ and $-\infty < \beta < 0$. The spectral measure H does not have point masses on 0 or 1, although its density h on $(0, 1)$ can only be expressed in terms of the root of a certain equation. A little reflection shows that one could even consider (9.9) with $0 < \alpha < 1$ and $-\infty < \beta < 0$ or vice versa, thereby obtaining some kind of hybrid between the bilogistic and negative bilogistic model.

The general multivariate version of (9.13) is

$$l(\boldsymbol{v}) = \sum_{j=1}^{d} v_j - \sum_{c \in C_d : |c| \geq 2} (-1)^{|c|} \left\{ \sum_{j \in c} (\psi_{c,j} v_j)^{1/\alpha_c} \right\}^{\alpha_c}$$

where C_d is the collection of non-empty subsets of $\{1, \ldots, d\}$; the parameter ranges are $-\infty \leq \alpha_c \leq 0$, $\psi_{c,j} \geq 0$, and $\sum_{c \ni j, |c| \geq 2} (-1)^{|c|} \psi_{c,j} \leq 1$ for all $j = 1, \ldots, d$. Also for this model, formula (8.34) can be used to find the spectral densities of the corresponding spectral measure H, see Coles and Tawn (1991). A related multivariate extension is proposed in Joe (1994); see also Kotz and Nadarajah (2000), p. 130.

Polynomial Pickands dependence function

Klüppelberg and May (1999) describe the class of Pickands dependence functions A that have a polynomial form,

$$A(t) = \psi_0 + \psi_1 t + \psi_2 t^2 + \cdots + \psi_m t^m, \qquad 0 \leq t \leq 1, \qquad (9.14)$$

with m a positive integer. The conditions $A(0) = 1$, $A(1) = 1$, $0 \geq A'(0) \geq -1$, $0 \leq A'(1) \leq 1$, $A''(0) \geq 0$ and $A''(1) \geq 0$ imply the necessary restrictions

$$\begin{cases} \psi_0 = 1 \\ \psi_1 = -(\psi_2 + \cdots + \psi_m) \\ 0 \leq \psi_2 + \cdots + \psi_m \leq 1 \\ \psi_2 \geq 0 \\ 0 \leq \psi_2 + 2\psi_3 + \cdots + (m - 1)\psi_m \leq 1 \\ \psi_2 + 3\psi_3 + \cdots + \binom{m}{2}\psi_m \geq 0 \end{cases} \qquad (9.15)$$

which, however, are not sufficient, in general, to guarantee that $A(t)$ in (9.14) is a Pickands dependence function [for instance, the function $A(t) = 1 - t^3 + t^4$ does satisfy (9.15) but is not convex].

The spectral measure H can be easily computed from A through (8.47). In particular, $H(\{0\}) = 1 - (\psi_2 + \cdots + \psi_m)$ and $H(\{1\}) = 1 - \{\psi_2 + 2\psi + \cdots + (m-1)\psi_m\}$, while its density on $(0, 1)$ is $h = A''$. Since A is a polynomial, complete dependence, $A(t) = \max(t, 1-t)$, can only be attained as $m \to \infty$. The linear case, $m = 1$, admits as only solution $A(t) = 1$, corresponding to independence. Most relevant for statistical purposes are the quadratic and the cubic case, corresponding to the mixed and asymmetric mixed model, respectively.

Quadratic case: mixed model. If $m = 2$ in (9.14), then we must have $-\psi_1 = \psi = \psi_2 \in [0, 1]$, leading to the (symmetric) mixed model

$$A(t) = 1 - \psi t + \psi t^2, \qquad 0 \leq t \leq 1, \tag{9.16}$$

appearing already in Gumbel (1962). Observe that this model also arises as a special case of the negative logistic: in (9.13), take $\alpha = -1$ and $\psi_1 = \psi_2 = \psi$. Independence arises for $\psi = 0$, but complete dependence is not possible in this model. For a random pair with dependence structure (9.16), the correlation coefficient is $6\pi^{-2}\{\arccos(1 - \psi/2)\}^2 \in [0, 2/3]$ if the margins are transformed to the Gumbel distribution as in (8.58) (Tiago de Oliveira 1980), and an even more complicated expression if the margins are transformed to the exponential distribution as in (8.59) (Tawn 1988a; Tiago de Oliveira 1989b).

Cubic case: asymmetric mixed model. If $m = 3$ in (9.14), the Pickands dependence function takes the form

$$A(t) = 1 - (\psi_2 + \psi_3)t + \psi_2 t^2 + \psi_3 t^3, \qquad 0 \leq t \leq 1, \tag{9.17}$$

see Figure 9.3(a). The conditions (9.15) reduce to

$$\psi_2 \geq 0, \quad \psi_2 + 3\psi_3 \geq 0, \quad \psi_2 + \psi_3 \leq 1, \quad \psi_2 + 2\psi_3 \leq 1,$$

which are also sufficient to guarantee that $A(t)$ in (9.17) is a Pickands dependence function. Independence occurs at $\psi_2 = \psi_3 = 0$, a corner of the parameter space, while, again, complete dependence is not possible.

Gaussian model

The Gaussian model is defined by its stable tail dependence function

$$l(v_1, v_2) = v_1 \Phi\{\lambda + (2\lambda)^{-1} \log(v_1/v_2)\} + v_2 \Phi\{\lambda + (2\lambda)^{-1} \log(v_2/v_1)\} \tag{9.18}$$

with parameter $\lambda \in [0, \infty]$, and with Φ the standard normal distribution function, see Figure 9.3(b). The cases $\lambda = 0$ and $\lambda = \infty$ correspond to complete dependence and independence, respectively. The extremal coefficient in (8.56) is $l(1, 1) = 2\Phi(\lambda)$, that is, dependence decreases as λ increases. By (8.35), the spectral measure H does not put mass on 0 or 1, while its density on $(0, 1)$ can be easily computed from (8.36).

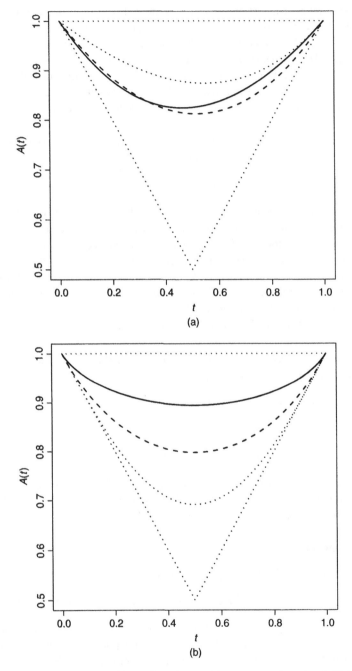

Figure 9.3 Pickands dependence functions: (a) asymmetric mixed (9.17) with (ψ_2, ψ_3) equal to $(1, -0.2)$ (——), $(0.6, 0.1)$ (- - - -), $(0.2, 0.2)$ ($\cdots\cdots$); (b) Gaussian (9.18) with λ equal to 1.25 (——), 0.83 (- - - -), 0.5 ($\cdots\cdots$).

Hüsler and Reiss (1989) characterized the model as the limit dependence structure of the suitably normalized component-wise maximum in a triangular array
$X_{1n}, \ldots, X_{nn}$ of independent, centred, unit-variance bivariate normal random pairs
with correlation ρ_n such that $(1 - \rho_n) \log(n) \to \lambda^2$ as $n \to \infty$. Hooghiemstra and
Hüsler (1996) prove a related characterization in terms of projections of standard
normal pairs in directions in the neighbourhood of a given fixed direction. Coles
and Pauli (2001) find comparable results for a class of bivariate Poisson distributions, the heuristic being that a Poisson distribution with a large intensity can be
approximated well by a normal distribution.

The model can also be obtained by the method of max-stable processes as in
(9.2), incidentally yielding a generalization to higher dimensions. Let both the index
set V and classification space S be $\mathbb{R}$, the frequency measure ν be the Lebesgue
measure, and the event profile function $f_\sigma(s, v)$ be the probability density function
in s of the normal distribution with mean v and variance σ^2. Then the stable tail
dependence function of the pair $(Z_v, Z_{v'})$ is equal to (9.18) with $\lambda = |v - v'|/(2\sigma)$.
A further extension of the model to $V = \mathbb{R}^2$ is used in Smith (1991) to describe
spatial dependence between storms in function of the distance between the storm
locations; see also Coles and Tawn (1996a) and Schlather and Tawn (2003). An
alternative multivariate extension is proposed in Joe (1994).

Circular model

Using the technique of max-stable processes in (9.2), Coles and Walshaw (1994)
construct a model for describing dependence between annual maxima of wind
speeds recorded at a fixed location across continuous directional space. A typical
storm has a principal direction $s \in S = (0, 2\pi]$ and strength $0 < r < \infty$. Its relative
strength at direction $v \in V = (0, 2\pi]$ is $r f_\zeta(s, v)$, where

$$f_\zeta(s, v) = \frac{1}{2\pi I_0(\zeta)} \exp\{\zeta \cos(s - v)\},$$

with $0 = \zeta < \infty$ and $I_0(\zeta) = (2\pi)^{-1} \int_0^{2\pi} \exp\{\zeta \cos(s)\} ds$ the modified Bessel
function of order 0. The function $f_\zeta(\cdot, v)$ is the von Mises circular density with
location and concentration parameters v and ζ respectively.

By (9.2), the joint distribution of the maximal wind speeds $(Z_v : v \in V_0)$
recorded in a given year at a collection $V_0 \subset (0, 2\pi]$ of directions and transformed
to standard-Fréchet margins is then given by

$$P[Z_v \leq z_v, \forall v \in V_0] = \exp\left[-\int_0^{2\pi} \max_{v \in V_0}\{z_v^{-1} f_\zeta(s, v)\} ds \right]$$

for $0 < z_v < \infty$. Large values of ζ correspond to profiles that are highly concentrated around a single direction, the limit $\zeta \to \infty$ being that of independent Z_v.
On the other hand, $\zeta = 0$ gives a constant profile and complete dependence.

Dirichlet model

The following model is an example of the construction of (9.4). For positive numbers α_1 and α_2, let

$$h^*(u) = \frac{\Gamma(\alpha_1 + \alpha_2)}{\Gamma(\alpha_1)\Gamma(\alpha_2)} u^{\alpha_1 - 1}(1 - u)^{\alpha_2 - 1}, \qquad 0 < u < 1;$$

observe that this is the probability density function of a Beta distribution. We have $m_j = \alpha_j/(\alpha_1 + \alpha_2)$ for $j = 1, 2$ in (9.3), so that by (9.4), we obtain after some calculation,

$$h(\omega) = \frac{\alpha_1^{\alpha_1}\alpha_2^{\alpha_2}\Gamma(\alpha_1 + \alpha_2 + 1)}{\Gamma(\alpha_1)\Gamma(\alpha_2)} \frac{\omega^{\alpha_1 - 1}(1 - \omega)^{\alpha_2 - 1}}{\{\alpha_1\omega + \alpha_2(1 - \omega)\}^{\alpha_1 + \alpha_2 + 1}}.$$

for $0 < \omega < 1$, the density of a measure H on $(0, 1)$ satisfying (8.29).

In d dimensions, we start from the Dirichlet density

$$h^*(\boldsymbol{u}) = \frac{\Gamma(\alpha_1 + \cdots + \alpha_d)}{\Gamma(\alpha_1) \cdots \Gamma(\alpha_d)} u_1^{\alpha_1 - 1} \cdots u_d^{\alpha_d - 1}$$

on $\boldsymbol{u} \in S_d$, with parameters $\alpha_j > 0$ for $j = 1, \ldots, d$. As its jth moment is $m_j = \alpha_j/(\alpha_1 + \cdots + \alpha_d)$, we obtain from (9.5) the spectral density

$$h(\boldsymbol{\omega}) = \frac{\Gamma(\alpha_1 + \cdots + \alpha_d + 1)}{\alpha_1\omega_1 + \cdots + \alpha_d\omega_d} \prod_{j=1}^{d} \frac{\alpha_j^{\alpha_j}}{\Gamma(\alpha_j)} \frac{\omega_j^{\alpha_j - 1}}{(\alpha_1\omega_1 + \cdots + \alpha_d\omega_d)^{\alpha_j}}$$

for $\boldsymbol{\omega} \in S_d$, which is called the *Dirichlet model* in Coles and Tawn (1991).

Piecewise algebraic spectral density

For the bivariate case, Nadarajah (1999) proposes a spectral measure H in (8.28) with point masses at 0, $0 < \theta < 1$, and 1, and with spectral density h on $(0, \theta)$ and $(\theta, 1)$:

$$H(\{x\}) = \gamma_x, \qquad x \in \{0, \theta, 1\},$$

$$h(\omega) = \begin{cases} \alpha\omega^r & \text{if } 0 < \omega < \theta, \\ \beta(1 - \omega)^s & \text{if } \theta < \omega < 1. \end{cases}$$

The parameter ranges are $\alpha \geq 0$, $\beta \geq 0$, $r > -1$, $s > -1$, and $\gamma_x \geq 0$ for $x \in \{0, \theta, 1\}$. The condition $\alpha\theta^r = \beta(1 - \theta)^s$ ensures that h can be continuously extended in θ, whereas the requirement (8.29) is met as soon as $\gamma_\theta \leq 1/$

$\{\theta \vee (1 - \theta)\}$ and

$$\gamma_0 = 1 - (1 - \theta)\gamma_\theta - \frac{\beta}{s+2}(1 - \theta)^{s+2} + \alpha\theta^{r+1}\left(\frac{\theta}{r+2} - \frac{1}{r+1}\right),$$

$$\gamma_1 = 1 - \theta\gamma_\theta - \frac{\alpha}{r+2}\theta^{r+2} + \beta(1 - \theta)^{s+1}\left(\frac{1-\theta}{s+2} - \frac{1}{s+1}\right).$$

In total, the number of free model parameters is five. The model can accommodate for a wide range of characteristics of bivariate extremal dependence structures.

9.3 Component-wise Maxima

Let $\{Y_1, \ldots, Y_k\}$ be an independent sample from a d-variate extreme value distribution function G. In this section, we explain how to estimate G. We also consider the more general situation where the dependence structure of G is that of a multivariate extreme value distribution, whereas the margins are arbitrary, that is,

$$G(y) = \exp[-l\{-\log G_1(y_1), \ldots, -\log G_d(y_d)\}],$$

for some stable tail dependence function l and arbitrary continuous distribution functions G_j.

These estimation problems may arise in a number of situations. The most typical one is where the Y_i can be viewed as component-wise maxima over blocks of variables of some underlying series, $\{X_1, \ldots, X_n\}$, that is,

$$Y_i = \bigvee_{r=(i-1)m+1}^{im} X_r, \qquad i = 1, \ldots, k, \qquad (9.19)$$

where $km = n$. The X_r may be observed or not. For instance, suppose X_{rj} denotes the maximal water height at day $r = 1, \ldots, n$ on location $j = 1, \ldots, d$ of a certain river. If blocks correspond to years ($m = 365$), then Y_{ij} is the maximal water height in year i at location j. With y_j the height of a dike or dam at location j, the probability that there will not be a flood in year i at any of the d locations is $G(y) = P[Y_i \leq y]$. This methodology is in fact the multivariate generalization of the annual maximum approach already advocated by Gumbel (1958); see also section 5.1. Historically, it marks the beginning of multivariate extreme value statistics based on the probabilistic theory of multivariate extremes (Gumbel and Goldstein 1964).

In (9.19), the X_r need not be independent or identically distributed. In the water height example, the X_r will certainly feature within-year seasonality as well as temporal dependence, with high water levels persisting for a number of days in a row. Still, in the absence of long-range dependence, the operation of taking maxima over observations aggregated over a whole year may be reasonably expected to produce an approximately independent sample from a multivariate extreme value distribution, see section 8.4.

The problem of estimating a multivariate extreme value dependence structure may be relevant outside extreme value statistics as well. Specifically, extreme value copulas form a large, non-parametric but still manageable subclass of the class of copulas with positive association. In this sense, they may be useful in modelling the dependence structure of random vectors for which positive association is a reasonable assumption.

Of course, retaining only the component-wise maxima over large blocks of observations, is, like in the univariate case, rather wasteful of data. In the multivariate set-up, the additional problem arises that the vector of component-wise maxima is typically not an observation itself as the maximal observations in each of the variables need not occur at the same moment. In section 9.4, we will therefore consider the more realistic problem of estimating a multivariate extreme value distribution based on a random sample from a distribution in its domain of attraction, thereby extending the familiar threshold approaches in the univariate case of Chapters 4 and 5 to the multivariate case.

Broadly speaking, there are two approaches: non-parametric (section 9.3.1) and parametric (section 9.3.2). In the non-parametric approach, we focus on the bivariate case, the estimation problem usually being formulated as how to estimate the Pickands dependence function A, introduced in section 8.2.5. In the parametric approach, the unknown copula is assumed to belong to a certain parametric family, usually one of the families described in section 9.2. In section 9.3.3, we will illustrate both approaches with the Loss-ALAE data of Figure 9.1.

In both approaches, the complication arises that, in practice, the margins of G are unknown. One option is to model the margins by univariate extreme value distributions. If, on the other hand, we do not want to make any assumptions on the margins, then the alternative consists of estimating the margins by the empirical distributions. In any case, proper credit should be given to the statistical uncertainty arising from having to estimate the margins, although it is not clear how to do this in a semi- or non-parametric context.

9.3.1 Non-parametric estimation

Let the random pair (Y_1, Y_2) have distribution function G with continuous margins G_1 and G_2 and with extreme value dependence structure, that is,

$$G(y_1, y_2) = \exp\left[\log\{G_1(y_1)G_2(y_2)\}A\left(\frac{\log\{G_2(y_2)\}}{\log\{G_1(y_1)G_2(y_2)\}}\right)\right], \qquad (9.20)$$

where A is a Pickands dependence function, see section 8.2.5. The joint survival function of the pair $\xi = -\log G_1(Y_1)$ and $\eta = -\log G_2(Y_2)$ is

$$P[\xi > x, \eta > y] = \exp\left\{-(x+y)A\left(\frac{y}{x+y}\right)\right\}, \qquad x \geq 0, \ y \geq 0. \quad (9.21)$$

Observe that ξ and η have a standard exponential distribution.

How to estimate A from a random sample $\{(\xi_i, \eta_i) : i = 1, \ldots, k\}$ from (9.21)? We will consider two families of estimators. The first one consists of refinements of and improvements over an estimator due to Pickands (1981), while the second one originates in a more recent proposal by Capéraà et al. (1997). A third approach, not discussed further in this section, consists of an estimator by Tiago de Oliveira (1989a), elaborated upon in Deheuvels and Tiago de Oliveira (1989) and Tiago de Oliveira (1992), the convergence rate of which is too slow to be practically applicable.

In practice, we do not observe the (ξ_i, η_i) but merely the (Y_{i1}, Y_{i2}). We cannot perform the required transformation as the margins are unknown. Hence, we have to replace the (ξ_i, η_i) by $\hat{\xi}_i = -\log \hat{G}_1(Y_{i1})$ and $\hat{\eta}_i = -\log \hat{G}_2(Y_{i2})$; here $\hat{G}_j$ is an estimate of G_j, for instance, the (modified) empirical distribution function $\hat{G}_j(y) = (k + 1)^{-1} \sum_{i=1}^{k} \mathbf{1}(Y_{ij} \leq y)$ or a member of a certain parametric family, typically that of the univariate extreme value distributions.

Pickands estimator

Let the random pair (ξ, η) be as in (9.21). For $t \in [0, 1]$,

$$P\left[\min\left(\frac{\xi}{1-t}, \frac{\eta}{t} \right) > x \right] = P[\xi > (1 - t)x, \eta > tx]$$

$$= \exp\{-x A(t)\}, \qquad x \geq 0.$$

In words, the random variable $\min\{\xi/(1 - t), \eta/t\}$ has an exponential distribution with mean $1/A(t)$. Pickands (1981, 1989) proposed to estimate $A(t)$ by the reciprocal of the sample mean of the $\min\{\xi_i/(1 - t), \eta_i/t\}$,

$$\frac{1}{\hat{A}_k^P(t)} = \frac{1}{k} \sum_{i=1}^{k} \min\left(\frac{\xi_i}{1-t}, \frac{\eta_i}{t} \right), \qquad t \in [0, 1]. \tag{9.22}$$

The Pickands estimator, $\hat{A}_n^P$, is conceptually simple and easy to compute, but has the drawback of not satisfying the necessary constraints to be itself a Pickands dependence function. This was the motivation for a number of modifications of the estimator. Denote the sample means of the ξ_i and the η_i by $\bar{\xi}_k = k^{-1} \sum_{i=1}^{k} \xi_i$ and $\bar{\eta}_k = k^{-1} \sum_{i=1}^{k} \eta_i$, respectively. Deheuvels (1991) proposed the variant

$$\frac{1}{\hat{A}_k^D(t)} = \frac{1}{k} \sum_{i=1}^{k} \min\left(\frac{\xi_i}{1-t}, \frac{\eta_i}{t} \right) - (1 - t)\bar{\xi}_k - t\bar{\eta}_k + 1, \qquad t \in [0, 1], \tag{9.23}$$

while Hall and Tajvidi (2000b) suggested

$$\frac{1}{\hat{A}_k^{HT}(t)} = \frac{1}{k} \sum_{i=1}^{k} \min\left(\frac{\xi_i/\bar{\xi}_k}{1-t}, \frac{\eta_i/\bar{\eta}_k}{t} \right), \qquad t \in [0, 1]. \tag{9.24}$$

The estimator of Deheuvels verifies $\hat{A}_k^{\mathrm{D}}(0) = \hat{A}_k^{\mathrm{D}}(1) = 1$, and the estimator of Hall and Tajvidi satisfies $\hat{A}_k^{\mathrm{HT}}(0) = \hat{A}_k^{\mathrm{HT}}(1) = 1$ as well as $\hat{A}_k^{\mathrm{HT}}(t) \geq \max(t, 1 - t)$.

Still, neither of the three estimators satisfies the constraint that a Pickands dependence function is convex. An obvious remedy is to replace an initial estimator $\hat{A}_k$ by its convex minorant, that is, the largest convex function on the interval $[0, 1]$ that is bounded by $\hat{A}_k$. Only in case of the Hall–Tajvidi estimator does the resulting modification satisfy all the constraints of a Pickands dependence function; for the Pickands and Deheuvels estimators, some further modifications are required to meet the constraint $\max(t, 1 - t) \leq A(t) \leq 1$ for $t \in [0, 1]$.

A final method that has been investigated to improve estimation is through smoothing. This might be a particularly good idea if the objective is to estimate the second derivative of A, which, in case of a differentiable model, is equal to the density of the spectral measure H on the interior of the unit interval, see (8.47). Smith *et al.* (1990) investigate various kinds of kernel estimators based on the original Pickands estimator, but conclude that this offers little gain over the usual finite-difference approximation of the second derivative based on the Pickands estimator itself. Alternatively, Hall and Tajvidi (2000b) suggest to approximate an arbitrary initial estimator by a polynomial smoothing spline of degree three or more, the knots being equally spaced on the unit interval. For the choice of the smoothing parameter, they suggest a cross-validation method. They illustrate by means of a small simulation study that this form of smoothing may lead to better estimation of A, although they do not mention the effect of smoothing on estimating the second derivative of A. Finally, the ideas of smoothing and taking convex minorants can be combined, in either order.

Deheuvels (1991) showed convergence of the stochastic processes

$$\delta_k^{\mathrm{P}}(t) = k^{1/2}[\{\hat{A}_k^{\mathrm{P}}(t)\}^{-1} - \{A(t)\}^{-1}],$$

$$\delta_k^{\mathrm{D}}(t) = k^{1/2}[\{\hat{A}_k^{\mathrm{D}}(t)\}^{-1} - \{A(t)\}^{-1}],$$

in $t \in [0, 1]$ to centred Gaussian processes with covariance structures depending on A. On the basis of this result, Deheuvels and Martynov (1996) proposed to use the Cramér-von Mises type statistic $T_k = \int_0^1 \{\delta_k^{\mathrm{P}}(t)\}^2 \mathrm{d}t$ to test the hypothesis of independence, $A \equiv 1$. To implement the test, they compute and tabulate the critical values of the limit distribution of the test statistic, T_k, under the null hypothesis. The use of the convergence results and the proposed test are hampered in practice because the fact is ignored that, prior to the estimation of A, the marginal distributions have to be estimated as well. Although it seems reasonable to conjecture that this preliminary marginal estimation will not affect the root-k consistency of the proposed estimators for A, it seems equally probable that the asymptotic distribution of the estimators will be different from when the marginal distributions are known.

Capéràa–Fougères–Genest estimator

Another type of estimator of A was proposed by Capéraà et al. (1997). The motivation and description we give here differ greatly from those in the cited paper and are intended to be slightly simpler.

Let (ξ, η) be a bivariate standard exponential pair with joint survival function given by (9.21). Then

$$P[\max\{t\xi, (1-t)\eta\} > x]$$

$$= P[\xi > x/t] + P[\eta > x/(1-t)] - P[\xi > x/t, \eta > x/(1-t)]$$

$$= \exp(-x/t) + \exp\{-x/(1-t)\} - \exp[-xA(t)/\{t(1-t)\}]$$

for $t \in [0, 1]$ and $x > 0$, so that

$$E[\log \max\{t\xi, (1-t)\eta\}] = \log A(t) + \int_0^\infty \log(x)e^{-x}dx.$$

This suggests estimating $A(t)$ by the empirical version of the previous equation, that is,

$$\log \hat{A}_k(t) = \frac{1}{k}\sum_{i=1}^{k} \log \max\{t\xi_i, (1-t)\eta_i\} - \int_0^\infty \log(x)e^{-x}dx.$$

However, $\hat{A}_k$ does not satisfy the constraints $A(0) = 1 = A(1)$. This is the motivation for the following modification, leading to the estimator of Capéraà et al. (1997):

$$\log \hat{A}_k^C(t) \tag{9.25}$$

$$= \log \hat{A}_k(t) - t \log \hat{A}_k(1) - (1-t)\log \hat{A}_k(0)$$

$$= \frac{1}{k}\sum_{i=1}^{k} \log \max\{t\xi_i, (1-t)\eta_i\} - t\frac{1}{k}\sum_{i=1}^{k}\log \xi_i - (1-t)\frac{1}{k}\sum_{i=1}^{k}\log \eta_i.$$

If required, further modifications are possible to make the estimator meet the constraints of convexity and $\max(t, 1-t) \le A(t) \le 1$. One such modification is proposed by Jiménez et al. (2001), although it leads to a consistent estimator only if A is also log-convex.

Capéraà et al. (1997) also conduct an extensive simulation study comparing $\hat{A}^P$, $\hat{A}^D$ and $\hat{A}^C$ for a wide range of dependence structures. Their results strongly indicate that in general, $\hat{A}^C$ performs better than $\hat{A}^D$, which in turn is preferable over $\hat{A}^P$. In a more restricted simulation study, Hall and Tajvidi (2000b) demonstrate that further improvements can be made by taking the convex hull of either $\hat{A}^C$ or $\hat{A}^{HT}$ and applying constrained spline smoothing.

9.3.2   Parametric estimation

Let $l(\cdot; \boldsymbol{\theta})$ be a parametric family of d-variate stable tail dependence functions indexed by the parameter vector $\boldsymbol{\theta}$; see section 9.2 for a list of popular parametric families. Assume that the d-variate distribution function G has an extreme value dependence structure with stable tail dependence function $l(\cdot; \boldsymbol{\theta})$ for some unknown $\boldsymbol{\theta}$:

$$G(y) = \exp[-l\{-\log G_1(y_1), \ldots, -\log G_d(y_d); \boldsymbol{\theta}\}]. \qquad (9.26)$$

How can we estimate $\boldsymbol{\theta}$ from an independent sample $\{Y_i : i = 1, \ldots, k\}$ from G? The answer differs according to the nature of our assumptions on the margins of G.

If the Y_i arise as component-wise maxima over large blocks of variables, it is natural to model the margins G_j $(j = 1, \ldots, d)$ as generalized extreme value (GEV) distributions

$$G_j(y_j) = \exp\left\{ -\left(1 + \gamma_j \frac{y_j - \mu_j}{\sigma_j} \right)_+^{-1/\gamma_j} \right\}. \qquad (9.27)$$

Here, γ_j is the extreme value index, while μ_j and $\sigma_j > 0$ are location and scale parameters, respectively. The combination of (9.26) and (9.27) now leads to a fully parametric model for G. The marginal and dependence parameters can be estimated simultaneously by maximum likelihood. Moreover, such joint modelling allows transfer of information from one margin to the other (Barão and Tawn 1999). Recall from section 5.1 that for the margin parameters, the estimation problem is regular for $\gamma_j > -1/2$. For the dependence parameter $\boldsymbol{\theta}$, complications may arise for parameter values corresponding to independence, and these have to be dealt with on a case-by-case basis, see below.

However, jointly modelling the margins and the dependence structure may not always be desirable. For instance, goodness-of-fit tests may cast doubts on the hypothesis of extreme value margins, although we may still believe in (9.26). Conversely, Dupuis and Tawn (2001) show that mis-specifying the dependence structure may have large adverse effects on the estimates of the margin parameters.

A more prudent approach, then, consists of the following. Write (9.26) as

$$G(y; \boldsymbol{\theta}) = C\{G_1(y_1), \ldots, G_d(y_d); \boldsymbol{\theta}\}, \qquad (9.28)$$

where

$$C(\boldsymbol{u}; \boldsymbol{\theta}) = \exp[-l\{-\log(u_1), \ldots, -\log(u_d); \boldsymbol{\theta}\}]$$

is the extreme value copula corresponding to $l(\cdot; \boldsymbol{\theta})$. The copula density is

$$c(\boldsymbol{u}; \boldsymbol{\theta}) = \frac{\partial^d}{\partial u_1 \cdots \partial u_d} C(\boldsymbol{u}; \boldsymbol{\theta}).$$

If we would know the margins G_j, then (9.28) would specify a parametric model for the distribution of the vector $(G_1(Y_1), \ldots, G_d(Y_d))$. Now, as we do not know the margins, we replace them by the (modified) empirical distribution functions:

$$\hat{G}_j(y) = \frac{1}{k+1} \sum_{i=1}^{k} \mathbf{1}(Y_{ij} \leq y).$$

Acting as if the $\hat{G}_j$ were the true margins, we can estimate θ by maximizing the pseudo-likelihood

$$L(\theta) = \prod_{i=1}^{k} c\{\hat{G}_1(Y_{i1}), \ldots, \hat{G}_d(Y_{id}); \theta\}. \tag{9.29}$$

The resulting estimator for θ is in fact a special case of the one considered in Genest et al. (1995). They establish asymptotic normality of the pseudo-maximum likelihood estimator for the parameter of a family of copulas in case the margins are unknown and are estimated by the empirical distribution functions. In particular, they show that the estimator is efficient in case the true parameter corresponds to independence. They also give an explicit expression for the variance-covariance matrix and propose a consistent estimator.

Specific models

Logistic model. Despite the large number of *ad hoc* methods for statistical inference on the parameter $\alpha \in (0, 1]$ in the symmetric bivariate logistic model (9.6) (Gumbel and Mustafi 1967; Hougaard 1986; Shi 1995b; Tiago de Oliveira 1980, 1984, 1989b; Yue 2001), maximum likelihood estimation is nevertheless the most efficient. In case $0 < \alpha < 1$, the estimation problem is regular. Oakes and Manatunga (1992) compute the information matrix in case of two-parameter Weibull margins, whereas Shi (1995a) computes the information matrix in case of generalized extreme value margins and symmetric multivariate logistic dependence structure (9.11). Robust estimation within the bivariate logistic model is considered in Dupuis and Morgenthaler (2002).

In the special case $\alpha = 1$, corresponding to independence, the estimation problem is non-regular because of two reasons: the parameter is on the boundary of the parameter space, and the variance of the score statistic is infinite. Tawn (1988a) investigates this case more closely and comes to the following conclusions. Assume first that the margins are known. Then we can transform the observations to standard exponential margins with joint survival function as in (8.51) for l as in (9.6); denote the transformed sample by (ξ_i, η_i), $i = 1, \ldots, k$. The score statistic at $\alpha = 1$ is equal to

$$U_k = \sum_{i=1}^{k} u(\xi_i, \eta_i), \tag{9.30}$$

where $u(\xi, \eta) = \xi \log \xi + \eta \log \eta - \log(\xi \eta)$
$$- (\xi + \eta - 2) \log(\xi + \eta) - (\xi + \eta)^{-1},$$

that is, $u(\xi, \eta)$ is the derivative of the log-likelihood evaluated at $\alpha = 1$. The asymptotic distribution of the score statistic is

$$\left(2^{-1}k \log k\right)^{-1/2} U_k \overset{\mathcal{D}}{\to} \mathrm{N}(0, 1), \qquad k \to \infty. \tag{9.31}$$

Large negative values of the score statistic lead to rejection of the null hypothesis $\alpha = 1$ versus the alternative $\alpha < 1$; in particular, the asymptotic p-value is $p = \Phi\{(2^{-1}k \log k)^{-1/2}U_k\}$, with Φ the standard normal distribution function. Unfortunately, the convergence in (9.31) is rather slow, so that the asymptotic p-value may be far from the true one; therefore, Tawn (1988a) suggests to compute small sample critical values at the desired significance levels by simulation. Alternatively, if λ_k denotes the likelihood ratio, that is, the ratio between the likelihoods at the maximum likelihood estimate $\hat{\alpha}_k$ and at $\alpha = 1$, then

$$\lim_{k \to \infty} P[2 \log \lambda_k \leq x] = \begin{cases} \Phi(x^{1/2}), & \text{if } x \geq 0, \\ 0, & \text{if } x < 0, \end{cases} \tag{9.32}$$

leading to a likelihood ratio test for $\alpha = 1$ versus $\alpha < 1$.

Typically, the marginal distributions are unknown and have to be estimated. Suppose that the margins follow an extreme value distribution and that the shape parameters are such that the maximum likelihood estimators are regular, that is, the extreme value indices are larger than $-1/2$, see section 5.1. Tawn (1988a) shows that if $\alpha = 1$, then the maximum likelihood estimator for α is asymptotically independent from the maximum likelihood estimators for the margin parameters. In particular, having to estimate the margin parameters does not change the asymptotic behaviour of the score and likelihood ratio tests for independence.

In the asymmetric bivariate logistic model (9.7), the problems already encountered for the symmetric bivariate logistic model are aggravated by the fact that the parameters ψ_j are non-identifiable when $\alpha = 1$. Tawn (1988a) pragmatically suggests to accept independence in the asymmetric model if it is accepted in the symmetric case.

In the multivariate case, testing for independence is certainly not simpler than in the bivariate case. By way of example, Tawn (1990) mentions the rather non-standard asymptotic behaviour of the score statistics at independence for the multivariate symmetric model (9.11) and the nested model (9.12). As pairwise independence implies independence, a simpler approach consists of applying just the relevant bivariate tests.

Finally, choosing between all the different logistic models is not easy. A proper understanding of the physical process generating the data should assist in identifying the appropriate structure, see, for instance, Tawn (1990).

Mixed model. Also for the mixed model (9.16), despite the abundance of *ad hoc* methods for statistical inference on ψ (Gumbel and Mustafi 1967; Posner *et al.* 1969; Tiago de Oliveira 1980, 1989b; Yue 2000; Yue *et al.* 1999), maximum likelihood estimation is the most efficient. The estimation problem is regular if

$0 < \psi < 1$, while at independence, $\psi = 0$, the situation is completely parallel with the one for the bivariate symmetric logistic model (9.6) at independence, the only differences being that now the score function at $\psi = 0$ is

$$u(\xi, \eta) = \frac{\xi\eta}{\xi + \eta} + 2\frac{\xi\eta}{(\xi + \eta)^3} - \frac{\xi^2 + \eta^2}{(\xi + \eta)^2},$$

the asymptotic distribution of the score statistic $U_k = \sum_{i=1}^{k} u(\xi_i, \eta_i)$ is

$$\left(15^{-1}k\log k\right)^{-1/2} U_k \overset{\mathcal{D}}{\to} \mathrm{N}(0, 1), \qquad k \to \infty,$$

and the null hypothesis $\psi = 0$ can be rejected in favour of the alternative $\psi > 0$ in case of large positive values of U_k, with p-value $1 - \Phi\{(15^{-1}k\log k)^{-1/2}U_k\}$; see Tawn (1988a), who also proposes a method to discriminate between the mixed and logistic model.

For the asymmetric mixed model (9.17), Tawn (1988a) reports that the score vector at independence converges to a bivariate normal distribution, although he does not give any details.

9.3.3 Data example

We applied the methods of this section to the 1500 Loss-ALAE data of Figure 9.1. As the data do not arise from a time series, there seems no obvious way to partition the data into groups. Therefore, we randomly permutated the data and formed $k = 50$ groups of size $m = 30$, seeking a compromise between the conflicting criteria of large group sizes and a large number of groups. Figure 9.4(a) shows, on a log-scale base 10, a scatterplot of the original data with, superimposed, the component-wise group maxima (y_{i1}, y_{i2}) $(i = 1, \ldots, k)$ computed in (9.19). We transformed the block maxima to standard exponential margins by $\xi_i = -\log \hat{G}_1(y_{i1})$ and $\eta_i = -\log \hat{G}_2(y_{i2})$, where $\hat{G}_j(y) = (k + 1)^{-1} \sum_{i=1}^{k} \mathbf{1}(y_{ij} \leq y)$ for $j = 1, 2$ are the (modified) empirical distribution functions, see Figure 9.4(b).

Next, we estimated Pickands dependence function by the various parametric and non-parametric estimators of this section. Figure 9.5(a) shows the non-parametric estimators by Pickands (9.22), Deheuvels (9.23), Hall–Tajvidi (9.24), and Capéràa–Fougères–Genest (9.25). The Pickands estimator does not satisfy the requirements $A(0) = A(1) = 1$; the estimators of Deheuvels and Hall–Tajvidi are modifications of the Pickands estimator to enforce this constraint. All estimators are clearly below the upper boundary of the triangle corresponding to independence. There also seems some evidence of asymmetry. Unfortunately, none of the estimators is convex, as a Pickands dependence function should be. A possible remedy (not shown) would be to replace the estimators by their convex minorants.

Estimates that do satisfy all requirements of a Pickands dependence function result from fitting parametric models as in section 9.3.2. We decided to employ the semi-parametric likelihood of (9.29) with margins estimated by the empirical distribution functions rather than to fit the fully parametric model consisting of

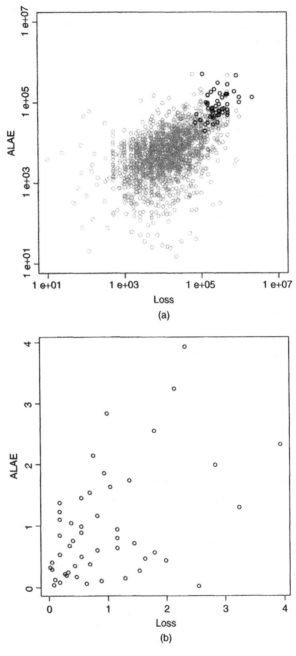

Figure 9.4   Loss-ALAE data: (a) Scatterplot of the data with, superimposed, the component-wise maxima corresponding to a random partition of the data in 50 blocks of size 30. (b) Scatterplot of these 50 component-wise maxima transformed to standard exponential margins.

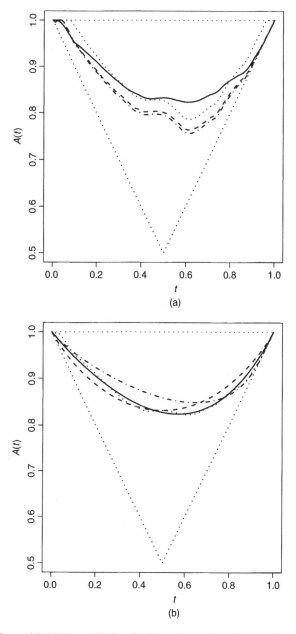

(a)

(b)

Figure 9.5   Loss-ALAE data: Pickands dependence function estimates based on block maxima of Figure 9.4. (a) Non-parametric estimates by Pickands ($\cdots\cdots$), Deheuvels (-----), Hall–Tajvidi (–·–·–), and Capéràa–Fougères–Genest (——). (b) Semi-parametric estimates based on (9.29) for asymmetric logistic (——), logistic (-----), and bilogistic (–·–·–) models together with non-parametric estimate by Capéràa–Fougères–Genest ($\cdots\cdots$).

(9.26) and (9.27) because the fit of the GEV to the margins was unsatisfactory (not shown). For simplicity, we only considered the logistic (9.6), asymmetric logistic (9.7), and bilogistic (9.9) models, although of course the other models in section 9.2 might have been tried as well. Since $A'(1) = \psi_1$ for the asymmetric logistic model, the non-parametric estimates in Figure 9.5 strongly suggest $\psi_1 = 1$, and indeed, imposing this constraint did not lead to a significant decrease in likelihood. The parameter estimates are given in Table 9.1, and the corresponding Pickands functions are shown in Figure 9.5(b). For comparison, we also show the Capéràa–Fougéres–Genest estimate, which is close to the asymmetric logistic one.

The p-value of Tawn's score statistic for independence (9.30) is equal to 0.005, clearly rejecting independence. The likelihood ratio test of the logistic against the bilogistic model gives a p-value of 0.12, showing only weak evidence against symmetry. Alternatively, in the case of the asymmetric logistic model with $\psi_1 = 1$, symmetry corresponds to the boundary value $\psi_2 = 1$. This time, the likelihood ratio statistic should be compared with a one-half chi-squared distribution with one degree of freedom (Tawn 1988a), also resulting in the p-value $P[\chi^2 > 1.37]/2 = 0.12$. Note that in all these similar tests and confidence intervals, the estimation uncertainty arising from having to estimate the margins is not taken into account.

An interesting way to visualize the estimated distribution function of component-wise maxima

$$\hat{G}(y_1, y_2) = \exp\left[\log\{\hat{G}_1(y_1)\hat{G}_2(y_2)\}\hat{A}\left(\frac{\log\{\hat{G}_2(y_2)\}}{\log\{\hat{G}_1(y_1)\hat{G}_2(y_2)\}} \right) \right]$$

is by quantile curves,

$$Q(\hat{G}, p) = \{(y_1, y_2) : \hat{G}(y_1, y_2) = p\}, \qquad 0 < p < 1. \tag{9.33}$$

Table 9.1 Loss-ALAE data: Parameter estimates, standard errors, and negative log-likelihoods for semi-parametric likelihood (9.29) with logistic, asymmetric logistic (constrained at $\psi_1 = 1$) and bilogistic models fitted to the block maxima of Figure 9.4.

Model	Parameter	(Standard error)	NLLH
Logistic	$\alpha = 0.73$	(0.08)	92.55
Asymmetric logistic	$\alpha = 0.61$	(0.13)	91.87
	$\psi_2 = 0.58$	(0.30)	
Bilogistic	$\alpha = 0.23$	(0.23)	91.38
	$\beta = 0.90$	(0.06)	

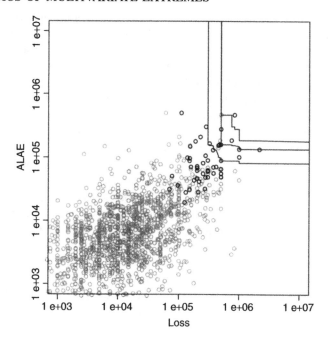

Figure 9.6   Loss-ALAE data: Estimated quantile curves $Q(\hat{F}, p)$ (9.34) for $p = 0.98, 0.99, 0.995$ based on block maxima of Figure 9.4(a) (non-parametric estimates for margins and Capéràa–Fougéres–Genest estimate for Pickands dependence function).

As $\hat{G}(y_1, y_2) = p$ if and only if there exists $w \in [0, 1]$ such that $\hat{G}_1(y_1) = p^{(1-w)/\hat{A}(w)}$ and $\hat{G}_2(y_2) = p^{w/\hat{A}(w)}$, the above quantile curve consists of the points

$$Q(\hat{G}, p) = \left\{ \left(\hat{G}_1^{\leftarrow}\{p^{(1-w)/\hat{A}(w)}\}, \hat{G}_2^{\leftarrow}\{p^{w/\hat{A}(w)}\} \right) : w \in [0, 1] \right\}.$$

Exploiting the relationship $F^m \approx G$, with m the block size ($m = 30$ for the Loss-ALAE data), quantile curves of $\hat{G}$ can be interpreted as quantile curves of $\hat{F}$ by

$$Q(\hat{F}, p) := Q(\hat{G}, p^m). \tag{9.34}$$

Figure 9.6 shows the quantile curves $Q(\hat{F}, p)$ for $p = 0.98, 0.99, 0.995$ with the margins estimated non-parametrically and with Pickands dependence function estimated by the Capéràa–Fougères–Genest estimator (9.25).

9.4   Excesses over a Threshold

Let F be a d-variate distribution function and let $X_1, \ldots, X_n$ be an independent sample from F. Let $x \in \mathbb{R}^d$ be such that $1 - F_j(x_j)$ is of the order $1/n$ for all

j. How to estimate $1 - F(x)$? Because we want to control the relative estimation error, the empirical distribution function is hardly of any use here. For example, for $x = x_n$ such that $1 - F(x_n) = 1/n$ and with $\hat{F}_n$ the empirical distribution function of the sample, the asymptotic distribution of $\{1 - \hat{F}_n(x_n)\}/\{1 - F(x_n)\}$ is Poisson(1), whereas in fact we want it to converge to 1.

In order to make any progress, we need to make some regularity assumptions on F that will allow us to extrapolate from within the sample region to its border or even beyond. Of course, we could assume a parametric model for F, but this assumption comes at a high price: how reliable do you believe the model to be outside the sample region? What we need instead is a more flexible assumption that still allows us to make the necessary jump out of the data.

Therefore, we will assume that F is in the domain of attraction of a d-variate extreme value distribution function G, the dependence structure of which is described by the stable tail dependence function l, see (8.14). Observe that this assumption is much more realistic than the one of section 9.3, where we assumed the data to come from a multivariate extreme value distribution itself, rather than from a distribution in its domain of attraction.

Recall that the condition $F \in D(G)$ motivates the approximation (8.81). Therefore, basically our estimator will take the form

$$\hat{F}(x) = \exp[-\hat{l}\{-\log \hat{F}_1(x_1), \ldots, -\log \hat{F}_d(x_d)\}]. \qquad (9.35)$$

Here, the $\hat{F}_j(x_j)$ are estimators for the marginal tails, typically by one of the methods of Chapters 4–5. So the task considered in this section is the estimation of the stable tail dependence function l, or, equivalently, of the exponent measure μ_*, of the spectral measure S w.r.t. any two norms on $\mathbb{R}^d$, or, in the bivariate case, of Pickands dependence function A, amongst others.

We have seen that there does not exist a finite-dimensional parametrization of the class of dependence structures for multivariate extreme value distributions. To facilitate statistical inference, we may still assume a parametric model, preferably one that combines parsimony, analytical tractability and flexibility, and, if possible, is motivated from the data; see section 9.4.2. But first, we consider in section 9.4.1 the more principled approach of not making such an assumption at all and estimating the extreme value dependence structure in its full generality. In section 9.4.3, we will apply the techniques to the Loss-ALAE data of Figure 9.1.

9.4.1 Non-parametric estimation

Estimation principle

In order to estimate the stable tail dependence function, l, we treat the limit relation (8.90) connecting l and C_F, the copula of F, as an approximate equality for small enough s. Denote the coordinates of X_i by X_{ij} for $j = 1, \ldots, d$. Setting $s = k/n$

with $k = k_n \to \infty$ and $k/n \to 0$ leads to an estimator of the form

$$\tilde{l}(v) = \frac{1}{k} \sum_{i=1}^{n} \mathbf{1}\{\exists j = 1, \ldots, d : \hat{F}_j(X_{ij}) > 1 - (k/n)v_j\}$$

$$= \frac{1}{k} \sum_{i=1}^{n} \mathbf{1}\left\{\frac{k}{n}\hat{X}_{*i} \not\leq (1/v_1, \ldots, 1/v_d)\right\}. \tag{9.36}$$

Here, $\hat{F}_j$ is an estimator of the marginal distribution F_j (see below), whereas

$$\hat{X}_{*ij} = 1/\{1 - \hat{F}_j(X_{ij})\}, \qquad j = 1, \ldots, d.$$

As we will see later, the estimator $\tilde{l}$ of (9.36) is not directly suited to be substituted in (9.35), and will have to be modified. Still, formula (9.36) contains the gist of all estimators to come.

Since the convergence in (8.90) is locally uniform in $v \in [0, \infty)$, we may replace $1 - sv_j$ by any function of the form $1 - sv_j + o(s)$ as $s \downarrow 0$. Taking empirical versions leads to variants of the estimators considered. For instance, the choice e^{-sv_j} leads to $\tilde{l}$ as in (9.36) but with $\hat{X}_{*ij} = -1/\log \hat{F}_j(X_{ij})$, see, for instance, Capéraà and Fougères (2000a). Alternatively, Abdous et al. (1999) prefer $(1 - s)^{v_j}$. Also, rather than taking a single s, Abdous et al. (1999) propose to integrate over $s > 0$ with respect to a suitable kernel, thereby replacing the problem of how to choose s by how to choose the kernel and, more importantly, the bandwidth.

Estimating the margins

We still have to specify the $\hat{F}_j(X_{ij})$ in the definition of $\hat{X}_{*ij}$. There are two options: non-parametric or parametric. In the first option, we estimate $\hat{F}_j$ by the empirical distribution function or a variant of it. Denote the rank of X_{ij} among $X_{1j}, \ldots, X_{nj}$ by $R_{ij} = \sum_{s=1}^{n} \mathbf{1}(X_{sj} \leq X_{ij})$, that is, $R_{ij} = r$ if and only if $X_{ij} = X_{(r),j}$, where $X_{(1),j} \leq \cdots \leq X_{(n),j}$ denote the order statistics of $X_{1j}, \ldots, X_{nj}$. Then possible estimators are

$$\hat{F}_j(X_{ij}) = \begin{cases} n^{-1}(R_{ij} - \alpha), & \text{for } \alpha \in \{0, 1/2, 1\}, \\ (n+1)^{-1}R_{ij}. \end{cases} \tag{9.37}$$

For instance, the choice $\alpha = 1$ leads to the so-called tail empirical dependence function

$$\tilde{l}(v) = \frac{1}{k} \sum_{i=1}^{n} \mathbf{1}\{\exists j = 1, \ldots, d : R_{ij} > n + 1 - kv_j\} \tag{9.38}$$

(Drees and Huang 1998; Huang 1992). The motivation to modify the ordinary empirical distribution function $\hat{F}_j(X_{ij}) = n^{-1}R_{ij}$ in the way described is to improve the relative estimation accuracy of $1 - F_j$ evaluated at high order statistics: for instance, if

F_j is continuous, then $E[1 - F_j(X_{(n),j})] = 1/(n + 1)$, whereas $1 - \hat{F}_j(X_{(n),j}) = 0$ in case $\hat{F}_j$ is the ordinary empirical distribution function.

The second option of estimating $\hat{F}_j(X_{ij})$ starts from the Generalized Pareto (GP) approximation to the tail function $1 - F_j$ (Pickands 1975), see Chapter 5. Choose the threshold $u_j = X_{(n-k),j}$, the $(k + 1)$th largest observation in the jth variable, and set

$$1 - \hat{F}_j(X_{ij}) = \frac{k}{n}\left(1 + \hat{\gamma}_j \frac{X_{ij} - u_j}{\hat{\sigma}_j}\right)_+^{-1/\hat{\gamma}_j} \tag{9.39}$$

where $(\hat{\gamma}_j, \hat{\sigma}_j)$ are estimators of the parameters of the approximating GP distribution to the excess distribution over the threshold u_j. In that case,

$$\tilde{l}(v) = \frac{1}{k}\sum_{i=1}^{n}\mathbf{1}\left\{\exists j = 1,\ldots,d : \left(1 + \hat{\gamma}_j \frac{X_{ij} - u_j}{\hat{\sigma}_j}\right)_+^{1/\hat{\gamma}_j} > \frac{1}{v_j}\right\}. \tag{9.40}$$

The GP parameters can, for instance, be estimated by

$$\hat{\gamma}_j = M_{1j} + 1 - \frac{1}{2}\left(1 - \frac{M_{1j}^2}{M_{2j}}\right)^{-1}, \tag{9.41}$$

$$\hat{\sigma}_j = u_j \frac{(3M_{1j}^2 - M_{2j})^{1/2}}{1 + (\hat{\gamma}_j)_-}\left(\frac{1 + 4(\hat{\gamma}_j)_-}{1 + 2(\hat{\gamma}_j)_-}\right)^{1/2}, \tag{9.42}$$

where

$$M_{rj} = \frac{1}{k}\sum_{i=0}^{k-1}\{\log X_{(n-i),j} - \log X_{(n-k),j}\}^r, \qquad r = 1, 2,$$

$$(\hat{\gamma}_j)_- = (-\hat{\gamma}_j) \vee 0$$

(de Haan and Resnick 1993). These estimators can only be used if the data are positive; in particular, they are not translation invariant. An alternative that does not share these drawbacks consists of estimating the GP parameters by maximum likelihood like in Smith (1987), see section 5.3.

Whereas the marginal distributions do not influence the non-parametric version of $\hat{F}_j(X_{ij})$, the performance of the parametric version (9.39) strongly depends on the quality of the GP approximation to the excess distribution and also on the performance of the estimators of the GP parameters. The estimator (9.40) may perform poorly even if the convergence in (8.90) is fast.

Exponent and spectral measure

The estimator $\tilde{l}$ of (9.36) with margins specified as in (9.37) or (9.39) leads to estimators of the exponent measure μ_*, the spectral measure S w.r.t. two arbitrary norms $\| \cdot \|_i$ ($i = 1, 2$) on $\mathbb{R}^d$, and, in the bivariate case, Pickands dependence function A. First of all,

$$\tilde{l}(v) = \tilde{\mu}_*([0, \infty] \setminus [0, (1/v_1, \ldots, 1/v_d)]), \tag{9.43}$$

where

$$\tilde{\mu}_*(\cdot) = \frac{1}{k} \sum_{i=1}^{n} \mathbf{1}\left(\frac{k}{n} \hat{X}_{*i} \in \cdot\right) \tag{9.44}$$

is the empirical version of (8.94), treated as an equality at $t = n/k$. Important special cases are the *tail empirical measure*

$$\tilde{\mu}_*(\cdot) = \frac{1}{k} \sum_{i=1}^{n} \mathbf{1}\left\{ \left(\frac{k}{n+1-R_{ij}}\right)_{j=1}^{d} \in \cdot \right\}, \tag{9.45}$$

derived from (9.38), and its semi-parametric variant

$$\tilde{\mu}_*(\cdot) = \frac{1}{k} \sum_{i=1}^{n} \mathbf{1}\left\{ \left(\left\{1 + \hat{\gamma}_j \frac{X_{ij} - u_j}{\hat{\sigma}_j}\right\}_+^{1/\hat{\gamma}_j}\right)_{j=1}^{d} \in \cdot \right\}, \tag{9.46}$$

derived from (9.40), see de Haan and Resnick (1993).

Second, by (8.16), we can turn $\tilde{\mu}_*$ into an estimator of the spectral measure S: set

$$\tilde{S}(\cdot) = \frac{1}{k} \sum_{i=1}^{n} \mathbf{1}(\hat{R}_i > n/k, \, \hat{W}_i \in \cdot), \tag{9.47}$$

where, with T as in (8.15),

$$(\hat{R}_i, \hat{W}_i) = T(\hat{X}_{*i}) = (\|\hat{X}_{*i}\|_1, \hat{X}_{*i}/\|\hat{X}_{*i}\|_2). \tag{9.48}$$

A useful choice of the two norms is the sum-norm, $\|x\| = |x_1| + \cdots + |x_d|$, in which case the transformation T simplifies to

$$\hat{R}_i = \hat{X}_{*i1} + \cdots + \hat{X}_{*id} \quad \text{and} \quad \hat{W}_{ij} = \hat{X}_{*ij}/\hat{R}_i. \tag{9.49}$$

The estimator of the spectral measure in (9.47) is based on all observations for which the radial component exceeds n/k. Their number is random, although by (8.97), there must be approximately $kS(\Xi)$ such observations. If we want to have exactly k observations involved in the estimation, we could choose $s = 1/\hat{R}_{(n-k)}$, the $(k + 1)$th largest of the $\hat{R}_i$, rather than $s = k/n$ as we did until now, giving

$$\tilde{S}(\cdot) = \hat{R}_{(n-k)} \frac{1}{n} \sum_{i=1}^{n} \mathbf{1}\{\hat{R}_i > \hat{R}_{(n-k)}, \, \hat{W}_i \in \cdot\}, \tag{9.50}$$

or equivalently

$$\tilde{S}(\Xi) = \frac{k}{n}\hat{R}_{(n-k)}, \tag{9.51}$$

$$\tilde{S}(\cdot)/\tilde{S}(\Xi) = \frac{1}{k}\sum_{i=1}^{n}\mathbf{1}\{\hat{R}_i > \hat{R}_{(n-k)}, \hat{W}_i \in \cdot\}. \tag{9.52}$$

The above estimator of $\tilde{S}(\Xi)$ may be rather volatile in k, and a good idea is to take the average over a range of k-values (Capéraà and Fougères 2000a).

However, an even better idea might be to choose the two norms equal to the sum-norm: in that case, the total mass of the spectral measure (denoted now by H) on the unit simplex S_d is by (8.26) always equal to the number of dimensions, d. Replacing the estimate $\hat{H}(S_d)$ by its true value d in (9.52) leads to the estimator

$$\hat{H}(\cdot) = \frac{d}{k}\sum_{i=1}^{n}\mathbf{1}\{\hat{R}_i > \hat{R}_{(n-k)}, \hat{W}_i \in \cdot\}, \tag{9.53}$$

with $\hat{R}_i$ and $\hat{W}_{ij}$ as in (9.49). If needed, the estimator of H can be turned into an estimator of the spectral measure w.r.t. two general norms through (8.38).

Pickands dependence function

In the bivariate case, we can transform the above estimators into estimators of Pickands dependence function A. Starting from (9.36), we get

$$\tilde{A}(t) = \tilde{l}(1-t, t) = \frac{1}{k}\sum_{i=1}^{n}\mathbf{1}[\max\{(1-t)\hat{X}_{*i1}, t\hat{X}_{*i2}\} > n/k]. \tag{9.54}$$

In particular, the extremal coefficient $\theta = 2A(1/2)$ in (8.56) may be estimated by setting $t = 1/2$: replacing k by $2k$ and letting $\tilde{l}$ be the tail empirical dependence function (9.38) yields

$$\tilde{\theta} = \frac{1}{k}\sum_{i=1}^{n}\mathbf{1}\{\max(R_{i1}, R_{i2}) > n - k\},$$

variants of which are considered in Falk and Reiss (2001, 2003). Alternatively, since $A(t) = tl\{(1-t)/t, 1\}$ we could use $t\tilde{l}\{(1-t)/t, 1\}$ to estimate $A(t)$ (Joe et al. 1992), although this estimator has the drawback of vanishing at $t = 0$, whereas in fact $A(0) = 1$.

The above estimator for A is not convex, and this property can be ensured if we start from an estimate of the spectral measure rather than from the stable tail dependence function. If $\tilde{S}$ denotes the denotes the estimator (9.47) of the spectral measure S, we obtain from (8.49),

$$\hat{A}(t) = \frac{1}{k}\sum_{i=1}^{n}\mathbf{1}(\hat{R}_i > n/k)\hat{R}_i^{-1}\max\{(1-t)\hat{X}_{*i1}, t\hat{X}_{*i2}\}, \tag{9.55}$$

with $\hat{R}_i$ as in (9.48). Estimator $\tilde{A}$ was proposed by Capéraà and Fougères (2000a) in case both norms are equal to the sum-norm. Finally, the estimator $\hat{H}$ in (9.53) leads via (8.46) to

$$\hat{A}(t) = \frac{2}{k} \sum_{i=1}^{n} \mathbf{1}\{\hat{R}_i > \hat{R}_{(n-k)}\} \max\{(1-t)\hat{W}_{i1}, t\hat{W}_{i2}\}, \qquad (9.56)$$

with $\hat{R}_i$ and $\hat{W}_{ij}$ as in (9.49).

The estimator of the extremal coefficient $\theta = 2A(1/2)$ corresponding to (9.56) is

$$\hat{\theta} = \frac{2}{k} \sum_{i=1}^{n} \mathbf{1}\{\hat{R}_i > \hat{R}_{(n-k)}\} \max(\hat{W}_{i1}, \hat{W}_{i2}). \qquad (9.57)$$

Observe that this estimator is always smaller than two, that is, even if the margins are perfectly independent, the estimator will still point to asymptotic dependence. The origin of this deficiency can be traced back to the approximation (8.93) whereupon the estimator is based: as discussed already, the approximation tends to undervalue the true probability of joint occurrences of extremes, the consequence of which is an inherent bias towards stronger asymptotic dependence for estimators that are based on it.

Finally, observe that the estimators $\hat{A}$ in (9.55) and (9.56) do not satisfy the constraint $\max(t, 1-t) \leq A(t) \leq 1$. A possible solution consists in the modification

$$\bar{A}(t) = \max\{t, 1-t, \hat{A}(t) + 1 - (1-t)\hat{A}(0) - t\hat{A}(1)\}. \qquad (9.58)$$

Via the usual transformation formulae, for instance, (8.44) or (8.47), we can turn $\bar{A}$ into estimators of the stable tail dependence function or the spectral measure that satisfy all the relevant constraints as well. Still, the modification (9.58) is rather *ad hoc*, and it is not clear what the consequences are for the performance of the estimator. Moreover, the procedure cannot be generalized to higher dimensions. The problem of constructing truly non-parametric estimators of the spectral measure that satisfy all the necessary constraints remains open.

Estimating F

Now let us return to the problem of estimating $1 - F(x)$ for large x_j as in (9.35). Typically, the marginal tail probabilities $1 - F_j(x_j)$ will be of the order $O(1/n)$ or even smaller, so that the estimator $\tilde{l}$ given in (9.36) is not suited to be substituted into (9.35), basically for the same reason why the empirical distribution is not a very good estimator in the first place: the estimator would involve a region of the sample space with (almost) no data. A possible remedy is to exploit the homogeneity of l: since $l(tv) = tl(v)$ for $t \geq 0$ and $v \geq 0$, we can put

$$\hat{l}(v) = \|v\| \tilde{l}(v/\|v\|), \qquad v \in [0, \infty) \setminus \{0\}, \qquad (9.59)$$

with $\tilde{l}$ as in (9.36), while $\|\cdot\|$ denotes an arbitrary norm on $\mathbb{R}^d$. Typical choices for the norm are the Euclidean norm $\|v\| = (v_1^2 + \cdots + v_2^2)^{1/2}$ (de Haan and de Ronde 1998) and the max-norm $\|v\| = |v_1| \vee \cdots \vee |v_d|$ (Drees 2001).

The estimator $\hat{l}(v)$ of (9.59) has the advantage that for any non-zero v, the number of observations used is of the order k. It also inherits the homogeneity property of l. However, $\hat{l}$ in (9.59) is not connected in a natural way to an exponent measure $\hat{\mu}_*$ or a spectral measure $\hat{S}$ like $\tilde{l}$ is connected to $\tilde{\mu}_*$ and $\tilde{S}$ in (9.43) and (9.47).

An alternative is to start from $\tilde{S}$ in (9.47) and exploit the connection between l and S in (8.23) to define instead, for $v \geq 0$,

$$\hat{l}(v) = \frac{1}{k} \sum_{i=1}^{n} \mathbf{1}(\hat{R}_i > n/k)\hat{R}_i^{-1} \bigvee_{j=1}^{d} (v_j \hat{X}_{*ij}), \qquad (9.60)$$

with $\hat{R}_i = \|\hat{X}_{*i}\|_1$. Similarly, from (8.17), we can define $\hat{\mu}_*$ from $\tilde{S}$ by

$$\hat{\mu}_* \circ T^{-1}(\mathrm{d}r, \mathrm{d}\omega) = r^{-2}\mathrm{d}r\,\tilde{S}(\mathrm{d}\omega),$$

with T the mapping (8.15). Both $\hat{l}$ from (9.60) and $\hat{\mu}_*$ above satisfy the required homogeneity properties; moreover, $\hat{l}$ is convex.

Alternatively, taking both norms equal to the sum-norm, we can estimate l starting from the estimator $\hat{H}$ of (9.53) rather than from $\tilde{S}$, leading to

$$\hat{l}(v) = \frac{d}{k} \sum_{i=1}^{n} \mathbf{1}\{\hat{R}_i > \hat{R}_{(n-k)}\} \bigvee_{j=1}^{d} (v_j \hat{W}_{ij}), \qquad (9.61)$$

with $\hat{R}_i$ and $\hat{W}_{ij}$ as in (9.49).

In the bivariate case, we can combine (9.35) with the definition of Pickands dependence function to find the estimator

$$\hat{F}(x_1, x_2) = \exp\left[\log\{\hat{F}_1(x_1)\hat{F}_2(x_2)\}\hat{A}\left(\frac{\log\{\hat{F}_2(x_2)\}}{\log\{\hat{F}_1(x_1)\hat{F}_2(x_2)\}}\right)\right]. \qquad (9.62)$$

This estimator coincides with the one of (9.35) if we set $\hat{A}(t) = \hat{l}(1 - t, t)$ for one of the choices of $\hat{l}$ above. Observe that this $\hat{A}$ is the same as the one in (9.55) or (9.56) for $\hat{l}$ as in (9.60) or (9.61), respectively.

Literature overview

The tail empirical dependence function (9.38) and tail empirical measure (9.45) were introduced by DM Mason in an unpublished 1991 manuscript and Huang (1992). Drees and Huang (1998) showed that the tail empirical dependence function attains the optimal rate of convergence for estimators of the stable tail dependence function.

The estimator (9.46) of the exponent measure was first considered by de Haan and Resnick (1993). They proposed to estimate the Generalized Pareto parameters as in (9.41) and (9.42). The paper is one of the few ones in the literature on multivariate extremes that is written down for arbitrary dimension.

For bivariate data, the estimator (9.47) of the spectral measure has been considered in a number of papers. The first hint was given in de Haan (1985) for both norms equal to the Euclidean norm as in (8.31). The idea was taken up further by Einmahl *et al.* (1993) under the simplifying assumption that the marginal distributions are the same and heavy-tailed. The restriction of identical margins was removed in Einmahl *et al.* (1997), who, for the combination of max-norm and Euclidean norm (8.33), proposed $\tilde{S}$ as in (9.47) with $\tilde{\mu}_*$ being the estimator (9.46) of de Haan and Resnick (1993). This estimator for S was modified into a fully non-parametric one in Einmahl *et al.* (2001) by choosing for $\tilde{\mu}_*$ the tail empirical measure (9.45). Alternatively, Capéraà and Fougères (2000a) considered the case where both norms are equal to the sum-norm; their estimator is computed as in (9.47) and (9.49) and with margins transformed to the standard Fréchet distribution. Still in the bivariate case, Abdous *et al.* (1999) replaced $1 - sv_j$ by $(1 - s)^{v_j}$ in (8.90) and consider kernel variants of (9.36).

The asymptotic theory for estimators of the dependence structure of extremes is rather involved, a major difficulty being the fact that the margins are unknown and are to be estimated as well. Useful tools in the area of local empirical processes can be found in Stute (1984) and Einmahl (1997).

Estimation of $1 - F(x)$ can be paraphrased as estimation of the probability of the 'failure region' $\mathbb{R}^d \setminus (\infty, x]$. More general regions are considered in de Haan and de Ronde (1998) and de Haan and Sinha (1999). In de Haan and Huang (1995), estimators of $1 - F(x)$ are turned into estimators of quantile curves $Q(F, p) = \{x \in \mathbb{R}^d : 1 - F(x) = p\}$ for small failure probabilities p.

9.4.2 Parametric estimation

We consider again the setting of d-variate observations $x_1, \ldots, x_n$ that can be assumed to be realizations of independent random vectors with common distribution F, the aim being to estimate $F(x)$ for x such that $F_j(x_j)$ is close to one. We assume that F is in the domain of attraction of some extreme value distribution function G, of which the stable tail dependence function belongs to some parametric family, $l(\cdot; \theta)$, indexed by a parameter (vector) θ, usually one of the families described in section 9.2.

The domain-of-attraction condition together with the parametric specification of the stable tail dependence function leads by the theory in section 8.3 to parametric models for F in regions of its support where all coordinates are large. The model parameters can be estimated by maximum likelihood, leading then to the desired estimates of $F(x)$. Still, different formulations of the domain-of-attraction condition lead to different models and hence to different estimators. The two most popular methods are the so-called point-process method (Coles and Tawn 1991;

Joe *et al.* 1992) and the censored-likelihood method (Ledford and Tawn 1996; Smith 1994; Smith *et al.* 1997), which we will discuss in turn in this section. For completeness, we mention that Tajvidi (1996) developed a procedure based on multivariate generalized Pareto distributions as in (8.68).

Point-process method

Coles and Tawn (1991) and Joe *et al.* (1992) found a way to turn the point process characterizations (8.73) and (8.98) into an estimation method. The method was applied to oceanographic data in Coles and Tawn (1994) and Morton and Bowers (1996). We present a derivation of the point-process likelihood by a quite different but simpler argument than the above authors, incidentally avoiding the point-process machinery.

By (8.93), we find

$$F(x) \approx 1 - l\{1 - F_1(x_1), \ldots, 1 - F_d(x_d); \theta\}, \tag{9.63}$$

provided all $1 - F_j(x_j)$ are sufficiently small. Univariate theory suggests to model the margins by generalized Pareto distributions: for $j = 1, \ldots, d$ and a high threshold u_j, we model F_j on $[u_j, \infty)$ by

$$F_j(x_j) \approx 1 - \lambda_j \left(1 + \gamma_j \frac{x_j - u_j}{\sigma_j}\right)_+^{-1/\gamma_j}, \qquad x_j \geq u_j. \tag{9.64}$$

with $\lambda_j = 1 - F_j(u_j)$. In terms of the function V_* of (8.8), we arrive at the model

$$F(x) \approx 1 - V_*\{z; \theta\}, \qquad x \in \mathbb{R}^d \setminus (-\infty, u], \tag{9.65}$$

$$z_j = z_j(x_j) = \begin{cases} \lambda_j^{-1} \left(1 + \gamma_j \dfrac{x_j - u_j}{\sigma_j}\right)_+^{1/\gamma_j} & \text{if } x_j > u_j, \\ 1/\{1 - F_j(x_j)\} & \text{if } x_j \leq u_j. \end{cases}$$

Since λ_j is close to zero, one can use the asymptotically equivalent marginal transformations

$$z_j = \begin{cases} -1 \Big/ \log\left\{1 - \lambda_j\left(1 + \gamma_j \dfrac{x_j - u_j}{\sigma_j}\right)_+^{-1/\gamma_j}\right\} & \text{if } x_j > u_j, \\ -1/\log F_j(x_j) & \text{if } x_j \leq u_j. \end{cases}$$

We use (9.65) to jointly estimate the marginal and dependence parameters from a sample $x_1, \ldots, x_n$. First, we simply estimate F_j on the region $(-\infty, u_j]$ by the marginal empirical distribution function and assume it to be known in the subsequent analysis. Then, we estimate the parameters (γ_j, σ_j), $j = 1, \ldots, d$, and θ by maximum likelihood, the likelihood contribution of an observation x_i depending on whether $x_i \leq u$ or not. On the one hand, if $x_i \leq u$, then the likelihood contribution is simply

$$L(x_i) = F(u) \approx 1 - l(\lambda; \theta).$$

On the other hand, if $x_i \nleq u$, then the likelihood contribution is

$$L(x_i) = \frac{\partial^d}{\partial x_1 \cdots \partial x_d} F(x_i)$$

$$\propto -\frac{\partial^d}{\partial z_1 \cdots \partial z_d} V_*(z_i; \theta) \prod_{j : x_{ij} > u_j} \frac{\mathrm{d}z_{ij}}{\mathrm{d}x_j}$$

where $z_{ij} = z_j(x_{ij})$. Defining $r_i = z_{i1} + \cdots + z_{id}$ and $w_i = r_i^{-1} z_i$, we can use (8.34) to rewrite the latter likelihood as

$$L(x_i) \propto r_i^{-(d+1)} h(w_i; \theta) \prod_{j : x_{ij} > u_j} \frac{\mathrm{d}z_{ij}}{\mathrm{d}x_j}$$

where $h(\cdot; \theta)$ is the spectral density of the spectral measure $H(\cdot; \theta)$ on the interior of the unit simplex S_d. With $N = \{i = 1, \dots, n : x_i \nleq u\}$, the total likelihood of the parameters given the sample is then

$$L\{(x)_{i=1}^n; (\gamma_j, \sigma_j)_{j=1}^d, \theta\}$$

$$\propto \{1 - l(\lambda; \theta)\}^{n - |N|} \prod_{i \in N} r_i^{-(d+1)} h(w_i; \theta) \prod_{j : x_{ij} > u_j} \frac{\mathrm{d}z_{ij}}{\mathrm{d}x_j}.$$

Since $l(\lambda; \theta) \leq \lambda_1 + \cdots + \lambda_d$ is small and since $|N|$ is small in comparison to n, we can approximate the former by the simpler

$$L\{(x)_{i=1}^n; (\gamma_j, \sigma_j)_{j=1}^d, \theta\}$$

$$\propto \exp\{-l(n\lambda; \theta)\} \prod_{i \in N} r_i^{-(d+1)} h(w_i; \theta) \prod_{j : x_{ij} > u_j} \frac{\mathrm{d}z_{ij}}{\mathrm{d}x_j}. \tag{9.66}$$

This is indeed the likelihood obtained in Coles and Tawn (1991) and Joe *et al.* (1992).

Optimization of the above likelihood is to be done numerically. A good initial guess for the optimizers can be found as follows. First estimate each pair of marginal parameters (γ_j, σ_j) separately by maximum likelihood from (9.64). For these estimates, compute z_i, and from z_i compute r_i and w_i. Now by (8.97), the probability density function of those w_i for which the corresponding r_i exceeds some high threshold is approximately $d^{-1} h(\cdot; \theta)$. Maximum likelihood estimation then yields an initial guess for θ.

Unfortunately, the point-process method suffers from a number of defects. First of all, it uses (9.63) for x such that *some* $1 - F_j(x_j)$ are small, whereas in fact the stated approximation is valid only if *all* $1 - F_j(x_j)$ are small. This improper use of (9.63) might corrupt the estimates of dependence parameters that are related to mass of the spectral measure not in the interior but on the lower-dimensional faces of the unit simplex. Joe *et al.* (1992) suggest a possible modification of the likelihood that should remedy the problem but do not pursue the issue further.

A second defect of the above method is that approximation (9.63) itself is not without its own worries: in the text following equation (8.93), we explained already that the right-hand side of (9.63) need not define a proper distribution and that it tends to undervalue the probability of joint extremes in several coordinates simultaneously. The result of this undervaluation is that estimates of dependence parameters will show a tendency to be biased towards stronger dependence. In particular, asymptotic independence will be rejected too often. All these drawbacks are avoided by the censored-likelihood method, to be discussed next.

Censored-likelihood method

Let u be a multivariate threshold such that $F_j(u_j) = \exp(-\lambda_j)$ for some small, positive λ_j. Equations (8.63) and (8.64) suggest the following parametric model for F on the region $[u, \infty)$:

$$F(x) \approx \exp\{-l(v; \theta)\}, \qquad x \geq u, \qquad (9.67)$$

$$v_j = \lambda_j \left(1 + \gamma_j \frac{x_j - u_j}{\sigma_j}\right)_+^{-1/\gamma_j}, \qquad j = 1, \ldots, d.$$

Observe that (9.67) entails the following model for the margin F_j on the region $[u_j, \infty)$, $j = 1, \ldots, d$:

$$F_j(x_j) \approx \exp\left\{-\lambda_j \left(1 + \gamma_j \frac{x_j - u_j}{\sigma_j}\right)_+^{-1/\gamma_j}\right\}, \qquad x_j \geq u_j. \qquad (9.68)$$

For small λ_j, this is approximately the same as the Generalized Pareto model (9.64) for the excess distribution over the threshold u_j.

The marginal parameters $(\lambda_j, \gamma_j, \sigma_j)$, $j = 1, \ldots, d$, and dependence parameters θ can be estimated jointly by maximum likelihood. Observe that model (9.67) is only specified on the region $[u, \infty)$, and hence does not apply directly to observations outside that region. The solution consists of considering the observation in a coordinate j that is smaller than u_j to be censored from below at u_j, hence the name 'censored likelihood'.

So, the likelihood of the parameters given a sample $x_1, \ldots, x_n$ is

$$L\{(x_i)_{i=1}^n; (\lambda_j, \gamma_j, \sigma_j)_{j=1}^d, \theta\} = \prod_{i=1}^n L(x_i), \qquad (9.69)$$

with the form of the likelihood contribution, $L(x)$, of an observation x, depending on which of its coordinates exceed the corresponding threshold coordinates. For $J \subset \{1, \ldots, d\}$, let R_J be the region in $\mathbb{R}^d$ of all x such that $x_j > u_j$ for $j \in J$ and $x_j \leq u_j$ for the other j. Then for $J = \{j_1, \ldots, j_m\}$, the likelihood contribution of an observation x in the region R_J is proportional to

$$L(x) \propto P[X_j \in dx_j, j \in J; X_j \leq u_j, j \notin J]$$

$$\propto \frac{\partial^m F}{\partial x_{j_1} \cdots \partial x_{j_m}}(x \vee u)$$

with F as in the right-hand side of (9.67). For instance, in the bivariate case, the plane is partitioned into four regions, depending on whether x_j ($j = 1, 2$) exceeds u_j or not. The likelihood contributions are

$$L(x_1, x_2) \propto \begin{cases} F(u_1, u_2) & \text{if } x_1 \leq u_1, x_2 \leq u_2, \\[2mm] \dfrac{\partial F}{\partial x_1}(x_1, u_2) & \text{if } x_1 > u_1, x_2 \leq u_2, \\[2mm] \dfrac{\partial F}{\partial x_2}(u_1, x_2) & \text{if } x_1 \leq u_1, x_2 > u_2, \\[2mm] \dfrac{\partial^2 F}{\partial x_1 \partial x_2}(x_1, x_2) & \text{if } x_1 > u_1, x_2 > u_2, \end{cases} \quad (9.70)$$

with F and its partial derivatives computed according to the right-hand side of (9.67).

Joint estimation of the marginal and dependence parameters has several advantages: transfer of information between variables, leading to better inference of the marginal parameters; proper assessment of the estimation uncertainty of the dependence parameters because of having to estimate the marginal parameters; possibility to incorporate connections between marginal parameters over different margins, for instance, a common shape parameter $\gamma_j = \gamma$. A drawback of the method is the computational complexity, growing worse as the dimension increases. A good idea might therefore be to include a preliminary step, estimating marginal and dependence parameters separately, and then using these estimates as starting values for the optimization procedure leading to the joint estimates.

The censored-likelihood method is first mentioned in Smith (1994). Ledford and Tawn (1996) give it its full development, focusing especially on testing for independence in the bivariate symmetric logistic model (9.6), for which the point-process method of the previous paragraph is known to perform badly for the reasons mentioned there. The method is useful as well in the analysis of extremes of univariate Markov chains (Smith *et al.* 1997), see section 10.4.5.

9.4.3 Data example

We continue the study of the Loss-ALAE data described in section 9.1. Whereas in section 9.3.3 we artificially partitioned the sample into blocks of equal size and extracted from each block the pair of component-wise maxima, now we will use all bivariate observations which are in some sense large.

To apply the non-parametric techniques of section 9.4.1, we transform the data to standard Fréchet margins by

$$x_{*ij} = -1/\log u_{ij}, \qquad i = 1, \ldots, n, \quad j = 1, 2,$$

with u_{ij} as in (9.1), the alternative consisting of transforming to standard Pareto margins by $x_{*ij} = 1/(1 - u_{ij})$. The transformation (9.49) of the pair (x_{*i1}, x_{*i2}) to pseudo-polar coordinates with both norms equal to the sum-norm takes the simple form

$$r_i = x_{*i1} + x_{*i2}, \qquad w_{ij} = x_{*ij}/r_i,$$

for $i = 1, \ldots, n$ and $j = 1, 2$. Let $r_{(1)} \leq \cdots \leq r_{(n)}$ be the radial coordinates r_i in ascending order.

If we are to construct estimates from the observations corresponding to the k largest r_i, then a sensible choice of k might be found by inspecting the plot of $(k/n)r_{(n-k)}$ as function of $k = 1, \ldots, n - 1$, see Figure 9.7(a). Recall that the estimator (9.50) of the spectral measure H may be written as

$$\tilde{H}(\cdot) = \frac{r_{(n-k)}}{n} \sum_{i=1}^{n} \mathbf{1}\{r_i > r_{(n-k)}, \ w_{i1} \in \cdot\}.$$

Hence, $\tilde{H}([0, 1]) = (k/n)r_{(n-k)}$ is an estimator of $H([0, 1]) = 2$. Therefore, we propose to choose the largest k for which $(k/n)r_{(n-k)}$ is close to two. Obviously, this is not more than a heuristic and should be formalized in some way. Also, it is not known if it leads to an optimal choice according to some criterion. Anyway, the plot suggests $k_0 = 337$ as a reasonable choice. Replacing $\tilde{H}([0, 1])$ by its true value then leads to the estimator

$$\hat{H}(\cdot) = \frac{2}{k_0} \sum_{i=1}^{n} \mathbf{1}\{r_i > r_{(n-k_0)}, \ w_{i1} \in \cdot\}, \tag{9.71}$$

see (9.53). Figure 9.7(b) shows a plot of $\hat{H}([0, w])$ as a function of $w \in [0, 1]$. The Pickands dependence function corresponding to $\hat{H}$ is

$$\hat{A}(t) = \frac{2}{k_0} \sum_{i=1}^{n} \mathbf{1}\{r_i > r_{(n-k_0)}\} \max\{(1 - t)w_{i1}, tw_{i2}\}, \tag{9.72}$$

see (9.56). As in (9.58), $\hat{A}$ can be modified into $\bar{A}$ to obtain an estimate satisfying all the requirements to be a Pickands dependence function, although in this case

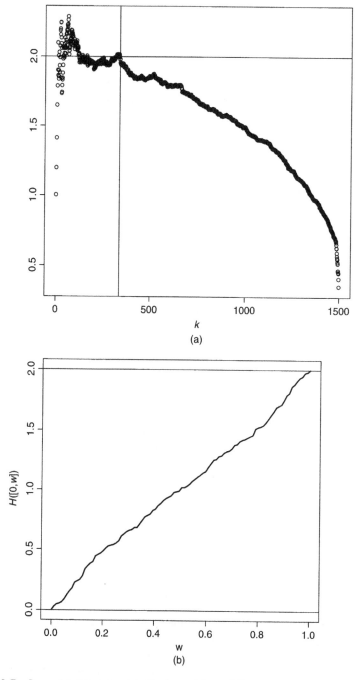

Figure 9.7 Loss-ALAE data: (a) Choice of $k_0 = 337$ via plot of $(k/n)r_{(n-k)}$ as function of k. (b) Plot of $\hat{H}_{k_0}([0, w])$ in (9.71) as function of $w \in [0, 1]$.

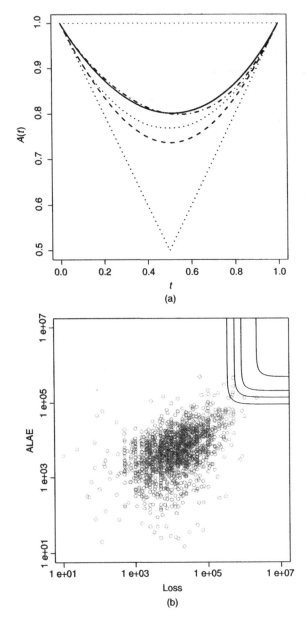

Figure 9.8   Loss-ALAE data: (a) Estimates of Pickands dependence function: asymmetric logistic model via censored likelihood (———), bilogistic model via censored likelihood (– · — · –) and point-process likelihood (- - - - -), and non-parametric estimate (· · · · · ·) obtained by modification of (9.72) via (9.58). (b) Quantile curves $Q(\hat{F}, p)$ of (9.73) for $p = 0.98, 0.99, 0.995, 0.999$ for asymmetric logistic model and GP margins estimated jointly via censored likelihood.

the difference between $\hat{A}$ and $\bar{A}$ turned out to be negligible. The function $\bar{A}$ is represented by the dotted line in Figure 9.8(a).

Alternatively, we may fit one of the parametric models of section 9.2 via the point-process or censored likelihoods of section 9.4.2. For both losses and ALAEs, we need to choose a threshold so that the approximations (9.65) and (9.67) are valid. Recall that the approximations entail that the marginal distributions of excesses above the corresponding thresholds are modelled by a GP or GEV distribution as in (9.64) or (9.68) and the dependence structure by that of a multivariate extreme value distribution. Sometimes, marginal and dependence considerations point to different thresholds; the required modifications of the methods are described Dixon and Tawn (1995).

For the Loss-ALAE data, we propose to choose the thresholds (u_1, u_2) in such a way that the total number of observations for which there is an exceedance in at least one coordinate is approximately $k_0 = 337$, the k-value found in Figure 9.7(a). Simplifying further, we propose $u_j = x_{(n-k_1),j}$ for $k_1 = \lfloor (k+1)/2 \rfloor = 169$; here $x_{(1),j} \le \cdots \le x_{(n),j}$ denote the observations in the jth coordinate in ascending order. The resulting thresholds are $u_1 = 88\,803$ for Loss and $u_2 = 23\,586$ for ALAE. Marginally fitting the GP by maximum likelihood to the threshold excesses led to $(\hat{\sigma}_1, \hat{\gamma}_1) = (79\,916, 0.52)$ for Loss and $(\hat{\sigma}_2, \hat{\gamma}_2) = (20\,897, 0.47)$ for ALAE. The goodness-of-fit was confirmed by W-plots (not shown) as in section 5.3.2.

We fitted the asymmetric logistic model (9.7) with the censored likelihood (9.69)–(9.70) and the bilogistic model (9.9) with the censored likelihood and the point-process likelihood (9.66). As for the component-wise maxima in section 9.3.3, imposing the constraint $\psi_1 = 1$ did not significantly decrease the likelihood. The parameter estimates are summarized in Table 9.2 and the Pickands dependence functions are shown in Figure 9.8(a).

Comparing the estimated Pickands dependence functions in Figure 9.8(a) with those in Figure 9.5(b) confirms our earlier findings about the inaccuracy of the

Table 9.2 Loss-ALAE data: Estimates (standard errors – ** if observed information matrix was near-singular) for marginal and dependence parameters for asymmetric model ($\psi_1 = 1$) with censored likelihood and bilogistic model with censored and point-process likelihoods.

Model	Loss $\hat{\sigma}_1/1000$	$\hat{\gamma}_1$	ALAE $\hat{\sigma}_2/1000$	$\hat{\gamma}_2$	Dependence
Asymmetric logistic	82	0.58	23	0.51	$\alpha = 0.66$ (0.04)
(Censored)	(1.3)	(0.10)	(0.4)	(0.09)	$\psi_2 = 0.89$ (0.15)
Bilogistic	84	0.59	25	0.47	$\alpha = 0.55$ (0.09)
(Censored)	(2.1)	(0.10)	(2.5)	(0.10)	$\beta = 0.76$ (0.05)
Bilogistic	84	0.79	25	0.64	$\alpha = 0.54$ (**)
(Point-process)	(**)	(**)	(**)	(**)	$\beta = 0.57$ (**)

approximation $1 - F(x_1, x_2) \approx l\{1 - F_1(x_1), 1 - F_2(x_2)\}$. Recall that this approximation underlies both the non-parametric methods as well as the point-process likelihood. It appears now that the non-parametric estimate (9.72), as well as the one from the bilogistic model fitted with the point-process likelihood, is biased towards stronger dependence. This is a consequence of the undervaluation of the probability of joint extremes by the mentioned approximation, see the explanation after (8.93). On the other hand, the censored likelihood is based on the more accurate approximation $F(x_1, x_2) \approx \exp[-l\{-\log F_1(x_1), -\log F_2(x_2)\}]$, and the resulting estimates, both with the asymmetric logistic and the bilogistic models, are much closer to the ones obtained from component-wise maxima. The standard errors of the censored-likelihood estimates are much smaller than their component-wise maxima counterparts, reflecting the more efficient use of information of the threshold approach.

We conclude with a picture of the quantile curves

$$Q(\hat{F}, p) = \{(x_1, x_2) : \hat{F}(x_1, x_2) = p\}, \qquad 0 < p < 1, \qquad (9.73)$$

with $\hat{F}$ as in the model (9.67) underling the censored likelihood and with asymmetric logistic dependence structure. The quantile curves are shown in Figure 9.8(b) for $p = 0.98, 0.99, 0.995, 0.999$. Since the model (9.67) can be written as

$$\hat{F}(x_1, x_2) = \exp\left[\log\{\hat{F}_1(x_1)\hat{F}_2(x_2)\}\hat{A}\left(\frac{\log\{\hat{F}_2(x_2)\}}{\log\{\hat{F}_1(x_1)\hat{F}_2(x_2)\}}\right)\right]$$

with marginal estimates $\hat{F}_j(x_j)$ as in (9.68) and with $\hat{A}(w) = A(w, \hat{\theta})$, the Pickands dependence function corresponding to the estimated dependence parameter vector $\hat{\theta}$, we have $\hat{F}(x_1, x_2) = p$ if and only if there exists $w \in [0, 1]$ such that $\hat{F}_1(x_1) = p^{(1-w)/\hat{A}(w)}$ and $\hat{F}_2(x_2) = p^{w/\hat{A}(w)}$. Therefore, the quantile curve can be computed from

$$Q(\hat{F}, p) = \left\{\left(\hat{F}_1^{\leftarrow}\{p^{(1-w)/\hat{A}(w)}\}, \hat{F}_2^{\leftarrow}\{p^{w/\hat{A}(w)}\}\right) : w \in [0, 1]\right\}.$$

For fixed $w \in [0, 1]$, point-wise confidence intervals could be added (not shown) to the quantile curves from the observed information matrix and the delta method.

9.5 Asymptotic Independence

Everything so far in this chapter was based on multivariate extreme value distributions. The justification is to be found in the theory of Chapter 8. Still, within the class of max-stable distributions, the only possible type of asymptotic independence is, in fact, perfect independence. This makes the class rather inappropriate for modelling data that exhibit positive or negative association that only gradually disappears at more and more extreme levels. To properly handle such cases, we are obliged to leave the by-now familiar framework of extreme value distributions and look for a class of models describing the tails of asymptotically independent distributions in a more refined way.

In section 9.5.1, we introduce a number of coefficients of extremal dependence useful in assessing whether a bivariate distribution is asymptotically dependent and, within each case, in giving a relative measure of strength of dependence (Coles *et al.* 1999). In particular, we find that the so-called coefficient of tail dependence (Ledford and Tawn 1996) is most useful in distinguishing asymptotic dependence from asymptotic independence and, within the class of asymptotically independent distributions, positive from negative association. Several methods to estimate this coefficient are described in section 9.5.2. Finally, section 9.5.3 describes a general model, due to Ledford and Tawn (1997), for the joint survivor function of a bivariate distribution, encompassing both asymptotic dependence as well as various types of asymptotic independence. We also discuss a number of inference techniques for this joint tail model, some of which are new.

9.5.1 Coefficients of extremal dependence

Asymptotic dependence

Let (X_1, X_2) be a bivariate random vector with distribution function F and marginal distribution functions F_1 and F_2. For simplicity, we will assume throughout that F_1 and F_2 are continuous. Assuming first that F_1 and F_2 are identical, a quite natural coefficient of extremal dependence between X_1 and X_2 at extreme levels is

$$\chi = \lim_{x \uparrow x_*} P[X_2 > x \mid X_1 > x], \qquad (9.74)$$

provided the limit exists; here, x_* denotes the right end-point of the common marginal distribution (Coles *et al.* 1999). Definition (9.74) can be generalized to the case where the marginal distribution functions F_1 and F_2 are non-identical. The variables $U_j = F_j(X_j)$ $(j = 1, 2)$ are uniformally distributed on $(0, 1)$. Now define

$$\chi = \lim_{u \uparrow 1} P[U_2 > u \mid U_1 > u], \qquad (9.75)$$

again provided that the limit exists. Observe that (9.74) is indeed a special case of (9.75).

The number χ can be interpreted as the tendency for one variable to be extreme given that the other is extreme. When $\chi = 0$, the variables are said to be asymptotically independent, whereas if $0 < \chi \leq 1$, they are said to be asymptotically dependent. Observe that the condition for asymptotic independence, that is, $\chi = 0$, coincides with the necessary and sufficient condition (8.100) for F to be asymptotically independent in the sense described there. Hence, if F_1 and F_2 are in the domain of attraction of univariate extreme value distributions G_1 and G_2 respectively, then $\chi = 0$ if and only if F is in the domain of attraction of the bivariate extreme value distribution $G(x, y) = G_1(x)G_2(y)$.

Recall from section 8.2.6 that the copula function of F, denoted by $C = C_F$, is equal to the distribution function of the pair (U_1, U_2), that is,

$$C(u_1, u_2) = P[U_1 \leq u_1, U_2 \leq u_2] = F\{F_1^{\leftarrow}(u_1), F_2^{\leftarrow}(u_2)\} \qquad (9.76)$$

for $(u_1, u_2) \in [0, 1]^2$, where, as usual, the arrow denotes the left-continuous inverse of a function. As the copula contains all information about the joint distribution of X_1 and X_2 except for the marginal information, it can be interpreted as the dependence structure associated with X_1 and X_2.

Now, defining

$$\chi(u) = 2 - \frac{\log C(u, u)}{\log u}, \qquad 0 < u < 1, \qquad (9.77)$$

we have

$$\chi(u) = 2 - \frac{1 - C(u, u)}{1 - u} + o(1) = P[U_2 > u \mid U_1 > u] + o(1), \qquad u \to 1,$$

whence

$$\lim_{u \to 1} \chi(u) = \chi. \qquad (9.78)$$

In general, the function $\chi(u)$ is bounded from below and above by

$$2 - \frac{\log\{\max(2u - 1, 0)\}}{\log u} \le \chi(u) \le 1, \qquad 0 < u < 1. \qquad (9.79)$$

These bounds follow from the respective bounds

$$\max(2u - 1, 0) \le C(u, u) \le u, \qquad 0 < u < 1, \qquad (9.80)$$

the left-hand side corresponding to perfect negative dependence and the right-hand side to perfect positive dependence.

Next to providing the limit χ, the function $\chi(u)$ also provides some insight in the dependence structure of the variables at lower quantile levels. In particular, $\chi(u)$ is less than, equal to or greater than 0 if and only if $C(u, u)$ is less than, equal to or greater than u^2 respectively. Since $C(u, u) = u^2$ corresponds to the case of exact independence, we find that the sign of $\chi(u)$ determines whether the variables are positively or negatively associated at quantile level u.

In the special case that C is a bivariate extreme value copula with Pickands dependence function A as in (8.54), we have $C(u, u) = u^\theta$ with $\theta = 2A(1/2) \in [1, 2]$ the extremal coefficient of (8.56). In particular, $\chi(u) = 2 - \theta \in [0, 1]$, constant in $0 < u < 1$. As a consequence, estimates of $\chi(u)$ can be used not only to gain information on the limiting behaviour as $u \to 1$ or the dependence structure at lower quantile levels but also as a diagnostic for membership to the bivariate extreme value class. More generally, if C is in the domain of attraction of a bivariate extreme value copula in the sense of (8.80), then by (8.92) we also have $\chi = 2 - \theta$.

Asymptotic independence

Within the class of asymptotically dependent variables ($0 < \chi \le 1$) the value of χ increases with increasing degree of dependence at extreme levels. The measure fails, however, to discriminate between the degrees of relative strength of dependence for asymptotically independent variables ($\chi = 0$). For that purpose, a quite natural alternative measure of dependence $\bar{\chi}$ has been defined, analogous to χ, but based on a comparison of joint and marginal survivor functions of U_1 and U_2 (Coles *et al.* 1999).

With the copula survivor function defined as

$$\bar{C}(u_1, u_2) = P[U_1 > u_1, U_2 > u_2] = 1 - u_1 - u_2 + C(u_1, u_2)$$

for $(u_1, u_2) \in [0, 1]^2$, let

$$\bar{\chi}(u) = \frac{2\log(1 - u)}{\log \bar{C}(u, u)} - 1, \qquad 0 < u < 1, \tag{9.81}$$

the precise definition being chosen for scaling convenience. From (9.80), we get

$$\frac{2\log(1 - u)}{\log\{\max(1 - 2u, 0)\}} - 1 \leq \bar{\chi}(u) \leq 1, \qquad 0 < u < 1. \tag{9.82}$$

Then, as a second limiting dependence measure, we define

$$\bar{\chi} = \lim_{u \to 1} \bar{\chi}(u), \tag{9.83}$$

provided the limit exists. By (9.82), we have $-1 \leq \bar{\chi} \leq 1$.

For asymptotically dependent variables, we have $\bar{\chi} = 1$; for asymptotically independent variables, we have $-1 \leq \bar{\chi} < 1$, and $\bar{\chi}$ provides a limiting measure that increases with relative dependence strength within this class. As a result, the pair $(\bar{\chi}, \chi)$ can be used as a one-dimensional summary of extremal dependence: if $\bar{\chi} = 1$ and $0 < \chi \leq 1$, the variables are asymptotically dependent and χ is a measure for strength of dependence within the class of asymptotically dependent distributions; if $-1 \leq \bar{\chi} < 1$ and $\chi = 0$, the variables are asymptotically independent, and $\bar{\chi}$ is a measure for strength of dependence within the class of asymptotically independent distributions.

The coefficient of tail dependence

Rather than to transform the original random variables X_1 and X_2 to uniform margins, it is also convenient to transform them to standard Fréchet margins by $Z_j = -1/\log U_j$ for $j = 1, 2$. Clearly, this leaves the copula invariant and hence does not affect the discussed dependence measures. The joint survival function of (Z_1, Z_2) can be found in terms of $\bar{C}$ through

$$P[Z_1 > z_1, Z_2 > z_2] = \bar{C}(e^{-1/z_1}, e^{-1/z_2}) \tag{9.84}$$

for $0 < z_j < \infty$ ($j = 1, 2$). Since $P[Z_j \leq z] = \exp(-1/z)$ for $z > 0$ and $j = 1, 2$, we have $P[Z_j > z] \sim 1/z$ as $z \to \infty$.

Next to χ and $\bar{\chi}$, Ledford and Tawn (1996) introduce a third dependence coefficient by assuming that the joint survivor function of Z_1 and Z_2 is a regularly varying function:

$$P[Z_1 > z, Z_2 > z] = \mathcal{L}(z)z^{-1/\eta}, \qquad z > 0. \tag{9.85}$$

Here, η is a positive constant, called the *coefficient of tail dependence*, and $\mathcal{L}$ is a slowly varying function, that is, $\mathcal{L}(xz)/\mathcal{L}(z) \to 1$ as $z \to \infty$ for all $0 < x < \infty$. The rate of decay in (9.85) is primarily controlled by η. Since $P[Z_1 > z, Z_2 > z] \leq 1 - \exp(-1/z) \sim 1/z$, we must have $\eta \leq 1$. Exploiting the fact that $P[Z_1 > z, Z_2 > z] = P[\min(Z_1, Z_2) > z]$, we can identify η as the tail index of the univariate variable $T = \min(Z_1, Z_2)$. Ledford and Tawn (1996) motivate their model through examples. The wide applicability of (9.85) is demonstrated by the extensive list of examples in Heffernan (2000). Still, the (somewhat pathological) counterexamples in Schlather (2001) show that (9.85) neither implies nor is implied by the familiar domain-of-attraction condition.

In (9.85), if $\mathcal{L}(z)z^{1-1/\eta} \sim P[Z_2 > z \mid Z_1 > z]$ converges as $z \to \infty$, the limit is equal to χ. Moreover, from (9.84), it follows that

$$\bar{C}(u, u) = \mathcal{L}(-1/\log u)(-\log u)^{1/\eta}, \qquad 0 < u < 1,$$

and thus, by (9.81),

$$\bar{\chi} = \lim_{u \to 1} \bar{\chi}(u) = 2\eta - 1.$$

As a consequence, if $\eta = 1$ and $\lim_{z \to \infty} \mathcal{L}(z) = c$ for some $0 < c \leq 1$, then $\bar{\chi} = 1$ and the variables are asymptotically dependent of degree $\chi = c$. On the other hand, if $0 < \eta < 1$ or if $\eta = 1$ and $\lim_{z \to \infty} \mathcal{L}(z) = 0$, then $\chi = 0$ and the variables are asymptotically independent of degree $\bar{\chi} = 2\eta - 1$.

Within the class of asymptotically independent variables, three types of independence can be identified according to the sign of $\bar{\chi} = 2\eta - 1$ (Heffernan 2000). First, when $1/2 < \eta < 1$ or $\eta = 1$ and $\mathcal{L}(z) \to 0$ as $z \to \infty$, observations for which both Z_1 and Z_2 exceed a large threshold z occur more frequently than under exact independence (positive association). Second, when $\eta = 1/2$, extremes of Z_1 and Z_2 are near independent and even exactly independent in case $\mathcal{L}(z) = 1$. Finally, when $0 < \eta < \frac{1}{2}$, observations for which both Z_1 and Z_2 exceed a large threshold z occur less frequently than under exact independence (negative association). All in all, the degree of dependence between large values of Z_1 and Z_2 is determined by η, with increasing values of η corresponding to stronger association. For a given η, the relative strength of dependence is characterized by $\mathcal{L}$.

Finally, remark that the whole story can be repeated if we transform the variables X_j to standard Pareto margins by $Z_j = 1/\{1 - F_j(X_j)\}$ rather than to standard Fréchet margins. The joint survivor of (Z_1, Z_2) is then given by

$$P[Z_1 > z_1, Z_2 > z_2] = \bar{C}(1 - 1/z_1, 1 - 1/z_2). \tag{9.86}$$

The only difference will be that the slowly varying function $\mathcal{L}$ corresponding to Pareto margins will have a different second-order behaviour than the one corresponding to Fréchet margins.

Example 9.1 The bivariate extreme value copula with logistic dependence structure (9.6) is given by

$$C(u_1, u_2) = \exp[-\{(-\log u_1)^{1/\alpha} + (-\log u_2)^{1/\alpha}\}^\alpha]$$

with parameter $0 < \alpha \le 1$. Perfect independence arises as $\alpha = 1$, while $0 < \alpha < 1$ leads to asymptotic dependence. The bivariate survivor function corresponding to standard Fréchet margins (9.84) satisfies

$$P[Z_1 > z, Z_2 > z] = (2 - 2^\alpha)z^{-1} + (2^{2\alpha-1} - 1)z^{-2} + o(z^{-2}) \qquad (9.87)$$

as $z \to \infty$, while transforming to standard Pareto margins (9.86) gives

$$P[Z_1 > z, Z_2 > z] = (2 - 2^\alpha)z^{-1} + (2^{2\alpha-1} - 2^{\alpha-1})z^{-2} + o(z^{-2}) \qquad (9.88)$$

as $z \to \infty$. If $0 < \alpha < 1$, then in both cases we find, as expected, a coefficient of tail dependence $\eta = 1$ and a slowly varying function $\mathcal{L}$ converging to $\chi = 2 - 2^\alpha$.

Example 9.2 The bivariate Farlie-Gumbel-Morgenstern copula is given by

$$C(u_1, u_2) = u_1 u_2 \{1 + \alpha(1 - u_1)(1 - u_2)\},$$

with parameter $-1 \le \alpha \le 1$. For $\alpha = 0$, $\alpha > 0$, and $\alpha < 0$, we get exact independence, positive dependence, and negative dependence, respectively. Complete dependence cannot be achieved under this model. As

$$\chi(u) = 2 - \frac{\log[u^2\{1 + \alpha(1 - u)^2\}]}{\log u}, \qquad 0 < u < 1,$$

we get $\chi(u) \to \chi = 0$ as $u \to 1$, that is, all distributions in this family are asymptotically independent. Examining the relative strength of dependence within the class of asymptotically independent variables, we notice that

$$\bar{\chi}(u) = \frac{2\log(1 - u)}{\log[1 - 2u + u^2\{1 + \alpha(1 - u)^2\}]} - 1, \qquad 0 < u < 1,$$

so $\bar{\chi}$ equals 0 for $\alpha > -1$ (near independence) and $-1/3$ for $\alpha = -1$ (negative association). Transforming to standard Fréchet margins leads to the joint survivor function (9.84)

$$P[Z_1 > z, Z_2 > z] = \frac{\alpha + 1}{z^2} - \frac{3\alpha + 1}{z^3} + \frac{55\alpha + 7}{12z^4} + o\left(\frac{1}{z^4}\right), \qquad z \to \infty.$$

This expansion allows us to identify η and $\mathcal{L}$ in (9.85) as a function of α. In case $\alpha > -1$, we have $\eta = 1/2$ and $\mathcal{L}(z) = (\alpha + 1) - (3\alpha + 1)z^{-1} + o(z^{-1})$ as $z \to \infty$ (near independence); in case $\alpha = -1$, we have $\eta = 1/3$ and $\mathcal{L}(z) = 2 - 4z^{-1} + o(z^{-1})$ as $z \to \infty$ (negative association). Notice that $\chi = \lim_{z\to\infty} \mathcal{L}(z)z^{1-1/\eta}$ and $\bar{\chi} = 2\eta - 1$, as expected.

Example 9.3 The bivariate normal distribution with correlation $\rho < 1$ is a prime example of asymptotic independence (Sibuya 1960). The joint survivor function (9.84) for margins transformed to the standard Fréchet distribution satisfies

$$P[Z_1 > z, \, Z_2 > z] \sim c_\rho (\log z)^{-\rho/(1+\rho)} z^{-2/(1+\rho)}, \qquad z \to \infty,$$

where $c_\rho = (1 + \rho)^{3/2} (1 - \rho)^{-1/2} (4\pi)^{-\rho/(1+\rho)}$ (Reiss 1989). In particular, the distribution is asymptotically independent ($\chi = 0$) with $\bar{\chi} = \rho$ and $\eta = (1 + \rho)/2$. Within the class of asymptotic independence, the cases of positive association, near independence, and negative association arise as $\rho > 0$, $\rho = 0$, and $\rho < 0$, respectively.

The bivariate normal distribution with correlation $0 < \rho < 1$ illustrates that dependence at intermediate levels, however strong, does not necessarily imply asymptotic dependence. This may lead to problems when we apply techniques based on bivariate extreme value dependence structures, where choices are limited to asymptotic dependence or exact independence. For instance, Ledford and Tawn (1996) show that the score test for independence using the censored likelihood (9.69)–(9.70) with logistic model will nearly always reject independence in case data are generated from a bivariate normal distribution with positive correlation. Still, extrapolating from an asymptotically dependent model fitted to the tail of the bivariate normal distribution will lead to overestimation of the probability of the occurrence of joint extremes.

Data example

For the Loss-ALAE data of Figure 9.1, an informal picture of the dependence functions $\chi(u)$ and $\bar{\chi}(u)$ can be created simply by plugging in empirical estimates

$$\hat{C}(u, u) = \frac{1}{n} \sum_{i=1}^{n} \mathbf{1}\{u_{i1} < u, \, u_{i2} < u\}$$

$$\hat{\bar{C}}(u, u) = \frac{1}{n} \sum_{i=1}^{n} \mathbf{1}\{u_{i1} > u, \, u_{i2} > u\}$$

in expressions (9.77) and (9.81) (Coles *et al.* 1999). Analyzing the behaviour of the empirical versions of $\chi(u)$ and $\bar{\chi}(u)$ as u tends to 1 can give an idea of the form of extremal dependence between the variables. Figure 9.9 shows estimates and 95% point-wise confidence intervals for $\chi(u)$ and $\bar{\chi}(u)$. The confidence intervals are based on bootstrap samples obtained by sampling with replacement from the original data (x_{i1}, x_{i2}), $i = 1, \ldots, n$, as suggested in Fermanian *et al.* (2004). Also shown are the cases of perfect positive dependence, exact independence and perfect negative dependence.

As $\chi(u) > 0$ for $u < 1$, there is evidence for dependence of the variables at lower quantile levels. It appears that $\chi(u) \approx 0.4$ for all u, even for u close to 1, suggesting an asymptotically dependent distribution that is possibly of the bivariate

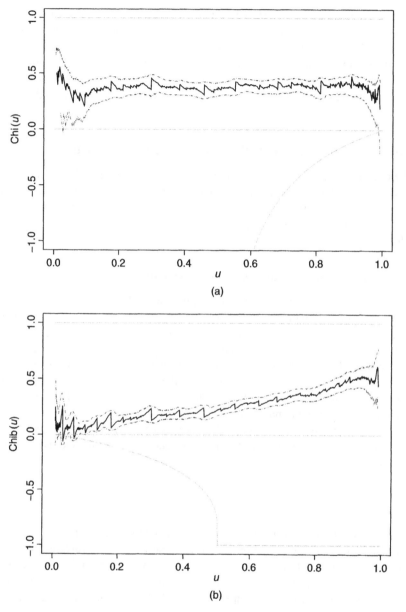

Figure 9.9 Loss-ALAE data: dependence measures (a) $\chi(u)$ and (b) $\bar{\chi}(u)$. Estimates (———), 95% point-wise confidence intervals (– · – · –) and cases corresponding to perfect positive dependence, exact independence and perfect negative dependence (- - - - -).

extreme value type. However, notice that for u close to 1, the point-wise confidence intervals cover a large range of possible limits, including 0. Moreover, $\bar{\chi}(u)$ seems to be smaller than 1, which is in contradiction with the hypothesis of an asymptotically dependent distribution. As a result, on the basis of the above informal analysis only, it is difficult to make a decision between asymptotic dependence and asymptotic independence for the insurance data. This shows the need for more formal diagnostics.

9.5.2 Estimating the coefficient of tail dependence

The coefficient of tail dependence, η, was found to be most useful in distinguishing between asymptotic dependence or asymptotic independence, and, within the latter class, between positive association, near independence, or negative association. This makes the problem of estimating η, the topic of this section, a particularly relevant one.

Hill estimator and maximum likelihood estimator

Assumption (9.85) entails that the univariate variable $T = \min(Z_1, Z_2)$ has a regularly varying tail with index $-1/\eta$; here, Z_j can be either $-1/\log F_j(X_j)$ (standard Fréchet margins) or $1/\{1 - F_j(X_j)\}$ (standard Pareto margins). Therefore, η and hence $\bar{\chi} = 2\eta - 1$ can be estimated as the tail index of T, for instance, by the univariate techniques in Chapters 4–5. Notice also that, subject to convergence of $\mathcal{L}(z)$ as $z \to \infty$ and η being equal to 1, the dependence parameter χ can be estimated as the scale parameter of T for large values of z, as in that case $\mathcal{L}(z)$ is approximately constant and equal to χ.

Given a sample of independent observations (X_{i1}, X_{i2}), $i = 1, \ldots, n$, Ledford and Tawn (1996) propose transforming the data to have approximate standard Fréchet margins by

$$Z_{ij} = -1/\log \hat{F}_j(X_{ij}), \qquad i = 1, \ldots, n, \quad j = 1, 2, \qquad (9.89)$$

with $\hat{F}_j$ estimates of the marginal distribution functions F_j, typically by empirical marginal distribution functions and incorporating extreme value estimators for the marginal tails. Alternatively, we may transform to standard Pareto margins by $Z_{ij} = 1/\{1 - \hat{F}_j(X_{ij})\}$. In any case, the $T_i = \min(Z_{i1}, Z_{i2})$, $i = 1, \ldots, n$, approximately form an independent sample distributed like T. Denote the order statistics of the T_i by $T_{1,n} \leq \cdots \leq T_{n,n}$.

We can use the T_i to estimate η, for example, by the Hill (1975) estimator (see section 4.2)

$$\hat{\eta} = \frac{1}{k} \sum_{i=1}^{k} \log T_{n-k+i,n} - \log T_{n-k,n} \qquad (9.90)$$

or by the maximum likelihood estimator in a peaks-over-threshold setting, where exceedances of T above a high-enough threshold u are assumed to follow a GP distribution

$$P[T > u + z \mid T > u] = (1 + \eta z/\sigma)^{-1/\eta}, \qquad 0 \le z < \infty, \qquad (9.91)$$

with shape parameter $0 < \eta \le 1$ and scale parameter $\sigma = \sigma(u) > 0$.

Under the model (9.91), Ledford and Tawn (1996) suggest to test for asymptotic independence ($\chi = 0$) by testing $\eta = 1$ against the alternative $0 < \eta < 1$. As mentioned before, observe that under model (9.85), the hypothesis $\eta = 1$ is implied by, but is not equivalent to, asymptotic dependence. Note, however, that the special case $\eta = 1$ and $\mathcal{L}(z) \to 0$ as $z \to \infty$ tends only to have theoretical value, so that, in practice, it is safe to assume that $\eta = 1$ is equivalent to asymptotic dependence.

So, let L_1 be the maximized likelihood for (9.91) for a given threshold u and L_0 be the corresponding maximized likelihood under the restriction $\eta = 1$. Since the null hypothesis corresponds to a boundary value of the parameter space, the likelihood ratio statistic $D = 2(\log L_1 - \log L_0)$ should be compared to a one-half chi-squared distribution with one degree of freedom (Self and Liang 1987), resulting in the p-value $P[\chi^2 > D]/2$. Still, it is likely that the true estimation uncertainty is larger than the one reflected in the likelihood ratio tests or profile likelihood-based confidence intervals as we falsely assumed that the T_i are independent, that each marginal distribution is estimated exactly and that the parametric specification of model (9.85) is correct.

Estimators of Peng (1999) and Draisma *et al.* (2002)

In order to avoid underestimation of the true uncertainty in the estimates of η as a consequence of the uncertainty introduced by possible marginal transformations, which is, for example, not accounted for in the above procedure, Peng (1999) and Draisma *et al.* (2004) propose estimating η in (9.85) through certain non-parametric alternatives that do not depend on the marginal distributions.

Assumption (9.85) formulated for standard Pareto $Z_j = 1/\{1 - F_j(X_j)\}$ implies

$$\lim_{t \to 0} \frac{P[X_1 > F_1^{\leftarrow}(1 - ts), \, X_2 > F_2^{\leftarrow}(1 - ts)]}{P[X_1 > F_1^{\leftarrow}(1 - t), \, X_2 > F_2^{\leftarrow}(1 - t)]} = s^{1/\eta} \qquad (9.92)$$

for $s > 0$. In both Peng (1999) and Draisma *et al.* (2004), this limiting relation is used to construct a non-parametric estimator for η on the basis of the empirical distribution function of the original observations (X_{i1}, X_{i2}), $i = 1, \ldots, n$. With $X_{(i,n),j}$ the ith ascending order statistic of the jth coordinate sample $(X_{ij})_{i=1}^n$ ($j = 1, 2$), define

$$S_n(k) = \sum_{i=1}^{n} \mathbf{1}\{X_{i1} > X_{(n-k,n),1}, \, X_{i2} > X_{(n-k,n),2}\} \qquad (9.93)$$

for $k = 0, \ldots, n - 1$. Notice that $S_n(k)$ depends on the data through their ranks only and that $S_n(k)/n$ can be seen as the empirical counterpart of $P[X_1 > F_1^{\leftarrow}(1 - k/n), X_2 > F_2^{\leftarrow}(1 - k/n)]$.

Taking logarithms on both sides of (9.92) with $s = 2$ leads quite naturally to the estimator

$$\hat{\eta}_1 = \frac{\log 2}{\log\{S_n(2k)/S_n(k)\}} \tag{9.94}$$

proposed by Peng (1999). Integrating both sides of (9.92) with respect to s from 0 to 1 gives the estimator

$$\hat{\eta}_2 = \frac{\sum_{j=1}^{k} S_n(j)}{kS_n(k) - \sum_{j=1}^{k} S_n(j)} \tag{9.95}$$

as introduced by Draisma *et al.* (2004). Note that $\hat{\eta}_1$ is based on $S_n(k)$ and $S_n(2k)$, while $\hat{\eta}_2$ is constructed from the $S_n(j)$ for j only up to k.

Peng (1999) and Draisma *et al.* (2004) establish asymptotic normality of their estimators under certain second-order conditions on the limiting behaviour of $P[Z_1 > x_1z, Z_2 > x_2z]$ for $0 < x_j < \infty$ as $z \to \infty$. The second-order conditions by Peng (1999) prohibit the slowly varying function $\mathcal{L}(z)$ in (9.85) to converge to zero as $z \to \infty$, so that the hypothesis of asymptotic independence ($\chi = 0$) is equivalent to $\eta < 1$. A drawback is that the distributions such as the bivariate normal (Example 9.3) are excluded. The second-order conditions by Draisma *et al.* (2004) are less restrictive in that they do allow for $\mathcal{L}(z) \to 0$ as $z \to \infty$.

Draisma *et al.* (2004) prove asymptotic normality not only for their estimator $\hat{\eta}_2$ (9.95) but also for the estimator $\hat{\eta}_1$ (9.94) by Peng (1999), the Hill estimator $\hat{\eta}_3$ (9.90), and the maximum likelihood estimator $\hat{\eta}_4$ arising from (9.91) with threshold $u = T_{n-k,n}$. They transform the data to standard Pareto margins by

$$Z_{ij} = \frac{1}{1 - \hat{F}_j(X_{ij})} \quad \text{where} \quad \hat{F}_j(x) = \frac{1}{n+1} \sum_{i=1}^{n} \mathbf{1}(X_{ij} \leq x). \tag{9.96}$$

Observe that the Z_{ij} depend on the original data X_{ij} through the ranks only. Draisma *et al.* (2004) show that under certain growth conditions for $k = k_n$, the standardized estimators $\{S_n(k)\}^{1/2}(\hat{\eta}_i - \eta)$ ($i = 1, 2$) and $k^{1/2}(\hat{\eta}_i - \eta)$ ($i = 3, 4$) are asymptotically normal with mean 0 and certain asymptotic variances σ_i^2 ($i = 1, \ldots, 4$). The expressions for the σ_i are rather complicated, but Draisma *et al.* (2004) propose a way to estimate them from the Z_{ij} as well. Denoting such estimates by $\hat{\sigma}_i$, asymptotic confidence intervals for η can easily be constructed, and $\eta = 1$ can be tested against the alternative $\eta < 1$. For instance, denoting by $\hat{\sigma}_{(i)}$ the estimated root variances $\{S_n(k, k)\}^{-1/2}\hat{\sigma}_i$ ($i = 1, 2$) and $k^{-1/2}\hat{\sigma}_i$ ($i = 3, 4$) in case $\eta = 1$, we can reject $\eta = 1$ in favour of $\eta < 1$ if $\hat{\eta}_i \leq 1 - z_\alpha\hat{\sigma}_{(i)}$, with z_α the $(1 - \alpha)$-quantile of the standard normal distribution.

Estimator of Beirlant and Vandewalle (2002)

The bias of estimators as introduced by Ledford and Tawn (1996), Peng (1999) and Draisma *et al.* (2004) depends heavily on the underlying bivariate distribution. If, for instance, the dependence parameter α in the logistic model of Example 9.1 is close to, but smaller than, one, the first-order terms in (9.87) and (9.88) will be small and dominated by the second-order terms unless z is large (note that the situation is worse for Pareto margins (9.88) than for Fréchet margins (9.87), as the second-order terms are smaller in the latter than in the former, although for other distributions the situation may be reversed). As a result, thresholds will need to be chosen high enough in order not to get estimates of η that are biased towards asymptotic independence. In view of the critical difference between asymptotic dependence and asymptotic independence regarding out-of-sample inference, it is therefore highly desirable to have available estimation methods for η that can cope with a slow rate of convergence in the model assumption $\mathcal{L}(xz)/\mathcal{L}(z) \to 1$ as $z \to \infty$.

In this respect, Beirlant and Vandewalle (2002) suggest an estimator based on scaled log-ratios

$$\tilde{Y}_j = j \log \left(\frac{T_{n-j+1,n} - T_{n-k,n}}{T_{n-j,n} - T_{n-k,n}} \right), \qquad j = 1, \ldots, k-1,$$

of excesses over a large threshold $T_{n-k,n}$; here, the T_i are defined as above, constructed from the data transformed to either standard Pareto or standard Fréchet margins. The coefficient of tail dependence, η, is then estimated by maximum likelihood from the exponential regression model

$$\tilde{Y}_j \stackrel{\mathcal{D}}{=} \frac{\eta}{1 - (j/k)^\eta} E_j, \qquad j = 1, \ldots, k-1, \tag{9.97}$$

with E_j independent standard exponential random variables, see section 5.4.

Beirlant and Vandewalle (2002) prove asymptotic normality of this estimator under the same second-order conditions as in Draisma *et al.* (2004) but restricted to the case of asymptotic independence ($\chi = 0$). The estimator has a smaller bias than other well-known estimators, whatever the marginal transformations and underlying distribution. Under a second-order refinement of (9.97), minimization of the estimated asymptotic mean squared error leads to a diagnostic for selecting the optimal k to be used in estimating η.

Data example

We transform the Loss-ALAE data of Figure 9.1 to approximate standard Fréchet margins by

$$z_{ij} = -1/\log u_{ij} \qquad i = 1, \ldots, n, \quad j = 1, 2,$$

with u_{ij} as in (9.1). The sample of minima, $t_i = \min(z_{i1}, z_{i2})$, serves to construct maximum likelihood estimates and profile likelihood confidence intervals for η

based on the GP likelihood derived from (9.91), see Figure 9.10(a). Here, the threshold varies along the entire range of threshold probabilities for T. Transforming to standard Pareto margins by $z_{ij} = 1/(1 - u_{ij})$ gives the same qualitative results, see Figure 9.10(b).

In view of the probable underestimation of the estimation uncertainty as reflected in the confidence intervals, we notice that although estimates of η for almost all threshold probabilities between 0.5 and 0.9 seem to be close to 0.9, corresponding to a strongly positively associated form of asymptotic independence, the value $\eta = 1$, consistent with asymptotic dependence, is still covered by almost all confidence intervals. Neither do likelihood ratio tests consistently permit us to reject asymptotic dependence against asymptotic independence.

Alternatively, Figure 9.11 shows point-wise estimates for η using the estimator (9.94) of Peng (1999), together with critical values under which $\eta = 1$ is rejected in favour of $\eta < 1$ based on a 5% one-sided test (left) and two-sided 95% confidence intervals (right). Again, we cannot consistently reject asymptotic dependence.

9.5.3 Joint tail modelling

Modelling the tail of a multivariate distribution by an extreme value distribution as in section 9.4 limits the options for the extremal dependence structure to either asymptotic dependence or exact independence. In the bivariate case, this means that the probability $P[X_1 > F_1^{\leftarrow}(1 - 1/z), X_2 > F_2^{\leftarrow}(1 - 1/z)]$ of joint exceedances of the respective $1/z$ tail quantiles in the two margins is of the order $O(z^{-1})$ (asymptotic dependence) or $O(z^{-2})$ (exact independence) as $z \to \infty$. However, for asymptotically independent distributions with positive association, that is, $1/2 < \eta < 1$ in (9.85), this probability is in fact of the order $O(z^{-1/\eta})$. Hence, for such distributions, the probability of joint extremes will be evaluated either too large, in case of an asymptotically dependent model, or too small, in case of the exactly independent model.

The model of Ledford and Tawn (1997)

A versatile model bridging the gap between asymptotic dependence and exact independence was introduced by Ledford and Tawn (1997). Before we can describe their model, we need some technical preliminaries.

A function $\mathcal{L} : (0, \infty)^2 \to (0, \infty)$ is called *bivariate slowly varying* if there exists a function $g : (0, \infty)^2 \to (0, \infty)$ such that

$$\lim_{t \to \infty} \frac{\mathcal{L}(tz_1, tz_2)}{\mathcal{L}(t, t)} = g(z_1, z_2), \qquad 0 < z_j < \infty \quad (j = 1, 2) \tag{9.98}$$

and if this function g is homogenous of order zero, that is,

$$g(sz_1, sz_2) = g(z_1, z_2), \qquad 0 < s < \infty, \quad 0 < z_j < \infty \quad (j = 1, 2)$$

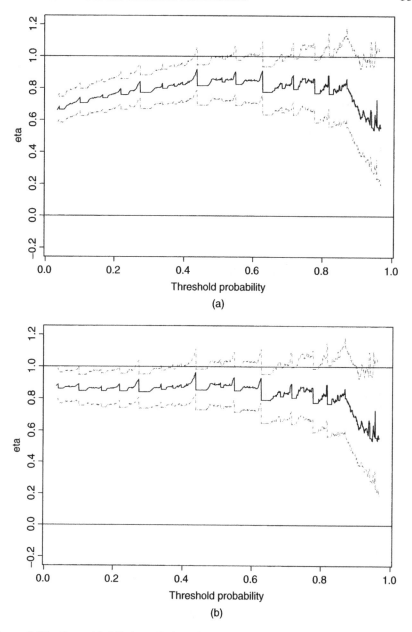

Figure 9.10   Loss-ALAE data: Point-wise maximum likelihood estimates (——)
and profile likelihood–based confidence intervals (- - - - -) for η based on (9.91) with
data transformed to (a) standard Fréchet margins and (b) standard Pareto margins.

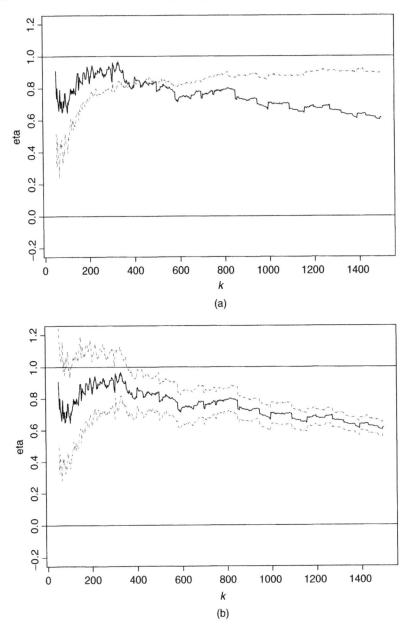

Figure 9.11 Loss-ALAE data: Point-wise η estimates (———) using (9.94) by Peng (1999) as function of $2k$ with (a) critical values (- - - - -) under which $\eta = 1$ is rejected in favour of $\eta < 1$ based on a 5% one-sided test and (b) 95% confidence intervals (- - - - -) (Draisma *et al.* 2004).

(Bingham *et al.* 1987). The homogeneity of g implies that there exists a function $g_* : (0, 1) \to (0, \infty)$ such that $g(z_1, z_2) = g_*\{z_1/(z_1 + z_2)\}$ for all $0 < z_j < \infty$ ($j = 1, 2$). We call $\mathcal{L}$ *ray independent* if g_* is constant and *ray dependent* otherwise. Furthermore, $\mathcal{L}$ is called *quasi-symmetric* if the function $g_*(w)/g_*(1 - w)$ is slowly varying at $w \to 0$ and $w \to 1$.

Now as in the previous sections, let (X_1, X_2) be a random pair with distribution function F and continuous marginal distribution functions F_1 and F_2. Transform the vector to standard Fréchet margins by $Z_j = -1/\log F_j(X_j)$ for $j = 1, 2$. Ledford and Tawn (1997) propose to model the joint survivor function of (Z_1, Z_2) as

$$P[Z_1 > z_1, Z_2 > z_2] = \mathcal{L}(z_1, z_2)z_1^{-c_1}z_2^{-c_2}, \tag{9.99}$$

with $c_j > 0$ for $j = 1, 2$ and $\mathcal{L}$ a quasi-symmetric, bivariate slowly varying function. Clearly, (9.99) implies (9.85) with $1/\eta = c_1 + c_2$. In this sense, the model of Ledford and Tawn (1997) provides an extension of the one of Ledford and Tawn (1996). All in all, (9.99) provides a smooth family of dependence models, incorporating asymptotically dependent distributions as well as positively or negatively associated asymptotically independent distributions.

The quasi-symmetry condition on $\mathcal{L}$ is imposed to identify c_1 and c_2. For, denoting $c_2 - c_1 = \kappa$, we also have

$$P[Z_1 > z_1, Z_2 > z_2] = \tilde{\mathcal{L}}(z_1, z_2)(z_1 z_2)^{-1/(2\eta)} \tag{9.100}$$

for $0 < z_j < \infty$ ($j = 1, 2$), where the function

$$\tilde{\mathcal{L}}(z_1, z_2) = (z_1/z_2)^{\kappa/2}\mathcal{L}(z_1, z_2) \tag{9.101}$$

is bivariate slowly varying with limit function $\tilde{g}(z_1, z_2) = (z_1/z_2)^{\kappa/2}g(z_1, z_2)$ and ray dependence function $\tilde{g}_*(w) = \{w/(1 - w)\}^{\kappa/2}g_*(w)$ for $0 < w < 1$. As $g_*(w)/g_*(1 - w)$ is slowly varying at 0 and 1, the function $\tilde{g}_*(w)/\tilde{g}_*(1 - w)$ is regularly varying at 0 and 1 with indices $-\kappa$ and κ, respectively. Observe that an alternative and perhaps simpler and less restrictive way to define the joint tail model is via (9.100) but without imposing regular variation of $\tilde{g}_*(w)/\tilde{g}_*(1 - w)$ at 0 or 1. This is in fact the approach taken in Ramos (2003).

Exponent measure

If the joint tail model (9.99) holds, then there exist proper analogues of the exponent and spectral measures of a max-stable distribution. Starting point is the simple observation that

$$\lim_{t \to \infty} \frac{P[Z_1 > tz_1, Z_2 > tz_2]}{P[Z_1 > t, Z_2 > t]} = g_*\{z_1/(z_1 + z_2)\}z_1^{-c_1}z_2^{-c_2}, \tag{9.102}$$

for $0 < z_j < \infty$ ($j = 1, 2$). This shows that $g(z, z) = g_*(1/2) = 1$ for $0 < z < \infty$. Now for $0 < t < \infty$, define a positive measure $\Lambda_t(\cdot)$ on $(0, \infty)^2$ by

$$\Lambda_t(B) = \frac{P[t^{-1}(Z_1, Z_2) \in B]}{P[Z_1 > t, Z_2 > t]}$$

for Borel sets B in $(0, \infty)^2$. Then (9.102) states convergence of $\Lambda_t\{(z_1, \infty) \times (z_2, \infty)\}$ as $t \to \infty$. Now by similar arguments as those leading to (8.71) or (8.94), this implies that there exists a positive measure Λ on $(0, \infty)^2$ given by

$$\Lambda\{(z_1, \infty) \times (z_2, \infty)\} = g_*\{z_1/(z_1 + z_2)\}z_1^{-c_1}z_2^{-c_2} \tag{9.103}$$

for $0 < z_j < \infty$ $(j = 1, 2)$ and such that

$$\Lambda_t(\cdot) \overset{v}{\to} \Lambda \text{ as } t \to \infty \text{ in } (0, \infty] \times (0, \infty], \tag{9.104}$$

with '$\overset{v}{\to}$' denoting vague convergence (Kallenberg 1983; Resnick 1987). Observe that in (9.104), the coordinate axes are excluded, in contrast to (8.94). The reason is that in case $\eta < 1$, the normalizing factor $1/P[Z_1 > t, Z_2 > t] = \{\mathcal{L}(t, t)\}^{-1}t^{1/\eta}$ in the definition of Λ_t is of larger order than the factor t in the definition of μ_{*t} in (8.94). In other words, for sets B hugging one or both of the axes, $\Lambda_t(B)$ may blow up to infinity if $\eta < 1$. Finally, observe that (9.104) suggests the approximation

$$P[(Z_1, Z_2) \in \cdot] \approx P[Z_1 > t, Z_2 > t]\Lambda(t^{-1} \cdot) \tag{9.105}$$

for large enough $0 < t < \infty$. This approximation forms the basis of statistical inference procedures on the bivariate tail of (Z_1, Z_2), see below.

Clearly, equation (9.103) implies that

$$\Lambda\{(sz_1, \infty) \times (sz_2, \infty)\} = s^{-1/\eta}\Lambda\{(z_1, \infty) \times (z_2, \infty)\},$$

for $0 < s < \infty$ and $0 < z_j < \infty$ $(j = 1, 2)$. Since rectangles of the kind $(z_1, \infty) \times (z_2, \infty)$ form a measure-determining class in $(0, \infty)^2$, we obtain

$$\Lambda(s \cdot) = s^{-1/\eta}\Lambda(\cdot), \qquad 0 < s < \infty. \tag{9.106}$$

Property (9.106) should be compared with the corresponding homogeneity property (8.11) of the exponent measure μ_*.

Spectral measure

Define the measure H_Λ on $(0, 1)$ by

$$H_\Lambda(B) = \Lambda\left(\left\{(z_1, z_2) \in (0, \infty)^2 : z_1 + z_2 > 1, \frac{z_1}{z_1 + z_2} \in B\right\}\right) \tag{9.107}$$

for Borel sets B in $(0, 1)$. By homogeneity in (9.106),

$$\Lambda\left(\left\{(z_1, z_2) \in (0, \infty)^2 : z_1 + z_2 > r, \frac{z_1}{z_1 + z_2} \in B\right\}\right) = r^{-1/\eta}H_\Lambda(B) \tag{9.108}$$

for $0 < r < \infty$ and Borel sets B in $(0, 1)$. Equation (9.108) implies that the measure Λ factorizes as a product measure when expressed in pseudo-polar coordinates $T(z_1, z_2) = (r, w)$ with $r = z_1 + z_2$ and $w = z_1/(z_1 + z_2)$, that is,

$$\Lambda \circ T^{-1}(dr\,dw) = \eta^{-1}r^{-1/\eta-1}dr\,H_\Lambda(dw). \tag{9.109}$$

This is the spectral decomposition of Λ, to be compared with the spectral decomposition for the exponent measure μ_* in (8.17). The measure H_Λ is the *spectral measure* of Λ.

The spectral decomposition (9.109) implies for $0 < z_j < \infty$ ($j = 1, 2$),

$$\Lambda\{(z_1, \infty) \times (z_2, \infty)\}$$

$$= \int_{(0,1)} \int_0^\infty \mathbf{1}\{rw > z_1, r(1-w) > z_2\} \eta^{-1} r^{-1/\eta-1} dr \, H_\Lambda(dw)$$

$$= \int_{(0,1)} \left\{ \min\left(\frac{w}{z_1}, \frac{1-w}{z_2}\right) \right\}^{1/\eta} H_\Lambda(dw). \tag{9.110}$$

Comparing this with (9.103) gives

$$g_*(w) w^{-c_1} (1-w)^{-c_2} = \int_{(0,1)} \left\{ \min\left(\frac{v}{w}, \frac{1-v}{1-w}\right) \right\}^{1/\eta} H_\Lambda(dv), \tag{9.111}$$

for $0 < w < 1$. As $g_*(1/2) = 1$, we find that the spectral measure H_Λ must satisfy the constraint

$$\int_{(0,1)} \{\min(w, 1-w)\}^{1/\eta} H_\Lambda(dw) = 1. \tag{9.112}$$

If Λ is absolutely continuous with density $\lambda(z_1, z_2)$, then H_Λ is absolutely continuous as well, and its density h_Λ can be calculated from g_* and (c_1, c_2) as follows. The Jacobian of the transformation $(z_1, z_2) \mapsto (r, w) = (z_1 + z_2, z_1/(z_1 + z_2))$ is equal to r^{-1}. Therefore, by the multivariate changes-of-variable formula,

$$\lambda(z_1, z_2) = \eta^{-1} r^{-2-1/\eta} h_\Lambda(w).$$

Since moreover $\lambda(z_1, z_2) = \partial^2 \Lambda\{(z_1, \infty) \times (z_2, \infty)\}/\partial z_1 \partial z_2$, we obtain

$$h_\Lambda(w) = \frac{c_1 c_2 g_*(w) + w(1-w) g_*'(w)(2w - 1 + c_1 - c_2) - g_*''(w) w^2 (1-w)^2}{(c_1 + c_2) w^{1+c_1} (1-w)^{1+c_2}}$$

$$\tag{9.113}$$

for $0 < w < 1$. This derivation shows that when specifying parametric models for $\mathcal{L}$ in (9.99) and hence for g_*, care has to be taken that the resulting spectral density h_Λ is indeed positive.

A simpler way to specify parametric models satisfying (9.99) is directly via the spectral measure (Ramos 2003). If $0 < \eta \leq 1$ and if H is a positive measure on $(0, 1)$ satisfying (9.107), then we can define a probability distribution with joint survivor function as in the right-hand side of (9.110) restricted to $1 \leq z_j < \infty$ ($j = 1, 2$). This survivor function can be written as in (9.100) with bivariate slowly varying function

$$\tilde{\mathcal{L}}_{\eta,H}(z_1, z_2) = \int_{(0,1)} \left[\min\left\{ \frac{z_2^{1/2} w}{z_1^{1/2}}, \frac{z_1^{1/2}(1-w)}{z_2^{1/2}} \right\} \right]^{1/\eta} H(dw)$$

for $1 \le z_j < \infty$ $(j = 1, 2)$, whose limit function, $\tilde{g}_{\eta,H}$, is equal to the expression in the right-hand side of the above equation extended to all $0 < z_j < \infty$ $(j = 1, 2)$. If, moreover, the corresponding ray dependence function $\tilde{g}_{*\eta,H}$ is regularly varying at 0 and 1 with indices κ and $-\kappa$, then we can define $\mathcal{L}_{\eta,H}$ by turning around (9.101), leading finally to the representation (9.100).

By (9.104), the spectral measure H_Λ can be related to the distribution of $Z_1/(Z_1 + Z_2)$ given that both Z_1 and Z_2 are large in the sense that

$$\frac{P[Z_1 + Z_2 > t,\ Z_1/(Z_1 + Z_2) \in \cdot\,]}{P[Z_1 > t,\ Z_2 > t]} \xrightarrow{v} H_\Lambda(\cdot),\qquad t \to \infty, \tag{9.114}$$

in the open interval $(0, 1)$. In particular, if H_Λ is absolutely continuous with spectral density h_Λ, then

$$\lim_{t \to \infty} \frac{P[Z_1 + Z_2 > t,\ w_0 \le Z_1/(Z_1 + Z_2) \le w_1]}{P[Z_1 > t,\ Z_2 > t]} = \int_{w_0}^{w_1} h_\Lambda(w)dw \tag{9.115}$$

for all $0 < w_0 \le w_1 < 1$. Observe that we do not allow the w_j to be 0 or 1, as in case $\eta < 1$ the limit would be infinity.

Example 9.4 If $\mathcal{L}$ is ray independent, that is, if $g_* = 1$, then the exponent measure is $\Lambda\{(z_1, \infty) \times (z_2, \infty)\} = z_1^{-c_1} z_2^{-c_2}$, while by (9.113), the spectral density is given simply by

$$h(w; c_1, c_2) = \frac{c_1 c_2}{c_1 + c_2}\, \frac{1}{w^{1+c_1}(1 - w)^{1+c_2}},\qquad 0 < w < 1.$$

If, moreover, $c_1 = c_2$, then, as $c_1 + c_2 = 1/\eta$,

$$h(w; \eta) = \frac{1}{4\eta\{w(1 - w)\}^{1+1/(2\eta)}},\qquad 0 < w < 1. \tag{9.116}$$

It is not hard to check (9.112) directly for $h(w; c_1, c_2)$.

Example 9.5 If Z_1 and Z_2 are independent standard Fréchet random variables, then

$$P[Z_1 > z_1,\ Z_2 > z_2] = \{1 - \exp(-1/z_1)\}\{1 - \exp(-1/z_2)\}$$
$$= \mathcal{L}(z_1, z_2)(z_1 z_2)^{-1},$$

where $\mathcal{L}(z_1, z_2)$ is a ray independent, bivariate slowly varying function. In particular, $c_1 = c_2 = 1$ and $\eta = 1/2$. By (9.116), the spectral density is given by $h(w; 1/2) = 2^{-1}\{w(1 - w)\}^{-2}$.

Example 9.6 Let the random pair (X_1, X_2) have a bivariate normal distribution with standard normal margins and correlation $-1 < \rho < 1$. Transform the margins to the standard Fréchet distribution by $Z_j = -1/\log \Phi(X_j)$ for $j = 1, 2$, where Φ

is the standard normal distribution function. Ledford and Tawn (1997) show that the bivariate survivor function of (Z_1, Z_2) can be written as

$$P[Z_1 > z_1, Z_2 > z_2] = \mathcal{L}(z_1, z_2; \rho)(z_1 z_2)^{-1/(1+\rho)},$$

with $\mathcal{L}(z_1, z_2; \rho)$ a ray independent, bivariate slowly varying function. Hence $c_1 = c_2 = 1/(1 + \rho)$, $\eta = (1 + \rho)/2$, and the spectral density is given by $h\{w; (1 + \rho)/2\}$ as in (9.116).

Example 9.7 Let (Z_1, Z_2) have a bivariate extreme value distribution with standard Fréchet margins and exponent measure μ_*, that is, $P[Z_1 \leq z_1, Z_2 \leq z_2] = \exp\{-V_*(z_1, z_2)\}$ with $V_*(z_1, z_2) = \mu_*\{[0, \infty)^2 \setminus [0, z_1] \times [0, z_2]\}$ for $0 < z_j < \infty$ ($j = 1, 2$), see (8.8). The joint survivor function of (Z_1, Z_2) is given by

$$P[Z_1 > z_1, Z_2 > z_2] = 1 - \exp(-1/z_1) - \exp(-1/z_2) + \exp\{-V_*(z_1, z_2)\}.$$

Assume that the Z_1 and Z_2 are not independent, that is, $\chi = 2 - V_*(1, 1) > 0$. Recalling from (8.11) that μ_* is homogenous of order -1, we obtain

$$P[Z_1 > t, Z_2 > t] \sim t^{-1}\chi, \qquad t \to \infty,$$

and hence, for $0 < z_j < \infty$ ($j = 1, 2$),

$$\lim_{t \to \infty} \frac{P[Z_1 > z_1 t, Z_2 > z_2 t]}{P[Z_1 > t, Z_2 > t]} = \chi^{-1}\{z_1^{-1} + z_2^{-1} - V_*(z_1, z_2)\}$$

$$= \chi^{-1}\mu_*\{(z_1, \infty) \times (z_2, \infty)\}. \qquad (9.117)$$

Therefore

$$P[Z_1 > z_1, Z_2 > z_2] = \mathcal{L}(z_1, z_2)(z_1 z_2)^{-1/2},$$

where the function $\mathcal{L}(z_1, z_2) = (z_1 z_2)^{1/2} P[Z_1 > z_1, Z_2 > z_2]$ is bivariate slowly varying with limit function

$$g(z_1, z_2) = \chi^{-1}(z_1 z_2)^{1/2} \mu_*\{(z_1, \infty) \times (z_2, \infty)\}.$$

The function g is homogenous of order zero, and $g(z_1, z_2) = g_*\{z_1/(z_1 + z_2)\}$ with

$$g_*(w) = \chi^{-1}\{w(1 - w)\}^{1/2} \mu_*\{(w, \infty) \times (1 - w, \infty)\}$$

$$= \frac{1 - A(w)}{\chi\{w(1 - w)\}^{1/2}}, \qquad 0 < w < 1,$$

where $A(w) = V_*\{(1 - w)^{-1}, w^{-1}\}$ is Pickands dependence function. Denoting the spectral measure of μ_* by H as in (8.28), we have by (8.48),

$$\frac{g_*(w)}{g_*(1 - w)} = \frac{1 - A(w)}{1 - A(1 - w)} \to \frac{-A'(0)}{A'(1)} = \frac{H((0, 1])}{H([0, 1))}, \qquad w \to 0,$$

confirming that $g_*(w)/g_*(1 - w)$ is slowly varying at 0 and 1, that is, $\mathcal{L}$ is quasi-symmetric and $c_1 = c_2 = 1/2$. From (9.117), we obtain that the exponent measure Λ is $\Lambda = \chi^{-1}\mu_*$ with spectral measure $H_\Lambda = \chi^{-1}H$, indeed satisfying (9.107). In the special case of complete dependence, H_Λ degenerates to a point mass of size two at $w = 1/2$.

Point processes

Convergence of point processes as in (8.98) under the domain-of-attraction condition can be formulated in the joint tail model (9.99) too. Let (Z_{n1}, Z_{n2}), $n = 1, 2, \ldots$ be an independent sequence of random pairs distributed as (Z_1, Z_2) in (9.99). Define the sequence of point processes

$$N_n(\cdot) = \sum_{i=1}^{n} \mathbf{1}\{t_n^{-1}(Z_{i1}, Z_{i2}) \in \cdot\}.$$

Rather than normalizing by n^{-1} as in (8.98), we normalize here by a sequence $(t_n)_n$ of positive numbers such that $P[Z_1 > t_n, Z_2 > t_n] \sim 1/n$ as $n \to \infty$. Since the function $0 < t \mapsto P[Z_1 > t, Z_2 > t] = \mathcal{L}(t, t)t^{-1/\eta}$ is regularly varying at infinity with index $-1/\eta$, we must have $t_n = n^\eta \mathcal{L}^\sharp(n)$ for some slowly varying function $\mathcal{L}^\sharp$ (Bingham *et al.* 1987). In particular, if $\eta < 1$, then $t_n = o(n)$ as $n \to \infty$.

Since, by (9.104),

$$nP[t_n^{-1}(Z_1, Z_2) \in \cdot] \xrightarrow{v} \Lambda(\cdot), \qquad n \to \infty$$

in $(0, \infty]^2$, Proposition 3.21 of Resnick (1987) implies that

$$N_n \xrightarrow{\mathcal{D}} N, \qquad n \to \infty, \tag{9.118}$$

where N is a non-homogenous Poisson process on $(0, \infty]^2$ with intensity measure Λ. Note again that we excluded the coordinate axes from the state space. The reason is that if $\eta < 1$, the normalization by t_n is too weak and can only control the (Z_{i1}, Z_{i2}) for which both coordinates are large (recall that the maximum of n independent standard Fréchet variables is of order n). Therefore, the number of points in N_n close to the axes will converge to infinity. Normalizing, on the other hand, by n rather than by t_n would indeed control the points near the axes, but since the limiting measure in case of asymptotic independence is concentrated on the axes, there would remain in the limit no points in the interior.

By (9.118), for $0 < z_j < \infty$ $(j = 1, 2)$,

$$P\left[\forall i = 1, \ldots, n : Z_{i1} \leq t_n z_1 \text{ or } Z_{i2} \leq t_n z_2\right]$$

$$= P[N_n\{(z_1, \infty) \times (z_2, \infty)\} = 0]$$

$$\to \exp[-\Lambda\{(z_1, \infty) \times (z_2, \infty)\}] = \exp\left\{-g_*\left(\frac{z_1}{z_1 + z_2}\right)z_1^{-c_1}z_2^{-c_2}\right\}$$

as $n \to \infty$. This relation can also be obtained directly from (9.99). More interestingly, we can find the limit distribution of the component-wise maximum of the sub-sample consisting of those pairs (Z_{i1}, Z_{i2}), $i = 1, \ldots, n$, that fall in the region

$(t_n, \infty)^2$: for $1 < z_j < \infty$ $(j = 1, 2)$,

$$P[\max\{(Z_{i1}, Z_{i2}) : i = 1, \ldots, n \text{ with } (Z_{i1}, Z_{i2}) > (t_n, t_n)\} \leq (t_n z_1, t_n z_2)]$$
$$= P\big[N_n\big(\{(1, \infty) \times (1, \infty)\} \setminus \{(1, z_1] \times (1, z_2]\}\big) = 0\big]$$
$$\rightarrow \exp\big[-\Lambda\big(\{(1, \infty) \times (1, \infty)\} \setminus \{(1, z_1] \times (1, z_2]\}\big)\big]$$
$$= \exp\{-g(z_1, 1)z_1^{-c_1} - g(1, z_2)z_2^{-c_2} + g(z_1, z_2)z_1^{-c_1} z_2^{-c_2}\}$$

as $n \rightarrow \infty$.

Statistical inference

The joint tail model (9.99) may be used for statistical inference on a bivariate distribution in that region of its support where both components are large. As before, the analysis splits into inference on the margins and inference on the joint dependence structure. Estimates $\hat{F}_j$ $(j = 1, 2)$ of the marginal distributions F_j are used to transform the original data (X_{i1}, X_{i2}) to approximate standard Fréchet margins by $\hat{Z}_{ij} = -1/\log \hat{F}_j(X_{ij})$, and these transformed data are then assumed to follow the joint tail model (9.99). The margins may be estimated non-parametrically or semi-parametrically as explained in section 9.4.1. Alternatively, under a parametric specification of (9.99), marginal and dependence parameters may be estimated jointly by maximum likelihood. In the text, we assume for simplicity that the margins are known, so that we dispose of independent, identically distributed random pairs (Z_{i1}, Z_{i2}) following the joint tail model (9.99).

We will not apply the methods to the Loss-ALAE data, since in section 9.5.2 we found insufficient proof for asymptotic independence. However, multivariate extreme value methods will come into play in section 10.4.6 again when we analyse the extremes of certain Markov processes. There we will illustrate some parametric techniques for asymptotic independence with suitable adaptations to Markov processes as in Bortot and Tawn (1998).

Non-parametric inference. Combining (9.106) and (9.104), we find that the distribution of (Z_1, Z_2) satisfies the following scaling relation: For a Borel set B in $(t, \infty)^2$ with t large and for $0 < s < \infty$, we have by successive applications of (9.105),

$$P[(Z_1, Z_2) \in B] \approx P[Z_1 > t, Z_2 > t]\Lambda(t^{-1}B)$$
$$= P[Z_1 > t, Z_2 > t]s^{1/\eta}\Lambda(st^{-1}B)$$
$$\approx s^{1/\eta}P[(Z_1, Z_2) \in sB]. \tag{9.119}$$

[For the approximations to work, B needs to be a continuity set of Λ, that is, $\Lambda(\partial B) = 0$, with ∂B the topological boundary of B.] Hence, if the set B does not contain any or only very few of the observations (Z_{i1}, Z_{i2}), $i = 1, \ldots, n$, we can still estimate the probability $p = P[(Z_1, Z_2) \in B]$ by

$$\hat{p} = s^{1/\hat{\eta}} \frac{1}{n} \sum_{i=1}^{n} \mathbf{1}\{(Z_{i1}, Z_{i2}) \in sB\}.$$

Here $\hat{\eta}$ is an estimator of the coefficient of tail dependence as in section 9.5.2, and $0 < s < 1$ is a scaling factor to be chosen such as to meet the two conflicting criteria of sufficiently scaling down the failure set B (small s) and keeping the approximation (9.119) sufficiently accurate (large s). The asymptotic properties of $\hat{p}$ are discussed in Draisma *et al.* (2004).

Alternatively, we may estimate the tail of (Z_1, Z_2) by first estimating the exponent measure Λ and secondly using the approximation (9.105). A naive estimator for the exponent measure Λ arises from replacing probabilities by empirical counts in that same approximation,

$$\tilde{\Lambda}(\cdot) = \frac{\sum_{i=1}^{n} \mathbf{1}\{t^{-1}(Z_{i1}, Z_{i2}) \in \cdot\}}{\sum_{i=1}^{n} \mathbf{1}\{\min(Z_{i1}, Z_{i2}) > t\}}. \tag{9.120}$$

Here, t acts as a threshold, the choice of which should strike a balance between a close approximation in (9.104) and a sufficient number of observations in the region $(t, \infty)^2$. The estimator $\tilde{\Lambda}(\cdot)$ does not satisfy the homogeneity property (9.106) and is therefore not directly suited to approximate the tail of (Z_1, Z_2) through (9.105). However, we can turn $\tilde{\Lambda}$ into an estimator for the spectral measure,

$$\hat{H}_{\Lambda}(\cdot) = \tilde{\Lambda}\left(\left\{(z_1, z_2) \in (0, \infty)^2 : z_1 + z_2 > 1, z_1/(z_1 + z_2) \in \cdot\right\}\right)$$

$$= \frac{\sum_{i=1}^{n} \mathbf{1}\{Z_{i1} + Z_{i2} > t, Z_{i1}/(Z_{i1} + Z_{i2}) \in \cdot\}}{\sum_{i=1}^{n} \mathbf{1}\{\min(Z_{i1}, Z_{i2}) > t\}}$$

Observe that replacing probabilities by empirical counts in (9.114) leads to $\hat{H}_{\Lambda}$ as well. Now, combine this $\hat{H}_{\Lambda}$ with an estimate $\hat{\eta}$ of the coefficient of tail dependence to find an estimator of Λ that does satisfy the required homogeneity property:

$$\hat{\Lambda}\{(z_1, \infty) \times (z_2, \infty)\}$$

$$= \int_{(0,1)} \left\{\min\left(\frac{w}{z_1}, \frac{1-w}{z_2}\right)\right\}^{1/\hat{\eta}} \hat{H}_{\Lambda}(dw)$$

$$= \frac{\sum_{i=1}^{n} \mathbf{1}(Z_{i1} + Z_{i2} > t)(Z_{i1} + Z_{i2})^{-1/\hat{\eta}}\{\min(Z_{i1}/z_1, Z_{i2}/z_2)\}^{1/\hat{\eta}}}{\sum_{i=1}^{n} \mathbf{1}\{\min(Z_{i1}, Z_{i2}) > t\}}.$$

The finite-sample or asymptotic properties of this estimator remain to be investigated.

Parametric inference.   Another possibility to perform statistical inference on the joint tail model (9.99) is within a parametric sub-model, analytically tractable but still sufficiently flexible. As for models based on multivariate extreme value distributions, inference can be done by the censored-likelihood approach, see section 9.4.2. The case for asymptotic dependence or asymptotic independence is not always clear-cut, however, making it useful to quantify the uncertainty in a Bayesian set-up through the posterior distribution on the parameters of a model that allows for both asymptotic dependence and independence (Coles and Pauli 2002).

It is not completely trivial to construct useful parametric models for the joint tail model (9.99). The modelling strategy adopted in Ledford and Tawn (1997) and Bruun and Tawn (1998) is to take a specification of the form $\mathcal{L}(z_1, z_2) = \mathcal{L}_*\{z_1/(z_1 + z_2)\}$, where $\mathcal{L}_*(w) = Kg_*(w)$ for a positive constant K and a quasi-symmetric ray dependence function g_*, see section 10.4.6. As a model selection diagnostic, one can first estimate Λ by the non-parametric estimator $\tilde{\Lambda}$ given in (9.120) and then identify g_* and (c_1, c_2) by evaluating (9.103) at $(w, 1 - w)$ and $(w - 1, w)$ and assuming that g_* is symmetric around $1/2$. It is not always obvious that a certain parametric form for g_* leads to a valid distribution; in particular, the spectral density h_Λ in (9.113) should be checked to be non-negative over its whole range. A more natural approach, therefore, is to specify a parametric form for the spectral density h_Λ itself (Ramos 2003).

9.6   Additional Topics

Only some components are extreme

Until now, we have only considered the tail function $1 - F(x)$ for x such that all marginal tail probabilities $1 - F_j(x_j)$ are small. In practice, however, we might want to perform estimation and extrapolation in a region of the support of the distribution where some, but not all, components are large. This, however, falls outside the scope of both the traditional approach based on extreme value distributions and the more recent approach of asymptotic independence in section 9.5. All in all, there seems to be a huge gap in the theory and practice of multivariate extremes in dire need of being filled in.

This need was already recognized in Maulik et al. (2002) in the analysis of certain internet traffic data. The size of a transmitted file is equal to the product of the transmission rate and the transmission time. The distributions of both the transmission rate and the transmission time are heavy-tailed, the first one having the heavier tail, and their joint distribution is asymptotically independent. However, this information is insufficient to characterize the tail of the distribution of the file length. To tackle the problem, the authors develop a more refined model, implicitly assuming a limit distribution for one variable given that the other one is large.

Heffernan and Tawn (2004) develop a comparable approach in a general d-variate setting. In Gumbel coordinates, they assume that conditionally on one variable being extreme, the distribution of the $d - 1$ other variables, properly centred and scaled, converges to a limit. Inductively proceeding from a number of analytical examples, they propose a parametric model for the normalizing constants. In this way, they end up with a multivariate semi-parametric regression model, which they then apply to five-dimensional air quality monitoring data recorded at the city centre of Leeds, UK.

Spatial extremes

In most studies of multivariate extremes, the number of variables is small, often just two. Many environmental phenomena, however, have a spatial dimension, and the aim is then to model the spatial dependence within extreme events in continuous space based on observations recorded at a grid. The basic modelling tool is formed by so-called max-stable processes (de Haan 1984), stochastic processes of which all finite-dimensional distributions are multivariate extreme value distributions.

In the context of coastal flood prevention, Coles and Tawn (1990) consider sea-level annual maxima along the British coast, assuming a bivariate logistic model for neighbouring sites. Coles (1993) constructs models for the spatial dependence of daily rainfall amounts recorded at 11 sites in the south-west of England; see also Coles (1994). The same data are considered again in Coles and Tawn (1996a), who extend the analysis to the aggregated rainfall over the whole region, and in Schlather and Tawn (2003), who construct non-parametric estimators for the extremal dependence between sites as a function of the inter-site distance. The issue of asymptotic dependence versus asymptotic independence in spatial processes is explored in Ancona-Navarrete and Tawn (2002).

A new impetus is the work by de Haan and Lin (2001). They develop an extension of the classical multivariate extreme value theory as developed in Chapter 8 to component-wise maxima of independent, identically distributed stochastic processes of a continuous variable. Within the same framework, Einmahl and Lin (2003) treat the simultaneous estimation of the tails of the marginal distributions.

9.7   Summary

Analysing multivariate extremes involves a number of choices: parametric models or not, block maxima or multivariate-threshold exceedances, asymptotic dependence or asymptotic independence, just to mention the most important ones. Unfortunately, the current state of the art does not seem developed far enough to provide the user with a fully automatic, universally applicable methodology. Rather than that, intelligent judgement of the user is, and probably will always be, necessary. To assist the reader in making wise decisions, we provide here an overview of all the methods, together with their drawbacks and benefits. We also sketch some avenues for further research.

Statistical inference on the class of multivariate extreme value distributions is hampered by the lack of a finite-dimensional parametrization for the dependence structure. A natural option then is to construct parametric sub-families that are, on the one hand, sufficiently flexible to satisfactorily approximate any given member from the general class and, on the other hand, still analytically and computationally tractable. The benefits of parametric modelling are easier statistical inference through likelihood machinery, joint estimation of marginal and dependence parameters, the possibility to tackle complex high-dimensional problems through careful model building and natural ways to include covariate information. Main drawback of course is the inherent risk of model mis-specification.

Like in the univariate case, historically, the first and conceptually the simplest multivariate extreme value method is based on block maxima. Observations are partitioned into blocks, each block is reduced to the vector of componentwise maxima, and the collection of block maxima is modelled as an independent sample from a common multivariate extreme value distribution. In the bivariate case, there is a range of direct non-parametric estimators for the Pickands dependence function to choose from. The efficiency of these estimators is still an open issue. The alternative consists of postulating a parametric model for the spectral measure and estimating the parameters by maximum likelihood, possibly jointly with the marginal parameters. A common critique to block maxima methods, univariate or multivariate, is that they throw away many relevant observations.

More efficient are so-called *thresholds methods*, as these employ all observations for which at least one coordinate exceeds a corresponding high threshold. The aim is now to estimate a distribution in a region where there are almost no observations. Starting point is the approximate relation

$$F(x) \approx \exp[-l\{-\log F_1(x_1), \dots, -\log F_d(x_d)\}] \tag{9.121}$$

where F is a distribution function in the domain of attraction of an extreme value distribution with stable tail dependence function l, and where x is such that $1 - F_j(x_j)$ is small for all $j = 1, \dots, d$. The task therefore can be torn apart into estimating the marginal tails and estimating the stable tail dependence function.

Non-parametric methods are essentially based on the *tail empirical dependence function*, which arises if we take the empirical version of the related approximation $F(x) \approx 1 - l\{1 - F_1(x_1), \dots, 1 - F_d(x_d)\}$. The tail empirical dependence function forms the starting point for fully non-parametric estimators for the spectral measure and the Pickands dependence function. We conjecture that existing non-parametric methods can still be improved if they take as a starting point the more accurate approximation (9.121).

Alternatively, the above approximations can be turned into fully parametric models for F by assuming a parametric model for l and by modelling the marginal tails by GP or GEV distributions. Marginal and dependence parameters can now be estimated jointly by maximum likelihood, benefits being transfer of information between coordinates, correct assessment of the global estimation uncertainty, possibility to exploit common features in the different marginal tails, and natural extensions to include covariate information. Some care needs to be taken, however, in the construction of the likelihood, as the model only specifies the form of F in a certain region of its support. Two possible ways to deal with this are the so-called *point-process* method and the *censored-likelihood* method. We prefer the latter as it yields the more accurate estimates of dependence parameters in case the dependence structure is close to asymptotic independence.

Methods derived from multivariate extreme value distributions only allow for models in which the components at extreme levels are either exactly independent or asymptotically dependent in the sense that joint extremes occur with a probability of the same order of magnitude as a single extreme. This is unsatisfactory for asymptotically independent distributions for which the components are still positively or negatively associated at extreme levels, as quantified by the coefficient of tail dependence. The deficiency is overcome by a model for the bivariate survivor function that bridges the gap between exact independence and asymptotic independence. The merits of the few available parametric and non-parametric inference techniques remain to be assessed.

A common feature of all models for multivariate extremes is that they describe the distribution only in that part of its support where all coordinates are extreme. The case where only some coordinates are extreme is relatively unexplored.

10

EXTREMES OF STATIONARY TIME SERIES

co-authored by Chris Ferro

10.1 Introduction

The extremes of time series can be very different to those of independent sequences. Serial dependence affects not only the magnitude of extremes but also their qualitative behaviour. This necessitates both a modification of standard methods for analysing extremes and a development of additional tools for describing these new features. In this chapter, we present mathematical characterizations for the extremes of stationary processes and statistical methods for their estimation.

The effect of serial dependence on extremes can be illustrated with a simple example. The moving-maximum process (Deheuvels 1983) is defined by

$$X_i = \max_{j \geq 0} \alpha_j Z_{i-j}, \qquad i \in \mathbb{Z}, \tag{10.1}$$

where the coefficients $\alpha_j \geq 0$ satisfy $\sum_{j \geq 0} \alpha_j = 1$ and the Z_i are independent, standard Fréchet random variables, that is, $P[Z \leq x] = \exp(-1/x)$ for $0 < x < \infty$; the marginal distribution of $\{X_i\}_{i \geq 1}$ is also standard Fréchet. A partial realization of the process when $\alpha_0 = \alpha_1 = 1/2$ (Newell 1964) is reproduced in Figure 10.1. The serial dependence causes large values to occur in pairs. This affects the distribution of order statistics: for example, the two largest order statistics have the same asymptotic distribution. More general clustering is possible with other choices for the coefficients. The presence of clusters of extremes is a phenomenon that is not experienced for independent sequences.

Statistics of Extremes: Theory and Applications   J. Beirlant, Y. Goegebeur, J. Segers, and J. Teugels
© 2004 John Wiley & Sons, Ltd   ISBN: 0-471-97647-4

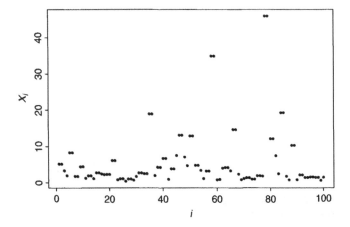

Figure 10.1 A partial realization of the moving-maximum process $X_i = \max(Z_i, Z_{i-1})/2$.

Extreme events in the physical world are often synonymous with clusters of large values: for example, a flood might be caused by several days with heavy rainfall. A single extreme event such as a flood can impact the environment, man-made structures and public health and generate a spate of insurance claims. It is therefore of great interest to know the rate at which such events can be expected to occur and what they might look like when they do.

There are two approaches to analysing the extremes of time series. One is to choose a time-series model for the complete process, fit it to the data and then determine its extremal behaviour either analytically or by simulation. This topic has been well treated elsewhere, by Embrechts *et al.* (1997) for instance, and we shall touch on it only briefly in section 10.6. The second approach is to choose a model for the process at extreme levels only and fit it to the extremes in the data. This alternative is attractive because, as we have seen elsewhere in this book, models for extremes can be derived under very weak conditions on the process. It is on this approach that we shall concentrate.

We begin in section 10.2 by considering the sample maximum, which can be modelled, as for independent sequences, with the generalized extreme value (GEV) distribution. In section 10.3, we achieve a characterization for all exceedances over a high threshold, which supplies a point-process model for clusters of extremes. Models for the extremes of Markov processes are established in section 10.4. Up to this point, we shall only deal with univariate sequences, for which both theory and methods are well developed. In section 10.5, we summarize some key results for the extremes of multivariate processes. Finally, in section 10.6, we provide the reader with some key references about additional topics that, despite their importance, did not make it to the core of the chapter.

Many of the statistical methods are illustrated for a series of daily maximum temperatures recorded at Uccle, Belgium; see also section 1.3.2. The data analysis was performed using R (version 1.6.1), freely available from www.r-project.org/. The routines for performing the computations in this chapter were written by Chris Ferro and most of them are being incorporated into the R package 'evd' written by Alec Stephenson, freely available from cran.r-project.org.

10.2 The Sample Maximum

Let $X_1, X_2, \ldots$ be a (strictly) stationary sequence of random variables with marginal distribution function F. The assumption entails that for integer $h \geq 0$ and $n \geq 1$, the distribution of the random vector $(X_{h+1}, \ldots, X_{h+n})$ does not depend on h. For the maximum $M_n = \max_{i=1,\ldots,n} X_i$, we seek the limiting distribution of $(M_n - b_n)/a_n$ for some choice of normalizing constants $a_n > 0$ and b_n. In Chapter 2, it was shown that for independent random variables, the only possible non-degenerate limits are the extreme value distributions. We shall see in section 10.2.1 that this remains true for stationary sequences if long-range dependence at extreme levels is suitably restricted. However, the limit distribution need not be the same as for the maximum $\tilde{M}_n = \max_{i=1,\ldots,n} \tilde{X}_i$ of the associated, independent sequence $\{\tilde{X}_i\}$ with the same marginal distribution as $\{X_i\}$. The distinction is due to the extremal index, introduced in section 10.2.3, which measures the tendency of extreme values to occur in clusters.

10.2.1 The extremal limit theorem

For a set J of positive integers, let $M(J) = \max_{i \in J} X_i$. For convenience, also set $M(\emptyset) = -\infty$. We shall partition the integers $\{1, \ldots, n\}$ into disjoint blocks $J_j = J_{j,n}$ and show that the block maxima $M(J_j)$ are asymptotically independent. Since $M_n = \max_j M(J_j)$, it follows as in Chapter 2 that the limit distribution of $(M_n - b_n)/a_n$, if it exists, must be an extreme value distribution.

Let $(r_n)_n$ be a sequence of positive integers such that $r_n = o(n)$ as $n \to \infty$, and put $k_n = \lfloor n/r_n \rfloor$. Partition $\{1, \ldots, n\}$ into k_n blocks of size r_n,

$$J_j = J_{j,n} = \{(j-1)r_n + 1, \ldots, jr_n\}, \qquad j = 1, \ldots, k_n, \qquad (10.2)$$

and, in case $k_n r_n < n$, a remainder block, $J_{k_n+1} = \{k_n r_n + 1, \ldots, n\}$. Now define thresholds u_n increasing at a rate for which the expected number of exceedances over u_n remains bounded: $\limsup n \bar{F}(u_n) < \infty$, with of course $\bar{F} = 1 - F$. We shall see that, under an appropriate condition,

$$P[M_n \leq u_n] = \prod_{j=1}^{k_n} P[M(J_j) \leq u_n] + o(1)$$

$$= (P[M_{r_n} \leq u_n])^{k_n} + o(1), \qquad n \to \infty. \qquad (10.3)$$

This is precisely the desired representation of M_n in terms of independent random variables, M_{r_n}.

To find out when (10.3) holds, observe that

$$P[M_n \leq u_n] = P\left[\bigcap_{j=1}^{k_n+1} \{M(J_j) \leq u_n\}\right].$$

Since $P[M(J_j) > u_n] \leq r_n \bar{F}(u_n) \to 0$, the remainder block can be omitted:

$$P[M_n \leq u_n] = P\left[\bigcap_{j=1}^{k_n} \{M(J_j) \leq u_n\}\right] + o(1), \qquad n \to \infty$$

A crucial point is that the events $\{X_i > u_n\}$ are sufficiently rare for the probability of an exceedance occurring near the ends of the blocks J_j to be negligible. Let $(s_n)_n$ be a sequence of positive integers such that $s_n = o(r_n)$ as $n \to \infty$, and let $J'_j = J'_{j,n} = \{jr_n - s_n + 1, \ldots, jr_n\}$ be the sub-block of size s_n at the end of J_j. The sub-blocks are asymptotically unimportant, as

$$P\left[\bigcup_{j=1}^{k_n} \{M(J'_j) > u_n\}\right] \leq k_n s_n \bar{F}(u_n) \to 0, \qquad n \to \infty.$$

This leaves us with

$$P[M_n \leq u_n] = P\left[\bigcap_{j=1}^{k_n} \{M(J^*_j) \leq u_n\}\right] + o(1), \qquad n \to \infty,$$

where the $J^*_j = \{(j-1)r_n + 1, \ldots, jr_n - s_n\}$ are separated from one another by a distance s_n. If the events $\{M(J^*_j) \leq u_n\}$ are approximately independent then we obtain, as required,

$$P[M_n \leq u_n] = \prod_{j=1}^{k_n} P[M(J^*_j) \leq u_n] + o(1)$$

$$= \prod_{j=1}^{k_n} P[M(J_j) \leq u_n] + o(1), \qquad n \to \infty,$$

using again $k_n P[M(J'_j) > u_n] \leq k_n s_n \bar{F}(u_n) \to 0$ as $n \to \infty$.

A mixing condition known as the $D(u_n)$ condition (Leadbetter 1974) suffices for the events $\{M(J^*_j) \leq u_n\}$ to become approximately independent as n increases. Let

$$\mathcal{I}_{j,k}(u_n) = \left\{\{M(I) \leq u_n\} : I \subseteq \{j, \ldots, k\}\right\}$$

be the set of all intersections of the events $\{X_i \leq u_n\}$, $j \leq i \leq k$.

Condition 10.1 $D(u_n)$. For all $A_1 \in \mathcal{I}_{1,l}(u_n)$, $A_2 \in \mathcal{I}_{l+s,n}(u_n)$ and $1 \le l \le n - s$,

$$|P(A_1 \cap A_2) - P(A_1)P(A_2)| \le \alpha(n, s)$$

and $\alpha(n, s_n) \to 0$ as $n \to \infty$ for some positive integer sequence s_n such that $s_n = o(n)$.

The $D(u_n)$ condition says that any two events of the form $\{M(I_1) \le u_n\}$ and $\{M(I_2) \le u_n\}$ can become approximately independent as n increases when the index sets $I_i \subset \{1, \ldots, n\}$ are separated by a relatively short distance $s_n = o(n)$. Hence, the $D(u_n)$ condition limits the long-range dependence between such events.

Now if the events $A_1, \ldots, A_k \in \mathcal{I}_{1,n}(u_n)$ are such that the corresponding index sets are separated from each other by a distance s, then, by induction on k, we get

$$\left| P\left(\bigcap_{j=1}^{k} A_j \right) - \prod_{j=1}^{k} P(A_j) \right| \le k\alpha(n, s).$$

Therefore, if $s_n = o(r_n)$ and $k_n\alpha(n, s_n) \to 0$, then

$$\left| P\left[\bigcap_{j=1}^{k_n} \{M(J_j^*) \le u_n\} \right] - \prod_{j=1}^{k_n} P[M(J_j^*) \le u_n] \right| \le k_n\alpha(n, s_n) \to 0,$$

as $n \to \infty$. When $\alpha(n, s_n) \to 0$ for some $s_n = o(n)$, it is indeed possible to find $r_n = o(n)$ such that $s_n = o(r_n)$ and $k_n\alpha(n, s_n) \to 0$; take, for instance, r_n to be the integer part of $[n \max\{s_n, n\alpha(n, s_n)\}]^{1/2}$. Together, we obtain the following fundamental result.

Theorem 10.2 (Leadbetter 1974) *Let* $\{X_n\}$ *be a stationary sequence for which there exist sequences of constants* $a_n > 0$ *and* b_n *and a non-degenerate distribution function* G *such that*

$$P\left[\frac{M_n - b_n}{a_n} \le x \right] \xrightarrow{\mathcal{D}} G(x), \qquad n \to \infty.$$

If $D(u_n)$ *holds with* $u_n = a_n x + b_n$ *for each* x *such that* $G(x) > 0$, *then* G *is an extreme value distribution function.*

Note that the $D(u_n)$ condition is required to hold for all sequences $u_n = a_n x + b_n$ for which $G(x) > 0$. The necessity of this requirement is shown by the process $X_i \equiv X_1$, for which $D(u_n)$ holds as soon as $F(u_n) \to 1$ as $n \to \infty$. Nevertheless, the condition is weak as it concerns events of the form $\{X_i \le u_n\}$ only. Compare this with strong mixing (Loynes 1965), for example, which requires Condition 10.1 to hold for classes of sets $\mathcal{I}_{j,k} = \sigma(X_i : j \le i \le k)$, the σ-algebra generated by the random variables $X_j, \ldots, X_k$. For Gaussian sequences with auto-correlation ρ_n at lag n, the $D(u_n)$ condition is satisfied as soon as $\rho_n \log n \to 0$ as $n \to \infty$ (Berman

1964). This is much weaker than the geometric decay assumed by auto-regressive models, for example.

In fact, Theorem 10.2 holds true for even weaker versions of the $D(u_n)$ condition as may be evident from our discussion. One example (O'Brien 1987) is asymptotic independence of maxima (AIM), which requires Condition 10.1 to hold when

$$\mathcal{I}_{j,k}(u_n) = \left\{ \{M(I) \leq u_n\} : I = \{i_1, \ldots, i_2\} \subseteq \{j, \ldots, k\} \right\},$$

comprising block maxima over intervals of integers rather than arbitrary sets of integers. This weakening admits a class of periodic Markov chains.

Example 10.3 The max-autoregressive process of order one, or ARMAX in short, is defined by the recursion

$$X_i = \max\{\alpha X_{i-1}, (1-\alpha)Z_i\}, \qquad i \in \mathbb{Z}, \tag{10.4}$$

where $0 \leq \alpha < 1$ and where the Z_i are independent standard Fréchet random variables. A stationary solution of the recursion is

$$X_i = \max_{j \geq 0} \alpha^j (1-\alpha) Z_{i-j}, \qquad i \in \mathbb{Z},$$

showing that the ARMAX process is a special case of the moving-maximum process of the introduction; in particular, the marginal distribution of the process is standard Fréchet. Furthermore, the $D(u_n)$ condition can be shown to hold for general moving-maximum processes, so we expect the limit distribution of M_n/n to be an extreme value distribution. Indeed, for $0 < x < \infty$, we have

$$
\begin{aligned}
P[M_n \leq x] &= P[X_1 \leq x, \ (1-\alpha)Z_2 \leq x, \ \ldots, \ (1-\alpha)Z_n \leq x] \\
&= P[X_1 \leq x]\{P[(1-\alpha)Z_1 \leq x]\}^{n-1} \\
&= \exp[-\{1 + (1-\alpha)(n-1)\}/x] \tag{10.5}
\end{aligned}
$$

so that

$$P[M_n/n \leq x] \to \exp\{-(1-\alpha)/x\} =: G(x), \qquad n \to \infty.$$

Compare this with the limit distribution $\tilde{G}(x) = \exp(-1/x)$ of $\tilde{M}_n/n$. We shall discover in section 10.2.3 that the relationship $G(x) = \tilde{G}(x)^{1-\alpha}$ is no coincidence.

If Theorem 10.2 holds, then we can fit the GEV distribution to block maxima from stationary sequences. For large n, we have

$$P[M_n \leq a_n x + b_n] \approx \exp\left[-\left\{ 1 + \gamma \left(\frac{x - \mu_0}{\sigma_0} \right) \right\}_+^{-1/\gamma} \right]$$

say. Therefore

$$P[M_n \leq x] \approx \exp\left[-\left\{1 + \gamma\left(\frac{x - \mu}{\sigma}\right)\right\}_+^{-1/\gamma}\right],$$

where the parameters $\mu = a_n\mu_0 + b_n$ and $\sigma = a_n\sigma_0$ assimilate the normalizing constants. The parameters (μ, σ, γ) can be estimated by maximum likelihood, for example, as in the following section.

10.2.2 Data example

The data plotted in Figure 10.2 are daily maximum temperatures recorded in degrees Celsius at Uccle, Belgium, during the years from 1901 to 1999. All days except those in July, which is generally the warmest month, have been removed in order to make our assumption of stationarity more reasonable. These data are freely available at www.knmi.nl/samenw/eca as part of the European Climate Assessment and Data set project (Klein Tank *et al.* 2002).

We begin our analysis of these data by fitting the GEV distribution to the July maxima. The maximum-likelihood estimates of the parameters, with standard errors in brackets, are $\hat{\mu} = 30.0$ (0.3), $\hat{\sigma} = 3.0$ (0.2) and $\hat{\gamma} = -0.34$ (0.07). The diagnostic plots in Figure 10.3 indicate a systematic discrepancy due perhaps to measurement error or non-stationary meteorological conditions, but the most extreme maxima are modelled well. The estimate of the upper limit for the distribution of July maximum temperature obtained from the GEV fit is $\hat{\mu} - \hat{\sigma}/\hat{\gamma} = 38.7°C$, with profile-likelihood 95% confidence interval (37.3, 43.9). The estimated 100,

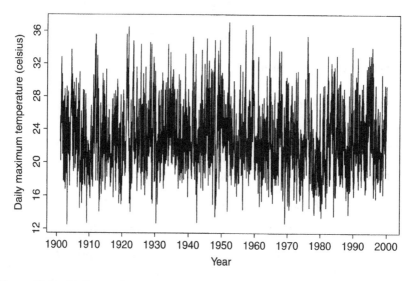

Figure 10.2 Daily maximum temperatures in July at Uccle from 1901 to 1999.

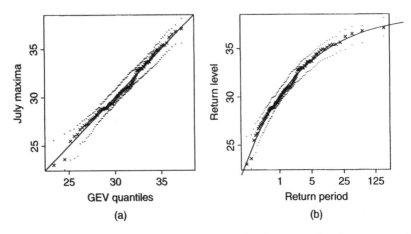

Figure 10.3 Quantile and return level plots for the generalized extreme value distribution fitted to July maximum temperatures.

1000 and 10,000 July return levels are 36.9 (36.2, 38.6), 37.9 (36.9, 40.5) and 38.3 (37.2, 41.8). We shall investigate other features of these data in later sections.

10.2.3 The extremal index

Theorem 10.2 shows that the possible limiting distributions for maxima of stationary sequences satisfying the $D(u_n)$ condition are the same as those for maxima of independent sequences. Dependence can affect the limit distribution, however, as illustrated by Example 10.3. We investigate the issue further in this section. First note that approximation (10.3) is also true for independent sequences. The effect of dependence is therefore to be found in the distribution of block maxima, M_{r_n}.

Choose thresholds u_n such that $n\bar{F}(u_n) \to \tau$ for some $0 < \tau < \infty$. For the associated, independent sequence,

$$P[\tilde{M}_n \le u_n] = \{F(u_n)\}^n = \left\{1 - \frac{1}{n}n\bar{F}(u_n)\right\}^n \to \exp(-\tau), \qquad n \to \infty.$$

For a general stationary process, however, $P[M_n \le u_n]$ need not converge and, if it does, the limit need not be $\exp(-\tau)$.

Suppose that u_n and v_n are two threshold sequences and that

$$n\bar{F}(u_n) \to \tau, \qquad P[M_n \le u_n] \to \exp(-\lambda),$$

$$n\bar{F}(v_n) \to \upsilon, \qquad P[M_n \le v_n] \to \exp(-\psi),$$

as $n \to \infty$, where $\tau, \upsilon \in (0, \infty)$ and $\lambda, \psi \in [0, \infty)$. We show that if $D(u_n)$ holds, then $\lambda/\tau = \psi/\upsilon =: \theta$. In other words, $P[M_n \le u_n] \to \exp(-\theta\tau)$ and the effect of dependence is expressed by the scalar θ, independently of τ.

Without loss of generality, assume that $\tau \geq \upsilon$ and define $n' = \lfloor (\upsilon/\tau)n \rfloor$. Clearly $n'\bar{F}(u_n) \to \upsilon$ so that

$$|P[M_{n'} \leq u_n] - P[M_{n'} \leq v_{n'}]| \leq n'|F(u_n) - F(v_{n'})| \to 0$$

and thus $P[M_{n'} \leq u_n] \to \exp(-\psi)$ as $n \to \infty$. Now suppose as in section 10.2.1 that $(r_n)_n$ and $(s_n)_n$ are positive integer sequences such that $r_n = o(n)$, $s_n = o(r_n)$, and $(n/r_n)\alpha(n, s_n) \to 0$ as $n \to \infty$. Since $n' \leq n$, we have by (10.3)

$$P[M_{n'} \leq u_n] = P[M_{r_n} \leq u_n]^{\lfloor n'/r_n \rfloor} + o(1),$$

$$P[M_n \leq u_n] = P[M_{r_n} \leq u_n]^{\lfloor n/r_n \rfloor} + o(1),$$

and thus

$$\frac{n'}{r_n} P[M_{r_n} > u_n] \to \psi, \qquad \frac{n}{r_n} P[M_{r_n} > u_n] \to \lambda,$$

as $n \to \infty$. Since $n' \sim (\upsilon/\tau)n$, we must have $\lambda/\tau = \psi/\upsilon$, as required, and

$$\theta = \frac{\lambda}{\tau} = \lim_{n \to \infty} \frac{P[M_{r_n} > u_n]}{r_n \bar{F}(u_n)}. \tag{10.6}$$

This argument is the basis for the following theorem.

Theorem 10.4 (Leadbetter 1983) *If there exist sequences of constants $a_n > 0$ and b_n and a non-degenerate distribution function $\tilde{G}$ such that*

$$P\left[\frac{\tilde{M}_n - b_n}{a_n} \leq x \right] \xrightarrow{\mathcal{D}} \tilde{G}(x), \qquad n \to \infty,$$

if $D(u_n)$ holds with $u_n = a_n x + b_n$ for each x such that $\tilde{G}(x) > 0$ and if $P[(M_n - b_n)/a_n \leq x]$ converges for some x, then

$$P\left[\frac{M_n - b_n}{a_n} \leq x \right] \xrightarrow{\mathcal{D}} G(x) := \tilde{G}^{\theta}(x), \qquad n \to \infty,$$

for some constant $\theta \in [0, 1]$.

The constant θ is called the *extremal index* and, unless it is equal to one, the limiting distributions for the independent and stationary sequences are not the same. If $\theta > 0$, then G is an extreme value distribution, but with different parameters than $\tilde{G}$. In particular, if (μ, σ, γ) are the parameters of G and $(\tilde{\mu}, \tilde{\sigma}, \tilde{\gamma})$ are the parameters of $\tilde{G}$, then their relationship is

$$\gamma = \tilde{\gamma}, \qquad \mu = \tilde{\mu} - \tilde{\sigma} \frac{1 - \theta^{\gamma}}{\gamma}, \qquad \sigma = \tilde{\sigma}\theta^{\gamma}, \tag{10.7}$$

or, if $\gamma = 0$, taking the limits $\mu = \tilde{\mu} + \sigma \log \theta$ and $\sigma = \tilde{\sigma}$. Observe that the extreme value index γ remains unaltered.

Example 10.5 The derivation in Example 10.3 shows that the extremal index of the ARMAX process is $\theta = 1 - \alpha$. More generally, for the moving-maximum process (10.1), we have

$$P(M_n \leq nx)$$

$$= P\left[\max_{j \geq 0}(\alpha_j Z_{1-j}) \leq nx, \ldots, \max_{j \geq 0}(\alpha_j Z_{n-j}) \leq nx\right]$$

$$= P\left[\max_{i \geq 0} \max_{1 \leq j \leq n} (\alpha_{i+j} Z_{-i}) \leq nx, \max_{1 \leq i \leq n} \max_{0 \leq j \leq n-i} (\alpha_j Z_i) \leq nx\right]$$

$$= \exp\left\{-\frac{1}{x}\left(\frac{1}{n}\sum_{i \geq 0}\max_{1 \leq j \leq n}\alpha_{i+j} + \frac{1}{n}\sum_{i=0}^{n-1}\max_{0 \leq j \leq i}\alpha_j\right)\right\}.$$

We treat both sums separately. The first sum can, for positive integer m, be bounded by

$$\frac{1}{n}\sum_{i \geq 0}\max_{1 \leq j \leq n}\alpha_{i+j} \leq \frac{m}{n} + \frac{1}{n}\sum_{i \geq m}\max_{1 \leq j \leq n}\alpha_{i+j} \leq \frac{m}{n} + \sum_{i \geq m+1}\alpha_i.$$

Let m tend to infinity to obtain that $n^{-1}\sum_{i \geq 0}\max_{1 \leq j \leq n}\alpha_{i+j} \to 0$ as $n \to \infty$. For the second sum, let $\alpha_{(1)} = \max_{j \geq 0}\alpha_j$. Since $\max_{0 \leq j \leq i} a_j \to a_{(1)}$ as $i \to \infty$, we have $n^{-1}\sum_{i=0}^{n-1}\max_{0 \leq j \leq i}\alpha_j \to \alpha_{(1)}$ as $n \to \infty$. Together, we obtain $\theta = \alpha_{(1)}$.

Asymptotic independence

The case $\theta = 1$ is true for independent processes, but it can be true for dependent processes too. The following condition (Leadbetter 1974) is sufficient when allied with $D(u_n)$.

Condition 10.6 $D'(u_n)$.

$$\lim_{k \to \infty} \limsup_{n \to \infty} n \sum_{j=2}^{\lfloor n/k \rfloor} P[X_1 > u_n, X_j > u_n] = 0.$$

To see the effect of $D'(u_n)$, apply the inclusion-exclusion formula to the event $\{M_{r_n} > u_n\} = \bigcup_{i=1}^{r_n}\{X_i > u_n\}$ to obtain

$$\sum_{i=1}^{r_n}\bar{F}(u_n) \geq P[M_{r_n} > u_n] \geq \sum_{i=1}^{r_n}\bar{F}(u_n) - \sum_{1 \leq i < j \leq r_n}P[X_i > u_n, X_j > u_n].$$

Therefore, $P[M_{r_n} > u_n] \sim r_n\bar{F}(u_n)$ and $\theta = 1$ by (10.6) if

$$\sum_{1 \leq i < j \leq r_n}P[X_i > u_n, X_j > u_n] = o\{r_n\bar{F}(u_n)\} = o(r_n/n).$$

as $n \to \infty$. Since the sum is not greater than $r_n \sum_{j=2}^{r_n} P(X_1 > u_n, X_j > u_n)$, this is satisfied if $D'(u_n)$ holds. In contrast to the $D(u_n)$ condition, which controls the long-range dependence, the $D'(u_n)$ condition limits the amount of short-range dependence in the process at extreme levels. In particular, it postulates that the probability of observing more than one exceedance in a block is negligible.

Example 10.7 When $\alpha = 0$, the ARMAX process (10.4) is independent and the $D'(u_n)$ condition holds. On the other hand,

$$P[X_1 > u_n, X_2 > u_n]$$

$$= 1 - P[X_1 \leq u_n] - P[X_2 \leq u_n] + P[X_1 \leq u_n, X_2 \leq u_n]$$

$$= 1 - 2\exp(-1/u_n) + P[X_1 \leq u_n, (1 - \alpha)Z_2 \leq u_n]$$

$$= 1 - 2\exp(-1/u_n) + \exp\{(\alpha - 2)/u_n\}$$

so that $nP[X_1 > u_n, X_2 > u_n] \to \alpha/x$ when $u_n = nx$ for some $0 < x < \infty$, that is, $D'(u_n)$ fails if $\alpha > 0$.

Positive extremal index

The case $\theta = 0$ is pathological, although not impossible, see Denzel and O'Brien (1975) or Leadbetter *et al.* (1983), p. 71. It entails that sample maxima M_n of the process are of smaller order than sample maxima $\tilde{M}_n$ of the associated independent sequence. Also, the expected number of exceedances in a block with at least one exceedance converges to infinity, see (10.10) below. For purposes of statistical inference, it will turn out to be convenient to assume that $0 < \theta \leq 1$. A sufficient condition is that the influence of a large value $X_1 > u_n$ reaches only finitely far over time, as in Condition 10.8 below. For integers $0 \leq j \leq k$, we denote $M_{j,k} = \max\{X_{j+1}, \ldots, X_k\}$ (with $\max \emptyset = -\infty$) and $M_k = M_{0,k}$.

Condition 10.8 *The thresholds u_n and the integers r_n are such that $F(u_n) < 1$, $\bar{F}(u_n) \to 0$, $r_n \to \infty$ and*

$$\lim_{m \to \infty} \limsup_{n \to \infty} P[M_{m,r_n} > u_n \mid X_1 > u_n] = 0. \tag{10.8}$$

For integer $m \geq 1$, by decomposing the event $\{M_{r_n} > u_n\}$ according to the time of the last exceedance,

$$P[M_{r_n} > u_n] \geq \sum_{i=1}^{\lfloor r_n/m \rfloor} P[X_{(i-1)m+1} > u_n, M_{im,r_n} \leq u_n]$$

$$\geq \lfloor r_n/m \rfloor \bar{F}(u_n) P[M_{m,r_n} \leq u_n \mid X_1 > u_n].$$

For large-enough m, therefore, Condition 10.8 guarantees that

$$\liminf_{n \to \infty} \frac{P[M_{r_n} > u_n]}{r_n \bar{F}(u_n)} \geq \liminf_{n \to \infty} \frac{1}{m} P[M_{m,r_n} \leq u_n \mid X_1 > u_n] > 0. \tag{10.9}$$

Hence, if also $r_n = o(n)$ and $n\alpha(n, s_n) = o(r_n)$ for some $s_n = o(r_n)$, then θ must indeed be positive by (10.6).

Blocks and runs

The extremal index has several interpretations. For example, $\theta = \lim \theta_n^B(u_n)$, where

$$\frac{1}{\theta_n^B(u_n)} = \frac{r_n \bar{F}(u_n)}{P[M_{r_n} > u_n]} = E\left[\sum_{i=1}^{r_n} \mathbf{1}(X_i > u_n) \,\middle|\, M_{r_n} > u_n\right] \qquad (10.10)$$

is the expected number of exceedances over u_n in a block containing at least one such exceedance. Therefore, the extremal index is the reciprocal of the limiting mean number of exceedances in blocks with at least one exceedance.

Another interpretation of the extremal index is due to O'Brien (1987). Assume again Condition 10.8 and let the integers $1 \le s_n \le r_n$ be such that $s_n \to \infty$ and $s_n = o(r_n)$ as $n \to \infty$; for instance, take s_n the integer part of $r_n^{1/2}$. On the one hand, we have

$$P[M_{r_n} > u_n] = \sum_{i=1}^{r_n} P[X_i > u_n, M_{i,r_n} \le u_n]$$

$$\ge r_n \bar{F}(u_n) P[M_{1,r_n} \le u_n \mid X_1 > u_n],$$

and on the other hand,

$$P[M_{r_n} > u_n] \le s_n \bar{F}(u_n) + (r_n - s_n)\bar{F}(u_n) P[M_{1,s_n} \le u_n \mid X_1 > u_n].$$

Moreover by (10.8)

$$0 \le P[M_{1,s_n} \le u_n \mid X_1 > u_n] - P[M_{1,r_n} \le u_n \mid X_1 > u_n]$$

$$\le P[M_{s_n,r_n} > u_n \mid X_1 > u_n] \to 0.$$

Writing

$$\theta_n^R(u_n) = P[M_{1,r_n} \le u_n \mid X_1 > u_n] \qquad (10.11)$$

we see that the upper and lower bounds on $P[M_{r_n} > u_n]$ give

$$\theta_n^R(u_n) = \theta_n^B(u_n) + o(1).$$

Therefore, $\theta = \lim \theta_n^R(u_n)$ represents the limiting probability that an exceedance is followed by a run of observations below the threshold. Both interpretations identify $\theta = 1$ with exceedances occurring singly in the limit, while $\theta < 1$ implies that exceedances tend to occur in clusters. Yet another interpretation of the extremal index, in terms of the times between exceedances over a high threshold, is given in section 10.3.4.

Example 10.9 For the ARMAX process of Example 10.3, we can derive the extremal index $\theta = 1 - \alpha$ by combining (10.5) with the block (10.10) or run (10.11) definitions, where $u_n = nx$ for some $0 < x < \infty$ and r_n is such that $r_n \to \infty$ but $r_n = o(n)$. Regarding the run definition (10.11), observe that by stationarity,

$$\theta_n^R(u_n) = \{P[M_{r_n-1} \leq u_n] - P[M_{r_n} \leq u_n]\}/\bar{F}(u_n).$$

Statistical relevance

Theorem 10.4 shows how the extremal index characterizes the change in the distribution of sample maxima due to dependence in the sequence. Suppose $0 < \theta \leq 1$. If $\tilde{G}^\leftarrow(p)$ is the quantile function for the limit $\tilde{G}$, then the quantile function for G is $G^\leftarrow(p) = \tilde{G}^\leftarrow(p^{1/\theta}) \leq \tilde{G}^\leftarrow(p)$. This inequality has implications for the estimation of quantiles from dependent sequences.

Suppose that we estimate the parameters (μ, σ, γ) of G by fitting, for example, an extreme value distribution to a sample of block maxima M_n. As before, the normalizing constants are assimilated into the location and scale parameters so that $P[M_n \leq x] \approx \{F(x)\}^{n\theta} \approx G(x)$, the latter being a GEV distribution with parameters (γ, μ, σ). We can exploit this relationship as in section 5.1.3 to approximate marginal quantiles by

$$F^\leftarrow(1 - p) \approx G^\leftarrow\{(1 - p)^{n\theta}\}$$

$$= \begin{cases} \mu + \sigma \dfrac{\{-n\theta \log(1 - p)\}^{-\gamma} - 1}{\gamma}, & \text{if } \gamma \neq 0, \\ \mu - \sigma \log\{-n\theta \log(1 - p)\}, & \text{if } \gamma = 0. \end{cases}$$

If we neglect the extremal index, then we risk underestimating the marginal quantiles. Conversely, suppose that we have an estimate of the tail of the marginal distribution F. Then the mn-observation return level is approximated by

$$G^\leftarrow(1 - 1/m) \approx F^\leftarrow\{(1 - 1/m)^{1/(n\theta)}\}.$$

If we neglect θ here, then we risk overestimating the return level. These two examples show why it is important to be able to estimate the extremal index. We discuss this problem in section 10.3.4, where the different interpretations that we have already seen for θ will motivate different estimators.

Finally, note that the frequency at which a process is sampled has consequences for the distribution of maxima. For example, let M_n' be the maximum from the sequence sampled every $m \geq 2$ time steps, with corresponding extremal index θ_m. Then

$$P[M_n \leq x] \approx \{F(x)\}^{n\theta} \approx \{P[M_n' \leq x]\}^{m\theta/\theta_m}.$$

Robinson and Tawn (2000) develop methods based on this approximation that enable inference for the distribution of M_n from data collected at the frequency of M_n'.

10.3   Point-Process Models

In this section, we broaden our outlook from the sample maximum to encompass all large values in the sequence, where 'large' means exceeding a high threshold. A particularly elegant and useful description of threshold exceedances is in terms of point processes. We shall see that these models are related to the distribution of large order statistics, describe the clustering of extremes and motivate statistical methods for the analysis of stationary processes at extreme levels. A brief and informal introduction to point processes is given in section 5.9.2; more detailed introductions focusing on relevant aspects for extreme value theory may be found in the appendix of Leadbetter *et al.* (1983), in Chapter 3 of Resnick (1987) and in Chapter 5 of Embrechts *et al.* (1997).

10.3.1   Clusters of extreme values

Let us seek the limit of the point process

$$N_n(\cdot) = \sum_{i \in \mathcal{I}} \mathbf{1}(i/n \in \cdot), \qquad \mathcal{I} = \{i : X_i > u_n, 1 \le i \le n\}, \tag{10.12}$$

which counts the times, normalized by n, at which the sample $\{X_i\}_{i=1}^n$ exceeds a threshold u_n. This process is related to order statistics by the relationship

$$P[X_{n-k,n} \le u_n] = P[N_n((0, 1]) \le k]. \tag{10.13}$$

If we can find the limit process of N_n, then we shall be able to derive the limiting distribution of the large order statistics.

Let the thresholds u_n be such that the expected number of exceedances remains finite, with $n\bar{F}(u_n) \to \tau \in (0, \infty)$, and reconsider the partition (10.2) of $\{1, \ldots, n\}$ into $k_n = \lfloor n/r_n \rfloor$ blocks J_j of length $r_n = o(n)$. The exceedances in a block are said to form a cluster. Now, because of the time normalization in N_n, the length of a block, r_n/n, converges to zero as $n \to \infty$, so that points in N_n making up a cluster converge to a single point in $(0, 1]$. In the limit, therefore, the points in N_n represent the positions of clusters and form a marked point process with marks equal to the number of exceedances in the cluster.

The distribution of the cluster size in N_n is given by

$$\pi_n(j) = P\left[\sum_{i=1}^{r_n} \mathbf{1}(X_i > u_n) = j \;\middle|\; M_{r_n} > u_n\right], \qquad j = 1, 2, \ldots, \tag{10.14}$$

and the mark distribution of the limit process, if it exists, will be $\pi = \lim \pi_n$. Recall that the events $\{M(J_j) \le u_n\}$ are approximately independent under $D(u_n)$, Condition 10.1. If we can say the same for the random variables $\mathbf{1}\{M(J_j) \le u_n\}$, then the number of clusters occurring in N_n during an interval $I \subseteq (0, 1]$ of length $|I|$ is approximately binomial, with probability $p_n = P[M_{r_n} > u_n]$ and mean $p_n k_n |I|$. If the process also has extremal index $\theta > 0$, then by (10.6), $p_n \sim \theta r_n \bar{F}(u_n) \to 0$

and $p_n k_n \to \theta\tau > 0$ as $n \to \infty$. Therefore, the number of clusters in I approaches a Poisson random variable with mean $\theta\tau|I|$. We might expect clusters to form a Poisson process with rate $\theta\tau$ and N_n to converge to a compound Poisson process $CP(\theta\tau, \pi)$.

Convergence in distribution of N_n to a $CP(\theta\tau, \pi)$ process N is equivalent to convergence in distribution, for all integer $m \geq 1$ and disjoint intervals $I_1, \ldots, I_m \subset (0, 1]$, of the random vector $(N_n(I_1), \ldots, N_n(I_m))$ to $(N(I_1), \ldots, N(I_m))$. A convenient way to check the latter is by proving convergence of Laplace transforms, that is, by showing that

$$\mathcal{L}_n(t_1, \ldots, t_m) = E\left[\exp\left\{-\sum_{i=1}^m t_i N_n(I_i)\right\}\right] \qquad (10.15)$$

converges for all $0 \leq t_i < \infty$ $(i = 1, \ldots, m)$ to

$$\mathcal{L}(t_1, \ldots, t_m) = \prod_{i=1}^m \exp\left[-\theta\tau|I_i|\left\{1 - \sum_{j\geq 1}\pi(j)e^{-jt_i}\right\}\right]. \qquad (10.16)$$

The limiting factorization of $\mathcal{L}_n$ is achieved in much the same way as the factorization (10.3) of $P[M_n \leq u_n]$ except that a mixing condition stronger than $D(u_n)$ is required (Hsing *et al.* 1988). Let $\mathcal{F}_{j,k}(u_n) = \sigma(\{X_i > u_n\} : j \leq i \leq k)$.

Condition 10.10 $\Delta(u_n)$. *For all* $A_1 \in \mathcal{F}_{1,l}(u_n)$, $A_2 \in \mathcal{F}_{l+s,n}(u_n)$ *and* $1 \leq l \leq n - s$,

$$|P(A_1 \cap A_2) - P(A_1)P(A_2)| \leq \alpha(n, s)$$

and $\alpha(n, s_n) \to 0$ *as* $n \to \infty$ *for some* $s_n = o(n)$.

The $\Delta(u_n)$ condition is more stringent than the $D(u_n)$ condition only in the number of events for which the long-range independence is required to hold; it is still weaker than strong mixing, for example. Lemma 2.1 of Hsing *et al.* (1988) tells us that we also have, for all $1 \leq l \leq n - s$, $\sup |E(B_1 B_2) - E(B_1)E(B_2)| \leq 4\alpha(n, s)$, where the supremum is over all random variables $0 \leq B_1 \leq 1$ measurable with respect to $\mathcal{F}_{1,l}(u_n)$ and $0 \leq B_2 \leq 1$ measurable with respect to $\mathcal{F}_{l+s,n}(u_n)$. This is precisely what we need to handle the Laplace transform (10.15).

Fix an interval $I \subseteq (0, 1]$ with positive length $|I|$. Let $(r_n)_n$ be a sequence of positive numbers such that $r_n/n \to 0$ as $n \to \infty$. Consider the partitioning $I = \bigcup_{i=1}^{m_n+1} J_i$ of I into disjoint, contiguous intervals J_i with lengths $|J_i| = r_n/n$ for $i = 1, \ldots, m_n$ and $|J_{m_n+1}| < r_n/n$. In particular, $m_n \sim (n/r_n)|I|$. Now, assume there exists a sequence $(s_n)_n$ of positive numbers such that $s_n = o(r_n)$ and $n\alpha(n, s_n) = o(r_n)$ as $n \to \infty$. Repeating the block-clipping technique that led to Theorem 10.2 yields

$$E\exp\{-tN_n(I)\} = [E\exp\{-tN_n(J_1)\}]^{(n/r_n)|I|} + o(1), \qquad n \to \infty.$$

Repeating a similar procedure for the Laplace transform (10.15), we obtain

$$\mathcal{L}_n(t_1, \ldots, t_m) = \prod_{i=1}^{m} [E \exp\{-t_i N_n(J_1)\}]^{(n/r_n)|I_i|} + o(1), \qquad n \to \infty.$$

It remains to check that each term in the product converges to the corresponding factor in the Laplace transform (10.16). If $\pi_n(j) \to \pi(j)$ for each integer $j \geq 1$, then the desired convergence is a consequence of

$$E \exp\{-t N_n(J_1)\} = P[M_{r_n} \leq u_n] + \sum_{j \geq 1} \pi_n(j) P[M_{r_n} > u_n] e^{-jt}$$

$$= 1 - (r_n/n)\theta\tau \left\{ 1 - \sum_{j \geq 1} \pi(j) e^{-jt} + o(1) \right\}.$$

Theorem 10.11 (Hsing et al. 1988) *Let $\{X_i\}$ be stationary with extremal index $\theta > 0$. Let there exist a sequence of thresholds u_n for which $\Delta(u_n)$ holds and $n\bar{F}(u_n) \to \tau \in (0, \infty)$. Let there exist positive sequences s_n and r_n and a distribution π such that $s_n = o(r_n)$, $r_n = o(n)$, $n\alpha(n, s_n) = o(r_n)$ and $\pi_n(j) \to \pi(j)$ for all integer $j \geq 1$ as $n \to \infty$. Then $N_n \overset{D}{\to} N$, where N is $CP(\theta\tau, \pi)$.*

A similar result was also obtained by Rootzén (1988). The rate of convergence for N_n and other point processes presented in this section has been investigated by Barbour et al. (2002) and Novak (2003) among others, where bounds are given for metrics such as the total variation distance.

Theorem 10.11 tells us that $\theta\tau$ clusters occur in $(0, 1]$ on average and that the cluster sizes are independent with distribution π. Since the expected number of exceedances in $(0, 1]$ is τ, this means that the average cluster size should be $1/\theta$. This was noted by Leadbetter (1983) and follows from our definition (10.10) of $\theta_n^B(u_n)$ since

$$\theta^{-1} = \lim_{n \to \infty} E \left[\sum_{i=1}^{r_n} \mathbf{1}(X_i > u_n) \, \middle| \, M_{r_n} > u_n \right] = \lim_{n \to \infty} \sum_{j \geq 1} j\pi_n(j). \tag{10.17}$$

By Fatou's lemma, we have $\theta^{-1} \geq \sum_{j \geq 1} j\pi(j)$, the mean of the limiting cluster size distribution. Smith (1988) shows by counterexample that not necessarily $\theta^{-1} = \sum_{j \geq 1} j\pi(j)$, although Hsing et al. (1988) give mild extra assumptions under which this is actually true. Note also that $\pi(1) = 1$ if $\theta = 1$.

Example 10.12 The cluster-size distribution of the ARMAX process (10.4) may be found intuitively as follows. Let $X_i > u_n$ be the first exceedance in a block. Subsequent values in the sequence will be $\alpha X_i, \alpha^2 X_i, \ldots$ with high probability, and the probability of observing another such run in the same block is negligible. With high probability, the number of exceedances in a block will therefore be j

provided $\alpha^j X_i \leq u_n < \alpha^{j-1} X_i$. Hence

$$\pi_n(j) = P\left[\alpha^j X_1 \leq u_n < \alpha^{j-1} X_1 \mid X_1 > u_n\right] + o(1)$$

$$= \frac{\exp\left(-\alpha^j/u_n\right) - \exp\left(-\alpha^{j-1}/u_n\right)}{1 - \exp(-1/u_n)} + o(1)$$

$$\to (1 - \alpha)\alpha^{j-1}, \qquad n \to \infty,$$

that is, the limiting cluster-size distribution is geometric with mean $(1 - \alpha)^{-1} = \theta^{-1}$.

Order statistics

Relation (10.13) allows us to derive from Theorem 10.11 the limiting distribution of order statistics; see Hsing *et al.* (1988) and Hsing (1988), for example. First, for blocks J_j in (10.2), let N_n^* be the point process of cluster positions,

$$N_n^*(\cdot) = \sum_{j \in \mathcal{I}} \delta_{jr_n/n}(\cdot), \qquad \mathcal{I} = \{j : M(J_j) > u_n, \, 1 \leq j \leq k_n\}, \qquad (10.18)$$

and let $P[M_n \leq u_n] \to G(x) = \exp(-\theta\tau)$. It follows from Theorem 10.11 that $N_n^* \overset{\mathcal{D}}{\to} N^*$, where N^* is a Poisson process on $(0, 1]$ with rate $\theta\tau = -\log G(x)$. If $K_1, K_2, \ldots$ are independent random variables with distribution π, then the limit of $P[X_{n-k,n} \leq u_n]$ is

$$P[N((0, 1]) \leq k]$$

$$= P[N^*((0, 1]) = 0] + \sum_{j=1}^{k} P[N^*((0, 1]) = j] P\left[\sum_{l=1}^{j} K_l \leq k\right]$$

$$= G(x)\left\{1 + \sum_{j=1}^{k}\sum_{i=j}^{k} \frac{\{-\log G(x)\}^j}{j!} P\left[\sum_{l=1}^{j} K_l = i\right]\right\}. \qquad (10.19)$$

For example,

$$P[X_{n-1,n} \leq u_n] \to G(x)\{1 - \pi(1)\log G(x)\},$$

$$P[X_{n-2,n} \leq u_n] \to G(x)\left[1 - \{\pi(1) + \pi(2)\}\log G(x) + \frac{1}{2}\{\pi(1)\log G(x)\}^2\right].$$

Setting $\pi(1) = 1$ and $\pi(j) = 0$ for $j \geq 2$ yields the limit distributions for the associated, independent sequence as in section 3.2.

The joint distribution of $X_{n,n}$ and $X_{n-k,n}$ for any $k \geq 1$, and indeed of any arbitrary set of extreme order statistics, can also be derived (Hsing 1988) although the class of limit distributions does not admit a finite-dimensional parametrization. Simpler characterizations are possible if stricter mixing conditions are imposed (Ferreira 1993).

10.3.2   Cluster statistics

Various properties of a cluster of exceedances may be of interest, such as the cluster size, the peak excess, or the sum of all excesses. In this section, we define a generic cluster statistic and give a characterization of its distribution that will be useful in section 10.4. We shall investigate point processes that focus on specific cluster statistics in the next section.

We study cluster statistics $c\{(X_i - u_n)_{i=1}^{r_n}\}$ for the following family of functions c.

Definition 10.13 (Yun 2000a) *A measurable map* $c : \mathbb{R} \cup \mathbb{R}^2 \cup \mathbb{R}^3 \cup \cdots \to \mathbb{R}$ *is called a* cluster functional *if for all integers* $1 \le j \le k \le r$ *and for all* $(x_1, \ldots, x_r)$ *such that* $x_i \le 0$ *whenever* $i = 1, \ldots, j - 1$ *or* $i = k + 1, \ldots, r$ *we have* $c(x_1, \ldots, x_r) = c(x_j, \ldots, x_k)$.

Example 10.14 Most cluster functionals of practical interest are of the form

$$c(x_1, \ldots, x_r) = \sum_{i=-m+2}^{r} \phi(x_i, \ldots, x_{i+m-1}),$$

where ϕ is a measurable function of m variables ($m = 1, 2, \ldots$) and $x_i = 0$ whenever $i \le 0$ or $i \ge r + 1$; the function ϕ should be such that $\phi(x_1, \ldots, x_m) = 0$ whenever $x_i \le 0$ for all $i = 1, \ldots, m$. Consider the following examples:

- $m = 1$ and $\phi(x) = \mathbf{1}(x > 0)$ gives the number of exceedances;

- $m = 1$ and $\phi(x) = \max(x, 0)$ gives the sum of all excesses;

- $m = 2$ and $\phi(x_1, x_2) = \mathbf{1}(x_1 \le 0 < x_2)$ gives the number of up-crossings over the threshold;

- $m = 1, 2, \ldots$ and $\phi(x_1, \ldots, x_m) = \mathbf{1}(x_1 > 0, \ldots, x_m > 0)$ gives the number of times, counting overlaps, there are m consecutive exceedances.

A cluster functional that is not of this type is the cluster duration

$$c(x_1, \ldots, x_r)$$
$$= \begin{cases} \max\{j - i + 1 : 1 \le i \le j \le r, \; x_i > 0, \; x_j > 0\} & \text{if } \max x_i > 0 \\ 0 & \text{otherwise.} \end{cases}$$

For general stationary processes, it turns out that the distribution of a cluster statistic can approximately be written in terms of the distribution of the process conditionally on the event that the first variable exceeds the threshold.

Proposition 10.15 (Segers 2003b) *Let* $\{X_i\}$ *be stationary. If the thresholds* u_n *and the positive integers* r_n *are such that Condition 10.8 holds, then, for every sequence* *of cluster functionals c_n and Borel sets $A_n \subset \mathbb{R}$,*

$$P[c_n\{(X_i - u_n)_{i=1}^{r_n}\} \in A_n \mid M_{r_n} > u_n] \qquad (10.20)$$

$$= \theta_n^{-1}\Big\{ P[c_n\{(X_i - u_n)_{i=1}^{r_n}\} \in A_n \mid X_1 > u_n]$$

$$- P[c_n\{(X_i - u_n)_{i=2}^{r_n}\} \in A_n, M_{1,r_n} > u_n \mid X_1 > u_n]\Big\} + o(1)$$

as $n \to \infty$, where θ_n can be either $\theta_n^R(u_n)$ or $\theta_n^B(u_n)$.

Specifying the c_n and A_n in (10.20) leads to interesting formulae, illustrating the usefulness of Proposition 10.15. For instance, with $c_n(x_1, \ldots, x_r) = \sum_{i=1}^{r} \mathbf{1}(x_i > 0)$ and $A_n = [j, \infty)$ for some integer $j \geq 1$, we obtain an approximation of the cluster-size distribution:

$$P\left[\sum_{i=1}^{r_n} \mathbf{1}(X_i > u_n) \geq j \;\middle|\; M_{r_n} > u_n \right]$$

$$= \theta_n^{-1} P\left[\sum_{i=2}^{r_n} \mathbf{1}(X_i > u_n) = j - 1 \;\middle|\; X_1 > u_n \right] + o(1).$$

This formula can be used to give a formal derivation of the limiting cluster-size distribution of the ARMAX process (Example 10.12).

Formula (10.20) also shows that the cluster maximum asymptotically has the same distribution as an arbitrary exceedance. For, setting $c_n(x_1, \ldots, x_r) = \sum_{i=1}^{r} \mathbf{1}(x_i > a_n x)$ and $A_n = [1, \infty)$, we obtain

$$P\left[\frac{M_{r_n} - u_n}{a_n} > x \;\middle|\; M_{r_n} > u_n \right]$$

$$= \theta_n^{-1} P\left[\frac{X_1 - u_n}{a_n} > x, \; \frac{M_{1,r_n} - u_n}{a_n} \leq x \;\middle|\; X_1 > u_n \right] + o(1)$$

$$= \frac{\theta_n^R(u_n + a_n x)}{\theta_n^R(u_n)} P\left[\frac{X_1 - u_n}{a_n} > x \;\middle|\; X_1 > u_n \right] + o(1).$$

Hence, if $\lim \theta_n^R(u_n + a_n x) = \lim \theta_n^R(u_n) = \theta > 0$, then indeed

$$P\left[\frac{M_{r_n} - u_n}{a_n} > x \;\middle|\; M_{r_n} > u_n \right] = P\left[\frac{X_1 - u_n}{a_n} > x \;\middle|\; X_1 > u_n \right] + o(1). \quad (10.21)$$

This notion is less surprising once it is realized that clusters with large maxima tend to contain other large exceedances.

10.3.3 Excesses over threshold

We have already seen a point process (10.12) with a limit that involves the cluster size. This corresponds to the first example of a cluster statistic in the previous section. The second example, concerning excesses over threshold, motivates the marked point process

$$Z_n(\cdot) = \sum_{i \in \mathcal{I}} \frac{X_i - u_n}{a_n} \delta_{i/n}(\cdot), \qquad \mathcal{I} = \{i : X_i > u_n, \ 1 \le i \le n\},$$

where each exceedance is marked with its excess. The normalizing constant a_n is used to ensure a non-degenerate limit for the distribution of the aggregate excess within a cluster,

$$\pi_n'(x) = P\left[a_n^{-1} \sum_{i=1}^{r_n} (X_i - u_n)_+ \le x \ \middle| \ M_{r_n} > u_n \right]. \tag{10.22}$$

In order to obtain limits for processes based on excesses, we require limiting independence of $(X_i - u_n)_+$ instead of $\mathbf{1}(X_i > u_n)$. Therefore define $\Delta'(u_n)$ to be the same as Condition 10.10 but with $\mathcal{F}_{j,k}(u_n) = \sigma\{(X_i - u_n)_+ : j \le i \le k\}$ and write $\alpha'(n, s)$ for the corresponding mixing coefficients.

Theorem 10.16 (Leadbetter 1995) *Let $\{X_i\}$ be stationary with extremal index $\theta > 0$. Let there exist a sequence of thresholds u_n for which $\Delta'(u_n)$ holds and $n\bar{F}(u_n) \to \tau \in (0, \infty)$. Let there exist positive integer sequences s_n and $r_n = o(n)$ and a distribution π' such that $s_n = o(r_n)$, $r_n = o(n)$, $n\alpha'(n, s_n) = o(r_n)$ and $\pi_n' \xrightarrow{D} \pi'$ as $n \to \infty$. Then $Z_n \xrightarrow{D} Z$, where Z is $\mathrm{CP}(\theta\tau, \pi')$.*

The limit process here is the same as that in Theorem 10.11 except that the mark distribution now describes the cluster excess; the method of proof is also similar. Results with different marks may be obtained analogously (Rootzén *et al.* 1998) as long as the appropriate mixing condition holds and the limiting mark distribution exists. One case is more substantial, that of the excess of just the cluster maximum, or peak, leading to the marked point process

$$Z_n^*(\cdot) = \sum_{j \in \mathcal{I}} \frac{M(J_j) - u_n}{a_n} \delta_{jr_n/n}(\cdot), \qquad \mathcal{I} = \{j : M(J_j) > u_n, \ 1 \le j \le k_n\},$$

for the blocks J_j in (10.2). The peak-excess distribution is

$$\pi_n^*(x) = P\left[\frac{M_{r_n} - u_n}{a_n} \le x \ \middle| \ M_{r_n} > u_n \right]$$

and, unlike π and π' above, here we are able to specify the form of $\pi^* = \lim \pi_n^*$ when it exists. If $\theta > 0$, then, by (10.21), we have

$$\pi_n^*(x) = P\left[\frac{X_1 - u_n}{a_n} \le x \ \middle| \ X_1 > u_n \right] + o(1), \qquad n \to \infty.$$

By Pickands (1975), the domain-of-attraction condition implies that the limit of the latter distribution is the Generalized Pareto (GP) distribution, that is,

$$\pi_n^*(x) \to \pi^*(x) = 1 - \left(1 + \frac{\gamma x}{\sigma}\right)_+^{-1/\gamma}, \qquad x > 0, \tag{10.23}$$

for a suitable choice of constants a_n; see also section 5.3.1.

Theorem 10.17 (Leadbetter 1991) *Let $\{X_i\}$ be stationary with extremal index $\theta >$ 0. Let there exist a sequence of thresholds u_n for which $\Delta'(u_n)$ holds and $n\bar{F}(u_n) \to \tau \in (0, \infty)$. Let there exist positive integer sequences r_n and s_n such that $s_n = o(r_n)$, $r_n = o(n)$, and $n\alpha'(n, s_n) = o(r_n)$ as $n \to \infty$. Then $Z_n^* \overset{\mathcal{D}}{\to} Z^*$, where Z^* is $CP(\theta\tau, \pi^*)$ and π^* is the GP distribution.*

Theorem 10.17 is the mathematical foundation of the so-called peaks-over-threshold (POT) method to be discussed in the next section.

10.3.4 Statistical applications

We have seen that the behaviour over high thresholds of certain stationary processes can be described by compound Poisson processes, where events corresponding to clusters occur at a rate $\upsilon = \theta\tau$ and the cluster statistics follow a mark distribution π. For a realization $\{x_i\}_{i=1}^n$, suppose that there are n_c clusters at times $\{t_j^*\}_{j=1}^{n_c}$ in $(0,1]$ and with marks $\{y_j^*\}_{j=1}^{n_c}$. We could fit the model by maximizing the likelihood,

$$L(\upsilon, \pi ; t^*, y^*) = e^{-\upsilon} \upsilon^{n_c} \prod_{j=1}^{n_c} \pi(y_j^*), \tag{10.24}$$

see, for example, section 4.4 of Snyder and Miller (1991). The form of the likelihood means that $\hat{\upsilon} = n_c$ independently of π. If we have a parametric model for π, then its maximum-likelihood estimate can be found, and it depends on the marks only. But the asymptotic theory specifies π only when the mark is the peak excess (Theorem 10.17), in which case π is the GP distribution. For other cluster statistics, we can either choose a parametric model for π or estimate it with the empirical distribution function of $\{y_j^*\}_{j=1}^{n_c}$.

Estimating υ and π relies on being able to identify clusters in the data. This problem, known as *declustering*, is not trivial because we observe only a finite sequence, and so clusters will not be defined at single points in time; rather, they will be slightly spread out and it may not always be clear whether a group of exceedances should form a single cluster or be split into separate clusters. Declustering is intrinsically linked to the extremal index, which we have seen is important also for its influence on marginal tail quantiles and return levels (section 10.2.3) and for its interpretation as the inverse mean cluster size (section 10.3.1). We continue this section by first discussing estimators for the extremal index and then exploring the connection with declustering before returning to estimation of the compound Poisson process. An alternative method for estimating cluster characteristics, which does not use the compound Poisson model, is described in section 10.4, where the evolution of the process over high thresholds is modelled by a class of Markov chains.

Estimating the extremal index

Our first characterization (10.10) of the extremal index was as the limiting ratio of $P[M_{r_n} > u_n]$ to $r_n \bar{F}(u_n)$. If we choose a threshold u and a block length r, then natural estimators for the quantities $P[M_r > u]$ and $\bar{F}(u)$ lead to the blocks estimator for the extremal index:

$$\hat{\theta}_n^B(u ; r) = \frac{k^{-1} \sum_{j=1}^{k} \mathbf{1}\{M_{(j-1)r, jr} > u\}}{rn^{-1} \sum_{i=1}^{n} \mathbf{1}(X_i > u)}, \tag{10.25}$$

where $k = \lfloor n/r \rfloor$. This can be improved by permitting overlapping blocks, giving the sliding-blocks estimator,

$$\tilde{\theta}_n^B(u ; r) = \frac{(n - r + 1)^{-1} \sum_{i=0}^{n-r} \mathbf{1}(M_{i,i+r} > u)}{rn^{-1} \sum_{i=1}^{n} \mathbf{1}(X_i > u)}.$$

Our second characterization (10.11) was in terms of the probability that an exceedance is followed by a run of observations below the threshold. If we choose a threshold u and a run length r, then we can estimate the quantities $P[X_1 > u, M_{1,r+1} \leq u]$ and $\bar{F}(u)$ to obtain the runs estimator for the extremal index:

$$\hat{\theta}_n^R(u ; r) = \frac{(n - r)^{-1} \sum_{i=1}^{n-r} \mathbf{1}(X_i > u, M_{i,i+r} \leq u)}{n^{-1} \sum_{i=1}^{n} \mathbf{1}(X_i > u)}. \tag{10.26}$$

The extremal index is also related to the times between threshold exceedances. We saw in Theorem 10.11 that the point process of exceedance times normalized by $1/n$ has a compound Poisson limit. Therefore, the corresponding times between consecutive exceedances are either zero, representing times between exceedances within the same cluster, or exponential with rate $\theta\tau$, representing times between exceedances in different clusters. Since we expect $\tau = \lim n\bar{F}(u_n)$ exceedances in total but only $\theta\tau$ clusters, the proportion of interexceedance times that are zero should be $1 - \theta$.

Formally, for u such that $F(u) < 1$, define the random variable $T(u)$ to be the time between successive exceedances of u, that is,

$$P[T(u) > r] = P[M_{1,1+r} \leq u \mid X > u].$$

Ferro and Segers (2003) showed that, under a slightly stricter mixing condition than $D(u_n)$, for $t > 0$,

$$P[\bar{F}(u_n)T(u_n) > t] = P[M_{1,1+\lfloor t/\bar{F}(u_n) \rfloor} \leq u_n \mid X_1 > u_n]$$

$$= P[M_{1,r_n} \leq u_n \mid X_1 > u_n]P[M_{\lfloor nt/\tau \rfloor} \leq u_n] + o(1)$$

$$= \theta_n^R(u_n) P[M_{r_n} \leq u_n]^{k_n t/\tau} + o(1)$$

$$\rightarrow \theta \exp(-\theta t), \qquad n \rightarrow \infty. \tag{10.27}$$

In other words, interexceedance times normalized by $\bar{F}(u_n)$ converge in distribution to a random variable T_θ with mass $1 - \theta$ at $t = 0$ and an exponential distribution with rate θ on $t > 0$. The reason that the rate is now θ and not $\theta\tau$ is that we have normalized by $\bar{F}(u_n) \sim \tau/n$ instead of $1/n$. In fact, the result also holds under $D(u_n)$, see Segers (2002).

The coefficient of variation, v, of a non-negative random variable is defined as the ratio of its standard deviation to its expectation. For T_θ,

$$1 + v^2 = E[T_\theta^2]/\{E[T_\theta]\}^2 = 2/\theta. \tag{10.28}$$

The interexceedance times are overdispersed compared to a Poisson process, that is, $v > 1$ and exceedances occur in clusters in the limit, if and only if $\theta < 1$. The case of underdispersion ($v < 1$), in which exceedances tend to repel one another requires long-range dependence and is prevented by the $D(u_n)$ condition.

Suppose that we observe $N = N_u = \sum_{i=1}^n \mathbf{1}(X_i > u)$ exceedances of u at times $1 \leq S_1 < \cdots < S_N \leq n$. The interexceedance times are $T_i = S_{i+1} - S_i$ for $i = 1, \ldots, N - 1$. Replacing the theoretical moments of T_θ in the ratio (10.28) with their empirical counterparts yields another estimator for the extremal index:

$$\hat{\theta}_n(u) = \frac{2\left(\sum_{i=1}^{N-1} T_i\right)^2}{(N-1)\sum_{i=1}^{N-1} T_i^2}.$$

Since the limiting distribution (10.27) models the small interexceedance times as zero, while the observed interexceedance times are always positive, a bias-adjusted version,

$$\hat{\theta}_n^*(u) = \frac{2\left\{\sum_{i=1}^{N-1}(T_i - 1)\right\}^2}{(N-1)\sum_{i=1}^{N-1}(T_i - 1)(T_i - 2)},$$

is preferable when $\max\{T_i : 1 \leq i \leq N - 1\} > 2$. Unlike the blocks and runs estimators, these two estimators are not guaranteed to lie in $[0, 1]$ so that the constraint must be imposed artificially. Doing so yields the intervals estimator for the extremal index:

$$\bar{\theta}_n^I(u) = \begin{cases} 1 \wedge \hat{\theta}_n(u) & \text{if } \max\{T_i : 1 \leq i \leq N - 1\} \leq 2, \\ 1 \wedge \hat{\theta}_n^*(u) & \text{if } \max\{T_i : 1 \leq i \leq N - 1\} > 2. \end{cases} \tag{10.29}$$

The blocks and runs estimators are used by Leadbetter *et al.* (1989) and Smith (1989); a variant of the blocks estimator is proposed by Smith and Weissman

(1994). Calculations of asymptotic bias by Smith and Weissman (1994) and Weissman and Novak (1998) suggest, however, that the runs estimator should be preferred. Asymptotic normality has been established under appropriate conditions by Hsing (1993) and Weissman and Novak (1998). The choice of auxiliary parameter, r, for both the blocks and runs estimators is largely arbitrary. It may be guided by physical reasoning about the likely range of dependence in the underlying process (Tawn 1988b) or parametric modelling of the evolution of extremes (Davison and Smith 1990). Alternatively, estimates with different r may be combined (Smith and Weissman 1994). The attraction of the intervals estimator (Ferro and Segers 2003) is its freedom from any auxiliary parameter.

Still more estimators can be found in the literature. For example, Ancona-Navarrete and Tawn (2000) derive estimators from Markov models fitted to the data (see also section 10.4). Gomes (1993) constructs an independent sequence by randomizing the data and then fits GEV distributions to sample maxima from both this and the original sequence. Since the parameters (10.7) of the two distributions are related by the extremal index, an estimator for θ may be obtained as a combination of the parameter estimates. A comparative study is made by Ancona-Navarrete and Tawn (2000).

The estimator of Gomes (1993) has the merit that it does not require the selection of a threshold, although it does require the selection of a block length to obtain a sample of maxima M_n. Threshold choice is a fundamental issue: the estimators presented in this section estimate a quantity $\theta(u)$ rather than $\theta = \lim \theta(u)$. Hsing (1993) considers threshold selection for the runs estimator and proposes an adaptive scheme to minimize mean square error under a model for the bias. A more common approach is simply to estimate the extremal index using several high thresholds and then assume that stability of estimates over a range of thresholds indicates that the limit has been reached.

Declustering the data

Recall that to estimate the limiting compound Poisson process, we need to decluster the data. Several schemes have been proposed in the literature, three of which relate to the blocks, runs and intervals estimators for the extremal index.

Blocks declustering (Leadbetter et al. 1989) is a natural application of the definition of clusters given in section 10.3.1. The data are partitioned into blocks of length r and exceedances of a threshold u are assumed to belong to the same cluster if they fall within the same block. The number of clusters identified in this way is the number of blocks with at least one exceedance. The example in Figure 10.4 identifies two clusters using block length $r = 6$. The number of clusters is precisely the quantity that appears in the numerator of the blocks estimator (10.25) for the extremal index, which is therefore the ratio of the number of clusters to the total number of exceedances, that is, the reciprocal of the average size of clusters found by blocks declustering.

The runs estimator (10.26) for the extremal index may also be interpreted as the ratio of the number of clusters to the number of exceedances, but where clusters

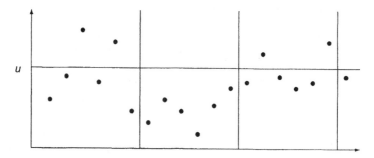

Figure 10.4   An illustration of blocks declustering with threshold u and block length $r = 6$.

are identified by runs declustering (Smith 1989). With this scheme, exceedances separated by fewer than r non-exceedances are assumed to belong to the same cluster; if $r = 0$, then each exceedance forms a separate cluster. In Figure 10.4, three clusters are identified if the run length is $r = 3$, but only two clusters are identified if $r = 4$.

As with the corresponding estimators for the extremal index, the troublesome issue for blocks and runs declustering is the choice of the auxiliary parameter, r. Diagnostic tools for selecting r have been proposed by Ledford and Tawn (2003), while the following scheme, intervals declustering (Ferro and Segers 2003), provides an alternative solution.

Recall that a proportion θ of normalized interexceedance times are non-zero in the limit (10.27), and that these represent times between clusters. If $\bar\theta$ is an estimate of the extremal index, then it is natural to take the largest $n_c - 1 = \lfloor (N - 1)\bar\theta \rfloor$ of the interexceedance times T_i, $1 \leq i \leq N - 1$, to be these intercluster times. This defines a partition of the remaining interexceedance times into sets of intracluster times. Note also that, because the point process of exceedance times is compound Poisson, the intercluster times are independent of one another, and the sets of intracluster times are independent both of one another and of the intercluster times. To be precise, if $T_{(n_c)}$ is the n_cth largest interexceedance time and T_{i_j} is the jth interexceedance time to exceed $T_{(n_c)}$, then $\{T_{i_j}\}_{j=1}^{n_c - 1}$ is a set of approximately independent intercluster times. In the case of ties, decrease n_c until $T_{(n_c - 1)}$ is strictly greater than $T_{(n_c)}$. Let also $\mathcal{T}_j = \{T_{i_{j-1}+1}, \ldots, T_{i_j - 1}\}$, where $i_0 = 0$, $i_{n_c} = N$ and $\mathcal{T}_j = \emptyset$ if $i_j = i_{j-1} + 1$. Then $\{\mathcal{T}_j\}_{j=1}^{n_c}$ is a collection of approximately independent sets of intracluster times. Furthermore, each set $\mathcal{T}_j$ has associated with it a set of threshold exceedances $\mathcal{X}_j = \{X_i : i \in \mathcal{S}_j\}$, where $\mathcal{S}_j = \{S_{i_{j-1}+1}, \ldots, S_{i_j}\}$ is the set of exceedance times. If we estimate θ with the intervals estimator (10.29), then this approach declusters the data into n_c clusters without requiring an arbitrary selection of auxiliary parameter. In fact, the scheme is equivalent to runs declustering but with run length $r = T_{(n_c)}$ estimated from the data and justified by the limiting theory.

Estimating the compound Poisson process

Once we have identified clusters $\mathcal{X}_j = \{x_i : i \in \mathcal{S}_j\}$ for $j = 1, \ldots, n_c$ over a high threshold u, we can compute the cluster statistics $y_j^* = c\{(x_i - u)_{i \in \mathcal{S}_j}\}$ corresponding to the marks of the limiting compound Poisson process. We have remarked already that $\hat{\upsilon} = n_c$, while π may be estimated by the empirical distribution function of the cluster statistics, if the theory does not supply a parametric model.

In the case of the peak excess, π is the GP distribution (Theorem 10.17) and may be estimated by maximum likelihood. This is known as POT modelling. Estimation methods, diagnostics and extensions of the model to handle seasonality and other regressors are described by Davison and Smith (1990); see also Chapter 7. An alternative POT approach is to fit the GP distribution to all of the excesses, not only those of the cluster maxima. The idea is justified by the fact (10.21) that, in the limit, the distribution of the excess of a cluster maximum is the same as that of an arbitrary exceedance, although the correspondence is often poor at finite thresholds. By fitting to all of the excesses, we avoid having to decluster the exceedances; on the other hand, the excesses can no longer be treated as though they were independent, which necessitates a modification of the estimation procedure. One approach is to adopt the estimation methods appropriate when the excesses are independent and adjust the standard errors, which will otherwise be underestimated. Several methods for obtaining standard errors in this case have been proposed: see Smith (1990a), Buishand (1993) and Drees (2000).

For any cluster statistic, a bootstrap scheme (Ferro and Segers 2003) that exploits the independence structure of the compound Poisson process may be used to obtain confidence limits on estimates of υ, π and derived quantities, ζ, such as the mean of π.

(i) Resample with replacement $n_c - 1$ intercluster times from $\{T_{i_j}\}_{j=1}^{n_c-1}$.

(ii) Resample with replacement n_c sets of intracluster times, some of which may be empty, and associated exceedances from $\{(T_j, \mathcal{X}_j)\}_{j=1}^{n_c}$.

(iii) Intercalate these interexceedance times and clusters to form a bootstrap replication of the process.

(iv) Compute N for the bootstrap process, estimate θ, and decluster accordingly.

(v) Estimate υ, π and ζ for the declustered bootstrap sample.

Forming B such bootstrap samples yields collections of estimates that may be used to approximate the distributions of the original point estimates. In particular, the empirical α- and $(1 - \alpha)$-quantiles of each collection define $(1 - 2\alpha)$-confidence intervals. Note that, when applied with intervals declustering, this scheme accounts for uncertainty in the run length used to decluster the data, as it is re-estimated for each sequence at step (iv).

Alternative confidence limits for the extremal index (Leadbetter et al. 1989) rely on the asymptotic normality and variance of the blocks estimator, which may be estimated (Hsing 1991; Smith and Weissman 1994) by

$$\frac{\{\bar{\theta}_n^B(u\,;\,r)\}^3 \hat{V}_s}{\sum_{i=1}^n \mathbf{1}(X_i > u)},\qquad(10.30)$$

where $\hat{V}_s$ is the sample variance of the cluster sizes, $\{\sum_{i=(j-1)r+1}^{jr} \mathbf{1}(X_i > u) : M_{(j-1)r,jr} > u, 1 \le j \le \lfloor n/r \rfloor \}$.

10.3.5   Data example

The intervals estimator for the extremal index of the Uccle temperature data (see section 10.2.2) is plotted against threshold in Figure 10.5. In this and subsequent plots, thresholds range from the 90% to the 99.5% empirical quantiles, and boot-strapped confidence intervals are based on the intervals declustering scheme of section 10.3.4 with 500 resamples. Note that in Figure 10.5, the lower confidence limits estimated by the bootstrap and the normal approximation (10.30) are similar, while the upper limits are higher with the bootstrap. The point estimates of the extremal index are stable up to the 97% threshold, with values just below 0.5. The increase of the estimates above the 97% threshold might indicate that the limit has not been reached, and possibly $\theta = 1$, or could be due to sampling variability. We shall return to this question in section 10.4.7; for now, we assume that the perceived stability indicates that the limit has been reached and that the limiting cluster characteristics of the data can be estimated by fixing a suitable threshold.

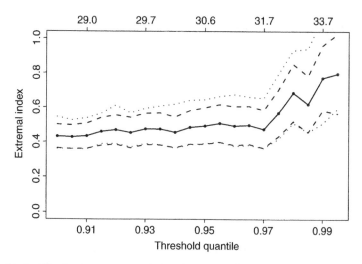

Figure 10.5   The intervals estimator for the extremal index (—○—) against thresh-old with 95% confidence intervals estimated by the bootstrap (······) and the normal approximation (- - - - -). The threshold is marked on the upper axis in degrees Celsius.

A cluster of hot days can have serious implications for public health and agriculture. By declustering the data, we can obtain estimates for the rate at which clusters occur and for the severity of clusters, which can be usefully measured with the distributions of statistics such as the cluster maximum, cluster size and cluster excess. We have seen already that the mean cluster size is $1/\theta \approx 2$.

The intervals declustering scheme, applied with the above estimates of the extremal index, enables the identification of clusters at different thresholds. The Poisson process rate at which clusters occur is approximately linearly decreasing with threshold exceedance probability according to the approximation $n_c \approx \theta n \bar{F}(u)$. On average, about 1.3 clusters occur over the 90% quantile every July, and the rate decreases by about 0.12 for every decrease of 0.01 in the threshold exceedance probability. Estimates of the declustering run length r are close to 4 for all thresholds, indicating that exceedances separated by about four days for which the temperature is below the threshold can be taken as independent.

For the POT model, we describe the excesses of cluster maxima by the GP distribution (10.23). The maximum-likelihood estimates of the GP parameters at different thresholds are represented in Figure 10.6. The model is fitted twice:

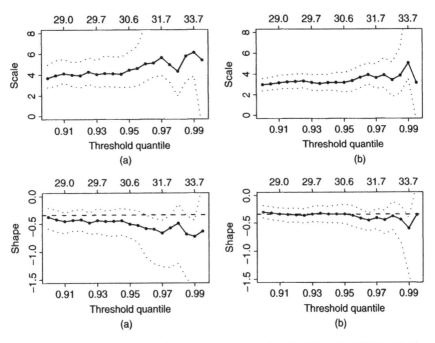

Figure 10.6   Parameter estimates (—o—) against threshold for the GP distribution fitted to cluster maxima (a) and all exceedances (b) with bootstrapped 95% confidence intervals (· · · · ·). The scale parameters have been normalized to $\sigma - \gamma u$. The estimate (- - - -) of the shape parameter from the GEV model is also indicated, and the threshold is marked on the upper axes in degrees Celsius.

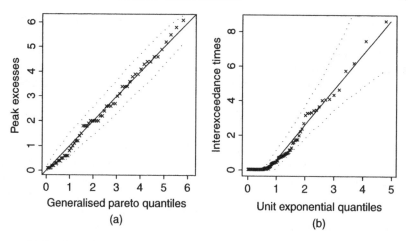

Figure 10.7   Quantile plots for excesses of cluster maxima (a) and for the normalized interexceedance times (b) over the 96% threshold with 95% confidence intervals obtained by simulating from the fitted models. For the interexceedance times, the continuous line has gradient $1/\hat{\theta}$ and breakpoint $-\log\hat{\theta}$, where $\hat{\theta} = 0.49$.

first to only the cluster maxima, and second, to all exceedances. As there is some disparity between the two fits, we should be wary of using the latter to model peaks. Note also that the estimate of the shape parameter is close to -0.5, below which the usual asymptotic properties of maximum-likelihood estimators do not hold (Smith 1985). Moment estimators give similar point estimates, however, and the bootstrap confidence intervals do not rely on asymptotic normality. For both fits, the parameter estimates are quite stable, perhaps with some bias below the 96% threshold, $31°C$, at which there are 120 exceedances and 59 identified clusters. The quantile plots in Figure 10.7 show that the GP model is a satisfactory fit at the 96% threshold and that the interexceedance times at this threshold are well modelled by their limit distribution (10.27). Furthermore, the mean-excess plot is approximately linear above the 96% threshold. We take the fit to cluster maxima at the 96% threshold, $\hat{\sigma} = 3.7$ and $\hat{\gamma} = -0.59$, for our POT model.

The marginal distribution of the temperature data is captured better by the fit to all exceedances, so we use the corresponding GP parameter estimates, $\hat{\sigma} = 2.8$ and $\hat{\gamma} = -0.42$, to describe the marginal tail. The 99%, 99.9% and 99.99% marginal quantiles, with bootstrapped 95% confidence intervals, are 33.9 (33.4, 34.3), 36.2 (35.6, 36.6) and 37.1 (36.2, 37.8). Compare the first two with the empirical quantiles, 33.7 and 36.2. Combining the estimate of the extremal index, 0.49, at the 96% threshold with this estimate of the GP distribution yields estimates of the 100, 1000 and 10000 July return levels: 36.5 (35.7, 36.9), 37.2 (36.2, 38.1) and 37.5 (36.3, 38.7). The confidence intervals are obtained by bootstrapping the extremal index and GP parameters with the scheme described in section 10.3.4.

The estimate of the upper end-point is 37.7 (36.4, 39.6). These estimates are lower, and the confidence intervals are narrower than the direct estimates from the GEV model of section 10.2.2. Bootstrapped confidence intervals have been preferred here to methods relying on asymptotic normality or profile likelihoods because they easily account for the dependence between threshold exceedances and for the uncertainty in the declustering scheme.

In addition to the cluster maxima, other statistics of interest are the numbers of exceedances and the sum of all excesses during a cluster. The empirical estimate of the cluster-excess distribution appears later in Figure 10.11. Estimates of the cluster-size distribution are presented in Figure 10.8. These appear stable, but again there is a hint that $\pi(1) \to 1$ as threshold increases. The point estimates at the 96% threshold are $\hat{\pi}(1) = 0.61$, $\hat{\pi}(2) = 0.15$, $\hat{\pi}(3) = 0.07$ and $\hat{\pi}(4) = 0.08$; 8% of clusters have more than four exceedances. These estimates can be combined with the GEV model to determine distributions (10.19) of large order statistics for July.

Inspecting the data reveals that clusters tend to comprise only consecutive exceedances, maximizing public health and agricultural impacts. This is reflected in the distribution, κ, of the maximum number of consecutive exceedances within a cluster: at the 96% threshold, the estimate is $\hat{\kappa}(1) = 0.64$, $\hat{\kappa}(2) = 0.17$, $\hat{\kappa}(3) = 0.05$ and $\hat{\kappa}(4) = 0.08$, which is very similar to the cluster-size distribution. The mean number of up-crossings per cluster is 1.17, with bootstrapped 95% confidence interval (1.00, 1.43).

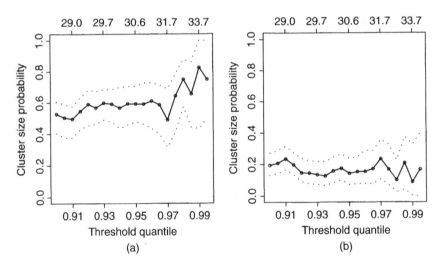

Figure 10.8   Cluster-size distribution estimates (—o—) against threshold for sizes 1 (a) and 2 (b) with bootstrapped 95% confidence intervals. The threshold is marked on the upper axes in degrees Celsius.

10.3.6 Additional topics

Two-dimensional point processes

In this section, we have considered one-dimensional point processes in which exceedance times are associated with marks defined by the exceeding random variables X_i. Another instructive approach is to consider two-dimensional processes recording time in the first dimension and X_i in the second. The process

$$V_n(\cdot) = \sum_{i=1}^n \delta_{(i/n,(X_i-b_n)/a_n)}(\cdot)$$

was studied by Hsing (1987), extending work of Pickands (1971) on independent sequences and Mori (1977) on strong-mixing sequences. When the normalizing constants are such that $(M_n - b_n)/a_n$ has a GEV limit, G, with lower and upper end-points $_*x$ and x_*, and the $\Delta(u_n)$ condition holds simultaneously at different thresholds, Hsing (1987) shows that any limit of V_n has the form

$$V(\cdot) = \sum_{i\geq 1}\sum_{j=1}^{K_i} \delta_{(S_i,X_{i,j})}(\cdot),$$

where S_i represents the occurrence time of a cluster of points $X_{i,1} \geq \cdots \geq X_{i,K_i}$. The times and heights, $\{(S_i, X_{i,1})\}_{i\geq 1}$, of cluster maxima occur according to a two-dimensional, nonhomogeneous Poisson process η on $(0, 1] \times (_*x, x_*)$ with intensity measure $-(b - a)\log G(x)$ on $(a, b) \times [x, x_*)$. This corresponds to our discussion of the process (10.18) of cluster maxima over a single threshold; see also section 5.3.1. Further insight is provided by the relationship between cluster maxima and the remaining points in a cluster. For each cluster, the points

$$Y_{i,j} = \frac{-\log G(X_{i,j})}{-\log G(X_{i,1})}, \qquad 1 \leq j \leq K_i,$$

occur according to an arbitrary point process η_i on $[1, \infty)$ with atom $Y_{i,1} = 1$, and these point processes are independent, identically distributed and independent of η.

More general normalizations than linear ones are considered in Novak (2002).

Tail array sums

Sometimes, we are interested in summaries of not just characteristics of individual clusters but also the cumulative effect of all exceedances over a longer-time period. Useful measures for such cumulative effects are tail array sums (Leadbetter 1995; Rootzén *et al.* 1998),

$$W_n = \sum_{i=1}^n \phi(X_i - u_n), \qquad (10.31)$$

for functions ϕ satisfying $\phi(x) = 0$ when $x \leq 0$ as in section 10.3.2. Note that we can decompose W_n as

$$W_n = \sum_{j=1}^{k_n} W_n(J_j),$$

where J_j are the blocks in (10.2) and $W_n(J_j) = \sum_{i \in J_j} \phi(X_i - u_n)$ are the block sums.

The tail array sum is related by $W_n = \Phi_n(0, 1]$ to the point process

$$\Phi_n(\cdot) = \sum_{i \in \mathcal{I}} \phi(X_i - u_n)\delta_{i/n}(\cdot), \qquad \mathcal{I} = \{i : X_i > u_n, \, 1 \leq i \leq n\},$$

of which we have seen examples in sections 10.3.1 and 10.3.3. Therefore, whenever Φ_n has a compound Poisson limit with mark distribution π_ϕ determined by the distribution of $W_n(J_1)$ conditional on $M(J_1) > u_n$, W_n will converge in distribution to $\sum_{j=1}^{N_c} W_j$, where N_c is a Poisson random variable representing the number of clusters and the W_j are independent random variables with distribution π_ϕ. The compound Poisson model does not provide a finite-parameter characterization for the limit distribution of W_n, except in cases where π_ϕ is known.

Previously, the number of clusters had a Poisson limit because its expectation was controlled by $n\bar{F}(u_n) \to \tau < \infty$. If, however, the thresholds are such that $n\bar{F}(u_n) \to \infty$, then we might hope to obtain a central limit theorem for W_n as the sum of a large number of block sums. To obtain non-degenerate limits, we normalize using

$$\sigma_n^2 = k_n \text{var}\{W_n(J_1)\} \tag{10.32}$$

and restrict the dependence with $\Delta^\phi(u_n)$, defined to be the same condition as $\Delta(u_n)$ but with $\mathcal{F}_{j,k}(u_n) = \sigma\{\phi(X_i - u_n)_+ : j \leq i \leq k\}$ and mixing coefficients $\alpha^\phi(n, s)$. With the usual moment conditions, we obtain the following result.

Theorem 10.18 (Leadbetter 1995) *Let there exist a sequence of thresholds u_n for which $\Delta^\phi(u_n)$ holds, $n\bar{F}(u_n) \to \infty$ and $E[\phi^2(X_i - u_n)] < \infty$. Let there exist a positive integer sequences r_n and s_n such that $s_n = o(r_n)$, $r_n = o(n)$, and $n\alpha^\phi(n, s_n) = o(r_n)$ as $n \to \infty$ and such that the Lindeberg condition, $k_n E\{\tilde{W}_{n1}^2 \mathbf{1}(|\tilde{W}_{n1}| > \varepsilon)\} \to 0$ as $n \to \infty$ for all $\varepsilon > 0$, holds with $\tilde{W}_{n1} = [W_n(J_1) - E\{W_n(J_1)\}]/\sigma_n$ and $k_n = \lfloor n/r_n \rfloor$. Then,*

$$\sigma_n^{-1}\{W_n - E(W_n)\} \overset{D}{\to} W,$$

where W has a standard normal distribution.

Theorem 10.18 says that we may model W_n by a normal distribution, reducing inference to estimation of its mean and variance. The mean may be estimated by the observed value of W_n and the variance by substituting the sample variance of the $W_n(J_j)$ into expression (10.32).

10.4   Markov-Chain Models

In the previous sections, we did not make any assumptions at all on the form of dependence among the variables X_i, except for a restriction on long-range dependence at extreme levels. This generality is of course attractive from a mathematical point of view, but leaves us with little means to analyse, for instance, the structure of clusters of high-threshold exceedances except for the usual empirical estimates obtained after application of a declustering scheme. As we saw earlier, the choice of such a scheme may be subjected to large uncertainty (which was quantified by our bootstrap scheme) and, moreover, if there are only a few clusters of extremes, then the empirical estimates are not very informative.

A possible way out of this problem is to make more detailed assumptions about the dependence structure in the series, for instance, by assuming some kind of (semi-)parametric model. In the present section, we focus on Markov chains for which the joint distribution of a pair of consecutive variables satisfies some regularity at extreme levels. Other time-series models are considered briefly in section 10.6.

The Markov-chain approach is successful because, under weak assumptions, the distribution of the chain given that it started at an extreme level, the so-called tail chain, can be represented in terms of a certain random walk, while the extremal index and, more generally, the distribution of clusters of extreme values can be written in terms of this tail chain (Perfekt 1994; Smith 1992). Moreover, an approximate likelihood can be constructed from which the Markov chain can be estimated, and the tail chain subsequently derived, given a set of data (Smith *et al.* 1997).

10.4.1   The tail chain

Let $\{X_n\}_{n\geq 1}$ be a stationary Markov chain. We assume that the joint distribution function $F(x_1, x_2)$ of (X_1, X_2) is absolutely continuous with joint density $f(x_1, x_2)$. Denote the marginal density of the chain by $f(x)$ and the marginal distribution function by $F(x)$, and let $x_* = \sup\{x \in \mathbb{R} : F(x) < 1\}$ be its right end-point. The Markov property entails that for every positive integer n, the joint density of the vector $(X_1, \ldots, X_n)$ is equal to

$$f(x_1, \ldots, x_n) = f(x_1) \prod_{i=2}^{n} f(x_i \mid x_{i-1})$$

$$= \prod_{i=2}^{n} f(x_{i-1}, x_i) \Big/ \prod_{i=2}^{n-1} f(x_i). \tag{10.33}$$

We shall model the extremes of the chain under the assumption that the joint distribution of (X_1, X_2) is in the domain of attraction of a bivariate extreme value distribution $G(x_1, x_2)$. Without loss of generality, we take the identical margins of

G to be the standard extreme value distribution with shape parameter $\gamma \in \mathbb{R}$:

$$G_\gamma(x) = \exp\{-(1 + \gamma x)^{-1/\gamma}\}, \qquad 1 + \gamma x > 0.$$

If the distribution of (X_1, X_2) is in the domain of attraction of G, then, by Pickands (1975) and Marshall and Olkin (1983), there exists a positive function $\sigma(u), u < x_*$, such that for x, x_1, x_2 with $1 + \gamma x > 0$ and $1 + \gamma x_i > 0$ $(i = 1, 2)$ we have

$$\frac{1 - F\{u + \sigma(u)x\}}{1 - F(u)} \to (1 + \gamma x)^{-1/\gamma}, \qquad (10.34)$$

$$\frac{1 - F\{u + \sigma(u)x_1, u + \sigma(u)x_2\}}{1 - F(u)} \to V(x_1, x_2), \qquad (10.35)$$

as $u \uparrow x_*$, where $V(x_1, x_2) = -\log G(x_1, x_2)$; see also equation (8.69).

Our model for the extremes of the chain and the methods of inference will be based on the limiting distribution of the vector $\{(X_i - u)/\sigma(u)\}_{i=1}^m$ conditionally on $X_1 > u$, where m is a positive integer. We shall show now that a non-trivial limit indeed exists provided we enforce conditions (10.34)–(10.35) to density convergence.

As a preliminary, we take a closer look at the extreme value distribution G. From section 8.2, we recall the following facts. The function

$$G_*(z_1, z_2) = G\left(\frac{z_1^\gamma - 1}{\gamma}, \frac{z_2^\gamma - 1}{\gamma}\right), \qquad 0 < z_i < \infty \quad (i = 1, 2)$$

is a bivariate extreme value distribution with standard Fréchet margins, and there exists a positive measure H on the unit interval $[0, 1]$ so that

$$V_*(z_1, z_2) = -\log G_*(z_1, z_2) = \int_{[0,1]} \max\{w/z_1, (1 - w)/z_2\} H(\mathrm{d}w). \quad (10.36)$$

The measure H is called the *spectral measure*, and it necessarily satisfies the constraints

$$\int_{[0,1]} w H(\mathrm{d}w) = 1 = \int_{[0,1]} (1 - w) H(\mathrm{d}w). \qquad (10.37)$$

For the sake of simplicity, we make the following assumption.

Condition 10.19 *The spectral measure H is absolutely continuous with continuous density function $h(w)$ for $0 < w < 1$.*

This condition poses a restriction indeed. For instance, it prohibits the margins of G to be independent, in which case H is concentrated on 0 and 1. Some parametric models, such as the asymmetric logistic (Tawn 1988a) in Example 8.1, also allow H to have non-zero mass at 0 and 1. The arguments below can be extended to cover these cases as well (Perfekt 1994; Yun 1998).

Under Condition 10.19, the function V_* is twice differentiable, and, denoting partial derivatives by appropriate subscripts, we have by equation (8.36)

$$V_{*12}(z_1, z_2) = -(z_1 + z_2)^{-3} h\{z_1/(z_1 + z_2)\} \tag{10.38}$$

for $0 < z_i < \infty$ $(i = 1, 2)$. As for (x_1, x_2) such that $1 + \gamma x_i > 0$ $(i = 1, 2)$,

$$V(x_1, x_2) = V_*(z_1, z_2), \qquad z_i = (1 + \gamma x_i)^{1/\gamma} \quad (i = 1, 2), \tag{10.39}$$

the function V is twice differentiable too, and we can formulate an assumption extending conditions (10.34)–(10.35) to densities.

Condition 10.20 *The function V is twice differentiable, and for x, x_1, x_2 such that $1 + \gamma x > 0$ and $1 + \gamma x_i > 0$ $(i = 1, 2)$ we have as $u \uparrow x_*$*

$$\frac{\sigma(u) f\{u + \sigma(u)x\}}{1 - F(u)} \to (1 + \gamma x)^{-1/\gamma - 1},$$

$$\frac{\sigma(u)^2 f\{u + \sigma(u)x_1, u + \sigma(u)x_2\}}{1 - F(u)} \to -V_{12}(x_1, x_2).$$

Under Condition 10.20, we can find the limit of the joint density of the vector $\{(X_i - u)/\sigma(u)\}_{i=1}^m$ conditionally on $X_1 > u$. For x_1 and x_2 such that $1 + \gamma x_i > 0$ for $i = 1, 2$, we find

$$\sigma(u) f\{u + \sigma(u)x_2 \mid u + \sigma(u)x_1\}$$

$$= \frac{\sigma^2(u) f\{u + \sigma(u)x_1, u + \sigma(u)x\}/\bar{F}(u)}{\sigma(u) f\{u + \sigma(u)x_1\}/\bar{F}(u)}$$

$$\to -(1 + \gamma x_1)^{1/\gamma + 1} V_{12}(x_1, x_2), \qquad u \uparrow x_*. \tag{10.40}$$

Hence by (10.33), the joint density of $\{(X_i - u)/\sigma(u)\}_{i=1}^m$ conditionally on $X_1 > u$ in $(x_1, \ldots, x_m)$ such that $x_1 > 0$ and $1 + \gamma x_i > 0$ for $i = 1, \ldots, m$ satisfies

$$\sigma^m(u) f\{u + \sigma(u)x_1, \ldots, u + \sigma(u)x_m\}/\bar{F}(u)$$

$$\to (1 + \gamma x_1)^{-1/\gamma - 1} \prod_{i=2}^m (1 + \gamma x_{i-1})^{1/\gamma + 1} \{-V_{12}(x_{i-1}, x_i)\}, \tag{10.41}$$

as $u \uparrow x_*$.

Now let T be a standard Pareto random variable, $P[T > t] = 1/t$ for $1 \le t < \infty$, and let $\{A_i\}_{i \ge 1}$ be independent, positive random variables, independent of T, and with common marginal distribution

$$P[A \le a] = \int_{1/(1+a)}^1 w h(w) dw = -V_{*1}(1, a), \qquad 0 \le a < \infty. \tag{10.42}$$

Let $\{Y_n\}_{n\geq 1}$ be the Markov chain given by the recursion

$$Y_1 = \frac{T^\gamma - 1}{\gamma}$$

$$Y_n = \frac{(1 + \gamma Y_{n-1})A_{n-1}^\gamma - 1}{\gamma}, \qquad n \geq 2, \tag{10.43}$$

or explicitly

$$Y_n = \frac{\left(T \prod_{i=1}^{n-1} A_i\right)^\gamma - 1}{\gamma}, \qquad n \geq 2. \tag{10.44}$$

The random variable Y_1 has a GP distribution with shape parameter γ. For $n \geq 2$ and (x_{n-1}, x_n) such that $1 + \gamma x_i > 0$ ($i = n - 1, n$), the density of Y_n conditionally on $Y_{n-1} = x_{n-1}$ is, denoting $z_i = (1 + \gamma x_i)^{1/\gamma}$, equal to

$$\frac{\mathrm{d}}{\mathrm{d}x_n} P[Y_n \leq x_n \mid Y_{n-1} = x_{n-1}]$$

$$= \frac{\mathrm{d}}{\mathrm{d}x_n} P[A_{n-1} \leq z_n/z_{n-1}]$$

$$= (1 + z_n/z_{n-1})^{-3} h\{(1 + z_n/z_{n-1})^{-1}\} z_{n-1}^{-1} \frac{\mathrm{d}z_n}{\mathrm{d}x_n}$$

$$= -V_{*12}(z_{n-1}, z_n) z_{n-1}^2 \frac{\mathrm{d}z_n}{\mathrm{d}x_n}$$

$$= -(1 + \gamma x_{n-1})^{1/\gamma+1} V_{12}(x_{n-1}, x_n) \tag{10.45}$$

where we used subsequently (10.43), (10.42), (10.38), and (10.39).

Combining (10.41) with (10.45), we obtain that under Conditions 10.19 and 10.20, for all positive integer m,

$$P\left[\left(\frac{X_1 - u}{\sigma(u)}, \ldots, \frac{X_m - u}{\sigma(u)}\right) \in \cdot \,\middle|\, X_1 > u\right] \xrightarrow{D} P[(Y_1, \ldots, Y_m) \in \cdot], \tag{10.46}$$

as $u \uparrow x_*$. The process $\{Y_n\}$ is called the *tail chain* of the Markov chain $\{X_n\}$. It describes the behaviour of the latter when started at a high value $X_1 > u$. Recall that the tail chain is completely determined by the extreme value index γ and the distribution of A; to find the approximate distribution of $(X_1, \ldots, X_m)$ conditional on $X_1 > u$, we also need the scaling parameter $\sigma(u)$. Finally, observe that (10.40) and (10.45) yield a convenient interpretation of the distribution of A in that

$$\lim_{n \to \infty} P\left[\left\{1 + \gamma \frac{X_2 - u}{\sigma(u)}\right\}^{1/\gamma} \leq a \,\middle|\, X_1 = u\right] \xrightarrow{D} P[A \leq a], \qquad u \uparrow x_*. \tag{10.47}$$

Example 10.21 A popular parametric model for V_* is the logistic model

$$V_*(z_1, z_2) = \left(z_1^{-1/\alpha} + z_2^{-1/\alpha}\right)^{\alpha}, \qquad 0 < z_j < \infty \quad (j = 1, 2)$$

with parameter $0 < \alpha \leq 1$, see (9.6). The case $\alpha = 1$ corresponds to independent margins, in which case the spectral measure H puts unit mass on 0 and 1, violating Condition 10.19. If $0 < \alpha < 1$, however, direct computation reveals that

$$P[A \leq a] = -V_{*1}(1, a) = (1 + a^{-1/\alpha})^{-(1-\alpha)}, \qquad 0 < a < \infty.$$

Without extra assumptions, we can find the limit behaviour of Y_n as $n \to \infty$. Observe first that by (10.42) and (10.37), we have

$$E(A) = \int_0^\infty P[A > a]\mathrm{d}a = \int_0^\infty \int_0^{1/(1+a)} wh(w)\mathrm{d}w\mathrm{d}a$$

$$= \int_0^1 \left(\frac{1}{w} - 1\right) wh(w)\mathrm{d}w = 1.$$

By Jensen's inequality, $-\infty \leq E\{\log(A)\} < 0$. Therefore, if $A_1, A_2, \ldots$ are independent copies of A, then by the law of large numbers $\sum_{i=1}^n \log(A_i) \to -\infty$ and thus $\prod_{i=1}^n A_i \to 0$ as $n \to \infty$. We obtain that

$$\lim_{n \to \infty} Y_n = \begin{cases} -1/\gamma & \text{if } \gamma > 0 \\ -\infty & \text{if } \gamma \leq 0 \end{cases} \tag{10.48}$$

with probability one. In particular, only a finite number of the Y_n are positive. The interpretation is that clusters of exceedances over a high threshold necessarily remain of finite length.

As mentioned before, Conditions 10.19 and 10.20 are not really necessary. A more general theory, formulated directly in terms of the transition kernel of the chain, is developed in Perfekt (1994). The main conclusions of this section remain valid in the more general framework: the representation (10.42) of the distribution of A in terms of V_*, the representation (10.43) of the tail chain $\{Y_n\}$, and the limit distribution (10.46). What changes is that the distribution of A need not be absolutely continuous anymore. In particular, A may have a point mass at zero, in which case an absorbing state for the tail chain is $-1/\gamma$ if $\gamma > 0$ and $-\infty$ if $\gamma \leq 0$. Also, it can happen that $P[A = 1] = 1$, corresponding to asymptotic complete dependence of the distribution of (X_1, X_2) (section 8.3.2), in which case $Y_n = Y_1$ for all $n \geq 1$, violating (10.48).

10.4.2 Extremal index

Suppose as in section 10.4.1 that $\{X_n\}$ is a stationary Markov chain with tail chain $\{Y_n\}$ satisfying (10.46). We want to express the extremal index θ of the Markov chain, provided it exists, in terms of the tail chain. This will allow us at a later stage to estimate the extremal index when we have estimated the tail chain from the data.

Recall our notation $M_{j,k} = \max\{X_{j+1}, \ldots, X_k\}$ (with $\max \emptyset = -\infty$) and $M_k = M_{0,k}$ for integers $0 \leq j \leq k$. In section 10.2.3, we saw that under suitable assumptions, the extremal index θ is the limit of $\theta_n^R(u_n) = P[M_{1,r_n} \leq u_n \mid X_1 > u_n]$. We shall find now that the limit of $\theta_n^R(u_n)$ is determined by the tail chain. Throughout, we assume Condition 10.8.

For $m = 2, 3, \ldots$, we have

$$\left| \theta_n^R(u_n) - P\left[\max_{i \geq 2} Y_i \leq 0\right] \right|$$

$$\leq P[M_{m,r_n} > u_n \mid X_1 > u_n] + P\left[\max_{i > m} Y_i > 0\right]$$

$$+ \left| P[M_{1,m} \leq u_n \mid X_1 > u_n] - P\left[\max_{2 \leq i \leq m} Y_i \leq 0\right] \right|. \qquad (10.49)$$

By (10.46), the last term on the right converges to zero as $n \to \infty$. Hence,

$$\limsup_{n \to \infty} \left| \theta_n^R(u_n) - P\left[\max_{i \geq 2} Y_i \leq 0\right] \right|$$

$$\leq \limsup_{n \to \infty} P[M_{m,r_n} > u_n \mid X_1 > u_n] + P\left[\max_{i > m} Y_i > 0\right].$$

Since m was arbitrary, we can let $m \to \infty$ to obtain, by (10.8) and (10.48),

$$\theta = \lim_{n \to \infty} \theta_n^R(u_n) = P\left[\max_{i \geq 2} Y_i \leq 0\right]. \qquad (10.50)$$

Observe that θ is indeed determined solely by the dependence structure in the chain: by (10.44),

$$\theta = P\left[\max_{i \geq 1} \prod_{j=1}^{i} A_i \leq U\right] \qquad (10.51)$$

(Perfekt 1994), where $U, A_1, A_2, \ldots$ are independent random variables with U uniformly distributed on $(0, 1)$ and the A_i distributed like A in (10.42).

10.4.3 Cluster statistics

Let c be a cluster functional (Definition 10.13) that is continuous almost everywhere. All the examples in section 10.3.2 satisfy this requirement. By Proposition 10.15, the distribution of the cluster statistic $c\{(X_i - u_n)/\sigma(u_n)\}_{i=1}^{r_n}$ conditional on $M_{r_n} > u_n$ converges to a limit that can be expressed in terms of the tail chain $\{Y_i\}$.

Using a similar decomposition as in (10.49), we obtain from (10.8), (10.48) and (10.20)

$$P\left[c\left\{\frac{X_i - u_n}{\sigma(u_n)}\right\}_{i=1}^{r_n} \in \cdot \;\middle|\; M_{r_n} > u_n\right] \tag{10.52}$$

$$\xrightarrow{\mathcal{D}} \theta^{-1}\left\{P[c(Y_1, Y_2, \ldots) \in \cdot] - P[c(Y_2, Y_3, \ldots) \in \cdot, \max_{i \geq 2} Y_i > 0]\right\}$$

(Yun 2000a). Here we have extended the domain of c to sequences $x_1, x_2, \ldots$ with only a finite number of positive members by setting $c(x_1, x_2, \ldots) = c(x_1, \ldots, x_r)$ where r is such that $x_i \leq 0$ for all $i > r$.

10.4.4   Statistical applications

In a practical data analysis, we might want to estimate the extremal index, for instance, to estimate high return levels as in section 10.2.3, or the distribution of a cluster statistic, for instance, the probability that the total amount of rainfall during a storm exceeds a high level. If we are willing to assume that the data $(x_1, \ldots, x_n)$ are a realization of a sample $(X_1, \ldots, X_n)$ from a stationary Markov chain satisfying the conditions of the previous sections, then we can use (10.51) and (10.52) to solve these problems.

Consider first the expression (10.51) for the extremal index. Given the bivariate extreme value distribution G, we can compute the distribution of the A_i, and then find θ in (10.51) by simulation or some other numerical technique. A fast method to compute the extremal index based on (10.51) that does not rely on direct simulation from the tail chain, but on the fast Fourier transform, is described in Hooghiemstra and Meester (1997).

For cluster statistics, we are usually interested in $c\{(X_i - u_n)\}_{i=1}^{r_n}$ without normalizing $\sigma(u_n)$. If c is invariant to scale, for example, if it depends only on $\mathbf{1}(X_i > u_n)$, then we can estimate the distribution of the cluster statistic by simulating the tail chain $\{Y_i\}$ for $1 \leq i \leq \max\{j \geq 1 : Y_j > 0\}$ according to the definition (10.43). In practice, we simulate $Y_1, \ldots, Y_r$, with r large enough such that the probability of a cluster being longer than r is negligible. Alternatively, if the distribution of the A_i has mass at $\{0\}$, an absorbing state, we can generate $r - 1$ from a geometric distribution with mean $1/P[A = 0]$. Simulating a large number of realizations of the tail chain allows the limit (10.52) to be approximated by a Monte Carlo average.

In cases where the normalization is needed, we must fix a threshold u and then, by (10.46), we can approximate the distribution of the cluster statistic conditional on the cluster maximum exceeding u by

$$\theta^{-1}\left\{P[c(\sigma Y_1, \sigma Y_2, \ldots) \in \cdot] - P[c(\sigma Y_2, \sigma Y_3, \ldots) \in \cdot, \max_{i \geq 2} Y_i > 0]\right\},$$

where $\sigma = \sigma(u)$.

A remarkable feature of these applications of the tail chain, which were invented by Yun (2000a), is that it requires knowledge of only the limiting forward transition probabilities. The sampling scheme of Smith *et al.* (1997) works differently: (1) generate a cluster maximum from the appropriate GP distribution as in (10.23); (2) generate the part of the cluster following the cluster maximum from the forward tail chain, rejecting samples that exceed the cluster maximum; (3) generate the part of the cluster preceding the cluster maximum from the backward tail chain, again rejecting those that exceed the cluster maximum. The backward tail chain, defined analogously to the forward tail chain, has transitions A_i with distribution function

$$P[A \le a] = \lim_{u \to x_*} P\left[\left\{1 + \gamma \frac{X_1 - u}{\sigma(u)}\right\}^{1/\gamma} \le a \,\middle|\, X_2 = u\right] = -V_{*2}(a, 1).$$

Although this scheme is intuitively straightforward, it is clearly less efficient than Yun's scheme, which only requires the forward tail chain and in which no samples need to be rejected. On the other hand, a benefit of the Smith *et al.* (1997) scheme is that it generates clusters directly, the empirical distribution of which can be used immediately as an estimate of the cluster distribution. A theoretical justification of the scheme is provided in Segers (2003b).

10.4.5 Fitting the Markov chain

It remains to estimate the marginal parameters, γ and $\sigma = \sigma(u)$, and the distribution of the A_i or, equivalently, the function V_* in (10.42). The estimation procedure basically consists of the censored-likelihood approach (section 9.4.2) as in Ledford and Tawn (1996), but now adapted to the Markov likelihood (10.33) as in Smith *et al.* (1997).

First we define our models for the marginal and joint distribution functions $F(x)$ and $F(x_1, x_2)$ in the regions $x > u$ and $x_i > u$ ($i = 1, 2$) for a sufficiently high threshold u. Denote $\lambda = \lambda(u) = 1 - F(u)$ and $\sigma = \sigma(u)$. Equation (10.34) suggests the approximation

$$F(x) \approx 1 - \lambda \left(1 + \gamma \frac{x - u}{\sigma}\right)_+^{-1/\gamma},$$

while from (10.35), using (10.39) and (10.36),

$$F(x_1, x_2) \approx 1 - \lambda V\left(\frac{x_1 - u}{\sigma}, \frac{x_2 - u}{\sigma}\right) = 1 - V_*(z_1, z_2), \tag{10.53}$$

$$\text{with} \quad z_i = \lambda^{-1}\left(1 + \gamma \frac{x_i - u}{\sigma}\right)_+^{1/\gamma}, \qquad i = 1, 2. \tag{10.54}$$

Slightly more accurate would be to use the tail equivalent models (9.67) and (9.68), but for simplicity we stick to the models above as in Smith *et al.* (1997).

As the models above are specified only for observations exceeding the threshold u, we must treat observations below the threshold as being censored at that threshold. Specifically, the marginal likelihood for a single observation x is set equal to

$$
f_u(x) = \begin{cases} \dfrac{\lambda}{\sigma}\left(1 + \gamma\dfrac{x-u}{\sigma}\right)_+^{-1/\gamma - 1} & \text{if } x > u, \\ 1 - \lambda & \text{if } x \le u, \end{cases}
$$

and the joint likelihood of a pair (x_1, x_2) is set equal to

$$
\begin{aligned}
&f_u(x_1, x_2) \\
&= \begin{cases} \dfrac{\partial^2}{\partial x_1 \partial x_2} F(x_1, x_2) \approx -\dfrac{\partial z_1}{\partial x_1}\dfrac{\partial z_2}{\partial x_2} V_{*12}(z_1, z_2) & \text{if } x_1 > u, x_2 > u \\[2mm] \dfrac{\partial}{\partial x_1} F(x_1, u) \approx -\dfrac{\partial z_1}{\partial x_1} V_{*1}(z_1, \lambda^{-1}) & \text{if } x_1 > u \ge x_2 \\[2mm] \dfrac{\partial}{\partial x_2} F(u, x_2) \approx -\dfrac{\partial z_2}{\partial x_2} V_{*2}(\lambda^{-1}, z_2) & \text{if } x_1 \le u < x_2 \\[2mm] F(u, u) \approx 1 - V_*(\lambda^{-1}, \lambda^{-1}) & \text{if } x_1 \le u, x_2 \le u, \end{cases}
\end{aligned}
$$

subscripts on V_* denoting partial derivatives and with (z_1, z_2) as in (10.54). Finally, the censored likelihood of a sample $(x_1, \ldots, x_n)$ is defined by replacing f with f_u in (10.33).

Usually we assume that the function V_* belongs to some parametric family, $V_*(\cdot \mid \boldsymbol{\theta})$ say, and estimate the unknown parameters $(\gamma, \sigma, \boldsymbol{\theta})$ by maximizing the censored likelihood; λ can be set equal to the ratio of the number of exceedances to n. Four such models for V_* are listed below; see section 9.2 for a more extensive list. Once we have estimated the model, we can implement the simulation schemes of the previous section to obtain estimates of the extremal index and properties of cluster statistics. Confidence intervals can be obtained by bootstrapping the observed Markov chain according to the scheme described in section 10.3.4 and refitting the model to each sequence. An alternative, more crude, approach could be to resample the maximum-likelihood parameter estimates from their estimated asymptotic multivariate normal distribution, assuming the usual properties of maximum-likelihood estimators hold.

Parametric models

For easy reference, we repeat here a couple of parametric models for V_* together with the corresponding distribution for A as in (10.42).

Asymmetric logistic model   (Tawn 1988a,b)

$$
V_*(z_1, z_2) = (1 - \psi_1)z_1^{-1} + (1 - \psi_2)z_2^{-1} + \{(\psi_1/z_1)^{1/\alpha} + (\psi_2/z_2)^{1/\alpha}\}^{\alpha}
$$

for $0 \leq \psi_i \leq 1$ ($i = 1, 2$) and $0 < \alpha \leq 1$, see (9.7). The logistic model arises as a special case, if $\psi_1 = \psi_2 = 1$. If $0 < \alpha < 1$, the associated transition distribution has $P[A = 0] = 1 - \psi_1$ and

$$P[A \leq a] = 1 - \psi_1 + \psi_1^{1/\alpha}(\psi_1^{1/\alpha} + \psi_2^{1/\alpha} a^{-1/\alpha})^{\alpha - 1}, \qquad a > 0.$$

In case $\alpha = 1$, we have $P[A = 0] = 1$ regardless of the ψ_i.

Asymmetric negative logistic model   (Joe 1990)

$$V_*(z_1, z_2) = z_1^{-1} + z_2^{-1} - \{(z_1/\psi_1)^r + (z_2/\psi_2)^r\}^{-1/r}$$

for $0 \leq \psi_1 \leq 1$ ($i = 1, 2$) and $r > 0$, see (9.13) where $\alpha = -1/r$. The associated transition distribution has $P[A = 0] = 1 - \psi_1$ and

$$P[A \leq a] = 1 - \psi_1^{-r}(\psi_1^{-r} + \psi_2^{-r} a^r)^{-1/r - 1}, \qquad a > 0.$$

In the limiting case $r = 0$, again $P[A = 0] = 1$.

Bilogistic model   (Smith 1990b)

$$V_*(z_1, z_2) = z_1^{-1} q^{1-\alpha} + z_2^{-1}(1 - q)^{1-\beta}$$

for $0 < \alpha < 1, 0 < \beta < 1$, and where $q = q(z_1, z_2)$ solves

$$(1 - \alpha)z_1^{-1}(1 - q)^\beta = (1 - \beta)z_2^{-1}q^\alpha, \qquad\qquad (10.55)$$

see (9.9). The associated transition distribution is

$$P[A \leq a] = q^{1-\alpha}, \qquad a > 0,$$

where q solves (10.55) when $z_1 = 1$ and $z_2 = a$.

Negative bilogistic model   (Coles and Tawn 1994)

$$V_*(z_1, z_2) = z_1^{-1} + z_2^{-1} - \{z_1^{-1}q^{1+\alpha} + z_2^{-1}(1 - q)^{1+\beta}\}$$

for $\alpha > 0, \beta > 0$, and where q solves

$$(1 + \alpha)z_1^{-1}q^\alpha = (1 + \beta)z_2^{-1}(1 - q)^\beta. \qquad\qquad (10.56)$$

The associated transition distribution is

$$P[A \leq a] = 1 - q^{1+\alpha}, \qquad a > 0,$$

where q solves (10.56) when $z_1 = 1$ and $z_2 = a$.

   Symmetric models are obtained from the first two models when $\psi_1 = \psi_2$ or from the last two models when $\alpha = \beta$.

10.4.6 Additional topics

Threshold dependence

Model (10.53) for the Markov chain assumes that the dependence between consecutive exceedances of a high threshold does not change as the threshold is increased. This is acceptable if we really are interested in the asymptotic properties of the process. Typically, however, we are interested in high, but finite, levels at which the process may behave very differently. For example, if the joint distribution of (X_1, X_2) is in the domain of attraction of an extreme value distribution with independent margins, that is, X_1 and X_2 are asymptotically independent, then $\theta = 1$ and there is no clustering in the limit. Clustering may occur at finite levels, however, and inferences such as return-level estimation can be improved, if we recognize that $\theta(u) < 1$. The asymptotically dependent model (10.53) is particularly inadequate in this situation because $\theta = 1$ can be achieved only if X_1 and X_2 are completely independent. In this section, we obtain threshold-dependent estimates of the extremal index and cluster statistics by extending the model (10.53) and using a penultimate approximation to the tail chain (10.43); see Bortot and Tawn (1998).

The model for the distribution of (X_1, X_2) in the joint-tail region $x_i \geq u$ ($i = 1, 2$) is taken from Ledford and Tawn (1997); see also section 9.5. Specifically,

$$\bar{F}(x_1, x_2) := P[X_1 > x_1, X_2 > x_2]$$

$$\approx \mathcal{L}(z_1, z_2) z_1^{-c_1} z_2^{-c_2}, \tag{10.57}$$

where $z_i \approx 1/\bar{F}(x_i)$ is the transformation (10.54), $\mathcal{L}$ is a bivariate slowly varying function, and c_1 and c_2 are positive parameters satisfying $c_1 + c_2 \geq 1$. The coefficient of tail dependence, η, defined by the limit

$$\lim_{t \to \infty} \bar{F}(tx, tx)/\bar{F}(t, t) = x^{-1/\eta}, \qquad 0 < x < \infty, \tag{10.58}$$

is $\eta = 1/(c_1 + c_2)$. If $c_1 + c_2 > 1$ then $\eta < 1$ and thus $P[X_2 > x \mid X_1 > x] \to 0$ as $x \to x_*$, that is, the pair (X_1, X_2) is asymptotically independent. In that case, we obtain $P[A = 0] = 1$ in (10.42), and the extremal index (10.51) is equal to unity, that is, there is no clustering in the limit.

Estimation proceeds with the censored likelihood of section 10.4.5 adapted to the new model, a possible parametric form for $\mathcal{L}$ being

$$\mathcal{L}(z_1, z_2) = a_0 + (z_1 z_2)^{-1/2} \{z_1 + z_2 - z_1 z_2 V_*(z_1, z_2)\}, \tag{10.59}$$

with $a_0 \geq 0$ and where V_* is one of the parametric models listed in section 10.4.4. The special case $c_1 = c_2 = 1/2$ and $a_0 = 0$ leads back to the previous model (10.53).

Suppose now that we want to find the extremal index or the distribution of a cluster statistic at some finite threshold $u_1 \geq u$. We can still use the tail-chain approximation (10.46), replacing u with u_1, and where Y_n is defined by (10.43).

However, instead of simulating the A_i from their degenerate limit distribution, we use (10.47) to simulate from the penultimate form

$$F_A(a; v) = P\left[\left\{1 + \gamma \frac{X_2 - v}{\sigma(v)}\right\}^{1/\gamma} \leq a \,\middle|\, X_1 = v\right]$$

$$\approx 1 - \lambda^{c_1 + c_2 - 1} a^{-c_2} \{c_1 \mathcal{L}(a\lambda^{-1}, \lambda^{-1}) - \lambda^{-1} \mathcal{L}_1(a\lambda^{-1}, \lambda^{-1})\},$$

with $\lambda = \bar{F}(v)$. Since this distribution depends on the particular value of the conditioning variable, v, the A_i are no longer identically distributed: given Y_i, we simulate A_i from $F_A\{\cdot; u_1 + \sigma(u_1)Y_i\}$. The tail chain can be simulated either for a fixed time r, as in section 10.4.4, or stopped when $X_i = u_1 + \sigma(u_1)Y_i$ falls below u, at which point the justification for the model is lost.

Non-parametric estimation

It is not necessary to fit a bivariate parametric model to obtain the distribution of the transitions A_i. The transitions satisfy

$$A_i = \left(\frac{1 + \gamma Y_{i+1}}{1 + \gamma Y_i}\right)^{1/\gamma}, \quad i = 1, 2, \ldots,$$

where Y_i approximates $(X_i - u)/\sigma(u)$ when $X_i > u$. In the special case that the X_i are standard exponentially distributed, we have $\gamma = 0$, $\sigma(u) = 1$, and $A_i = \exp(Y_{i+1} - Y_i) \approx \exp(X_{i+1} - X_i)$. For data $\{x_j\}_{1 \leq j \leq n}$, therefore, we can define the empirical values of A_i to be

$$\left\{\exp\left(\tilde{x}_{j+i} - \tilde{x}_{j+i-1}\right) : x_j > u, 1 \leq j \leq n - i\right\}, \tag{10.60}$$

where $\tilde{x}_j$ are the data transformed to standard exponential margins, for instance, by the empirical distribution function. The transition distribution can be estimated with a kernel density estimator based on these empirical values (Bortot and Coles 2000). Such an estimate also provides a method for assessing the fit of parametric models.

Higher-order Markov chains

Extremes of d-order Markov chains, $d \geq 1$, were considered in Yun (1998, 2000a). The ideas remain the same, but the appropriate higher-order transition probabilities lead to a tail chain that also has order d. Statistical modelling requires a $(d + 1)$-variate extreme value distribution, suitably restricted to ensure stationarity and fitted with the appropriate extension of the likelihood in section 10.4.5. To select between models of different order, it is advantageous for the lower-order model to be nested within the higher-order model. In this case, the models can be compared by evaluating both of them for the higher-order likelihood: the form of the censored likelihood means that likelihoods of different orders are not necessarily comparable.

10.4.7 Data example

In this section, we fit first-order Markov models to the Uccle data of section 10.2.2, consider the issue of asymptotic independence and compare the simulated cluster characteristics to the empirical estimates of section 10.3.5.

We fit Markov chains with the six asymptotic dependence structures listed in Table 10.1 at thresholds ranging from the 90% to the 99.5% empirical quantile. As with the compound Poisson models of section 10.3.5, parameter estimates are stable above the 96% threshold, and constraining the asymmetric logistic and asymmetric negative logistic models to $\psi_2 = 1$ causes almost no change in the maximum likelihood. The model fits at the 96% threshold are summarized in Table 10.1.

Symmetry corresponds to the hypothesis $\alpha = \beta$ in the case of the bilogistic and negative bilogistic models. Under this hypothesis, the models reduce to the logistic and negative logistic, and a likelihood-ratio test gives no indication of asymmetry. Note that we assume here and elsewhere that standard likelihood properties hold even though the censored likelihood is an approximation to the joint density. Simulating test statistics under the null hypothesis is an alternative, but computationally expensive, approach. In the case of the asymmetric logistic and asymmetric negative logistic models with $\psi_2 = 1$, symmetry corresponds to the boundary value $\psi_1 = 1$. This is one example of the nonregular problems encountered in multivariate extremes (Tawn 1988a, 1990); the likelihood-ratio statistic should be compared to a one-half chi-squared distribution with one degree of freedom. For the asymmetric logistic model, the statistic is 2.12 with p-value $P[\chi_1^2 \geq 2.12]/2 = 0.073$,

Table 10.1 Parameter estimates, standard errors, negative log-likelihoods and extremal indices for six asymptotically dependent Markov models. The asymmetric logistic and asymmetric negative logistic models are constrained to $\psi_2 = 1$, with as special cases for $\psi_1 = 1$ the logistic and negative logistic models, respectively.

Model	σ	γ	Dependence	NLLH	θ
Logistic	2.8	−0.30	$\alpha = 0.67$ (0.04)	597.15	0.54
	(0.4)	(0.11)			
Bilogistic	2.7	−0.29	$\alpha = 0.74$ (0.05)	595.89	0.55
	(0.4)	(0.11)	$\beta = 0.58$ (0.08)		
Asymmetric logistic	2.8	−0.30	$\alpha = 0.62$ (0.06)	596.09	0.56
	(0.4)	(0.12)	$\psi_1 = 0.76$ (0.14)		
Negative logistic	2.7	−0.28	$r = 0.77$ (0.09)	597.63	0.54
	(0.4)	(0.11)			
Negative bilogistic	2.7	−0.27	$\alpha = 0.89$ (0.04)	596.71	0.54
	(0.07)	(0.04)	$\beta = 1.81$ (0.07)		
Asymmetric	2.7	−0.28	$r = 0.92$ (0.16)	596.51	0.55
Negative logistic	(0.4)	(0.11)	$\psi_1 = 0.75$ (0.14)		

and for the asymmetric negative logistic model, the statistic is 2.24 with p-value
0.067. We conclude that there is only weak evidence for asymmetry and we proceed
with the symmetric logistic model.

We can assess how well the model fits the data with some diagnostic plots. The
estimated shape parameter is greater than that obtained from the marginal analysis
($\hat{\gamma} = -0.42$) and the quantile plot for threshold excesses is poor. That there is
little to choose between the models featured in Table 10.1 is exemplified by the
similarity of the estimates of the Pickands dependence function A in Figure 10.9.
Recall from (8.54) that the Pickands dependence function of a bivariate extreme
value distribution is defined by $A(w) = V_*\{(1 - w)^{-1}, w^{-1}\}$ for $0 < w < 1$. In
addition, the parametric estimates are close to the non-parametric one by Capéraà
and Fougères (2000a); see also section 9.4.1.

We also investigate how closely the data follow the asymptotic tail chain $\{Y_n\}$
of the model by comparing the empirical values (10.60) of the transitions with
their estimated distribution in Figure 10.10. The joint density plot shows that the
empirical values are negatively correlated, so we would need a higher threshold
to find the independence structure of the tail chain. On the other hand, the dis-
crepancies between the empirical and model marginal distributions are sufficiently

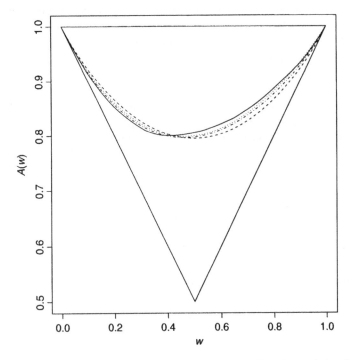

Figure 10.9 Estimates of the Pickands dependence function of the bivariate
Markov model: non-parametric (———), logistic (- - - - -), asymmetric logistic
($\cdots\cdots$) and bilogistic ($-\cdot-\cdot-$).

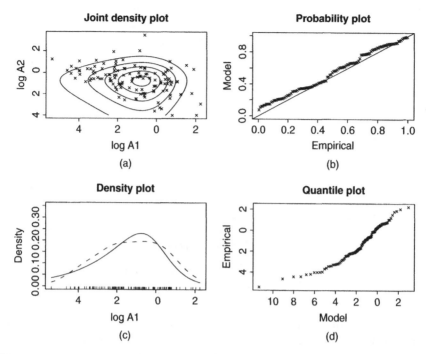

Figure 10.10   Diagnostic plots for the tail-chain transitions. The joint density plot shows the empirical transitions with model contours; the density plot shows the model estimate (———) and a kernel density estimate (- - - - -); the probability and quantile plots refer to the transitions A_1.

small that an empirical version of the tail chain could be obtained by simulating independent transitions from the kernel density estimate in Figure 10.10.

Extremal characteristics of the fitted logistic model are found from 10 000 simulations of the model tail chain with length $r = 100$. The extremal index is 0.54 with bootstrapped 95% confidence interval (0.42, 0.69), the mean cluster size is 1.84 (1.45, 2.37) and the mean number of up-crossings per cluster is 1.09 (1.04, 1.17). The cluster-size distribution is $\hat{\pi}(1) = 0.60$, $\hat{\pi}(2) = 0.20$, $\hat{\pi}(3) = 0.10$ and $\hat{\pi}(4) = 0.05$. Figure 10.11 exhibits the estimate of the distribution of the aggregate cluster excess that deviates from the empirical estimate mainly around $1°C$–$4°C$. The Markov model produces clusters that are smaller than, but in general agreement with, those found empirically at the same threshold. The choice of parametric model in fact has little influence on the extremal characteristics: witness the extremal indices from all six models displayed in Table 10.1.

The estimates of the 100, 1 000 and 10 000 July return levels with bootstrapped 95% confidence intervals are 37.6 (36.3, 39.0), 38.9 (37.0, 41.9) and 39.6 (37.2, 44.2); the estimated upper end-point is 40.3 (37.2, 53.9). These are larger than the estimates from the marginal analysis in section 10.2.2, due principally to the

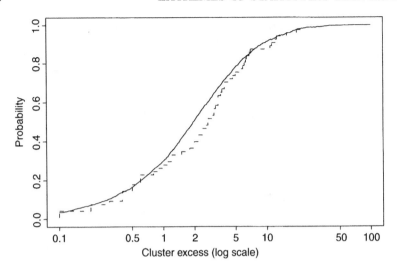

Figure 10.11   Empirical distribution function (- - - - -) and estimate from the Markov model (——) of cluster excess at the 96% threshold.

different shape parameters, and intimate a deficiency in this Markov model. This is in line with findings of Dupuis and Tawn (2001) that misspecification of the dependence model may corrupt estimates of marginal parameters.

We have noted some evidence for asymptotic independence, such as the empirical estimates of the extremal index in Figure 10.5 that increase at high thresholds. To assess this evidence more formally, we test $\eta = 1$, where η is the coefficient of tail dependence (10.58); see also section 9.5. First we transform the data $X_1, \ldots, X_n$ to approximate standard Pareto margins by $Z_i = 1/\{1 - \hat{F}_n(X_i)\}$, where $\hat{F}_n$ is the empirical distribution function; an alternative is to transform to standard Fréchet margins. Next, define

$$T_i = \min(Z_i, Z_{i+1}) \quad \text{for } i \in \{j : X_j \text{ and } X_{j+1} \text{ fall in the same year}\}.$$

In view of (10.58), the tail function of T_i is regularly varying with index η. Hence, if $T_{(1)} > T_{(2)} > \ldots$ are the T_i in descending order, then Hill's estimator for η is

$$\hat{\eta} = \frac{1}{k-1} \sum_{i=1}^{k-1} \log T_{(i)} - \log T_{(k)}$$

see, for instance, Ledford and Tawn (2003). Values of $\hat{\eta}$ for different k are reproduced in Figure 10.12, with bootstrapped 95% confidence intervals constructed by resampling the data blocked by year. The estimates are about 0.8 and are significantly less than 1 for all values of k. There is some evidence, therefore, that the series is asymptotically independent and we should be wary of extrapolating the results obtained from the previous Markov-chain model.

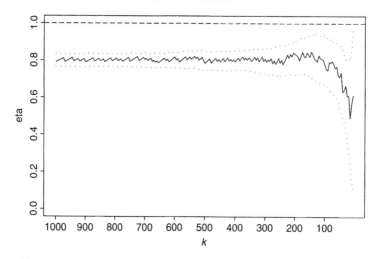

Figure 10.12   Hill's estimates (——) of the coefficient of tail dependence, η, against the number of order statistics, with bootstrapped 95% confidence intervals.

Asymptotic independence can be handled by model (10.57) and supports cluster characteristics that can change with threshold. We choose again the symmetric logistic model for V_* in (10.59) and find that parameter estimates are stable above the 96% threshold, although a_0 is poorly estimated. At the 96% threshold, $\hat{\eta} = 0.84$ and the p-value for the nonregular, likelihood-ratio test of $\eta = 1$ (Bortot and Tawn 1998) is 0.03, confirming our earlier conclusion of asymptotic independence. A likelihood-ratio test does not reject $c_1 = c_2$ so we refit the model with this constraint, obtaining $\hat{\sigma} = 2.7$ (0.4), $\hat{\gamma} = -0.35$ (0.10), $\hat{c}_1 = \hat{c}_2 = 0.59$ (0.07), $\hat{a}_0 = 0.2$ (0.3) and $\hat{\alpha} = 0.53$ (0.10).

The estimates of the extremal index from this model, obtained by simulating tail chains of length $r = 20$ and truncating once the chain falls below the model threshold, are reproduced in Figure 10.13. Other cluster characteristics were simulated too: the mean cluster size decreased from 1.73 at the 96% threshold to 1.47 by the 99.5% threshold; the mean number of up-crossings per cluster rose from 1.00 to 1.06; and $\hat{\pi}(1)$ increased from 0.60 to 0.69, which is consistent with the empirical estimates in Figure 10.8.

When the extremal index changes with threshold, return-level estimation is improved if the approximation $P[M_n \leq x] \approx \{F(x)\}^{n\theta}$ is used with $\theta = \theta(x)$. The return levels obtained in this way from our model are 37.1 (36.2, 38.0), 38.1 (36.8, 40.0) and 38.5 (37.0, 41.4), with upper end-point 38.8 (37.1, 44.7). These are close to the return levels estimated from the GEV model, principally because of the similar shape parameters.

This concludes our analysis of the Uccle data. We have found evidence for asymptotic independence, which means that cluster characteristics change with threshold. Within the data, the empirical estimates of section 10.3.5 provide a

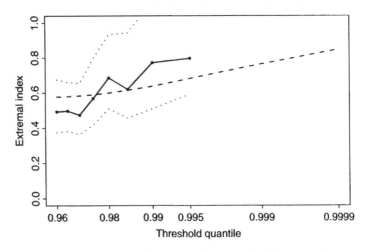

Figure 10.13   Extremal index estimates against threshold on complementary log-log scale: empirical (—o—) with bootstrapped 95% confidence intervals (······) and from the asymptotically independent Markov model (-----).

valuable description, but if inference is required for levels at which we have no data then the asymptotic independent Markov model of this section can be used.

Return levels from the different models are summarized in Table 10.2. Of the marginal models, we prefer the GP model for threshold exceedances to the GEV model for block maxima because the estimates are more precise. In section 10.3.5, the GP return levels were estimated with $\theta = 0.49$. In light of asymptotic independence, we should use $\theta = 1$, which yields estimates that are closer to the GEV estimates. The asymptotically dependent model is inconsistent with the other results because of its larger shape parameter. The asymptotically independent model, however, produces estimates similar to the GEV estimates and with similar confidence intervals. We can conclude with some confidence, therefore, that the point estimates from the GEV model are good estimates of the true July return levels.

Table 10.2   Return levels (°C) with 95% confidence intervals and shape parameters from five models: GP with $\theta = 0.49$, GP; with $\theta = 1$, GP1; GEV; asymptotically independent Markov chain, MCI; asymptotically dependent Markov chain, MCD.

Model	100	1 000	10 000	γ
GP	36.5 (35.7, 36.9)	37.2 (36.2, 38.1)	37.5 (36.3, 38.7)	−0.42
GP1	36.8 (35.9, 37.2)	37.3 (36.3, 38.3)	37.5 (36.4, 38.9)	−0.42
GEV	36.9 (36.2, 38.6)	37.9 (36.9, 40.5)	38.3 (37.2, 41.8)	−0.34
MCI	37.1 (36.2, 38.0)	38.1 (36.8, 40.0)	38.5 (37.0, 41.4)	−0.35
MCD	37.6 (36.3, 39.0)	38.9 (37.0, 41.9)	39.6 (37.2, 44.2)	−0.30

10.5 Multivariate Stationary Processes

Up to now the setting of this chapter consisted of a univariate stationary time series. Complementarily, the framework of Chapters 8 and 9 was that of independent multivariate observations. In this section, we join both lines to the study of extremes of multivariate stationary time series. Although this area is relatively unexplored, some theory is already available, mainly on the vector of component-wise maxima. In particular, we shall encounter an appropriate generalization of the extremal limit theorem (ELT) in section 10.5.1 and of the extremal index in section 10.5.2. These results, however, have so far hardly led to any practical statistical procedures. It is our hope, therefore, that the present overview of the theory might stimulate further research in the area.

10.5.1 The extremal limit theorem

Let $X_n = (X_{n,1}, \ldots, X_{n,d})$, $n \geq 1$, be a stationary sequence of random vectors in $\mathbb{R}^d$ with distribution function F. We seek to model the extremes of the process. A natural starting point is the sample maximum, defined as the vector of component-wise maxima,

$$M_n = \left(\max_{i=1,\ldots,n} X_{i,1}, \ldots, \max_{i=1,\ldots,n} X_{i,d} \right).$$

We shall investigate the asymptotic distribution of $a_n^{-1}(M_n - b_n)$, where $a_n > 0 = (0, \ldots, 0)$ and b_n are d-dimensional vectors. By convention, operations on and relations between such vectors are to be read component-wise.

The case of independent vectors X_n was treated in Chapter 8. A central problem there was to characterize the class of distribution functions G with non-degenerate margins that can arise as the limit in

$$P[a_n^{-1}(M_n - b_n) \leq x] \overset{\mathcal{D}}{\to} G(x), \qquad n \to \infty. \tag{10.61}$$

This gave rise to the class of multivariate extreme value distributions that were described in detail. In the stationary case now, we shall seek conditions so that any limit distribution G in (10.61) must be a d-variate extreme value distribution as well. This will provide a proper generalization of the univariate ELT (Theorem 10.2). As in the univariate case, the long-range dependence in the process will need to be restricted in some way.

At this stage it pays off to reflect a little on the structure of the arguments in the univariate case. Let $\{X_n\}$ be a stationary sequence of univariate random variables and recall the notation of section 10.2. For a sequence of thresholds u_n consider the events $A_{n,i} = \{X_i \leq u_n\}$. Observe that for fixed n the sequence of indicator variables $\{\mathbf{1}(A_{n,i})\}_{i \geq 1}$ is stationary.

The crucial step in the proof of Theorem 10.2 is the decomposition (10.3) $P[M_n \leq u_n] = \{P[M_{r_n} \leq u_n]\}^{\lfloor n/r_n \rfloor} + o(1)$ for a positive integer sequence r_n tending to infinity but at a slower rate than n. It is a useful exercise to rewrite the whole argument leading to (10.3) in terms of the events $A_{n,i}$. Explicitly, for a set I of positive integers we can write

$$P[M(I) \leq u_n] = P\left[\bigcap_{i \in I}\{X_i \leq u_n\}\right] = P\left[\bigcap_{i \in I} A_{n,i}\right].$$

The $D(u_n)$ condition required in the theorem can be expressed in terms of the events $A_{n,i}$ as well since

$$\alpha(n, s) = \max_{1 \leq l \leq n-s} \max_{I,J} \left| P\left[\bigcap_{i \in I \cup J} A_{n,i}\right] - P\left[\bigcap_{i \in I} A_{n,i}\right] P\left[\bigcap_{i \in J} A_{n,i}\right] \right| \quad (10.62)$$

the second maximum ranging over all $I \subseteq \{1, \ldots, l\}$ and $J \subseteq \{l + s, n\}$.

How does this help us in the multivariate case? Let u_n be a sequence of d-dimensional thresholds and consider the events $A_{n,i} = \{X_i \leq u_n\}$, the ordering of vectors being component-wise. Clearly, the translated version of the univariate argument goes through without change. In particular, define $\alpha(n, s)$ as in (10.62) and say that *Condition $D(u_n)$* holds if $\alpha(n, s_n) \to 0$ for some positive integer sequence s_n such that $s_n = o(n)$. We arrive at the multivariate version of the ELT, due to Hsing (1989) and Hüsler (1990).

Theorem 10.22 *Let $\{X_n\}$ be a stationary sequence for which there exist sequences of constant vectors $a_n > 0$ and b_n, and a distribution function G with non-degenerate margins such that*

$$P[a_n^{-1}(M_n - b_n) \leq x] \overset{\mathcal{D}}{\to} G(x), \quad n \to \infty.$$

If $D(u_n)$ holds with $u_n = a_n x + b_n$ for each x such that $G(x) > 0$, then G is a d-variate extreme value distribution function.

The dependence may affect the limiting distribution G in the sense that it can be different from the corresponding limit $\tilde{G}$ for the associated, independent sequence $\{\tilde{X}_n\}$ of random vectors with the same marginal distribution as X_1. So what is the connection between G and $\tilde{G}$ and when are they the same?

The latter question is the easier one to answer. *Condition $D'(u_n)$* holds if

$$\lim_{k \to \infty} \limsup_{n \to \infty} n \sum_{i=1}^{\lfloor n/k \rfloor} P[X_1 \not\leq u_n, X_i \not\leq u_n] = 0.$$

Observe that this is the direct translation of Condition $D'(u_n)$ via the $A_{n,i}$. The arguments in the univariate case go through here as well: the inclusion-exclusion formula and $D'(u_n)$ give

$$P[M_{r_n} \not\leq u_n] = r_n \bar{F}(u_n) + o(r_n/n)$$

whenever $r_n = o(n)$, so that

$$P[M_n \leq u_n] = \{P[M_{r_n} \leq u_n]\}^{\lfloor n/r_n \rfloor} + o(1) = \{F(u_n)\}^n + o(1),$$

provided $n\alpha(n, s_n) = o(r_n)$ for some $s_n = o(r_n)$. We obtain the following result.

Theorem 10.23 *Let G be a d-variate extreme value distribution and let $a_n > 0$ and b_n be d-dimensional vectors such that $D(u_n)$ and $D'(u_n)$ hold for every $u_n = a_n x + b_n$ with $x \in \mathbb{R}^d$ such that $G(x) > 0$. Then*

$$P[a_n^{-1}(M_n - b_n) \leq x] \xrightarrow{\mathcal{D}} G(x), \qquad n \to \infty,$$

if and only if

$$F^n(a_n x_n + b_n) \xrightarrow{\mathcal{D}} G(x), \qquad n \to \infty.$$

10.5.2 The multivariate extremal index

Recall that under the $D'(u_n)$ condition the asymptotic distribution of M_n is the same as in the case of an independent sequence. The reason is that the $D'(u_n)$ condition prevents local clustering of extremes, so that the temporal dependence becomes negligible at high-levels. Things become different, however, if we allow for local dependence at such high levels as well. Whereas in the univariate case, the effect of local dependence was summarized by a single number, the extremal index, the multivariate setting is more difficult: the analogue of the extremal index turns out to be a function (Nandagopalan 1994; Perfekt 1997; Smith and Weissman 1996).

Let again $\{X_n\}$ be a stationary sequence of random vectors in $\mathbb{R}^d$ with distribution function F. Assume that there are vectors $a_n > 0$ and b_n and d-variate extreme value distributions G and $\tilde{G}$ such that

$$P[a_n^{-1}(M_n - b_n) \leq x] \xrightarrow{\mathcal{D}} G(x),$$

$$F^n(a_n x + b_n) \xrightarrow{\mathcal{D}} \tilde{G}(x),$$

as $n \to \infty$. Assume also that the jth marginal series $\{X_{n,j}\}_n$ has extremal index $0 < \theta_j \leq 1$, so that the margins of G and $\tilde{G}$ are related by $G_j(x) = \{\tilde{G}_j(x)\}^{\theta_j}$ for $j = 1, \ldots, d$. The θ_j need not be the same, showing that the connection between G and $\tilde{G}$ may be more complicated than in the univariate case. We will also need the stable tail dependence functions l and $\tilde{l}$ of G and $\tilde{G}$, defined by

$$G(x) = \exp[-l\{-\log G_1(x_1), \ldots, -\log G_d(x_d)\}],$$

$$\tilde{G}(x) = \exp[-\tilde{l}\{-\log \tilde{G}_1(x_1), \ldots, -\log \tilde{G}_d(x_d)\}],$$

see (8.12).

Definition

To define the multivariate extremal index, it is convenient to make abstraction of the margins. For $v \in [0, \infty) \setminus \{0\}$, let $x = x(v)$ be such that $v_j = -\log \tilde{G}_j(x_j) = -\theta_j^{-1} \log G_j(x_j)$ for $j = 1, \ldots, d$. In case $v_j = 0$, we set $x_j = \sup\{x \in \mathbb{R} : \tilde{G}_j(x) < 1\}$. Let $x_n = x_n(v)$ be a sequence in $\mathbb{R}^d$ such that $x_n \to x$ as $n \to \infty$ and let $u_n = a_n x_n + b_n$. Clearly

$$\lim_{n \to \infty} n P[X_{1,j} > u_{n,j}] = v_j, \qquad j = 1, \ldots, d, \qquad (10.63)$$

together with

$$\lim_{n \to \infty} P[M_n \le u_n] = G(x), \qquad \lim_{n \to \infty} F^n(u_n) = \tilde{G}(x).$$

Now define the *extremal index function*, or *extremal index* in short, of the sequence $\{X_n\}$ by

$$\theta(v) = \frac{\log G(x)}{\log \tilde{G}(x)}, \qquad v \in [0, \infty) \setminus \{0\}. \qquad (10.64)$$

This is a straightforward extension of the definition in the univariate case (Theorem 10.4). In terms of the stable tail dependence functions, we have

$$\theta(v) = \frac{l(\theta_1 v_1, \ldots, \theta_d v_d)}{\tilde{l}(v_1, \ldots, v_d)}, \qquad v \in [0, \infty) \setminus \{0\}. \qquad (10.65)$$

Properties

The multivariate extremal index satisfies a number of properties.

  (i) $\theta(v)$ is a continuous function in v.

  (ii) $\theta(cv) = \theta(v)$ for $0 < c < \infty$ and $v \in [0, \infty) \setminus \{0\}$.

  (iii) for $j = 1, \ldots, d$ we have $\theta(e_j) = \theta_j$ where e_j is the jth unit vector.

  (iv) $0 \le \theta(\cdot) \le 1$.

Properties (i–iii) are immediate consequences of (10.65) and properties of stable tail dependence functions. To prove (iv), observe first that, with $x = x(v)$ and $u_n = a_n x_n + b_n$ as above,

$$P[M_n \le u_n] = 1 - P[M_n \not\le u_n] \ge 1 - n\{1 - F(u_n)\}$$

so that

$$G(x) = \lim_{n \to \infty} P[M_n \le u_n] \ge \lim_{n \to \infty} [1 - n\{1 - F(u_n)\}] = 1 + \log \tilde{G}(x),$$

and thus $\exp\{-l(\theta_1 v_1, \ldots, \theta_d v_d)\} \geq 1 - \tilde{l}(v)$. This inequality and property (ii) imply

$$\theta(v) = \lim_{s\downarrow 0} \frac{l(s\theta_1 v_1, \ldots, s\theta_d v_d)}{\tilde{l}(sv)} \leq \lim_{s\downarrow 0} \frac{-\log\{1 - \tilde{l}(sv)\}}{\tilde{l}(sv)} = 1,$$

whence (iv).

Property (iii) can be extended to a univariate characterization of the multivariate extremal index (Smith and Weissman 1996). Consider the random variables

$$Y_n(v) = \max_{j=1,\ldots,d} \frac{v_j}{1 - F_j(X_{n,j})}, \qquad v \in [0, \infty) \setminus \{0\}. \tag{10.66}$$

Denoting the quantile function of F_j by $F_j^{\leftarrow}(p) = \inf\{x \in \mathbb{R} : F_j(x) \geq p\}$ $(0 < p < 1)$, we have, assuming for simplicity that F_j is continuous,

$$P\left[\max_{i=1,\ldots,n} Y_i(v) \leq n\right] = P\left[\max_{i=1,\ldots,n} F_j(X_{i,j}) \leq 1 - \frac{v_j}{n}, \forall j = 1, \ldots, d\right]$$

$$= P\left[M_{n,j} \leq F_j^{\leftarrow}\left(1 - \frac{v_j}{n}\right), \forall j = 1, \ldots, d\right]$$

$$\to G(x), \qquad n \to \infty,$$

by (10.63). Similarly, $\{P[Y_1(v) \leq n]\}^n \to \tilde{G}(x)$ as $n \to \infty$. Hence

(v) $\theta(v)$ is the (univariate) extremal index of the sequence $\{Y_n(v)\}$.

Finally, we mention that the multivariate extremal index admits similar interpretations as the univariate one. For instance, under condition $D\{u_n(v)\}$ and for suitable integers $r_n = o(n)$ we have $\theta(v) = \lim \theta_n^B(v) = \lim \theta_n^R(v)$ where

$$\frac{1}{\theta_n^B(v)} = \frac{r_n\{1 - F(u_n)\}}{P[\exists k = 1, \ldots, r_n : X_k \not\leq u_n]}$$

$$= E\left[\sum_{k=1}^{r_n} \mathbf{1}(X_k \not\leq u_n) \,\middle|\, \exists k = 1, \ldots, r_n : X_k \not\leq u_n\right],$$

$$\theta_n^R(v) = P\left[\max_{k=2,\ldots,r_n} X_k \leq u_n \,\middle|\, X_1 \not\leq u_n\right].$$

The arguments are perfectly analogous to the univariate case and are omitted. In effect, the multivariate extremal index summarizes temporal dependence at extreme levels, but the strength of dependence can vary with direction.

Example 10.24 Let Z_i, $i \in \mathbb{Z}$, be independent, standard Fréchet random variables. Also, let α_{jk}, $j = 1, \ldots, d$ and $k = 0, 1, 2, \ldots$ be non-negative constants such that $\sum_{k\geq 0} \alpha_{jk} = 1$ for $j = 1, \ldots, d$. The *multivariate moving-maximum process* $\{X_n\}$ is defined by

$$X_{n,j} = \max_{k\geq 0} \alpha_{jk} Z_{n-k}, \qquad j = 1, \ldots, d.$$

Observe that the margins of X_n are standard Fréchet, and recall from Example 10.5 that the marginal extremal indices are $\theta_j = \max_{k \geq 0} a_{jk}$. Let F be the distribution function of X_n. For $v \in [0, \infty) \setminus \{0\}$, we have

$$F^n(n/v_1, \ldots, n/v_d) = \exp\left(- \sum_{k \geq 0} \max_{j=1,\ldots,d} \alpha_{jk} v_j\right).$$

Similarly to the univariate case, for $v \in [0, \infty) \setminus \{0\}$,

$$P[M_{n,j} \leq n/v_j, \forall j = 1, \ldots, d]$$

$$= \exp\left\{-\frac{1}{n}\left(\sum_{l=0}^{n-1} \max_{k=0,\ldots,l} \max_{j=1,\ldots,d} a_{jk} v_j + \sum_{l \geq 0} \max_{k=l+1,\ldots,l+n} \max_{j=1,\ldots,d} a_{jk} v_j\right)\right\}$$

$$\to \exp\left(-\max_{k \geq 0} \max_{j=1,\ldots,d} a_{jk} v_i\right), \qquad n \to \infty.$$

We conclude that the multivariate extremal index of $\{X_n\}$ is

$$\theta(v) = \frac{\max_{k \geq 0} \max_{j=1,\ldots,d} a_{jk} v_j}{\sum_{k \geq 0} \max_{j=1,\ldots,d} a_{jk} v_j}, \qquad v \in [0, \infty) \setminus \{0\}.$$

Estimation

How to estimate the multivariate extremal index? Observe that the blocks, runs and intervals estimators of the univariate extremal index can all be written in terms of the indicator variables $\mathbf{1}(X_k \leq u)$. In the multivariate case, then, we can choose a vector of thresholds, u, compute $\hat{v}$ where $\hat{v}_j = \sum_{i=1}^n \mathbf{1}(X_{i,j} > u_j)$ estimates $v_j = nP[X_{1,j} > u_j]$, and construct blocks, runs or intervals estimators of $\theta(\hat{v})$ from the indicator variables $\mathbf{1}(X_i \leq u)$, $i = 1, \ldots, n$. A related method would be to first compute $\hat{Y}_i(v)$ ($i = 1, \ldots, n$) by plugging in estimates of the unknown F_j into (10.66) and next to estimate the (ordinary) extremal index of this sequence.

Unfortunately, to estimate a function rather than a number is markedly more difficult: thresholds need to be chosen for every v, and the point-wise estimates $\hat{\theta}(v)$ need not necessarily satisfy (i to iv). Up to our knowledge, there is no literature yet on estimation of the multivariate extremal index, except for a manuscript of Smith and Weissman (1996), in which a less direct method based on Pickands dependence function is proposed.

10.5.3 Further reading

The multivariate extremal index was proposed in Nandagopalan (1994). The same paper also discusses multivariate extensions of some point-process results in the spirit of section 10.3.

Smith and Weissman (1996) and Zhang (2002) introduced a class of processes called *multivariate maxima of moving maxima*, or M_4 in short. These processes constitute a generalization of the multivariate moving-maximum processes of Example 10.24. The multivariate extremal indices of M_4 processes turn out to form a rich subclass of those of general multivariate stationary processes. In this sense, the problem of modelling extremes of multivariate stationary processes can be stylized to the study of extremes of M_4 processes.

Extremes of multivariate Markov chains are treated in Perfekt (1997). The multivariate extremal index is studied first for general multivariate stationary processes and next for multivariate Markov chains, with special attention to a multivariate version of the tail chain.

A few declustering schemes have been proposed for multivariate sequences (Coles and Tawn 1991; Nadarajah 2001). These schemes are designed to extract independent observations from a multivariate, stationary sequence: clusters are identified and then summarized by a single value, such as the component-wise maximum of the observations in the cluster. The approach of Coles and Tawn (1991) is a multivariate version of blocks declustering; that of Nadarajah (2001) is a complicated extension of runs declustering. Both methods require the choice of one or more declustering parameters. The intervals declustering scheme (Ferro and Segers 2003) can be applied without arbitrary choice of declustering parameters by considering the return times to a 'failure set', membership of which defines an observation as extreme. Such a general formulation, already alluded to by Nandagopalan (1994), is developed in Segers (2002).

10.6 Additional Topics

Heavy-tailed time series

Efforts to model financial time series have led to the development of various time-series models, extending the classical framework of linear processes (Brockwell and Davis 1991)

$$X_t = \sum_{i=1}^{\infty} \psi_i Z_{t-i}, \qquad t \in \mathbb{Z},$$

in particular, of auto-regressive moving-average (ARMA) processes; here the innovations Z_t are independent, identically distributed with finite second moment, while the parameters ψ_i satisfy a certain summability constraint. Deficiencies of these ARMA processes are that they do not satisfactorily model the more extreme observations of financial time series with respect to both the magnitude and the serial dependence of such extremes. For a financial risk manager, such shortcomings are particularly grave because the financial risk involved in holding a certain portfolio may be underestimated.

A natural extension of the classical framework is to allow the innovations Z_t to be heavy-tailed, leading to heavy-tailed linear time series. Extremal characteristics of such processes, like the extreme value index, the extremal index, and the limiting distribution of clusters of extremes, can be expressed in terms of the tail of the innovation distribution and the parameters ψ_i. Moreover, for ARMA(p, q) processes

$$X_t - \sum_{i=1}^{p} \phi_i X_{t-i} = Z_t + \sum_{j=1}^{q} \theta_j Z_{t-j}, \qquad t \in \mathbb{Z},$$

with innovation distribution in the domain of attraction of a stable distribution, it is known how to estimate the coefficients ϕ_i and θ_j (Mikosch et al. 1995). This allows reconstruction of the innovations, leading, after estimation of the innovation distribution, to estimates of characteristics of clusters of extremes. A recommendable overview with numerous references of extreme value theory for heavy-tailed linear time series is Chapter 7 of Embrechts et al. (1997).

Particularly popular in finance are the auto-regressive conditionally heteroscedastic (ARCH) process (Engle 1982) and its numerous ramifications, in particular, generalized ARCH or GARCH (Bollerslev 1986). Not surprisingly, their extremal properties have been thoroughly investigated (Basrak et al. 2002; Borkovec 2000; Borkovec and Klüppelberg 2003; de Haan et al. 1989; Mikosch and Stărică 2000), even for multivariate versions (Stărică 1999).

Finally, replacing sums by maxima in the definition of linear time series and requiring the innovation distribution to be Fréchet leads to max-stable processes, in particular, max-ARMA processes, of which the ARMAX and moving-maximum processes considered in this chapter are special cases. The probability theory for such processes is well developed (Alpuim 1989; Alpuim et al. 1995; Davis and Resnick 1989, 1993; Deheuvels 1983; de Haan 1984; de Haan and Pickands 1986), although statistical applications have appeared only recently (Hall et al. 2002; Zhang 2002; Zhang and Smith 2001).

Tail estimation for the marginal distribution

How to estimate the tail of the marginal distribution of a random sample was the topic of Chapters 4 and 5. Unfortunately, the assumption of independence is all too often not very reasonable: hot summer days group together in heat waves, and large positive or negative returns of financial assets occur in periods of high volatility. Two questions arise: Are these estimation procedures still applicable? And what is the effect of dependence on estimation uncertainty?

The answer to the first question is affirmative: all familiar tail estimators, be it the Hill estimator (Hill 1975) or the maximum likelihood estimator in the POT model (Smith 1987) or indeed any other estimator, are consistent and even asymptotically normal provided the dependence between observations that are far apart in time is small. The second question, unfortunately, is more difficult to answer. Still, we can assert that typically, the effect of dependence is to increase the asymptotic variances of tail estimators, although it is not easy to say by how much. In particular, confidence intervals based on theory for independent variables risk being too narrow.

Broadly speaking, two strategies are conceivable: (1) Proceed with estimation as if the data were independent, but adapt the standard errors; (2) Extract from the original sample a new, approximately independent series, on which the inference procedures can then be applied as usual. The simplest example of the second strategy is the method of annual maxima, in which data are grouped in blocks and a GEV distribution is fitted to the block maxima. Recall from section 10.2 that under $D(u_n)$ type conditions such block maxima are indeed approximately independent. Alternatively, in the POT method we fit a GP distribution not to all excesses over a high threshold but only to the cluster maxima, a procedure motivated by the point process results of Section 10.3.

Which of the two strategies is the better one depends on the model assumptions one is willing to make, perhaps motivated by the problem at hand. In general, the more information one has about the model, the easier it becomes to extract approximately independent residuals, and the more successful will the second method become. For instance, Resnick and Stărică (1997) considered an autoregressive model

$$X_t = \sum_{i=1}^p \phi_i X_{t-i} + Z_t, \qquad t \in \mathbb{Z},$$

with independent, identically distributed innovations Z_t with positive extreme value index γ. They showed that to estimate γ with the Hill estimator on the sample $X_1, \ldots, X_n$ is inferior to first estimating the coefficients ϕ_i (for instance, as in Mikosch et al. (1995)) and second, applying the Hill estimator to the estimated residuals $\hat{Z}_t = X_t - \sum_{i=1}^p \hat{\phi}_i X_{t-i}$, the latter procedure attaining the efficiency of the case of independent data. Similarly, when studying extremes of a financial return series, McNeil and Frey (2000) propose to fit a GARCH model to the series and apply standard tail estimators to the estimated innovation sequence.

However, if there is no clear indication as to which model to use, basically the only approximately independent series to be extracted are, as mentioned already, block maxima or peaks over high thresholds. In both cases, potentially useful information is thrown away, rendering these methods less attractive. A more promising road then is to apply an appropriate estimator directly to the data and estimate its asymptotic variance. This presupposes that the asymptotic distribution of the estimator is known for dependent data as well.

Not surprisingly, the first tail estimator for which this program was carried out is the classical Hill estimator. Hsing (1991) proved asymptotic normality of the Hill estimator for stationary sequences satisfying certain mixing conditions and gave explicit estimators for its asymptotic variance. Also Resnick and Stărică (1995, 1998) gave general consistency results, with specializations to various specific models such as infinite order moving averages, bilinear processes, solutions of stochastic difference equations, and hidden semi-Markov models. Related to the Hill estimator is the ratio estimator (Goldie and Smith 1987), which was investigated in the setting of dependent variables by Novak (1999).

Unfortunately, all these methods are somewhat *ad hoc* in the sense that it is not clear how to generalize them to other estimators like, for instance, the popular maximum-likelihood estimator for the GP distribution fitted to excesses over a high threshold. A real breakthrough was achieved by Drees (2000, 2002, 2003). He established powerful convergence results for tail empirical quantile processes for certain stationary time series. Since most tail estimators can be written as smooth functionals of such processes, the classical delta-method immediately leads to asymptotic normality for a wide variety of estimators of the extreme value index and high quantiles. Moreover, the resulting expressions for the asymptotic variance lend themselves to data-driven methods for the construction of confidence intervals, the actual coverage probability of which improves considerably upon that of intervals constructed under the (false) assumption of independence.

Still, these methods deal only with the problem of estimating the marginal tail. But often, it is also the aggregate effect of extreme observations occurring one after the other that is of interest: although a single day with a large amount of rainfall may not cause much trouble, the succession of several such days definitely will. Therefore, we need to estimate appropriate summaries of the strength of temporal dependence as well. To assess the uncertainty on estimates of these summaries together with the marginal tail, we have in this chapter relied on bootstrap techniques motivated by point-process theory.

Non-stationary processes

In this chapter, we have relaxed the assumption of independent, identically distributed random variables to that of a stationary sequence. In practice, however, data are seldom stationary: meteorological data typically have a strong seasonal component, tick-by-tick financial data exhibit a clear daily pattern, while macro-economic data often show an upward or downward trend. For the Uccle temperature data, our solution, which was, by the way, only partially successful, was to extract from the whole series the July data. In other applications, however, the non-stationarity itself of extremes may be of interest. This was treated in Chapter 7 in case there is no serial dependence.

Exceedances of a non-stationary sequence $X_1, X_2, \ldots$ above a boundary function $u_{n,1}, u_{n,2}, \ldots$ define a point process,

$$N_n(\cdot) = \sum_{i \in \mathcal{I}} \delta_{i/n}(\cdot), \qquad \mathcal{I} = \{i : X_i > u_{n,i}, 1 \leq i \leq n\}.$$

Like in the stationary case (Section 10.3), N_n converges, under mild mixing conditions and assumptions on the marginal distributions, to a certain compound Poisson process (Hüsler 1993; Hüsler and Schmidt 1996). This result hints at the possibility of extending regression analysis for extremes to allow for serial dependence and clustering.

11

BAYESIAN METHODOLOGY IN EXTREME VALUE STATISTICS

co-authored by Daan de Waal

11.1 Introduction

The Bayesian paradigm provides a set of interesting additional statistical tools when carrying out an extreme value analysis. There are several good reasons for that.

- Given the low amount of information often available in extreme value analysis, it is natural to consider other sources of knowledge; these can occur in the form of known constraints, whether from physical, economical or other origin. For instance, an economist may want to specify a maximum value for a quantity or variable under study. There are, however, several other possible ways in which an expert with knowledge of the processes behind the data may deliver information that is relevant to extremal behaviour and which is independent of the available data.

- Prediction is also naturally incorporated in a Bayesian setting. The concept of posterior prediction matches with the fact that the principal inferential objective of an extreme value analysis is of predictive nature.

- Bayesian analysis is not dependent on regularity assumptions required by, for instance, the maximum likelihood and probability weighted moments methods.

Statistics of Extremes: Theory and Applications   J. Beirlant, Y. Goegebeur, J. Segers, and J. Teugels
© 2004 John Wiley & Sons, Ltd   ISBN: 0-471-97647-4

As with the Pickands type estimators, the moment estimator and others discussed in Chapter 5, Bayesian inference provides a viable alternative in cases when maximum likelihood and probability weighted moments break down.

At the other side, many statisticians argue that the problem of *prior elicitation* leads to subjectiveness. Without taking part in the discussion for and against Bayesian methodology, we aim at showing that a practical statistical analysis can indeed gain from this approach. Some important contributions to this subject are found in Pickands (1994), Coles and Powell (1996), Coles and Tawn (1996b), Smith (2000), Smith and Goodman (2000) and Coles (2001) (Chapter 9). Bayesian inference of extremes has only quite recently been discovered because of the availability of *Markov Chain Monte Carlo* (MCMC) *techniques*. These computer-intensive methods have opened up the field of extremes to complicated settings involving large parameter sets. So the methods described in this chapter appear to be alternatives that are of full value or even preferable to more conventional ones.

We will briefly review some of the basic characteristics of a Bayesian analysis here. Then we go over the statistical problems raised in Part I (see Chapters 4 and 5), to end with a more complex application from environmetrics.

11.2   The Bayes Approach

Let $y = (y_1, \ldots, y_m)$ denote the observed data of a random variable Y distributed according to a distribution with density function $f(y|\theta)$. For instance, y can represent a random sample of m independent observations consisting of maxima. θ denotes the vector of parameters. Let $\pi(\theta)$ denote the density of the prior distribution for θ. We write the likelihood for θ as $f(y|\theta)$, which equals $\prod_{i=1}^{m} f(y_i|\theta)$ in case of independence. According to Bayes' theorem,

$$\pi(\theta|y) = \frac{f(y|\theta)\pi(\theta)}{\int_\Omega f(y|\theta)\pi(\theta)d\theta} \propto f(y|\theta)\pi(\theta), \qquad (11.1)$$

where the integral is taken over the parameter space Ω. This well-known probabilistic result provides a framework for statisticians to convert an initial set of beliefs about θ, represented by the prior $\pi(\theta)$, into a posterior distribution $\pi(\theta|y)$ of θ that is proportional to the product of the likelihood and the prior. Estimates of θ will then be obtained through the mode or mean of the posterior, while the accuracy of an inference is described by the posterior distribution itself, for instance, through a *highest posterior density* (hpd) *region* according to a certain probability $1 - \alpha$, which is the region of values that contains $100(1 - \alpha)\%$ of the posterior probability and also has the characteristic that the density within the region is never lower than that outside. Here, there is no need to fall back to asymptotic theory.

Ease of prediction is another attractive characteristic of the Bayesian approach. If Y_{m+1} denotes a future observation with density function $f(y_{m+1}|\theta)$, then the *posterior predictive density* of a future observation Y_{m+1} given y is given by

$$f(y_{m+1}|y) = \int_\Omega f(y_{m+1}|\theta)\pi(\theta|y)d\theta. \qquad (11.2)$$

Compared to other approaches to prediction, the predictive density has the advantage that it reflects uncertainty in the model through $\pi(\theta|y)$ ánd uncertainty due to variability in future observations through $f(y_{m+1}|\theta)$. *The posterior predictive probability* of Y_{n+1} exceeding some high threshold y is accordingly given by

$$P(Y_{m+1} > y|\, y) = \int_\Omega P(Y_{m+1} > y|\,\theta)\pi(\theta|y)d\theta. \tag{11.3}$$

The posterior predictive distribution (11.3) most of the time is difficult to obtain analytically. However, it can be approximated if the posterior distribution has been estimated by simulation as discussed further on. Given a sample $\theta_1, \ldots, \theta_r$ from $\pi(\theta|y)$, then we can use the approximation

$$P(Y_{m+1} > y|y) \sim \frac{1}{r}\sum_{i=1}^{r} P(Y_{m+1} > y|\theta_i), \tag{11.4}$$

where $P(Y_{m+1} > y|\theta_i)$ follows immediately from the postulated density function $f(y|\theta)$. A *posterior predictive* $(1 - p)$ *quantile* is obtained by solving

$$P(Y_{m+1} > y|y) = p. \tag{11.5}$$

Most often, this solution cannot be found analytically, and then the solution y of (11.5) can be found using a standard numerical solver.

11.3   Prior Elicitation

The main objection against the use of Bayesian analysis is the need for specifying a prior $\pi(\theta)$. When available information is minimal, one can start an updating scheme with an *objective prior* distribution. Uniform priors are the simplest examples of this kind. Other proposals, for instance, are Jeffreys' prior and the maximal data information (MDI) prior. Advantages of using objective prior distributions are found in the fact that objective priors are sometimes used as a benchmark that will not reflect the particular biases of the analyst and that the use of such priors will yield statistical procedures that are analogous to those developed using classical (frequentist) procedures. In multiple parameter situations, the parameters should not be taken to be independent, which is sometimes the case with objective priors. Another point of concern is the invariance under certain groups of transformations and different parametrizations.

Jeffreys' prior
Jeffreys' prior (Jeffreys (1961)) is defined as $J(\theta) \propto \sqrt{|I(\theta)|}$ where $I(\theta)$ is Fisher's information matrix with (i, j)–th element

$$I_{ij}(\theta) = E\left\{-\frac{\partial^2 \log f(Y|\theta)}{\partial\theta_i\partial\theta_j}\right\}, \quad i, j = 1, \ldots, p,$$

where p denotes the dimension of θ. Jeffreys' prior is considered to be the standard starting rule for an objective Bayesian analysis. It is invariant under one-to-one transformations and takes the dependence between the parameters into account. When applied to the models appearing in extreme value methodology, Jeffreys' prior leads to the same restrictions on the parameter set as with the maximum likelihood approach. See, for instance, Bernardo and Smith (1994), Chapter 4, for more details.

MDI prior

Zellner (1971) defined the MDI prior to provide maximal average data information on θ. These priors are not invariant under reparametrization, but are usually easy to implement, however. The MDI prior for θ is defined as $\pi(\theta) \propto \exp E \{\log f(Y|\theta)\}$. Constraints on the parameters can be built into the prior. We refer to Zellner (1971) for more details.

On the other hand, *subjective prior distributions* represent an attempt to bring prior knowledge about the phenomenon under study into the problem. This always leads to proper priors, which means that they integrate to 1, and these priors are typically well behaving analytically. However, it is not always easy to translate the prior knowledge into a meaningful probability distribution. Also, the results of a Bayesian analysis that used a subjective prior are meaningful to the particular analyst whose prior distribution was used, but not necessarily to other researchers. Families of subjective distributions are the natural conjugate families, exponential power distributions and mixture prior distributions. Natural conjugates are most popular possibly due to their mathematical convenience: it is the class of distributions that is closed under transformation from prior to posterior; that is, the implied posterior distribution with respect to a natural conjugate prior is in the same family as the prior distribution.

Specifically in an extreme value context, authors have rather systematically advocated the specification of priors in terms of extreme quantiles of the underlying process rather than the extreme value model parameters themselves, see, for instance, Coles and Tawn (1996b). Of course, subject to self-consistency, a prior distribution on a set of two or three parameters can always be transformed to a prior distribution on the original model parameters themselves. An example in insurance comes from the fact that finite right end-points are sometimes specified to loss distributions, while the claim data appear to be of Pareto-type on the basis of the data analytic methods described in the first part of this book. Alternatively, in some specific contents, a prior can be designed so that the analysis can meet requirements set by experts. A well-known example of this kind is the requirement in reinsurance applications that the EVI γ should not to be larger than 1, or even 0.5, for the most common premium calculation methods to be valid. We will give some examples of this kind using conjugate priors, but mostly we will restrict ourselves to the use of objective priors. In this way, we hope to convince more people of the possible added value of a Bayesian approach to an extreme value analysis. Of course, the question of sensitivity of tail estimations to changes in prior specification should be posed.

11.4   Bayesian Computation

The main obstacle to the widespread use of Bayesian techniques for a long time was the difficulty of computation of the different integrals involved with posterior inference. Appropriate choice of prior families for certain models can avoid the necessity to calculate, for instance, the normalizing integral in (11.1) but this is rather exceptional, certainly in multi-parameter settings. This difficulty has been lifted with the development of simulation-based techniques. MCMC methods have popularized the use of Bayesian techniques to a great extent. We discuss here briefly two of the more popular MCMC methods: the Gibbs sampler and the Metropolis–Hastings algorithm. More details can, for instance, be found in Chapter 5 of Carlin and Louis (2000).

The Metropolis–Hastings algorithm

The basic idea of the Metropolis–Hastings algorithm is particularly simple. One simulates a sequence $\theta_1, \theta_2, \ldots$ in the following way: starting with an initial point θ_1, the next state $\theta^{(i+1)}$ is chosen by first sampling a candidate point θ^* from a proposal density $q(\theta^* | \theta^{(i)})$ that depends on the current state $\theta^{(i)}$. Examples of proposal densities can be the multivariate normal distribution with mean $\theta^{(i)}$ and a suitable chosen covariance matrix. The candidate θ^* is accepted with probability α_i where

$$\alpha_i = \min \left\{ \frac{\pi(\theta^* | y) q(\theta^{(i)} | \theta^*)}{\pi(\theta^{(i)} | y) q(\theta^* | \theta^{(i)})}, 1 \right\} \tag{11.6}$$

If the candidate is accepted, the next state becomes $\theta^{(i+1)} = \theta^*$, otherwise the chain remains at $\theta^{(i+1)} = \theta^{(i)}$. Both rejection and acceptance count as an iteration of the algorithm. When a candidate is sampled for which the posterior (11.1) is 0, we must continue sampling until we have a candidate with $f(y | \theta^*) > 0$. Remarkably, under some regularity conditions, the stationary distribution is exactly the posterior distribution $\pi(\theta | y)$, called the *target distribution of the Markov chain*. Although the proposal density can be arbitrarily chosen, the convergence largely depends on the proposal density. On the one hand, a proposal density with large jumps to places far from the support of the posterior has low acceptance rate and causes the Markov chain to stand still most of the time. On the other hand, a proposal density with small jumps and high acceptance rate may cause the chain to move slowly and to get stuck in one state. A great advantage, though, of this algorithm is that it only depends on the posterior density through ratios of the form $\pi(\theta^* | y) / \pi(\theta^{(i)} | y)$. Hence, the posterior density only needs to be known up to a proportionality constant.

The Gibbs sampler

The Gibbs sampler, introduced by Geman and Geman (1984), is an alternating conditional MCMC algorithm. It can be regarded as a special case of the Metropolis–Hastings algorithm as follows. Consider a parametric vector $\boldsymbol{\theta} = (\theta_1, \ldots, \theta_p)$. One iteration of the Gibbs sampler consists of cycling through the p coordinates $\theta_1, \ldots, \theta_p$ by drawing a sample of one coordinate conditional on the values of all the others. For every iteration step, say i, there are p steps. With a predetermined ordering of the p coordinates, each $\theta_j^{(i)}$ is sampled from the conditional distribution given all the other components of $\boldsymbol{\theta}$. So given a set of values $\{\theta_1^{(i)}, \ldots, \theta_p^{(i)}\}$, the algorithm proceeds as follows:

$$\text{Draw } \theta_1^{(i+1)} \sim \pi(\theta_1 | \theta_2^{(i)}, \ldots, \theta_p^{(i)}, \boldsymbol{y})$$

$$\text{Draw } \theta_2^{(i+1)} \sim \pi(\theta_2 | \theta_1^{(i+1)}, \theta_3^{(i)}, \ldots, \theta_p^{(i)}, \boldsymbol{y})$$

$$\vdots$$

$$\text{Draw } \theta_p^{(i+1)} \sim \pi(\theta_p | \theta_1^{(i+1)}, \ldots, \theta_{p-1}^{(i+1)}, \boldsymbol{y}).$$

It can be proven that the corresponding acceptance probability is equal to unity so that every jump is therefore accepted for Gibbs sampling. We also mention the possibility of Gibbs sampling combined with some Metropolis steps; see, for instance, Chapter 11 in Gelman *et al* (1995).

11.5   Univariate Inference

In this section, we revisit the most important models considered in Part I, namely, the fit of the GEV based on block maxima, followed by the different methods considering peaks over threshold data. At the end of the chapter, we also consider some extensions of these basic models that can be used to provide good global fits in addition to appropriate tail fits.

11.5.1   Inference based on block maxima

To illustrate the use of the Bayesian methodology described above when block maxima are available, we consider again the annual maximal discharges of the Meuse river in Belgium, which were already considered in sections 2.2 and 5.1. The likelihood model is

$$Y_i | \sigma, \gamma, \mu \sim GEV(\sigma, \gamma, \mu), \quad i = 1, \ldots, 85,$$

where Y_i denotes the maximum for the year indexed by i. Here, $\boldsymbol{\theta} = (\sigma, \gamma, \mu)$ and

$$f(y|\boldsymbol{\theta}) = \frac{1}{\sigma}\left(1 + \gamma \frac{y-\mu}{\sigma}\right)^{-1/\gamma - 1} \exp\left(-\left(1 + \gamma \frac{y-\mu}{\sigma}\right)^{-1/\gamma}\right).$$

As a prior distribution, we choose here the MDI prior

$$\pi(\boldsymbol{\theta}) = \exp E \{\log f(Y|\boldsymbol{\theta})\} \propto \frac{1}{\sigma} e^{-\psi(1)(1+\gamma)}$$

where $\psi(1)$ denotes Euler's constant. Jeffreys' prior in case of the GEV is quite complicated and only exists when $\gamma > -0.5$.

We use the Metropolis–Hastings algorithm with the proposal density q taken as a multivariate normal on $(\log \sigma, \mu, \gamma)$ with independent components and respective standard deviations $(0.01, 10, 0.01)$:

$$\log \sigma^* = \log \sigma^{(i)} + 0.01\epsilon_1$$

$$\mu^* = \mu^{(i)} + 10\epsilon_2$$

$$\gamma^* = \gamma^{(i)} + 0.01\epsilon_3$$

where $(\epsilon_1, \epsilon_2, \epsilon_3)$ denote independent standard normally distributed random variables. The values of the variances in the specification of q were chosen after a little trial and error in order to make the algorithm work more efficient. Initializing with $(\sigma, \mu, \gamma) = (500, 1200, 0.01)$, the values generated by 15,000 iterations of the chain are plotted in Figure 11.1. The convergence can be speeded up if instead q is taken to be a trivariate normal distribution with the covariance matrix determined by the information matrix from the log-posterior. In Figure 11.2, we show the estimated posterior densities of the GEV parameters (in the original scale for σ) and the 100-year return level $q_{Y,0.01}$. The estimated posterior density of the 100-year return level is obtained from the GEV quantile function

$$q_{Y,p} = \mu + \frac{\sigma}{\gamma} \left[(-\log(1-p))^{-\gamma} - 1 \right]$$

replacing σ, γ and μ by their respective posterior realizations. The mean posterior estimates together with the 95% hpd confidence regions are given by

$$\hat{\mu} = 1264 \ (1156, 1385), \quad \hat{\sigma} = 471 \ (400, 547), \quad \hat{\gamma} = -0.075 \ (-0.200, 0.072),$$

$$\hat{q}_{0.01} = 3109 \ (2711, 3809).$$

In Figure 11.3, we show the estimated posterior predictive distribution of a future observation Y_{m+1} given $\boldsymbol{y}$ and the corresponding posterior predictive 0.99 quantile. These estimates are obtained along (11.4) and (11.5) respectively.

11.5.2   Inference for Fréchet-Pareto-type models

As in the first part of the book, we again consider the estimation of extreme events within the Fréchet-Pareto-type framework, that is, $1 - F(x) = x^{-1/\gamma} \ell_F(x)$ with ℓ_F some slowly varying function at infinity. Recall that here there are mainly two approaches possible.

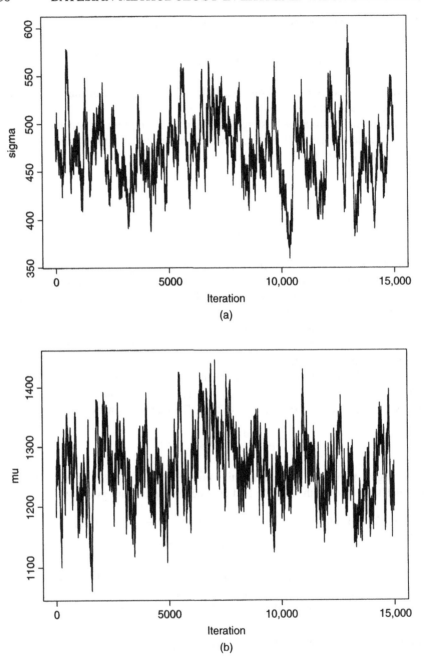

Figure 11.1   Annual maximal river discharges of the Meuse: Metropolis–Hastings realizations of (a) σ, (b) μ, (c) γ and (d) $q_{Y,0.01}$.

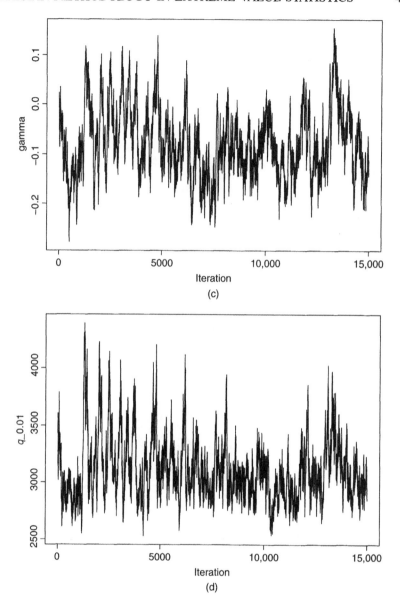

(c)

(d)

First, one can consider the relative excesses $Y_j = X/t$ $(X > t)$ for some appropriate threshold t and fit a strict Pareto distribution with distribution function $1 - y^{-1/\gamma}$ $(y > 1)$. As with Hill's estimator for γ, we choose $t = X_{n-k,n}$ so that the ordered excesses are given by $Y_j = X_{n-j+1,n}/X_{n-k,n}$, $j = 1, \dots, k$. Here, the likelihood model is given by

$$Y_j|\gamma \sim Pa(1/\gamma), \quad j = 1, \dots, k,$$

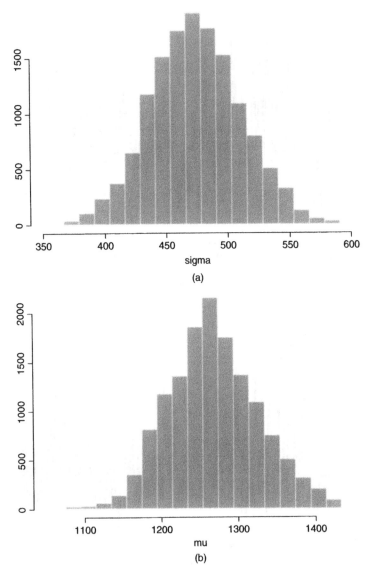

Figure 11.2   Annual maximal river discharges of the Meuse: estimated posterior density of (a) σ, (b) μ, (c) γ and (d) $q_{Y,0.01}$.

for some fixed k. So $\theta = \gamma$ and

$$f(y|\gamma) = \frac{1}{\gamma} y^{-1-1/\gamma}, \ y > 1.$$

Here, Jeffreys' prior turns out to be particularly simple, namely, $\pi(\gamma) \propto 1/\gamma$, while the MDI prior is proportional to $(1/\gamma) \exp(-\gamma)$. Continuing with Jeffreys' prior,

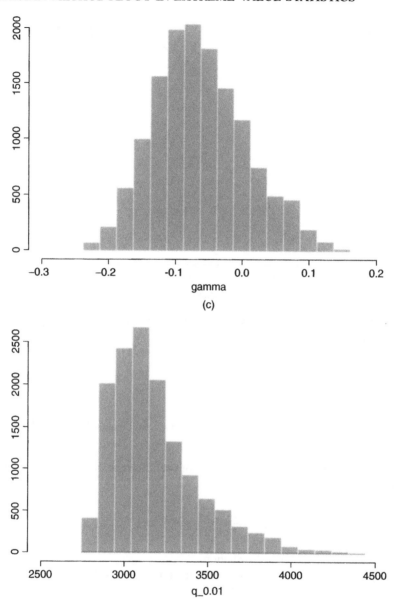

(c)

(d)

the posterior is given by

$$\pi(\gamma \,|\, \boldsymbol{y}) \propto \gamma^{-k-1} \Pi_{j=1}^{k} y_j^{-1-1/\gamma}$$

$$\propto e^{-\frac{1}{\gamma} \sum_{j=1}^{k} \log y_j} \gamma^{-k-1}$$

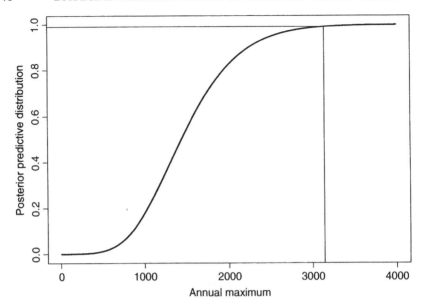

Figure 11.3   Annual maximal river discharges of the Meuse: $P(Y_{m+1} \leq y_{m+1}|\mathbf{y})$ and $q_{Y_{m+1},0.01}$.

leading to the posterior mode estimator

$$\frac{1}{k+1} \sum_{j=1}^{k} \log Y_j = \frac{1}{k+1} \sum_{j=1}^{k} (\log X_{n-j+1,n} - \log X_{n-k,n}) = \frac{k}{k+1} H_{k,n}$$

which is almost identical to the Hill estimator itself. Remark that when normalizing the posterior $\pi(\gamma|\mathbf{y})$, one obtains

$$\pi(\gamma|\mathbf{y}) = \frac{(kH_{k,n})^k}{(k-1)!} e^{-\frac{1}{\gamma}kH_{k,n}} \gamma^{-k-1}, \qquad (11.7)$$

which is an inverse gamma distribution.

The posterior predictive density (11.2) of a future excess $\tilde{Y}$ when using Jeffreys' prior is given by

$$f(\tilde{y}|\mathbf{y}) = \frac{(kH_{k,n})^k}{\tilde{y}(k-1)!} \int_0^\infty \gamma^{-k-2} e^{-(\log \tilde{y} + kH_{k,n})/\gamma} d\gamma$$

$$= \frac{(kH_{k,n})^k}{\tilde{y}(k-1)!} \int_0^\infty w^k e^{-(\log \tilde{y} + kH_{k,n})w} dw$$

$$= \frac{k(kH_{k,n})^k}{\tilde{y}} (\log \tilde{y} + kH_{k,n})^{-k-1}$$

where, in the second step, we use the substitution $w = 1/\gamma$. This entails that the posterior predictive density of $\log \tilde{Y} =: \tilde{V}$ is given by

$$f(\tilde{v}|y) = \frac{1}{H_{k,n}} \left(1 + \frac{\tilde{v}}{k H_{k,n}} \right)^{-k-1}$$

which turns out to be a GP distribution with scale equal to Hill's statistic $H_{k,n}$ and shape equal to $1/k$. This then leads to an interesting objective Bayesian alternative to the Weissman estimator (see section 4.6.1) for small tail probabilities of Pareto-type distributions through (11.3):

$$P(X > x|y) = \frac{k}{n} \left(1 + \frac{\log(x/X_{n-k,n})}{k H_{k,n}} \right)^{-k}. \tag{11.8}$$

The results of this estimator in case of the Secura Belgian Re insurance data set introduced in section 1.3.3 (i) with $x = 10,000,000$ are plotted in Figure 11.4 as a function of k together with the result of the original Weissman estimator.

Simulating γ-values from the inverse gamma distribution (11.7) and substituting them in the expression $(k/n)(x/X_{n-k,n})^{-1/\gamma}$ as suggested in (11.4) leads to an alternative approach and yields the possibility of calculating a 95% hpd region. The results at $k = 95$ are shown in Figure 11.5. The 95% hpd region is $(0.00041, 0.00208)$.

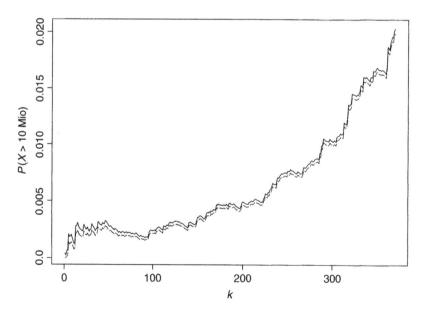

Figure 11.4   Secura data: $P(X > 10\text{ Mio}|y)$ (solid line) and $\hat{p}^{+}_{k,10\text{ Mio}}$ (broken line) as a function of k.

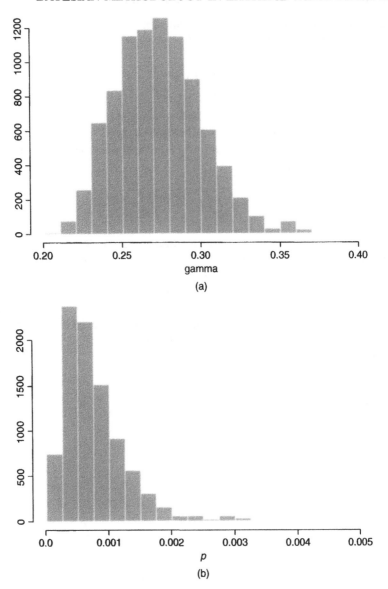

Figure 11.5 Secura data: simulated posterior density of (a) γ and (b) $(95/371)(10{,}000{,}000/2{,}580{,}025)^{-1/\gamma}$.

Setting (11.8) equal to p and solving for x leads to the estimator

$$q_{X,p} = X_{n-k,n}e^{kH_{k,n}((np/k)^{-1/k}-1)} \tag{11.9}$$

of extreme posterior predictive quantiles.

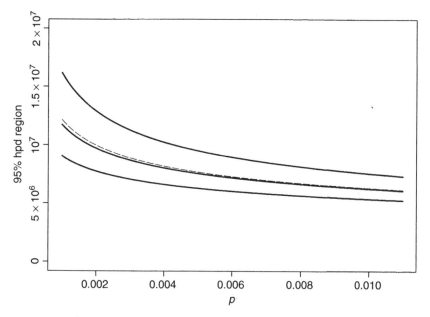

Figure 11.6   Secura data: $q_{X,p}$ (broken line) and posterior median and 95% hpd region of $Q(1 - p)$ (solid lines) at $k = 95$ as a function of p.

Besides the posterior predictive quantile function $q_{X,p}$, interest is often in the posterior distribution of the quantile function $Q(1 - p)$ associated with $f(y|\theta)$. Let us consider the estimation of $Q(1 - p)$ with $p \in [0.001, 0.01]$ in the insurance example using the threshold $t = x_{276,371}$ or equivalently $k = 95$. Substituting γ_i-values obtained by MCMC from the posterior into the expression $t(np/k)^{-\gamma}$ allows to construct a 95% hpd region for $Q(1 - p)$. In Figure 11.6, we show $q_{X,p}$ (broken line) and the posterior median and 95% hpd region of $Q(1 - p)$ (solid lines) at $k = 95$ as a function of p.

From a subjective Bayesian point of view, inverse gamma priors provide more possibilities to incorporate an expert's view. Inspired by Hsieh (2001), we consider here a three-parameter inverse gamma prior $IG(\lambda, \eta, \tau)$ as a prior for γ, defined by

$$\pi(\gamma) = \frac{\lambda^\eta}{\Gamma(\eta)} e^{-\lambda(\gamma^{-1} - \tau)} (\gamma^{-1} - \tau)^{\eta-1} \gamma^{-2}, \quad 0 < \gamma < 1/\tau, \tag{11.10}$$

with $\lambda, \eta > 0$ and $\tau \geq 0$. In case $\tau = 0$, we obtain back the classical inverse gamma distribution. The truncation parameter τ can be used to bound the possible values of γ. For instance, in insurance applications, the value $\tau = 1$ is an appropriate choice, since values $\gamma \geq 1$ (or rv's X possessing infinite mean) are not acceptable to (most) actuaries. Some would even argue for $\gamma < 1/2$ and hence $\tau = 2$, since in many insurance branches, variances are believed to be finite. The parameters λ and η together can be used to reflect the degree of uncertainty of an expert concerning the specification of γ. Further, the prior mode of γ is given by

$$\hat{\gamma} = \frac{\lambda\tau + \eta + 1 - D}{4\tau}$$

where

$$D = \sqrt{(\lambda\tau + \eta + 1)^2 - 8\lambda\tau}.$$

Now the posterior for γ is given by

$$\pi(\gamma|y) \propto \gamma^{-k-2} e^{-kH_{k,n}/\gamma} e^{-\lambda(\gamma^{-1}-\tau)} (\gamma^{-1} - \tau)^{\eta-1}$$

$$\propto \gamma^{-k-2} e^{-(\lambda+kH_{k,n})(\gamma^{-1}-\tau)} (\gamma^{-1} - \tau)^{\eta-1}. \qquad (11.11)$$

The posterior mode can be found analytically:

$$\hat{\gamma} = \frac{1}{2\tau} \left(1 + \frac{1}{k+2}[\tau(\lambda + kH_{k,n}) + \eta - 1] - D \right) \qquad (11.12)$$

where

$$D = \sqrt{(1 + \frac{1}{k+2}[\tau(\lambda + kH_{k,n}) + \eta - 1])^2 - 4\frac{\tau}{k+2}(\lambda + kH_{k,n}))}.$$

The Bayesian estimator (11.12) of a positive EVI γ constitutes an interesting alternative compared to the Hill estimator.

The exact normalization of the posterior (11.11) can be found using the substitution $\gamma^{-1} - \tau = u$

$$\int_0^{1/\tau} \gamma^{-k-2} e^{-(\lambda+kH_{k,n})(\gamma^{-1}-\tau)} (\gamma^{-1} - \tau)^{\eta-1} d\gamma$$

$$= \int_0^{\infty} (u + \tau)^k e^{-(\lambda+kH_{k,n})u} u^{\eta-1} du$$

$$= \sum_{j=0}^{k} \binom{k}{j} \tau^{k-j} \int_0^{\infty} u^{\eta+j-1} e^{-(\lambda+kH_{k,n})u} du$$

$$= \sum_{j=0}^{k} \binom{k}{j} \tau^{k-j} \Gamma(\eta + j)(\lambda + kH_{k,n})^{-(\eta+j)}.$$

Since the m-th inverse moment of a random variable with density given by (11.10) equals

$$E(\Gamma^{-m}(\lambda, \eta, \tau)) := \frac{1}{\Gamma(\eta)} \sum_{j=0}^{m} \binom{m}{j} \tau^{m-j} \Gamma(\eta + j)\lambda^{-j},$$

the posterior is given by

$$\pi(\gamma|y) = \frac{(\lambda + kH_{k,n})^\eta}{\Gamma(\eta)E(\Gamma^{-k}(\lambda + kH_{k,n}, \eta, \tau))} \gamma^{-k-2} e^{-(\gamma^{-1}-\tau)(\lambda+kH_{k,n})} (\gamma^{-1} - \tau)^{\eta-1}.$$

The posterior predictive distribution of a future excess $\tilde{Y}$ with a bounded generalized inverse gamma prior (11.10), is found to be, using again the substitution $\gamma^{-1} - \tau = u$

$$f(\tilde{y}|y) = \int_0^{1/\tau} \gamma^{-1} \tilde{y}^{-1-\gamma^{-1}} \pi(\gamma|y) d\gamma$$

$$= \tilde{y}^{-\tau-1} \frac{E(\Gamma^{-(k+1)}(\log \tilde{y} + \lambda + kH_{k,n}, \eta, \tau))}{E(\Gamma^{-k}(\lambda + kH_{k,n}, \eta, \tau))} \left(\frac{\log \tilde{y} + \lambda + kH_{k,n}}{\lambda + kH_{k,n}} \right)^{-\eta}$$

$$= \tilde{y}^{-\tau-1} \frac{\sum_{j=0}^{k+1} \binom{k+1}{j} \tau^{k+1-j} \Gamma(\eta + j)(\log \tilde{y} + \lambda + kH_{k,n})^{-\eta-j}}{\sum_{j=0}^{k} \binom{k}{j} \tau^{k-j} \Gamma(\eta + j)(\lambda + kH_{k,n})^{-\eta-j}},$$

while

$$P(X > x|y) = \frac{k}{n} \left(\frac{x}{X_{n-k,n}} \right)^{-\tau}$$

$$\times \frac{\sum_{j=0}^{k} \binom{k}{j} \tau^{-j} \Gamma(\eta + j)(\log(\frac{x}{X_{n-k,n}}) + \lambda + kH_{k,n})^{-\eta-j}}{\sum_{j=0}^{k} \binom{k}{j} \tau^{-j} \Gamma(\eta + j)(\lambda + kH_{k,n})^{-\eta-j}}.$$

$$(11.13)$$

We apply (11.11) and (11.13) with $x = 10$ Mio Euro to the insurance data from section 1.3.3 (i) with $\tau = 1$ and $\tau = 2$, and $(\lambda, \eta) = (8, 4)$, see Figure 11.7. The 95% hpd region for γ is $(0.31489, 0.40361)$ in case $\tau = 1$ and $(0.30110, 0.38417)$ in case $\tau = 2$. The corresponding hpd regions for $95/371(10,000,000/2,580,025)^{-1/\gamma}$ are respectively $(0.00155, 0.00526)$ and $(0.00121, 0.00422)$. Note that the posterior mode of γ is slightly larger than the Hill estimate obtained in section 6.2 ($H_{95,371} = 0.27109$), which can be understood from the values of the prior modes, $\hat{\gamma} = 0.68826$ when $(\lambda, \eta, \tau) = (8, 4, 1)$, respectively $\hat{\gamma} = 0.41352$ when $(\lambda, \eta, \tau) = (8, 4, 2)$.

11.5.3 Inference for all domains of attractions

Considering now the estimation of extreme events within a general extreme value (GEV) context as discussed in Chapter 5, and hence considering GP fits to the excesses $Y_j = X_i - t$ ($X_i > t$) for some appropriate threshold t. Choosing t again in an observation $X_{n-k,n}$, then the model

$$F(y|\sigma, \gamma) = 1 - \left(1 + \gamma \frac{y}{\sigma}\right)^{-1/\gamma}$$

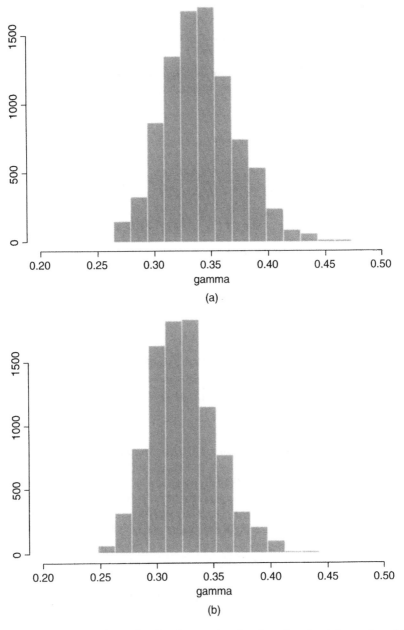

Figure 11.7   Secura data: simulated posterior density of γ for (a) $\tau = 1$ and (b) $\tau = 2$ and simulated posterior density of $95/371(10,000,000/2,580,025)^{-1/\gamma}$ for (c) $\tau = 1$ and (d) $\tau = 2$.

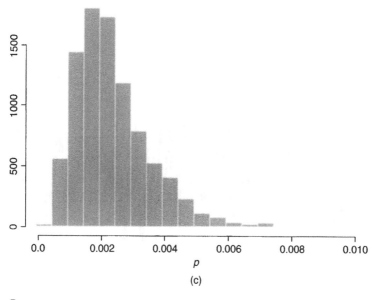

(c)

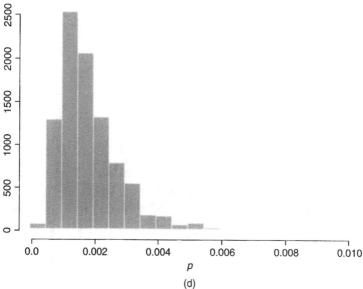

(d)

is fitted to the excesses $Y_j = X_{n-j+1,n} - X_{n-k,n}$, $j = 1, \ldots, k$. Jeffreys' prior for the GP distribution $\pi(\sigma, \gamma) = \dfrac{1}{\sigma(1+\gamma)\sqrt{1+2\gamma}}$, which is finite for $\gamma > -\frac{1}{2}$, is given in Smith (1984). The MDI prior, however, is given by

$$\pi(\sigma, \gamma) \propto \frac{1}{\sigma} e^{-\gamma}.$$

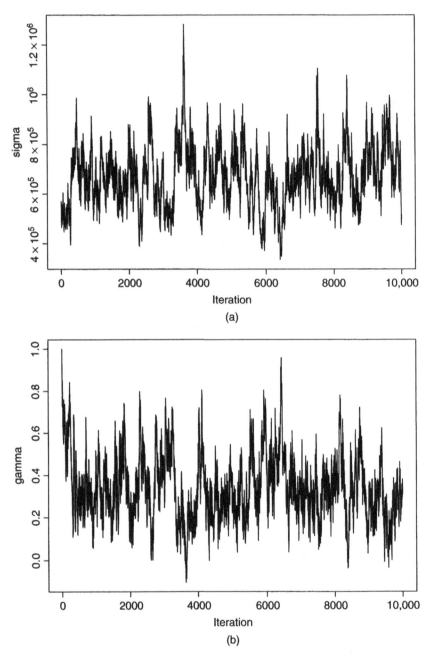

Figure 11.8   Secura data: Metropolis–Hastings realizations of (a) σ and (b) γ and simulated posterior density of (c) σ and (d) γ.

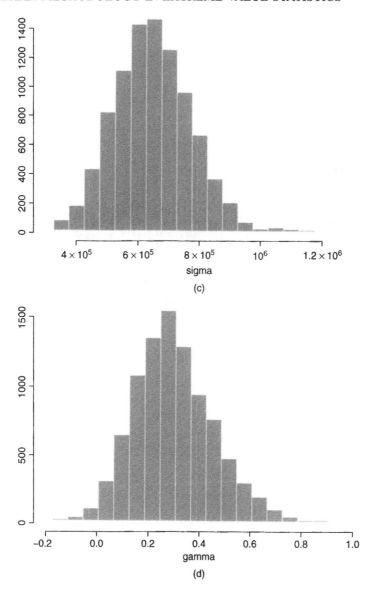

(c)

(d)

Using the MDI prior in the insurance example with t set at the 96th-largest observation, and setting up the Metropolis–Hastings algorithm with the proposal density q taken as a multivariate normal on $(\log \sigma, \gamma)$ with independent components and respective standard deviations $(0.04, 0.04)$, that is,

$$\log \sigma^* = \log \sigma^{(i)} + 0.04\epsilon_1$$
$$\gamma^* = \gamma^{(i)} + 0.04\epsilon_2$$

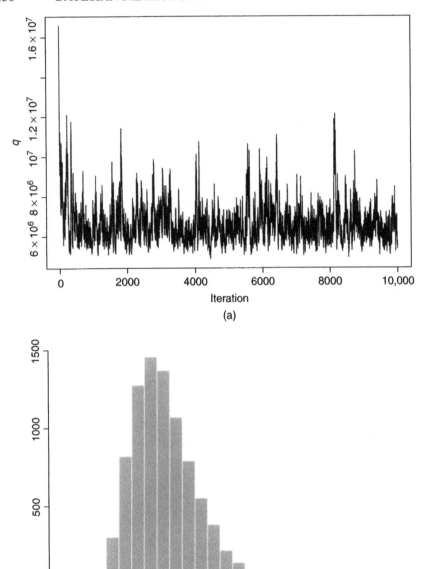

Figure 11.9   Secura data: (a) Metropolis–Hastings realizations of $q_{0.01}$ and (b) simulated posterior density of $q_{0.01}$.

where (ϵ_1, ϵ_2) denote independent standard normally distributed random variables, we obtain the estimated posterior densities of σ and γ given in Figure 11.8(c) and (d) respectively. Initializing with $(\sigma, \gamma) = (570000, 1)$ the values generated by 10000 iterations of the chain are plotted in Figure 11.8(a) and (b). The mean posterior estimates together with the 95% hpd confidence regions are given by

$$\hat{\sigma} = 664, 221.3 \ (449, 704.8; 917, 847.3), \ \hat{\gamma} = 0.32339 \ (0.07401, 0.68992),$$

The posterior distribution of the quantile function of the original X-distribution can be simulated on the basis of

$$q_p = t + \frac{\sigma}{\gamma}((np/k)^{-\gamma} - 1)$$

replacing σ and γ by their respective posterior realizations. In the insurance example with t set at the 96th-largest observation and $p = 0.01$, we obtain Figure 11.9. The posterior predictive distribution of a future claim given the past claims is given in Figure 11.10.

In a hydrological context, Coles and Tawn (1996a) consider prior elicitation on the basis of annual return levels rather than in terms of the GEV or GP parameters. It can indeed be argued that hydrological experts are probably more familiar with quantile specification rather than parameter specification. Then, in case of GEV modelling of maxima, one specifies prior information in terms of $(q_{p_1}, q_{p_2}, q_{p_3})$ with $p_1 > p_2 > p_3$ where

$$q_p = \mu + \frac{\sigma}{\gamma}\left([-\log(1 - p)]^{-\gamma} - 1\right). \tag{11.14}$$

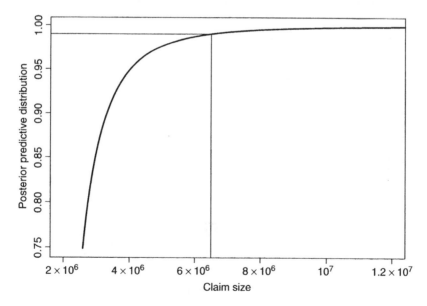

Figure 11.10   Secura data: $P(\tilde{Y} \le \tilde{y}|y)$ as a function of $\tilde{y}$ at $k = 95$.

Since the quantiles $q_{p_1}, q_{p_2}, q_{p_3}$ have to be ordered, Coles and Tawn considered independent gamma (α_i, β_i) $(i = 1, 2, 3)$ priors on

$$\tilde{q}_1 = q_{p_1}, \quad \tilde{q}_2 = q_{p_2} - q_{p_1}, \quad \tilde{q}_3 = q_{p_3} - q_{p_2}.$$

The hyperparameters α_i, β_i were determined by measures of location and variability in prior belief. The experts were asked to specify the median and 90% quantiles of each of the $\tilde{q}_i$, from which the gamma parameter estimates were obtained. In the rainfall example considered in Coles and Tawn (1996a), the values $p_1 = 0.1$, $p_2 = 0.01$ and $p_3 = 0.001$ were chosen. From the prior specification, the joint prior for the q_{p_i} is obtained as

$$\pi(q_{p_1}, q_{p_2}, q_{p_3}) \propto \Pi_{i=1}^{3}(q_{p_i} - q_{p_{i-1}})^{\alpha_i - 1} \exp(-\beta_i(q_{p_i} - q_{p_{i-1}}))$$

where $q_{p_0} = 0$ and with $0 \leq q_{p_1} \leq q_{p_2} \leq q_{p_3}$. Substituting the quantile expression (11.14) in this prior for $(q_{p_1}, q_{p_2}, q_{p_3})$ and multiplying by the Jacobian of the transformation $(q_{p_1}, q_{p_2}, q_{p_3}) \to \boldsymbol{\theta} = (\mu, \sigma, \gamma)$ leads directly to an expression for the prior in terms of the GEV parameters.

11.6   An Environmental Application

We end by illustrating the use of Bayesian modelling of extreme values with an example from an environmental context. It concerns the Bayesian modelling of wind-speed measurements from three different locations in Cape Town (South Africa). Figure 11.11 contains the boxplots of the monthly maximal wind gust

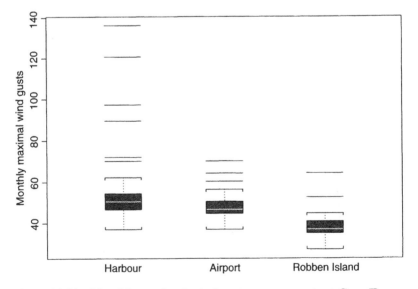

Figure 11.11   Monthly maximal wind gust measurements at Cape Town.

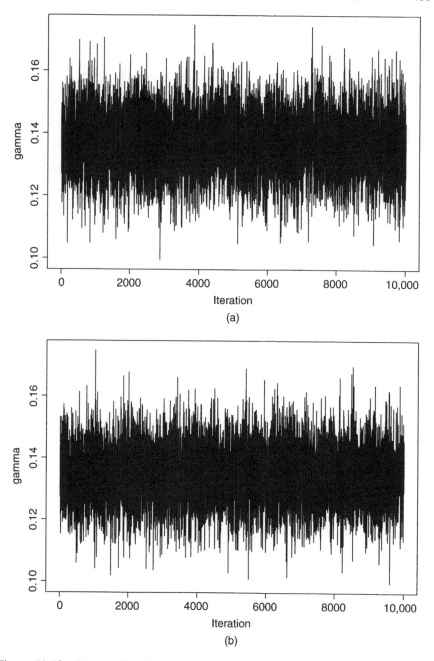

Figure 11.12   Metropolis–Hastings realizations of γ_i: (a) harbour, (b) airport, (c) Robben Island.

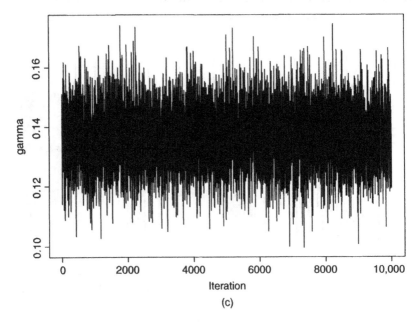

Figure 11.12   (*continued*)

measurements at Cape Town harbour, airport and Robben Island respectively. Let $Y_{i,j}$ denote the maximum wind gust measurement of month j, $j = 1, \ldots, 70$, at location i, $i = 1, 2, 3$. Following condition $(\mathcal{C}_\gamma)$

$$Y_{i,j} | \sigma_i, \gamma_i, \mu_i \sim GEV(\sigma_i, \gamma_i, \mu_i), \qquad j = 1, \ldots, 70, \ i = 1, 2, 3.$$

The parameter vectors $(\sigma_i, \gamma_i, \mu_i)$, $i = 1, 2, 3$, are assumed to be i.i.d. according to

$$(\log \sigma, \log \gamma, \mu) \sim N_3(\phi, \Sigma),$$

with $\phi' = (3, -2, 40)$ and $\Sigma = 10 I_3$. The proposed prior distribution reflects the beliefs of the harbour master. We use the Metropolis–Hastings algorithm to simulate the posterior distribution of the model parameters. The proposal density is taken as a multivariate normal on $(\log \sigma_i, \log \gamma_i, \mu_i)$, $i = 1, 2, 3$, with independent components and respective standard deviations $(0.03, 0.03, 0.03)$. Initializing with $(\log \sigma_i, \log \gamma_i, \mu_i) = (3, -2, 40)$, $i = 1, 2, 3$, the values generated by 10,000 iterations of the chain are plotted in Figures 11.12-11.14. Note that the three locations are quite similar with respect to the tail index γ. The (heavy-tailed) posterior distributions of γ_1, γ_2 and γ_3 have a median around 0.135. The differences between the monthly maximal wind gust distributions manifest themselves through the posterior distributions of the parameters μ and σ. Finally, in Figure 11.15, we show the posterior median and 95% hpd region of $q_{Y,p}$ as a function of p for the three locations. The quantiles of the maximal wind gust distribution clearly tend to be largest at the harbour.

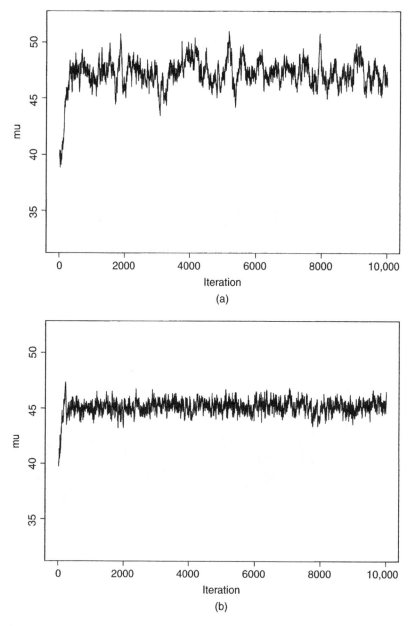

Figure 11.13   Metropolis–Hastings realizations of μ_i: (a) harbour, (b) airport, (c) Robben Island.

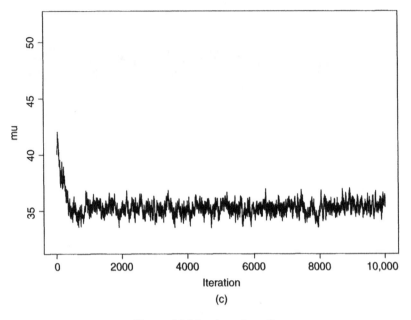

(c)

Figure 11.13   (*continued*)

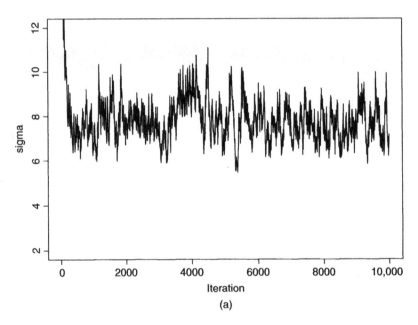

(a)

Figure 11.14   Metropolis–Hastings realizations of σ_i: (a) harbour, (b) airport, (c) Robben Island.

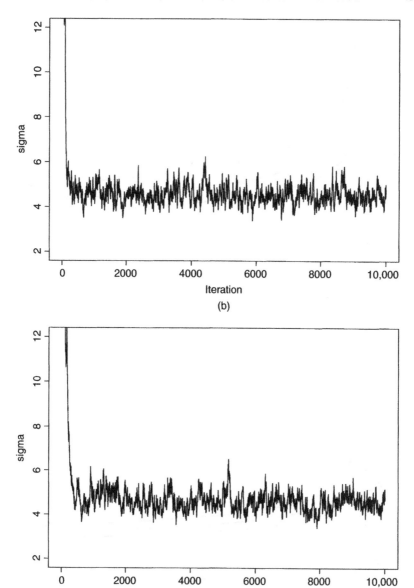

(b)

(c)

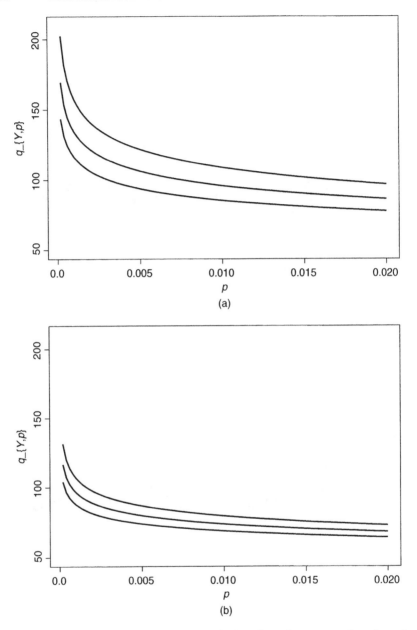

Figure 11.15   Posterior median and 95% hpd region of $q_{Y,p}$ as a function of p:
(a) harbour, (b) airport and (c) Robben Island.

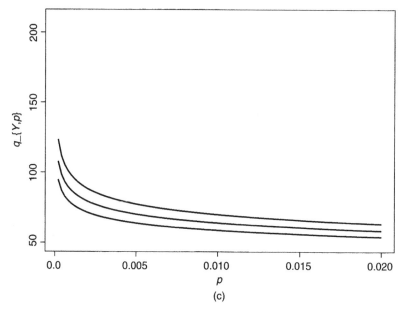

(c)

Figure 11.15

Bibliography

Abdous B, Ghoudi K and Khoudraji A 1999 Non-parametric estimation of the limit dependence function of multivariate extremes. *Extremes* **2**, 245–268.

Alpuim MT 1989 An extremal Markovian sequence. *Journal of Applied Probability* **26**, 219–232.

Alpuim MT, Catkan NA and Hüsler J 1995 Extremes and clustering of nonstationary max-AR(1) sequences. *Stochastic Processes and their Applications* **56**, 171–184.

Ancona-Navarrete MA and Tawn J 2000 A comparison of methods for estimating the extremal index. *Extremes* **3**, 5–38.

Ancona-Navarrete MA and Tawn J 2002 Diagnostics for pairwise extremal dependence in spatial processes. *Extremes* **5**, 271–285.

Anderson CW and Coles SG 2000 The largest inclusions in a piece of steel. *Extremes* **5**, 237–252.

Bacro JN and Brito M 1993 Strong limiting behaviour of a simple tail Pareto-index estimator. *Statistics and Decisions* **3**, 133–134.

Balakrishnan N and Chan PS 1992 Order statistics from extreme value distribution, II: best linear unbiased estimates and some other uses. *Communications in Statistics–Simulation and Computation* **21**, 1219–1246.

Balkema AA and Resnick SI 1977 Max-infinite divisibility. *Journal of Applied Probability* **14**, 309–319.

Barão MI and Tawn JA 1999 Extremal analysis of short series with outliers: sea-levels and athletic records. *Applied Statistics* **48**, 469–487.

Barbour AD, Novak SY and Xia A 2002 Compound Poisson approximation for the distribution of extremes. *Advances in Applied Probability* **24**, 223–240.

Barndorff-Nielsen OE and Cox DR 1994 *Inference and Asymptotics*. Chapman & Hall.

Barnett V 1976 The ordering of multivariate data (with discussion). *Journal of the Royal Statistical Society, Series A* **139**, 318–354.

Basrak B, Davis RA and Mikosch T 2002 Regular variation of GARCH processes. *Stochastic Processes and their Applications* **99**, 95–115.

Beirlant J, Bouquiaux C and Werker BJM 2002a Semiparametric lower bounds for tail index estimation. *Journal of Statistical Planning and Inference*; to appear.

Beirlant J, Dierckx G, Goegebeur Y and Matthys G 1999 Tail index estimation and an exponential regression model. *Extremes* **2**, 177–200.

Statistics of Extremes: Theory and Applications J. Beirlant, Y. Goegebeur, J. Segers, and J. Teugels
© 2004 John Wiley & Sons, Ltd ISBN: 0-471-97647-4

Beirlant J, Dierckx G and Guillou A 2002b Estimation of the extreme value index and regression on generalized quantile plots. *Bernoulli*; submitted.

Beirlant J, Dierckx G, Guillou A and Stărică C 2002c On exponential representations of log-spacings of extreme order statistics. *Extremes* **5**, 157–180.

Beirlant J and Goegebeur Y 2003 Regression with response distributions of Pareto-type. *Computational Statistics and Data Analysis* **42**, 595–619.

Beirlant J and Goegebeur Y 2004a Local polynomial maximum likelihood estimation for Pareto-type distributions. *Journal of Multivariate Analysis* **89**, 97–118.

Beirlant J and Goegebeur Y 2004b Simultaneous tail index estimation. *REVSTAT—Statistical Journal* **2**, 15–39.

Beirlant J, Goegebeur Y, Verlaak R and Vynckier P 1998 Burr regression and portfolio segmentation. *Insurance: Mathematics and Economics* **23**, 231–250.

Beirlant J, Joossens E and Segers J 2004 Discussion of "Generalized Pareto Fit to the Society of Actuaries' Large Claims Database" by A Cebrian, M Denuit and P Lambert. *North American Actuarial Journal* **8**, 108–111.

Beirlant J, Matthys G and Dierckx G 2001 Heavy-tailed distributions and rating. *ASTIN Bulletin* **31**, 41–62.

Beirlant J and Teugels JL 1995 A simple approach to classical extreme value theory. In *Festschrift for Academician V.S. Korolyuk* (ed. Skorokhod HI). VNU Publishers.

Beirlant J, Teugels JL and Vynckier P 1996a *Practical Analysis of Extreme Values*. Leuven University Press.

Beirlant J and Vandewalle B 2002 Some comments on the estimation of a dependence index in bivariate extreme value statistics. *Statistics & Probability Letters* **60**, 265–278.

Beirlant J, Vynckier P and Teugels JL 1996b Tail index estimation, Pareto quantile plots, and regression diagnostics. *Journal of the American Statistical Association* **91**, 1659–1667.

Beirlant J, Vynckier P and Teugels JL 1996c Excess functions and estimation of the extreme-value index. *Bernoulli* **2**, 293–318.

Beretta S and Anderson CW 2002 Extreme value statistics in metal fatigue. In *Società Italiana di Statistica: Atti della XLI Riunione Scientifica*, pp. 251–260.

Berman SM 1961 Convergence to bivariate limiting extreme value distribution. *Annals of the Institute for Statistical Mathematics* **13**, 217–223.

Berman SM 1964 Limit theorems for the maximum term in stationary sequences. *Annals of Mathematical Statistics* **35**, 502–516.

Bernardo JM and Smith AFM 1994 *Bayesian Theory*. Wiley.

Billingsley P 1995 *Probability and Measure*, 3rd edn. Wiley.

Bingham NH, Goldie CM and Teugels JL 1987 *Regular Variation*. Cambridge University Press.

Bollerslev T 1986 Generalized autoregressive conditional heteroskedasticity. *Journal of Econometrics* **31**, 307–327.

Bomas H, Mayr P and Linkewitz T 1999 Inclusion size distribution and endurance limit of a hard steel. *Extremes* **2**, 149–164.

Borkovec M 2000 Extremal behavior of the autoregressive process with ARCH(1) errors. *Stochastic Processes and their Applications* **85**, 189–207.

Borkovec M and Klüppelberg C 2003 The tail of the stationary distribution of an autoregressive process with ARCH(1) errors. *Annals of Applied Probability* **11**, 1220–1241.

Bortot P and Coles S 2000 A sufficiency property arising from the characterization of extremes of Markov chains. *Bernoulli* **6**, 183–190.

Bortot P and Tawn JA 1998 Models for the extremes of Markov chains. *Biometrika* **85**, 851–867.

Brockwell PJ and Davis RA (1991) *Time Series: Theory and Methods.* Springer.

Bruun T and Tawn JA 1998 Comparison of approaches for estimating the probability of coastal flooding. *Applied Statistics* **47**, 405–423.

Buishand TA 1993 Rainfall depth-duration-frequency curves; a problem of dependent extremes. In *Statistics for the Environment* (eds Barnett V and Turkman KF), pp. 183–197. Wiley.

Caers J, Beirlant J and Maes MA 1999a Statistics for modelling heavy tailed distributions in geology: Part I. Methodology. *Mathematical Geology* **31**, 391–410.

Caers J, Beirlant J and Maes MA 1999b Statistics for modelling heavy tailed distributions in geology: Part II. Applications. *Mathematical Geology* **31**, 411–434.

Capéraà P and Fougères AL 2000 Estimation of a bivariate extreme value distribution. *Extremes* **3**, 311–329.

Capéraà P, Fougères AL and Genest C 1997 A nonparametric estimation procedure for bivariate extreme value copulas. *Biometrika* **84**, 567–577.

Capéraà P, Fougères AL and Genest C 2000 Bivariate distributions with given extreme value attractor. *Journal of Multivariate Analysis* **72**, 30–49.

Carlin BP and Louis TA 2000 *Bayes and Empirical Bayes Methods for Data Analysis.* Chapman & Hall.

Castillo E and Hadi AS 1997 Fitting the generalized Pareto distribution to data. *Journal of the American Statistical Association* **92**, 1609-1620.

Chavez-Demoulin V 1999 *Two Problems in Environmental Statistics: Capture-Recapture Models and Smooth Extremal Models.* PhD Thesis, EPFL, Lausanne.

Chavez-Demoulin V and Davison AC 2001 Generalized additive models for sample extremes. *Journal of the Royal Statistical Society C*; to appear.

Chavez-Demoulin V and Embrechts P 2004 Smooth extremal models in finance. *Journal of Risk and Insurance* **71**, 183–199.

Cheng S, de Haan L and Yang J 1995 Asymptotic distributions of multivariate intermediate order statistics. *Theory of Probability and its Applications* **41**, 646–656.

Christopeit N 1994 Estimating parameters of an extreme value distribution by the method of moments. *Journal of Statistical Planning and Inference* **41**, 173–186.

Coles SG 1993 Regional modeling of extreme storms via max-stable processes. *Journal of the Royal Statistical Society, Series B* **55**, 797–816.

Coles SG 1994 Some aspects of spatial extremes. In *Extreme Value Theory and Applications* (eds Galambos J, Lechner J and Simiu E), pp. 269–282. Kluwer.

Coles SG 2001 *An Introduction to Statistical Modelling of Extreme Values.* Springer-Verlag.

Coles SG and Dixon MJ 1999 Likelihood-based inference of extreme value models. *Extremes* **2**, 5–23.

Coles S, Heffernan J and Tawn JA 1999 Dependence measures for extreme value analyses. *Extremes* **2**, 339–365.

Coles S and Pauli P 2001 Extremal limit laws for a class of bivariate Poisson vectors. *Statistics & Probability Letters* **54**, 373–379.

Coles S and Pauli P 2002 Models and inference for uncertainty in extremal dependence. *Biometrika* **89**, 183–196.

Coles SG and Powell EA 1996 Bayesian methods in extreme value modelling: a review and new developments. *International Statistical Review* **64**, 119–136.

Coles SG and Tawn JA 1990 Statistics of coastal flood prevention. *Philosophical Transactions of the Royal Society of London, Series A* **332**, 457–476.

Coles SG and Tawn JA 1991 Modelling extreme multivariate events. *Journal of the Royal Statistical Society, Series B* **53**, 377–392.

Coles SG and Tawn JA 1994 Statistical methods for multivariate extremes: an application to structural design. *Applied Statistics* **43**, 1–48.

Coles SG and Tawn JA 1996a Modelling extremes of the areal rainfall process. *Journal the Royal Statistical Society, Series B* **58**, 329–347.

Coles SG and Tawn JA 1996b Modelling extremes: a Bayesian approach. *Applied Statistics* **45**, 463–478.

Coles SG and Walshaw D 1994 Directional modelling of extreme wind speeds. *Applied Statistics* **43**, 139–157.

Csörgő S, Deheuvels P and Mason DM 1985 Kernel estimates of the tail index of a distribution. *Annals of Statistics* **13**, 1050–1077.

Csörgő S and Mason DM 1985 Central limit theorems for sums of extreme values. *Mathematical Proceedings of the Cambridge Philosophical Society* **98**, 547–558.

Csörgő S and Viharos L 1998 Estimating the tail index. In *Asymptotic Methods in Probability and Statistics* (ed. Szyszkowicz B), pp. 833–881. North Holland.

Danielsson J, de Haan L, Peng L and Vries CG de 1997 Using a bootstrap method to choose the sample fraction in tail index estimation. Technical Report TI 97-016/4, Tinbergen Institute, Rotterdam.

Danielsson J and de Vries CG 1997 Beyond the sample: extreme quantile and probability estimation. Technical Report, Tinbergen Institute, Rotterdam.

David HA 1994 Concomitants of extreme order statistics. In *Extreme Value Theory and Applications* (eds Galambos J, Lechner J and Simiu E), pp. 211–224. Kluwer.

Davis RA and Resnick ST 1984 Tail estimates motivated by extreme value theory. *Annals of Statistics* **12**, 1467–1487.

Davis RA and Resnick SI 1989 Basic properties and prediction of max-ARMA processes. *Advances in Applied Probability* **21**, 781–803.

Davis RA and Resnick SI 1993 Prediction of stationary max-stable processes. *Annals of Applied Probability* **3**, 497–525.

Davison AC and Ramesh NI 2000 Local likelihood smoothing of sample extremes. *Journal of the Royal Statistical Society* **62**, 191–208.

Davison AC and Smith RL 1990 Models for exceedances over high thresholds. *Journal of the Royal Statistical Society B* **52**, 393–442.

Deheuvels P 1983 Point processes and multivariate extreme values. *Journal of Multivariate Analysis* **13**, 257–272.

Deheuvels P 1984 Probabilistic aspects of multivariate extremes. In *Statistical Extremes and Applications* (ed Tiago de Oliveira J), pp. 117–130. Reidel.

Deheuvels P 1991 On the limiting behavior of the Pickands estimator for bivariate extreme-value distributions. *Statistics & Probability Letters* **12**, 429–439.

Deheuvels P, Haeusler E and Mason DM 1988 Almost sure convergence of the Hill estimator. *Mathematical Proceedings of the Cambridge Philosophical Society* **104**, 371–381.

Deheuvels P and Martynov GV 1996 Cramér-von Mises-type tests with applications to tests of independence for multivariate extreme-value distributions. *Communications in Statistics–Theory and Methods* **25**, 871–908.

Deheuvels P and Tiago de Oliveira J 1989 On the non-parametric estimation of the bivariate extreme-value distributions. *Statistics & Probability Letters* **8**, 315–323.

Dekkers ALM, Einmahl JHJ and de Haan L 1989 A moment estimator for the index of an extreme-value distribution. *Annals of Statistics* **17**, 1833-1855.

Dekkers ALM and de Haan L 1989 On the estimation of the extreme-value index and large quantile estimation. *Annals of Statistics* **17**, 1795-1832.

Denzel GE and O'Brien GL 1975 Limit theorems for extreme values of chain-dependent processes. *Annals of Probability* **3**, 773–779.

Dietrich D and Hüsler J 1996 Minimum distance estimators in extreme value distributions. *Communications in Statistics—Theory and Methods* **25**, 695–703.

Dixon MJ and Tawn JA 1995 A semiparametric model for multivariate extreme values. *Statistics and Computing* **5**, 215–225.

Draisma G, Drees H, Ferreira A and de Haan L 2004 Bivariate tail estimation: dependence in asymptotic independence. *Bernoulli* **10**, 251–280.

Drees H 1995 Refined Pickands estimators for the extreme value index. *Annals of Statistics* **23**, 2059–2080.

Drees H 1998 Optimal rates of convergence for estimates of the extreme value index. *Annals of Statistics* **26**, 434–448.

Drees H 2000 Weighted approximations of tail processes under mixing conditions. *Annals of Applied Probability* **10**, 1274–1301.

Drees H 2001 Exceedance over Threshold. In *Encyclopedia of Environmetrics*, Vol. 2 (eds El-Shaarawi AH and Piegorsch WW), pp. 715–728. Wiley.

Drees H 2002 Tail empirical processes under mixing conditions. In *Empirical Process Techniques for Dependent Data* (eds Dehling HG, Mikosch T and Sorensen M), pp. 325–342. Birkhäuser.

Drees H 2003 Extreme quantile estimation for dependent data with applications to finance. *Bernoulli* **9**, 617–657.

Drees H, Ferreira A and de Haan L 2002 On maximum likelihood estimation of the extreme value index. *Annals of Applied Probability*; to appear.

Drees H, de Haan L and Resnick SI 2000 How to make a Hill plot. *Annals of Statistics* **29**, 266–294.

Drees H and Huang X 1998 Best attainable rates of convergence for estimators of the stable tail dependence function. *Journal of Multivariate Analysis* **64**, 25–47.

Drees H and Kaufmann E 1998 Selecting the optimal sample fraction in univariate extreme value estimation. *Stochastic Processes and their Applications* **75**, 149–172.

Drees H and Reiss R-D 1992 Tail behaviour in Wicksell's corpuscle problem. In *Probability and Applications: Essays to the Memory of József Mogyoródi* (eds Galambos J and Katai I), pp. 205–220. Kluwer.

Dupuis DJ and Morgenthaler S 2002 Robust weighted likelihood estimators with an application to bivariate extreme value problems. *The Canadian Journal of Statistics* **30**, 17–36.

Dupuis DJ and Tawn JA 2001 Effects of mis-specification in bivariate extreme value problems. *Extremes* **4**, 315–330.

Efron B and Tibshirani RJ 1993 *An Introduction to the Bootstrap*. Chapman & Hall.

Einmahl JHJ 1997 Poisson and Gaussian approximation of weighted local empirical processes. *Stochastic Processes and their Applications* **70**, 31–58.

Einmahl JHJ, de Haan L and Huang X 1993 Estimating a multidimensional extreme-value distribution. *Journal of Multivariate Analysis* **47**, 35–47.

Einmahl JHJ, de Haan L and Piterbarg VI 2001 Nonparametric estimation of the spectral measure of an extreme value distribution. *The Annals of Statistics* **29**, 1401–1423.

Einmahl JHJ, de Haan L and Sinha AK 1997 Estimating the spectral measure of an extreme value distribution. *Stochastic Processes and their Applications* **70**, 143–171.

Einmahl JHJ and Lin T 2003 Asymptotic normality of extreme value estimators on $C[0, 1]$. Center Discussion Paper 2003–132.

Embrechts P, Klüppelberg C and Mikosch T 1997 *Modelling Extremal Events*. Springer-Verlag.

Engle RF 1982 Autoregressive conditional heteroscedastic models with estimates of the variance of United Kingdom inflation. *Econometrica* **50**, 987–1007.

Esary JD, Proschan F and Walup DW 1967 Association of random variables, with applications. *The Annals of Mathematical Statistics* **38**, 1466–1474.

Falk M 1994 Efficiency of convex combinations of Pickands' estimator of the extreme value index. *Journal of Nonparametric Statistics* **4**, 133–147.

Falk M, Hüsler J and Reiss R-D 1994 *Laws of Small Numbers: Extremes and Rare Events*. Birkhäuser.

Falk M and Reiss R-D 2001 Estimation of canonical dependence parameters in a class of bivariate peaks-over-threshold models. *Statistics & Probability Letters* **52**, 233–242.

Falk M and Reiss R-D 2002 A characterization of the rate of convergence in bivariate extreme value models. *Statistics & Probability Letters* **59**, 341–351.

Falk M and Reiss R-D 2003 Efficient estimators and LAN in canonical bivariate POT models. *Journal of Multivariate Analysis* **84**, 190–207.

Fan J and Gijbels I 1996 *Local Polynomial Modelling and its Applications*. Chapman & Hall.

Fermanian JD, Radulovic D and Wegkamp M 2004 Weak convergence of empirical copula processes. *Bernoulli*; to appear.

Ferreira H 1993 Joint exceedances of high levels under a local dependence condition. *Journal of Applied Probability* **30**, 112–120.

Ferreira A, de Haan L and Peng L 2003 On optimising the estimation of high quantiles of a probability distribution. *Statistics* **37**, 403–434.

Ferro CAT and Segers J 2003 Inference for clusters of extreme values. *Journal of the Royal Statistical Society B*, **65**, 545–556.

Feuerverger A and Hall P 1999 Estimating a tail exponent by modelling departure from a Pareto distribution. *Annals of Statistics* **27**, 760–781.

Fisher RA and Tippett LHC 1928 On the estimation of the frequency distributions of the largest or smallest member of a sample. *Proceedings of the Cambridge Philosophical Society* **24**, 180–190.

Fougères AL 2004 Multivariate extremes. In *Extreme Values in Finance, Telecommunications, and the Environment* (eds Finkelstädt B and Rootzén H) pp. 373–388. Chapman & Hall.

Fraga Alves MI 1995 Estimation of the tail parameter in the domain of attraction of an extremal distribution. *Journal of Statistical Planning and Inference* **45**, 143–173.

Fraga Alves MI 2001 A location invariant Hill-type estimator. *Extremes* **4**, 199–217.

Fraga Alves MI, Gomes MI and de Haan L 2003 A new class of semi-parametric estimators of the second order parameter. *Portugaliae Mathematica* **60**, 193–213.

Fréchet M 1927 Sur la loi de probabilité de l'écart maximum. *Annales de la Société Polonaise de Mathematique, Cracow* **6**, 93–117.

Frees EW and Valdez EA 1998 Understanding relationships using copulas. *North American Actuarial Journal* **2**, 1–15.

Galambos J 1978 *The Asymptotic Theory of Extreme Order Statistics*. Wiley.

Galambos J 1987 *The Asymptotic Theory of Extreme Order Statistics*. Krieger.

Garralda Guillem AI 2000 Structure de dépendance des lois de valeurs extrêmes bivariées. *Comptes Rendues de l'Académie des Sciences de Paris* **330**, 593–596.

Geffroy J 1958/59 Contribution à la théorie des valuers extrêmes. *Publications de l'Institut de Statistique de l'Université de Paris* **7**, 37–121 (1958); **8**, 123–184 (1959).

Gelman A, Carlin JB, Stern HS and Rubin DB 1995 *Bayesian Data Analysis*. Chapman & Hall.

Geman S and Geman D 1984 Stochastic relaxation, Gibbs distributions, and the Bayesian restoration of images. *IEEE Transactions on Pattern Analysis and Machine Intelligence* **6**, 721–741.

Genest C, Ghoudi K and Rivest LP 1995 A semiparametric estimation procedure of dependence parameters in multivariate families of distributions. *Biometrika* **82**, 543–552.

Genest C and MacKay RJ 1986 Copules archimédiennes et familles de lois bidimensionelles dont les marges sont données. *Canadian Journal of Statistics* **14**, 145–159.

Genest C and Rivest LP 1989 A characterization of Gumbel's family of extreme value distributions. *Statistics and Probability Letter* **8**, 207–211.

Gerritse G 1986 Supremum self-decomposable random vectors. *Probability Theory and Related Fields* **72**, 17–34.

Ghoudi K, Khoudraji A and Rivest LP 1998 Propriétés statistiques des copules de valeurs extrêmes bidimensionelles. *Canadian Journal of Statistics* **14**, 145–159.

Gnedenko BV 1943 Sur la distribution limite du terme maximum d'une série aléatoire. *Annals of Mathematics* **44**, 423–453.

Goegebeur Y, Planchon V, Beirlant J and Oger R 2004 Quality assessment of pedochemical data using extreme value methodology. *Journal of Applied Science*; to appear.

Goldie CM and Resnick SI 1995 Many multivariate records. *Stochastic Processes and their Applications* **59**, 185–216.

Goldie CM and Smith RL 1987 Slow variation with remainder: theory and applications. *Quarterly Journal of Mathematics, Oxford* **38**, 45–71.

Gomes MI 1993 On the estimation of parameters of rare events in environmental time series. In *Statistics for the Environment 2: Water Related Issues* (eds Barnett V and Turkman KF), pp. 225–241. Wiley.

Gomes MI, de Haan L and Peng L 2002 Semi-parametric estimators of the second order parameter in statistics of extremes. *Extremes* **5**, 387–414.

Gomes MI and Martins MJ 2002 "Asymptotically unbiased" estimators of the tail index based on external estimation of the second order parameter. *Extremes* **5**, 5–31.

Gomes MI, Martins MJ and Neves M 2000 Alternatives to a semi-parametric estimator of parameters of rare events—the Jackknife methodology. *Extremes* **3**, 207–229.

Gomes MI and Oliveira O 2001 The bootstrap methodology in statistics of extremes – choice of the optimal sample fraction. *Extremes* **4**, 331–358.

Gomes MI and Oliveira O 2003 Maximum likelihood revisited under a semi-parametric context—estimation of the tail index. *Journal of Statistical Planning and Inference* **73**, 285–301.

Grazier, KL and G'Sell Associates 1997 *Group Medical Insurance Large Claims Database and Collection.* SOA Monograph M-HB97-1, Society of Actuaries, Schaumburg.

Green PJ and Silverman BW 1994 *Nonparametric Regression and Generalized Linear Models: A Roughness Penalty Approach.* Chapman & Hall.

Greenwood JA, Landwehr JM, Matalas NC and Wallis JR 1979 Probability weighted moments: definition and relation to parameters of several distributions expressable in inverse form. *Water Resources Research* **15**, 1049–1054.

Guerin CA, Nyberg H, Perrin O, Resnick SI, Rootzén H and Stărică C 2000 Empirical testing of the infinite source Poisson data traffic model. Report 2000:4, Mathematical Statistics, Chalmers University of Technology.

Guillou A and Hall P 2001 A diagnostic for selecting the threshold in extreme value analysis. *Journal of the Royal Statistical Society B* **63**, 293–305.

Gumbel EJ 1958 *Statistics of Extremes.* Columbia University Press.

Gumbel EJ 1960a Distributions des valeurs extrêmes en plusieurs dimensions. *Publications de l'Institute de Statistique de l'Université de Paris* **9**, 171–173.

Gumbel EJ 1960b Bivariate exponential distributions. *Journal of the American Statistical Association* **55**, 698–707.

Gumbel EJ 1962 Multivariate extremal distributions. In *Bulletin de l'Institut International de Statistique, Actes de la 33e Session,* pp. 471–475.

Gumbel EJ and Goldstein N 1964 Analysis of empirical bivariate extremal distributions. *Journal of the American Statistical Association* **59**, 794–816.

Gumbel EJ and Mustafi CK 1967 Some analytical properties of bivariate extremal distributions. *Journal of the American Statistical Association* **62**, 569–588.

de Haan L 1970 *On Regular Variation and its Applications to the Weak Convergence of Sample Extremes.* Mathematical Centre Tract, 32, Amsterdam.

de Haan L 1984 A spectral representation for max-stable processes. *Annals of Probability* **12**, 1194–1204.

de Haan L 1985 Extremes in higher dimensions: the model and some statistics. In *Bulletin of the International Statistical Institute, Proceedings of the 45th Session,* Book 4, Paper 26.3.

de Haan L and Huang X 1995 Large quantile estimation in a multivariate setting. *Journal of Multivariate Analysis* **53**, 247–263.

de Haan L and Lin T 2001 On convergence towards an extreme value distribution in $C[0, 1]$. *Annals of Probability* **29**, 467–483.

de Haan L and Peng L 1997 Rates of convergence for bivariate extremes. *Journal of Multivariate Analysis* **61**, 195–230.

de Haan L and Peng L 1998 Comparison of tail index estimators. *Statistica Neerlandica* **52**, 60–70.

de Haan L and Pickands J 1986 Stationary min-stable stochastic processes. *Probability Theory and Related Fields* **72**, 477–492.

de Haan L and Resnick SI 1977 Limit theory for multivariate sample extremes. *Zeitschrift für Wahrscheinlichkeitstheorie und Verwandte Gebiete* **40**, 317–337.

de Haan L and Resnick SI 1987 On regular variation of probability densities. *Stochastic Processes and their Applications* **25**, 83–93.

de Haan L and Resnick SI 1993 Estimating the limit distribution of multivariate extremes. *Communications in Statistics–Stochastic Models* **9**, 275–309.

de Haan L and Resnick SI 1998 On asymptotic normality of the Hill estimator. *Stochastic Models* **14**, 849–867.

de Haan L, Resnick SI, Rootzén H and de Vries CG 1989 Extremal behaviour of solutions to a stochastic difference equation with applications to ARCH processes. *Stochastic Processes and their Applications* **32**, 213–224.

de Haan L and de Ronde J 1998 Sea and wind: multivariate extremes at work. *Extremes* **1**, 7–45.

de Haan L and Rootzén H 1993 On the estimation of high quantiles. *Journal of Statistical Planning and Inference* **35**, 1–13.

de Haan L and Sinha AK 1999 Estimating the probability of a rare event. *The Annals of Statistics* **27**, 732–759.

de Haan L and Stadtmüller U 1996 Generalized regular variation of second order. *Journal of the Australian Mathematical Society A* **61**, 381–395.

Habach L 1997 Comportement p.s. des extrêmes multivariées de suites i.i.d. de vecteurs gaussiens. *Comptes Rendues de l'Académie des Sciences de Paris* **325**, 541–544.

Haeusler E and Teugels JL 1985 On asymptotic normality of Hill's estimator for the exponent of regular variation. *Annals of Statistics* **13**, 743–756.

Hall P 1982 On some simple estimates of an exponent of regular variation. *Journal of the Royal Statistical Society B* **44**, 37–42.

Hall P 1990 Using the bootstrap to estimate mean squared error and select smoothing parameter in nonparametric problems. *Journal of Multivariate Analysis* **32**, 177–203.

Hall P, Peng L and Yao Q 2002 Moving-maximum models for extrema of time series. *Journal of Statistical Planning and Inference* **103**, 51–63.

Hall P and Tajvidi N 2000a Nonparametric analysis of temporal trend when fitting parametric models to extreme-value data. *Statistical Science* **15**, 153–167.

Hall P and Tajvidi N 2000b Distribution and dependence-function estimation for bivariate extreme-value distributions. *Bernoulli* **6**, 835–844.

Hall P and Welsh AH 1984 Best attainable rates of convergence for estimates of parameters of regular variation. *Annals of Statistics* **12**, 1079–1084.

Hall P and Welsh AH 1985 Adaptive estimates of parameters of regular variation. *Annals of Statistics* **13**, 331-341.

Heffernan JE 2000 A directory of coefficients of tail dependence. *Extremes* **3**, 279–290.

Heffernan JE and Tawn JA 2004 A conditional approach for multivariate extreme values. *Journal of the Royal Statistical Society, Series B* **66**, 1–34.

Hill BM 1975 A simple general approach to inference about the tail of a distribution. *Annals of Statistics* **3**, 1163–1174.

Hooghiemstra G and Hüsler J 1996 A note on maxima of bivariate random vectors. *Statistics & Probability Letters* **31**, 1–6.

Hooghiemstra G and Meester LE 1997 The extremal index in 10 seconds. *Journal of Applied Probability* **34**, 818–822.

Hougaard P 1986 A class of multivariate failure time distributions. *Biometrika* **73**, 671–678.

Hosking JRM 1984 Testing whether the shape parameter is zero in the generalized extreme-value distribution. *Biometrika* **71**, 367–374.

Hosking JRM 1985 Algorithm AS 215: Maximum likelihood estimation of the parameters of the generalized extreme value distribution. *Applied Statistics* **34**, 301–310.

Hosking JRM and Wallis JR 1987 Parameter and quantile estimation for the generalized Pareto distribution. *Technometrics* **29**, 339–349.

Hosking JRM, Wallis JR and Wood EF 1985 Estimation of the generalized extreme-value distribution by the method of probability-weighted moments. *Technometrics* **27**, 251–261.

Hsieh PH 2001 On Bayesian predictive moments of next record value using three-parameter gamma priors. *Communications in Statistics: Theory and Methods* **30**, 729-738.

Hsing T 1987 On the characterization of certain point processes. *Stochastic Processes and their Applications* **26**, 297–316.

Hsing T 1988 On the extreme order statistics for a stationary sequence. *Stochastic Processes and their Application* **29**, 155–169.

Hsing T 1989 Extreme value theory for multivariate stationary sequences. *Journal of Multivariate Analysis* **29**, 274–291.

Hsing T 1991 On tail index estimation using dependent data. *Annals of Statistics* **19**, 1547–1569.

Hsing T 1993 Extremal index estimation for a weakly dependent stationary sequence. *Annals of Statistics* **21**, 2043–2071.

Hsing T, Hüsler J and Leadbetter MR 1988 On the exceedance point process for a stationary sequence. *Probability Theory and Related Fields* **78**, 97–112.

Huang X 1992 *Statistics of Bivariate Extremes*. PhD Thesis, Erasmus University, Rotterdam, Tinbergen Institute Research series No. 22.

Hult H and Lindskog F 2002 Multivariate extremes, aggregation and dependence in elliptical distributions. *Advances in Applied Probability* **34**, 587–608.

Hürlimann W 2003 Hutchinson-Lai's conjecture for bivariate extreme value copulas. *Statistics & Probability Letters* **61**, 191–198.

Hüsler J 1989a Limit distributions of multivariate extreme values in nonstationary sequences of random vectors. In *Extreme Value Theory: Proceedings, Oberwolfach 1987* (eds Hüsler J and Reiss RD), pp. 234–245. Lecture Notes in Statistics 51, Springer-Verlag, Berlin.

Hüsler J 1989b Limit properties for multivariate extreme values in sequences of independent, non-identically distributed random vectors. *Stochastic Processes and their Applications* **31**, 105–116.

Hüsler J 1990 Multivariate extreme values in stationary random sequences. *Stochastic Processes and their Applications* **35**, 99–108.

Hüsler J 1993 A note on exceedances and rare events of non-stationary sequences. *Journal of Applied Probability* **30**, 877–888.

Hüsler J 1994 Maxima of bivariate random vectors: between independence and complete dependence. *Statistics & Probability Letters* **21**, 385–394.

Hüsler J and Reiss R-D 1989 Maxima of normal random vectors: between independence and complete dependence. *Statistics & Probability Letters* **7**, 283–286.

Hüsler J and Schmidt M 1996 A note on the point processes of rare events. *Journal of Applied Probability* **33**, 654–663.

Hutchinson TP and Lai CD 1990 *Continuous Bivariate Distributions, emphasizing Applications*. Rumsby Scientific, Adelaide.

Jeffreys H 1961 *Theory of Probability*. Oxford University Press.

Jiménez JR, Villa-Diharce E and Flores M 2001 Nonparametric estimation of the dependence function in bivariate extreme value distributions. *Journal of Multivariate Analysis* **76**, 159–191.

Joe H 1990 Families of min-stable multivariate exponential and multivariate extreme value distributions. *Statistics & Probability Letters* **9**, 75–81.

Joe H 1994 Multivariate extreme value distributions with applications to environmental data. *Canadian Journal of Statistics* **22**, 47–64.

Joe H, Smith RL and Weissman I 1992 Bivariate threshold methods for extremes. *Journal of the Royal Statistical Society, Series B* **54**, 171–183.

Kallenberg O 1983 *Random Measures*, 3rd edn. Akademie-Verlag, Berlin.

Kaufmann E and Reiss RD 1995 Approximation rates for multivariate exceedances. *Journal of Statistical Planning and Inference* **45**, 235–245.

Kaufmann E and Reiss RD 1998 Approximations of the Hill estimator process. *Statistics & Probability Letters* **39**, 347–354.

Kimeldorf G and Sampson AR 1987 Positive dependence orderings. *The Annals of the Institute for Statistics and Mathematics* **39**, 113–128.

Klein Tank AMG and co-authors 2002 Daily dataset of 20th-century surface air temperature and precipitation series for the European climate assessment. *International Journal of Climatology* **22**, 1441–1453.

Klugman SA and Parsa R 1999 Fitting bivariate loss distributions with copulas. *Insurance: Mathematics and Economics* **24**, 139–148.

Klüppelberg C and May A (1999) *The dependence function for bivariate extreme value distributions—a systematic approach*. Unpublished manuscript.

Koenker R and Bassett G 1978 Regression quantiles. *Econometrica* **46**, 33–50.

Kotz S and Nadarajah S 2000 *Extreme Value Distributions*. Imperial College Press.

Kratz M and Resnick SI 1996 The qq-estimator of the index of regular variation. *Communications in Statistics: Stochastic Models* **12**, 699–724.

Landwehr JM, Matalas NC and Wallis JR 1979 Probability weighted moments compared with some traditional techniques in estimating Gumbel parameters and quantiles. *Water Resources Research* **15**, 1055–1064.

Leadbetter MR 1974 On extreme values in stationary sequences. *Zeitschrift für Wahrscheinlichkeitstheorie und Verwandte Gebiete* **28**, 289–303.

Leadbetter MR 1983 Extremes and local dependence in stationary sequences. *Zeitschrift für Wahrscheinlichkeitstheorie und Verwandte Gebiete* **65**, 291–306.

Leadbetter MR 1991 On a basis for 'Peaks over Threshold' modeling. *Statistics & Probability Letters* **12**, 357–362.

Leadbetter MR 1995 On high level exceedance modeling and tail inference. *Journal of Statistical Planning and Inference* **45**, 247–260.

Leadbetter MR, Lindgren G and Rootzén H 1983 *Extremes and Related Properties of Random Sequences and Processes*. Springer-Verlag.

Leadbetter MR, Weissman I, de Haan L and Rootzén H 1989 On clustering of high values in statistically stationary series. In *Proceedings of the 4th International Meeting on Statistical Climatology* (ed. Sansom J), pp. 217–222. New Zealand Meteorological Service, Wellington.

Ledford AW and Tawn JA 1996 Statistics for near independence in multivariate extreme values. *Biometrika* **83**, 169–187.

Ledford AW and Tawn JA 1997 Modelling dependence within joint tail regions. *Journal of the Royal Statistical Society B* **59**, 475–499.

Ledford AW and Tawn JA 1998 Concomitant tail behaviour for extremes. *Advances in Applied Probability* **30**, 197–215.

Ledford AW and Tawn JA 2003 Diagnostics for dependence within time series extremes. *Journal of the Royal Statistical Society B* **65**, 521–543.

Lehmann EL 1966 Some concepts of dependence. *The Annals of Mathematical Statistics* **37**, 1137–1153.

Longin F and Solnik B 2001 Extreme correlation of international equity markets. *The Journal of Finance* **56**, 649–676

Loynes RM 1965 Extreme values in uniformly mixing stationary stochastic processes. *Annals of Mathematical Statistics* **36**, 993–999.

Lye LM, Hapuarachchi KP and Ryan S 1993 Bayes estimation of the extreme-value reliability function. *IEEE Transactions on Reliability* **42**, 641–644.

Macleod AJ 1989 AS R76 - A remark on algorithm AS 215: Maximum likelihood estimation of the parameters of the generalized extreme-value distribution. *Applied Statistics* **38**, 198–199.

Marohn F 1999 Testing the Gumbel hypothesis via the POT-method. *Extremes* **1**, 191–213.

Marshall AW and Olkin I 1967 A multivariate exponential distribution. *Journal of the American Statistical Association* **62**, 30–44.

Marshall AW and Olkin I 1983 Domains of attraction of multivariate extreme value distributions. *Annals of Probability* **11**, 168–177.

Mason DM 1982 Laws of large numbers for sums of extreme values. *Annals of Probability* **10**, 754–764.

Matthys G and Beirlant J 2000 Adaptive threshold selection in tail index estimation. In *Extremes and Integrated Risk Management* (ed. Embrechts P), pp. 37–57. UBS Warburg.

Matthys G and Beirlant J 2003 Estimating the extreme value index and high quantiles with exponential regression models. *Statistica Sinica* **13**, 853–880.

Maulik K, Resnick S, and Rootzén H (2002) Asymptotic independence and a network traffic model. *Journal of Applied Probability* **39**, 671–699.

McFadden D 1978 Modelling the choice of residential location. In *Spatial Interaction Theory and Planning Models* (eds Karlqvist A, Lundquist L, Snickers F and Weibull J), pp. 75–96. North Holland.

McNeil AJ and Frey R 2000 Estimation of tail-related risk measures for heteroscedastic financial time series: an extreme value approach. *Journal of Empirical Finance* **7**, 271–300.

Meerschaert MM and Scheffler HP 2001 *Limit Distributions for Sums of Independent Random Vectors*. Wiley.

Mikosch T 2004 Modeling dependence and tails of financial time series. In *Extreme Values in Finance, Telecommunications, and the Environment* (eds Finkelstädt B and Rootzén H) pp. 185–286. Chapman & Hall.

Mikosch T, Gadrich T, Klüppelberg C and Adler RJ (1995) Parameter estimation for ARMA models with infinite variance innovations. *Annals of Statistics* **23**, 305–326.

Mikosch T and Stărică C 2000 Limit theory for the sample autocorrelations and extremes of a GARCH(1,1) process. *Annals of Statistics* **28**, 1427–1451.

Mori T 1977 Limit distributions of two-dimensional point processes generated by strong-mixing sequences. *Yokohama Mathematical Journal* **25**, 155–168.

Morton ID and Bowers J 1996 Extreme value analysis in a multivariate offshore environment. *Applied Ocean Research* **18**, 303–317.

Murakami Y and Beretta S 1999 Small defects and inhomogeneities in fatigue strength: experiments, models and statistical implications. *Extremes* **2**, 123–147.

Nadarajah S 1999b A polynomial model for bivariate extreme value distributions. *Statistics & Probability Letters* **42**, 15–25.

Nadarajah S 2000 Approximations for bivariate extreme values. *Extremes* **3**, 87–98.

Nadarajah S 2001 Multivariate declustering techniques. *Environmetrics* **12**, 357–365.

Nadarajah S, Anderson CW and Tawn JA 1998 Ordered multivariate extremes. *Journal of the Royal Statistical Society, Series B* **60**, 473–496.

Nagaraja HN and David HA 1994 Distribution of the maximum of concomitants of selected order statistics. *The Annals of Statistics* **22**, 478–494.

Nandagopalan S 1994 On the multivariate extremal index. *Journal of Research of the National Institute of Standards and Technology* **99**, 543–550.

Nelder JA and Wedderburn RWM 1972 Generalized linear models. *Journal of the Royal Statistical Society, Series A* **153**, 370–384.

Nelsen RB 1999 *An Introduction to Copulas*. Springer, New York.

Newell GF 1964 Asymptotic extremes for m-dependent random variables. *The Annals of Mathematical Statistics* **35**, 1322–1325.

Novak SY 1999 Inference on heavy tails from dependent data. Technical Report 99-043, Eurandom.

Novak SY 2002 Multilevel clustering of extremes. *Stochastic Processes and their Applications* **97**, 59–75.

Novak SY 2003 On the accuracy of multivariate compound Poisson approximation. *Statistics & Probability Letters* **62**, 35–43.

Oakes D and Manatunga AK 1992 Fisher information for a bivariate extreme value distribution. *Biometrika* **79**, 827–832.

O'Brien GL 1987 Extreme values for stationary and Markov sequences. *Annals of Probability* **15**, 281–291.

Obrenetov A 1991 On the dependence function of Sibuya in multivariate extreme value theory. *Journal of Multivariate Analysis* **36**, 35–43.

Omey E and Rachev ST 1991 Rates of convergence in multivariate extreme value theory. *Journal of Multivariate Analysis* **38**, 36–50.

Pauli F and Coles S 2001 Penalized likelihood inference in extreme value analysis. *Journal of Applied Statistics* **28**, 547–560.

Peng L 1998 Asymptotically unbiased estimators for extreme value index. *Statistics & Probability Letters* **38**, 107–115.

Peng L 1999 Estimation of the coefficient of tail dependence in bivariate extremes. *Statistics & Probability Letters* **43**, 399–409.

Pereira TT 1994 Second order behaviour of domains of attraction and the bias of generalized Pickands' estimator. In *Extreme Value Theory and Applications III* (eds Galambos J, Lechner J and Simiu E), pp. 165–177. NIST.

Perfekt R 1994 Extremal behaviour of stationary Markov chains with applications. *Annals of Applied Probability* **4**, 529–548.

Perfekt R 1997 Extreme value theory for a class of Markov chains with values in $\mathbb{R}^d$. *Advances in Applied Probability* **29**, 138–164.

Pickands J 1971 The two-dimensional Poisson process and extremal processes. *Journal of Applied Probability* **8**, 745–756.

Pickands J 1975 Statistical inference using extreme order statistics. *Annals of Statistics* **3**, 119–131.

Pickands J 1981 Multivariate extreme value distributions. In *Bulletin of the International Statistical Institute, Proceedings of the 43rd Session*, Buenos Aires, pp. 859–878.

Pickands J 1989 Multivariate negative exponential and extreme value distributions. In *Extreme Value Theory: Proceedings, Oberwolfach 1987* (eds Hüsler J and Reiss R-D), pp. 262–274. Lecture Notes in Statistics 51, Springer-Verlag, Berlin.

Pickands J 1994 Bayes quantile estimation and threshold selection for the generalized Pareto family. In *Extreme Value Theory and Applications* (eds Galambos J, Lechner J and Simiu E), pp. 123–138. Kluwer.

Pisarenko VF and Sornette D 2003 Characterization of the frequency of extreme events by the generalized Pareto distribution. *Pure and Applied Geophysics* **160**, 2343–2364.

Posner EC, Rodemich ER, Asholock JC and Lurie S 1969 Application of an estimator of high efficiency in bivariate extreme value theory. *Journal of the American Statistical Association* **64**, 1403–1414.

Prescott P and Walden AT 1980 Maximum likelihood estimation of the parameters of the generalized extreme-value distribution. *Biometrika* **67**, 723–724.

Prescott P and Walden AT 1983 Maximum likelihood estimation of the parameters of the three-parameter generalized extreme value distribution from censored samples. *Journal of Statistical Computation and Simulation* **16**, 241–250.

Ramos A 2003 *Multivariate Joint Tail Modelling and Score Tests of Independence.* PhD Thesis, Department of Mathematics and Statistics, University of Surrey.

Reiss R-D 1989 *Approximate Distributions of Order Statistics.* Springer-Verlag.

Reiss R-D and Thomas M 2001 *Statistical Analysis of Extreme Values*, 2nd edn. Birkhäuser.

Resnick SI 1987 *Extreme Values, Regular Variation and Point Processes.* Springer-Verlag.

Resnick SI 1997 Heavy tail modeling and teletraffic data. *Annals of Statistics* **25**, 1805–1869.

Resnick SI and Rootzén H 2000 Self-similar communication models and very heavy tails. *Annals of Applied Probability* **10**, 753–778.

Resnick SI and Stărică C 1995 Consistency of Hill's estimator for dependent data. *Journal of Applied Probability* **32**, 239–267.

Resnick SI and Stărică C 1997a Smoothing the Hill estimator. *Advances in Applied Probability* **29**, 271-293.

Resnick SI and Stărică C 1997b Asymptotic behavior of Hill's estimator for auto-regressive data. *Communications in Statistics – Stochastic Models* **13**, 703–721.

Resnick SI and Stărică C 1998 Tail index estimation for dependent data. *Annals of Applied Probability* **8**, 1156–1183.

Robinson ME and Tawn JA 2000 Extremal analysis of processes sampled at different frequencies. *Journal of the Royal Statistical Society B* **62**, 117–135.

Rootzén H 1988 Maxima and exceedances of stationary Markov chains. *Advances in Applied Probability* **20**, 371–390.

Rootzén H, Leadbetter MR and de Haan L 1998 On the distribution of tail array sums for strongly mixing stationary sequences. *Annals of Applied Probability* **8**, 868–885.

Schlather M 2001 Examples for the coefficient of tail dependence and the domain of attraction of a bivariate extreme value distribution. *Statistics & Probability Letters* **53**, 325–329.

Schlather M and Tawn J 2002 Inequalities for the extremal coefficients of multivariate extreme value distributions. *Extremes* **5**, 87–102.

Schlather M and Tawn J 2003 A dependence measure for multivariate and spatial extreme values: properties and inference. *Biometrika* **90**, 139–156.

Schultze J and Steinebach J 1996 On least squares estimates of an exponential tail coefficient. *Statistics and Decisions* **14**, 353–372.

Segers J 2001 *Extremes of a Random Sample: Limit Theorems and Statistical Applications.* PhD Thesis, Department of Mathematics, K.U., Leuven.

Segers J 2002 Extreme events: dealing with dependence. EURANDOM report 2002-036.

Segers J 2003a Functionals of clusters of extremes. *Advances in Applied Probability* **35**, 1028–1045.

Segers J 2003b Approximate distributions of clusters of extremes. Center Discussion Paper 2003-91, Tilburg University, The Netherlands.

Segers J 2004 Generalized Pickands estimators for the extreme value index. *Journal of Statistical Planning and Inference*; to appear.

Self SG and Liang KY 1987 Asymptotic properties of maximum likelihood estimators and likelihood ratio tests under non-standard conditions. *Journal of the American Statistical Association* **82**, 605-610.

Shi D 1995a Fisher information for a multivariate extreme value distribution. *Biometrika* **82**, 644–649.

Shi D 1995b Moment estimation for multivariate extreme distributions. *Journal of Applied Mathematics* **10B**, 61–68.

Sibuya M 1960 Bivariate extreme statistics. *Annals of the Institute of Statistical Mathematics* **11**, 195–210.

Sklar A 1959 Fonctions de répartition à *n* dimensions et leurs marges. *Publications de l'Institut de Statistique de l'Université de Paris* **8**, 229–231.

Skorohod AV 1956 Limit theorems for stochastic processes. *Theory of Probability and its Applications* **1**, 261–290.

Smith RL 1984 Threshold methods for sample extremes. In *Statistical Extremes and Applications* (ed. Tiago de Oliveira J), pp. 621–638. Reidel.

Smith RL 1985 Maximum likelihood estimation in a class of nonregular cases. *Biometrika* **72**, 67–90.

Smith RL 1987 Estimating tails of probability distributions. *Annals of Statistics* **15**, 1174–1207.

Smith RL 1988 A counterexample concerning the extremal index. *Advances in Applied Probability* **20**, 681–683.

Smith RL 1989 Extreme value analysis of environmental time series: an application to trend detection in ground-level ozone. *Statistical Science* **4**, 367–393.

Smith RL 1990a Regional estimation from spatially dependent data. Unpublished.

Smith RL 1990b Extreme value theory. In *Handbook of Applicable Mathematics Supplement* (eds Ledermann W, Lloyd E, Vajda S and Alexander C), pp. 437–472. Wiley.

Smith RL 1991 Max-stable Processes and spatial extremes. Preprint, Department of Mathematics, University of Surrey, Guilford.

Smith RL 1992 The extremal index for a Markov chain. *Journal of Applied Probability* **29**, 37–45.

Smith RL 1994 Multivariate threshold methods. In *Extreme Value Theory and Applications* (eds Galambos J, Lechner J and Simiu E), pp. 225–248. Kluwer.

Smith RL and Goodman DJ 2000 Bayesian risk analysis. In *Extremes and Integrated Risk Management* (ed. Embrechts P), pp. 235–251. UBS, Warburg.

Smith RL, Tawn JA and Coles SG 1997 Markov chain models for threshold exceedances. *Biometrika*, **84**, 249–268.

Smith RL, Tawn JA and Yuen HK 1990 Statistics of multivariate extremes. *International Statistical Review* **58**, 47–58.

Smith RL and Weissman I 1994 Estimating the extremal index. *Journal of the Royal Statistical Society B* **56**, 515–528.

Smith RL and Weissman I 1996 *Characterization and Estimation of the Multivariate Extremal Index.* www.stat.unc.edu/faculty/rs/papers/RLS_Papers.html

Snyder DL and Miller MI 1991 *Random Point Processes in Time and Space*, 2nd edn. Springer-Verlag.

Stărică C 1999 Multivariate extremes for models with constant conditional correlations. *Journal of Empirical Finance* **6**, 515–553.

Stephenson A 2003 Simulating multivariate extreme value distributions of logistic type. *Extremes* **6**, 49–59.

Stute W 1984 The oscillation behavior of empirical processes: the multivariate case. *Annals of Probability* **12**, 361–379.

Svensson T and Maré J de 1999 Random features of the fatigue limit. *Extremes* **2**, 165–176.

Tajvidi N 1996 *Characterization and Some Statistical Aspects of Univariate and Multivariate Generalized Pareto Distributions.* PhD Thesis, University of Göteborg. Available at www.maths.lth.se/matstat/staff/nader/fullpub.html.

Takahashi R 1994 Asymptotic independence and perfect dependence of vector components of multivariate extreme statistics. *Statistics & Probability Letters* **19**, 19–26.

Takahashi R and Sibuya M 1996 The maximum size of the planar sections of random spheres and its application to metallurgy. *Annals of the Institute of Statistical Mathematics* **48**, 127–144.

Takahashi R and Sibuya M 1998 Prediction of the maximum size in Wicksell's corpuscle problem. *Annals of the Institute of Statistical Mathematics* **50**, 361–377.

Tawn JA 1988a Bivariate extreme value theory: models and estimation. *Biometrika* **75**, 397–415.

Tawn JA 1988b An extreme value theory model for dependent observations. *Journal of Hydrology* **101**, 227–250.

Tawn JA 1990 Modelling multivariate extreme value distributions. *Biometrika* **77**, 245–253.

Thatcher AR 1999 The long-term pattern of adult mortality and the highest attained age. *Journal of the Royal Statistical Society A* **162**, 5–43.

Tiago de Oliveira J 1958 Extremal distributions. *Revista Faculdade de Ciencias de Lisboa*, 2 Ser., A, Mat., **7**, 219–227.

Tiago de Oliveira J 1962 La représentation des distributions extrêmales bivariées. In *Bulletin de l'Institut International de Statistique, Actes de la 33e Session*, pp. 477–480.

Tiago de Oliveira J 1962/1963 Structure theory of bivariate extremes; extension. *Estudos de Matemática, Estatistica e Econometrica* **7**, 165–195.

Tiago de Oliveira J 1969 Biextremal distributions; statistical decision. *Trabajos de Estadistica y Investigación Operative* **21**, 107–117.

Tiago de Oliveira J 1971 A new model of bivariate extremes. In *Studi di Probabilitá Statistica e Ricerca Operativa in Onore de Guiuseppe Pompilj* (ed Istituto di Calcolo delle Probabilitá dell' Universitá degli Studi di Roma), pp. 437–449. Tipografia Oderisi.

Tiago de Oliveira J 1974 Regression in the nondifferentiable bivariate extreme models. *Journal of the American Statistical Association* **69**, 816–818.

Tiago de Oliveira J 1980 Bivariate extremes: foundations and statistics. In *Multivariate Analysis V* (ed Krishnaiah PR), pp. 349–366. North Holland.

Tiago de Oliveira J 1984 Bivariate models for extremes; statistical decision. In *Statistical Extremes and Applications* (ed Tiago de Oliveira J), pp. 131–153. Reidel.

Tiago de Oliveira J 1989a Intrinsic estimation of the dependence structure for bivariate extremes. *Statistics & Probability Letters* **8**, 213–218.

Tiago de Oliveira J 1989b Statistical decision for bivariate extremes. In *Extreme Value Theory: Proceedings, Oberwolfach 1987* (eds Hüsler J and Reiss RD), pp. 246–261. Lecture Notes in Statistics 51, Springer-Verlag, Berlin.

Tiago de Oliveira J 1992 Intrinsic estimation of the dependence function in bivariate extremes: a new and simpler approach. *Communications in Statistics–Theory and Methods* **21**, 599–611.

Vanroelen G 2003 *The Effect of Transformations on Second-Order Regular Variation.* PhD Thesis, Department of Mathematics, Catholic University, Leuven.

Wahba G 1978 Improper priors, spline smoothing, and the problem of guarding against model errors in regression. *Journal of the Royal Statistical Society B* **40**, 364–372.

Weissman I 1978 Estimation of parameters and large quantiles based on the k largest observations. *Journal of the American Statistical Association* **73**, 812–815.

Weissman I and Novak SY 1998 On blocks and runs estimators of the extremal index. *Journal of Statistical Planning and Inference* **66**, 281–288.

Wicksell SD 1925 The corpuscle problem: a mathematical study of a biometric problem. *Biometrika* **17**, 84–99.

Willis JC 1922 *Age and Area.* Cambridge University Press.

Wiśniewski M 1996 On extreme-order statistics and point processes of exceedances in multivariate stationary Gaussian sequences. *Statistics & Probability Letters* **29**, 55–59.

Yue S 2000 The Gumbel Mixed model applied to storm frequency analysis. *Water Resources Management* **14**, 377–389.

Yue S 2001 The Gumbel Logistic model for representing a multivariate storm event. *Advances in Water Resources* **24**, 179–185.

Yue S, Ouarda TBMJ, Boée B, Legendre P and Bruneau P 1999 The Gumbel Mixed model for flood frequency analysis. *Journal of Hydrology* **226**, 88–100.

Yun S 1997 On domains of attraction of multivariate extreme value distributions under absolute continuity. *Journal of Multivariate Analysis* **63**, 277–295.

Yun S 1998 The extremal index of a higher-order stationary Markov chain. *Annals of Applied Probability* **8**, 408–437.

Yun S 2000a The distributions of cluster functionals of extreme events in a dth-order Markov chain. *Journal of Applied Probability* **37**, 29–44.

Yun S 2000b A class of Pickands-type estimators for the extreme value index. *Journal of Statistical Planning and Inference* **83**, 113–124.

Yun S 2002 On a generalized Pickands estimator of the extreme value index. *Journal of Statistical Planning and Inference* **102**, 389–409.

Zellner A 1971 *An Introduction to Bayesian Inference in Econometrics*. Wiley.

Zhang Z 2002 *Multivariate Extremes, Max-Stable Process Estimation and Dynamic Financial Modeling*. PhD Thesis, University of North Carolina at Chapel Hill.

Zhang Z and Smith RL 2001 Modeling financial time series data as moving maxima processes. Working Paper.

Zipf GK 1935 *The Psycho-Biology of Language; An Introduction to Dynamic Philology*. Houghton Mifflin, Boston.

Zipf GK 1941 *National Unity and Disunity: The Nation as a Bio-Social Organism*. The Principia Press, Bloomington.

Zipf GK 1949 *Human Behavior and the Principle of Least Effort: An Introduction to Human Ecology*. Addison-Wesley.

Author Index

Statistics of Extremes: Theory and Applications   J. Beirlant, Y. Goegebeur, J. Segers, and J. Teugels
© 2004 John Wiley & Sons, Ltd   ISBN: 0-471-97647-4

Subject Index

Algorithms
  bootstrap, 128
  Drees-Kaufmann, 128
  Gibbs sampler, 434
  least squares, 6
  Metropolis-Hastings, 433
  optimal sample size, 205
  weighted least squares, 108
Angular component, 284
Asymptotic independence, 285,
    342–365
Automobile insurance, 25, 188–199
Auxiliary function, 49, 56

Bandwidth, 238
Bayesian methods, 429–459
  in Fréchet-Pareto case,
    435–445
  in general case, 445–452
Beirlant-Vandewalle estimator, 353
Benktander II distribution, 72
Bessel function, 311
Beta distribution, 68, 69
Bias reduction
  Fréchet-Pareto case, 113–119
  general max-domain of
    attraction, 165
  negative, 308
Block maxima
  in Bayesian methodology, 434
  in multivariate case, 313–325

in regression, 211–218
in time series, 376
in univariate case, 132–140
Bonferroni inequality, 285
Bootstrap, 126, 235, 409
Boundary, 48. *See also* Endpoint
de Bruyn conjugate, 57, 66, 79
Box-Cox transformation, 52
Burr distribution, 59, 60, 63
  reversed, 68

Capéraà-Fougères-Genest estimator,
    317
Cauchy distribution, 60
Censored likelihood estimation, 336,
    409
Central limit problem, 45
Cluster, 382
  functional, 386
  maximum, 387, 388, 399
  size, 387
  statistics, 386, 406
Coefficient of extremal dependence,
    343–345
Coefficient of tail dependence,
    346–354
Concomitant, 289
Condroz data, 34, 177–187,
    213–241
Continuity correction, 5, 19
Convergence in distribution, 47

Statistics of Extremes: Theory and Applications   J. Beirlant, Y. Goegebeur, J. Segers, and J. Teugels
© 2004 John Wiley & Sons, Ltd   ISBN: 0-471-97647-4

WILEY SERIES IN PROBABILITY AND STATISTICS

ESTABLISHED BY WALTER A. SHEWHART AND SAMUEL S. WILKS

Editors: *David J. Balding, Noel A. C. Cressie, Nicholas I. Fisher,*
Iain M. Johnstone, J. B. Kadane, Geert Molenberghs, Louise M. Ryan,
David W. Scott, Adrian F. M. Smith, Jozef L. Teugels
Editors Emeriti: *Vic Barnett, J. Stuart Hunter, David G. Kendall*

The *Wiley Series in Probability and Statistics* is well established and authoritative. It covers many topics of current research interest in both pure and applied statistics and probability theory. Written by leading statisticians and institutions, the titles span both state-of-the-art developments in the field and classical methods.

Reflecting the wide range of current research in statistics, the series encompasses applied, methodological and theoretical statistics, ranging from applications and new techniques made possible by advances in computerized practice to rigorous treatment of theoretical approaches.

This series provides essential and invaluable reading for all statisticians, whether in academia, industry, government, or research.

ABRAHAM and LEDOLTER · Statistical Methods for Forecasting
AGRESTI · Analysis of Ordinal Categorical Data
AGRESTI · An Introduction to Categorical Data Analysis
AGRESTI · Categorical Data Analysis, *Second Edition*
ALTMAN, GILL, and McDONALD · Numerical Issues in Statistical Computing
  for the Social Scientist
AMARATUNGA and CABRERA · Exploration and Analysis of DNA Microarray and Protein Array
  Data
ANDĚL · Mathematics of Chance
ANDERSON · An Introduction to Multivariate Statistical Analysis, *Third Edition*
*ANDERSON · The Statistical Analysis of Time Series
ANDERSON, AUQUIER, HAUCK, OAKES, VANDAELE, and WEISBERG ·
  Statistical Methods for Comparative Studies
ANDERSON and LOYNES · The Teaching of Practical Statistics
ARMITAGE and DAVID (editors) · Advances in Biometry
ARNOLD, BALAKRISHNAN, and NAGARAJA · Records
*ARTHANARI and DODGE · Mathematical Programming in Statistics
*BAILEY · The Elements of Stochastic Processes with Applications to the Natural
  Sciences
BALAKRISHNAN and KOUTRAS · Runs and Scans with Applications
BARNETT · Comparative Statistical Inference, *Third Edition*
BARNETT · Environmental Statistics: Methods & Applications
BARNETT and LEWIS · Outliers in Statistical Data, *Third Edition*
BARTOSZYNSKI and NIEWIADOMSKA-BUGAJ · Probability and Statistical Inference
BASILEVSKY · Statistical Factor Analysis and Related Methods: Theory and
  Applications
BASU and RIGDON · Statistical Methods for the Reliability of Repairable Systems
BATES and WATTS · Nonlinear Regression Analysis and Its Applications
BECHHOFER, SANTNER, and GOLDSMAN · Design and Analysis of Experiments for
  Statistical Selection, Screening, and Multiple Comparisons
BELSLEY · Conditioning Diagnostics: Collinearity and Weak Data in Regression
BELSLEY, KUH, and WELSCH · Regression Diagnostics: Identifying Influential
  Data and Sources of Collinearity

*Now available in a lower priced paperback edition in the Wiley Classics Library.

*Now available in a lower priced paperback edition in the Wiley Classics Library.

*Now available in a lower priced paperback edition in the Wiley Classics Library.

*Now available in a lower priced paperback edition in the Wiley Classics Library.

*Now available in a lower priced paperback edition in the Wiley Classics Library.

MORGENTHALER and TUKEY · Configural Polysampling: A Route to Practical Robustness

MUIRHEAD · Aspects of Multivariate Statistical Theory

MURRAY · X-STAT 2.0 Statistical Experimentation, Design Data Analysis, and Nonlinear Optimization

MURTHY, XIE, and JIANG · Weibull Models

MYERS and MONTGOMERY · Response Surface Methodology: Process and Product Optimization Using Designed Experiments, *Second Edition*

MYERS, MONTGOMERY, and VINING · Generalized Linear Models. With Applications in Engineering and the Sciences

NELSON · Accelerated Testing, Statistical Models, Test Plans, and Data Analyses

NELSON · Applied Life Data Analysis

NEWMAN · Biostatistical Methods in Epidemiology

OCHI · Applied Probability and Stochastic Processes in Engineering and Physical Sciences

OKABE, BOOTS, SUGIHARA, and CHIU · Spatial Tesselations: Concepts and Applications of Voronoi Diagrams, *Second Edition*

OLIVER and SMITH · Influence Diagrams, Belief Nets and Decision Analysis

PALTA · Quantitative Methods in Population Health: Extensions of Ordinary Regressions

PANKRATZ · Forecasting with Dynamic Regression Models

PANKRATZ · Forecasting with Univariate Box-Jenkins Models: Concepts and Cases

*PARZEN · Modern Probability Theory and It's Applications

PEÑA, TIAO, and TSAY · A Course in Time Series Analysis

PIANTADOSI · Clinical Trials: A Methodologic Perspective

PORT · Theoretical Probability for Applications

POURAHMADI · Foundations of Time Series Analysis and Prediction Theory

PRESS · Bayesian Statistics: Principles, Models, and Applications

PRESS · Subjective and Objective Bayesian Statistics, *Second Edition*

PRESS and TANUR · The Subjectivity of Scientists and the Bayesian Approach

PUKELSHEIM · Optimal Experimental Design

PURI, VILAPLANA, and WERTZ · New Perspectives in Theoretical and Applied Statistics

PUTERMAN · Markov Decision Processes: Discrete Stochastic Dynamic Programming

*RAO · Linear Statistical Inference and Its Applications, *Second Edition*

RAUSAND and HØYLAND · System Reliability Theory: Models, Statistical Methods and Applications, *Second Edition*

RENCHER · Linear Models in Statistics

RENCHER · Methods of Multivariate Analysis, *Second Edition*

RENCHER · Multivariate Statistical Inference with Applications

RIPLEY · Spatial Statistics

RIPLEY · Stochastic Simulation

ROBINSON · Practical Strategies for Experimenting

ROHATGI and SALEH · An Introduction to Probability and Statistics, *Second Edition*

ROLSKI, SCHMIDLI, SCHMIDT, and TEUGELS · Stochastic Processes for Insurance and Finance

ROSENBERGER and LACHIN · Randomization in Clinical Trials: Theory and Practice

ROSS · Introduction to Probability and Statistics for Engineers and Scientists

ROUSSEEUW and LEROY · Robust Regression and Outlier Detection

RUBIN · Multiple Imputation for Nonresponse in Surveys

RUBINSTEIN · Simulation and the Monte Carlo Method

RUBINSTEIN and MELAMED · Modern Simulation and Modeling

RYAN · Modern Regression Methods

RYAN · Statistical Methods for Quality Improvement, *Second Edition*

SALTELLI, CHAN, and SCOTT (editors) · Sensitivity Analysis

*SCHEFFE · The Analysis of Variance

*Now available in a lower priced paperback edition in the Wiley Classics Library.

*Now available in a lower priced paperback edition in the Wiley Classics Library.

Printed and bound by CPI Group (UK) Ltd, Croydon, CR0 4YY

16/04/2025

14658497-0003